# CATALYTIC NAPHTHA REFORMING

# CHEMICAL INDUSTRIES

A Series of Reference Books and Textbooks

*Consulting Editor*

HEINZ HEINEMANN
*Berkeley, California*

1. *Fluid Catalytic Cracking with Zeolite Catalysts,* Paul B. Venuto and E. Thomas Habib, Jr.
2. *Ethylene: Keystone to the Petrochemical Industry,* Ludwig Kniel, Olaf Winter, and Karl Stork
3. *The Chemistry and Technology of Petroleum,* James G. Speight
4. *The Desulfurization of Heavy Oils and Residua,* James G. Speight
5. *Catalysis of Organic Reactions,* edited by William R. Moser
6. *Acetylene-Based Chemicals from Coal and Other Natural Resources,* Robert J. Tedeschi
7. *Chemically Resistant Masonry,* Walter Lee Sheppard, Jr.
8. *Compressors and Expanders: Selection and Application for the Process Industry,* Heinz P. Bloch, Joseph A. Cameron, Frank M. Danowski, Jr., Ralph James, Jr., Judson S. Swearingen, and Marilyn E. Weightman
9. *Metering Pumps: Selection and Application,* James P. Poynton
10. *Hydrocarbons from Methanol,* Clarence D. Chang
11. *Form Flotation: Theory and Applications,* Ann N. Clarke and David J. Wilson
12. *The Chemistry and Technology of Coal,* James G. Speight
13. *Pneumatic and Hydraulic Conveying of Solids,* O. A. Williams
14. *Catalyst Manufacture: Laboratory and Commercial Preparations,* Alvin B. Stiles
15. *Characterization of Heterogeneous Catalysts,* edited by Francis Delannay
16. *BASIC Programs for Chemical Engineering Design,* James H. Weber
17. *Catalyst Poisoning,* L. Louis Hegedus and Robert W. McCabe
18. *Catalysis of Organic Reactions,* edited by John R. Kosak
19. *Adsorption Technology: A Step-by-Step Approach to Process Evaluation and Application,* edited by Frank L. Slejko
20. *Deactivation and Poisoning of Catalysts,* edited by Jacques Oudar and Henry Wise

21. *Catalysis and Surface Science: Developments in Chemicals from Methanol, Hydrotreating of Hydrocarbons, Catalyst Preparation, Monomers and Polymers, Photocatalysis and Photovoltaics,* edited by Heinz Heinemann and Gabor A. Somorjai
22. *Catalysis of Organic Reactions,* edited by Robert L. Augustine
23. *Modern Control Techniques for the Processing Industries,* T. H. Tsai, J. W. Lane, and C. S. Lin
24. *Temperature-Programmed Reduction for Solid Materials Characterization,* Alan Jones and Brian McNichol
25. *Catalytic Cracking: Catalysts, Chemistry, and Kinetics,* Bohdan W. Wojciechowski and Avelino Corma
26. *Chemical Reaction and Reactor Engineering,* edited by J. J. Carberry and A. Varma
27. *Filtration: Principles and Practices: Second Edition,* edited by Michael J. Matteson and Clyde Orr
28. *Corrosion Mechanisms,* edited by Florian Mansfeld
29. *Catalysis and Surface Properties of Liquid Metals and Alloys,* Yoshisada Ogino
30. *Catalyst Deactivation,* edited by Eugene E. Petersen and Alexis T. Bell
31. *Hydrogen Effects in Catalysis: Fundamentals and Practical Applications,* edited by Zoltán Paál and P. G. Menon
32. *Flow Management for Engineers and Scientists,* Nicholas P. Cheremisinoff and Paul N. Cheremisinoff
33. *Catalysis of Organic Reactions,* edited by Paul N. Rylander, Harold Greenfield, and Robert L. Augustine
34. *Powder and Bulk Solids Handling Processes: Instrumentation and Control,* Koichi Iinoya, Hiroaki Masuda, and Kinnosuke Watanabe
35. *Reverse Osmosis Technology: Applications for High-Purity-Water Production,* edited by Bipin S. Parekh
36. *Shape Selective Catalysis in Industrial Applications,* N. Y. Chen, William E. Garwood, and Frank G. Dwyer
37. *Alpha Olefins Applications Handbook,* edited by George R. Lappin and Joseph L. Sauer
38. *Process Modeling and Control in Chemical Industries,* edited by Kaddour Najim
39. *Clathrate Hydrates of Natural Gases,* E. Dendy Sloan, Jr.
40. *Catalysis of Organic Reactions,* edited by Dale W. Blackburn
41. *Fuel Science and Technology Handbook,* edited by James G. Speight
42. *Octane-Enhancing Zeolitic FCC Catalysts,* Julius Scherzer
43. *Oxygen in Catalysis,* Adam Bielański and Jerzy Haber
44. *The Chemistry and Technology of Petroleum: Second Edition, Revised and Expanded,* James G. Speight

45. *Industrial Drying Equipment: Selection and Application,* C. M. van't Land
46. *Novel Production Methods for Ethylene, Light Hydrocarbons, and Aromatics,* edited by Lyle F. Albright, Billy L. Crynes, and Siegfried Nowak
47. *Catalysis of Organic Reactions,* edited by William E. Pascoe
48. *Synthetic Lubricants and High-Performance Functional Fluids,* edited by Ronald L. Shubkin
49. *Acetic Acid and Its Derivatives,* edited by Victor H. Agreda and Joseph R. Zoeller
50. *Properties and Applications of Perovskite-Type Oxides,* edited by L. G. Tejuca and J. L. G. Fierro
51. *Computer-Aided Design of Catalysts,* edited by E. Robert Becker and Carmo J. Pereira
52. *Models for Thermodynamic and Phase Equilibria Calculations,* edited by Stanley I. Sandler
53. *Catalysis of Organic Reactions,* edited by John R. Kosak and Thomas A. Johnson
54. *Composition and Analysis of Heavy Petroleum Fractions,* Klaus H. Altgelt and Mieczyslaw M. Boduszynski
55. *NMR Techniques in Catalysis,* edited by Alexis T. Bell and Alexander Pines
56. *Upgrading Petroleum Residues and Heavy Oils,* Murray R. Gray
57. *Methanol Production and Use,* edited by Wu-Hsun Cheng and Harold H. Kung
58. *Catalytic Hydroprocessing of Petroleum and Distillates,* edited by Michael C. Oballah and Stuart S. Shih
59. *The Chemistry and Technology of Coal: Second Edition, Revised and Expanded,* James G. Speight
60. *Lubricant Base Oil and Wax Processing,* Avilino Sequeira, Jr.
61. *Catalytic Naphtha Reforming: Science and Technology,* edited by George J. Antos, Abdullah M. Aitani, and José M. Parera
62. *Catalysis of Organic Reactions,* edited by Michael Scaros

ADDITIONAL VOLUMES IN PREPARATION

# CATALYTIC NAPHTHA REFORMING

## Science and Technology

edited by

### George J. Antos
*UOP Research Center
Des Plaines, Illinois*

### Abdullah M. Aitani
*King Fahd University of
Petroleum and Minerals
Dhahran, Saudi Arabia*

### José M. Parera
*Instituto de Investigaciones en
Catálisis y Petroquímica (INCAPE)
Santa Fe, Argentina*

Marcel Dekker, Inc.          New York • Basel • Hong Kong

Library of Congress Cataloging-in-Publication Data

Catalytic naphtha reforming: science and technology / edited by
  George J. Antos, Abdullah M. Aitani, José M. Parera.
     p.   cm. — (Chemical industries; 61)
  Includes bibliographical references and index.
  ISBN 0-8247-9236-X (acid-free)
  1. Naphtha.  2. Catalytic reforming.  I. Antos, George J.
  II. Aitani, Abdullah M.  III. Parera, José M.
  IV. Series: Chemical industries; v. 61.
  TP692.4.NC37   1995
  665.5'3824—dc20                                            94-29127
                                                                 CIP

The publisher offers discounts on this book when ordered in bulk quantities. For more information, write to Special Sales/Professional Marketing at the address below.

This book is printed on acid-free paper.

**Copyright © 1995 by MARCEL DEKKER, INC. All Rights Reserved.**

Neither this book nor any part may be reproduced or transmitted in any form or by any means, electronic or mechanical, including photocopying, microfilming, and recording, or by any information storage and retrieval system, without permission in writing from the publisher.

MARCEL DEKKER, INC.
270 Madison Avenue, New York, New York  10016

Current printing (last digit):
10 9 8 7 6 5 4 3 2 1

**PRINTED IN THE UNITED STATES OF AMERICA**

# Preface

The use of catalytic naphtha reforming as a process to produce high-octane gasoline is as important now as it has been for over the 45 years of its commercial use. The catalytic reformer occupies a key position in a refinery, providing high value-added reformate for the gasoline pool; hydrogen for feedstock improvement by the hydrogen-consuming hydrotreatment processes; and frequently benzene, toluene, and xylene aromatics for petrochemical uses. The technology has even further impact in the refinery complex. The processes of hydrogenation, dehydrogenation, and isomerization have all benefited from the catalyst, reactor, and feed treatment technologies invented for catalytic reforming processes. The long-term outlook for the reforming catalyst market remains strong. The conditions of operation of catalytic reforming units are harsh and there is an increasing need for reformate. Presently, the catalytic reforming process is currently operated to produce research octane numbers of 100 and more.

Since its introduction, catalytic reforming has been studied extensively in order to understand the catalytic chemistry of the process. The workhorse for this process is typically a catalyst composed of minor amounts of several components, including platinum supported on an oxide material such as alumina. This simplification masks the absolute beauty of the chemistry involved in combining these components in just the proper manner to yield a high-performance, modern reforming catalyst. The difficulty in mastering this chemistry and of characterizing the catalyst to know what has been wrought is the driving force behind the many industrial and academic studies in reforming catalysis available today.

Several questions come to mind. Why are scientists continuing to research this area of catalysis? What have all the preceding studies taught us about these catalysts, and what remains unknown? Given the numerous studies reported in the patent literature and in technical journals, it is surprising that a survey aimed at answering these questions summarizing the preceding experiences is not readily found. All the editors and contributors of this book are experienced in the study of reforming catalysts, and each one of them would have employed such a survey in his own research program. This volume provides information not currently available from one single literature source. The chapters are written by well-known authorities in the fields encompassed by catalytic reforming, starting with the process chemistry and focusing on the preparation, characterization, evaluation, and operation of the catalyst itself. The unknown aspects of catalyst chemistry and fundamental studies attempting to provide an understanding are also presented. Some attempt is made to predict the future for this catalyst technology, a task made complicated by the conflicting demand for more transportation fuels and petrochemicals, and the resolution to reduce the pollution resulting from their use.

It has been our pleasure to work with the contributors involved in this book. Their effort in combining their own research with the recent literature in the field of catalytic naphtha reforming is highly appreciated. This effort would not have been possible without their willingness to share valuable knowledge and experience. Moreover, we express our gratitude for their responsiveness to deadlines and review comments.

The editors hope that veteran industrial researchers will recognize this volume as an important resource and that novice researchers in the field of reforming and related catalysts—industrial chemists assigned to their first major catalysis project, graduate students embarking on the study of catalysis, and chemical engineers in the refinery responsible for full-scale commercial catalytic reforming—will find this a valuable reference volume and tool for their future endeavors in this exciting area.

*George J. Antos*
*Abdullah M. Aitani*
*José M. Parera*

# Contents

*Preface* iii
*Contributors* vii

**Part I  Reaction Chemistry**

1  Chemistry and Processing of Petroleum  1
   *José M. Parera and Nora S. Fígoli*

2  Basic Reactions of Reforming on Metal Catalysts  19
   *Zoltán Paál*

3  Reactions in the Commercial Reformer  45
   *José M. Parera and Nora S. Fígoli*

**Part II  Catalyst Preparation and Characterization**

4  Catalyst Preparation  79
   *J. P. Boitiaux, J. M. Devès, B. Didillon, and C. R. Marcilly*

| 5 | Characterization of Naphtha Reforming Catalysts<br>*Burtron H. Davis and George J. Antos* | **113** |
|---|---|---|
| 6 | Evaluation of Catalysts for Catalytic Reforming<br>*S. Tiong Sie* | **181** |
| 7 | Structure and Performance of Reforming Catalysts<br>*K. R. Murthy, N. Sharma, and N. George* | **207** |

## Part III   Catalyst Deactivation and Regeneration

| 8 | Naphtha Hydrotreatment<br>*H. J. Lovink* | **257** |
|---|---|---|
| 9 | Deactivation by Coking<br>*Patrice Marécot and Jacques Barbier* | **279** |
| 10 | Deactivation by Poisoning and Sintering<br>*Jorge Norberto Beltramini* | **313** |
| 11 | Regeneration of Reforming Catalysts<br>*Jorge Norberto Beltramini* | **365** |
| 12 | Recovery of Pt and Re from Spent Reforming Catalysts<br>*J. P. Rosso and Mahmoud I. El Guindy* | **395** |

## Part IV   Technology and Applications

| 13 | Catalytic Reforming Processes<br>*Abdullah M. Aitani* | **409** |
|---|---|---|
| 14 | Modeling Commercial Reformers<br>*Lee E. Turpin* | **437** |
| 15 | Reforming for Gasoline and Aromatics: Recent Developments<br>*Subramanian Sivasanker and Paul Ratnasamy* | **483** |

*Index* — *509*

# Contributors

**Abdullah M. Aitani**   Research Institute, King Fahd University of Petroleum and Minerals, Dhahran, Saudi Arabia

**George J. Antos**   UOP Research Center, Des Plaines, Illinois

**Jacques Barbier**   Department of Chemistry, Laboratoire de Catalyse en Chimie Organique, Université de Poitiers, Poitiers, France

**Jorge Norberto Beltramini**   Department of Chemical Engineering, King Fahd University of Petroleum and Minerals, Dhahran, Saudi Arabia

**J. P. Boitiaux**   Institut Français du Pétrole, Rueil-Malmaison, France

**Burtron H. Davis**   Center for Applied Energy Research, University of Kentucky, Lexington, Kentucky

**J. M. Devès**   Department of Kinetics and Catalysis, Institut Français du Pétrole, Rueil-Malmaison, France

**B. Didillon**   Department of Kinetics and Catalysis, Institut Français du Pétrole, Rueil-Malmaison, France

**Mahmoud I. El Guindy**   Gemini Industries, Inc., Santa Ana, California

**Nora S. Fígoli** Instituto de Investigaciones en Catálisis y Petroquimica (INCAPE), Santa Fe, Argentina

**N. George** Research Centre, Indian Petrochemicals Corporation Ltd., Vadodara, India

**H. J. Lovink** AKZO Catalysts, Amersfoort, The Netherlands

**C. R. Marcilly** Department of Kinetics and Catalysis, Institut Français du Pétrole, Rueil-Malmaison, France

**Patrice Marécot** Department of Chemistry, Laboratoire de Catalyse en Chimie Organique, Université de Poitiers, Poitiers, France

**K. R. Murthy** Research Centre, Indian Petrochemicals Corporation Ltd., Vadodara, India

**Zoltán Paál** Institute of Isotopes of the Hungarian Academy of Sciences, Budapest, Hungary

**José M. Parera** Instituto de Investigaciones en Catálisis y Petroquímica (INCAPE), Santa Fe, Argentina

**Paul Ratnasamy** Catalysis Division, National Chemical Laboratory, Pune, India

**J. P. Rosso** Gemini Industries, Inc., Santa Ana, California

**N. Sharma** Research Centre, Indian Petrochemicals Corporation Ltd., Vadodara, India

**S. Tiong Sie** Faculty of Chemical Technology and Materials Science, Technical University Delft, Delft, The Netherlands

**Subramanian Sivasanker** Catalysis Division, National Chemical Laboratory, Pune, India

**Lee E. Turpin** Process Modeling Services, Profimatics, Inc., Thousand Oaks, California

# 1
# Chemistry and Processing of Petroleum

**José M. Parera and Nora S. Fígoli**
*Instituto de Investigaciones en Catálisis y Petroquímica (INCAPE), Santa Fe, Argentina*

## I. INTRODUCTION

This chapter gives a brief outline of the chemistry and processing of petroleum in order to show how the production of high-octane gasolines and aromatic hydrocarbons from crude oil is achieved. There are many publications on the subject; some of them are Refs. (1–5).

The main constituents of crude oils are hydrocarbons, with small amounts of impurities. There are different types of crude oils from which fuels, lubricants, and petrochemical raw materials are obtained by proper processing. Refinery processes are either simple, like distillation, or more complicated, where chemical reactions take place and the structure of the molecules is changed. Among the latter processes, hydroprocessing and reforming are important.

The objective of naphtha reforming is the transformation of low-octane naphthas (virgin or processed) into high-octane gasoline by increasing the concentration of aromatics and $i$-paraffins. Environmental regulations also have an impact on the composition of the gasoline.

## II. COMPOSITION OF PETROLEUM

Petroleum is a complex mixture of hydrocarbons (paraffins, naphthenes, and aromatics) plus small amounts of water and organic compounds of sulfur, oxygen, and nitrogen, as well as compounds with metallic constituents, particularly vanadium, nickel, and sodium. Although concentrations of nonhydrocarbon compounds are very small, their influence on catalytic petroleum processing is important. Proportions of the elements in petroleum from different origins vary only slightly within narrow limits, as shown in Table 1. However, physical properties and technical qualities of various petroleums may be quite variable.

### A. Hydrocarbons

Hydrocarbons are the major constituents of petroleum, and they belong to the paraffinic, naphthenic, and aromatic series. Olefins are not usually found in crude oils, but they are produced during petroleum processing.

The number of individual hydrocarbons in petroleum is enormous. Table 2 shows the hydrocarbons with boiling points lower than 111°C found in a particular crude oil. These 37 compounds represent only 16.5% of the total crude volume. It can be noticed that normal paraffins predominate. The most volatile fraction is formed only by paraffins, because the lightest naphthene is cyclopentane (boiling point [BP] 49°C) and the lightest aromatic is benzene (BP 80°C).

An increase in the number of carbon atoms in the paraffins increases the number of their isomers, as shown in Table 3. In a light gasoline fraction containing hydrocarbons of 4 to 10 carbon atoms, nearly 500 hydrocarbons can be present, most of them only in trace quantities.

The distribution of the hydrocarbon types throughout the boiling range of petroleum is shown in Figure 1.

The main properties of the three hydrocarbon series contained in petroleum are as follows.

**Table 1** Elemental Composition of Crude Oils, wt %

| Origin | Carbon | Hydrogen | Nitrogen | Oxygen | Sulfur |
|---|---|---|---|---|---|
| Romania | 86.6 | 12.1 | —0.7— | | 0.6 |
| Canada | 83.2 | 10.4 | 0.4 | 0.9 | 5.1 |
| Mexico | 83.0 | 11.0 | —1.7— | | 4.3 |
| United States | 85.7 | 11.0 | —2.6— | | 0.7 |
| Argentina | 86.7 | 12.1 | —1.0— | | 0.2 |
| Colombia | 85.6 | 11.9 | 0.5 | | |
| Venezuela | 82.5 | 10.4 | 0.6 | 0.8 | 5.7 |

*Source:* Extracted from Ref. 1, p. 50.

**Table 2** Boiling Point and Volume Fraction of Hydrocarbons Having a Boiling Point Lower Than 111 °C Contained in a Petroleum

| Hydrocarbon | Boiling point (°C) | Fraction of volume (%) |
|---|---|---|
| Methane | −151.49 | — |
| Ethane | −88.63 | 0.06 |
| Propane | −42.07 | 0.11 |
| Isobutane | −11.73 | 0.14 |
| n-Butane | 0.50 | 0.79 |
| 2-Methylbutane | 27.85 | 0.77 |
| n-Pentane | 36.07 | 1.43 |
| Cyclopentane | 49.26 | 0.05 |
| 2,2-Diemthylbutane | 49.74 | 0.04 |
| 2,3-Dimethylbutane | 57.99 | 0.08 |
| 2-Methylpentane | 60.27 | 0.37 |
| 3-Methylpentane | 63.28 | 0.35 |
| n-Hexane | 68.74 | 1.80 |
| Methylcyclopentane | 71.81 | 0.87 |
| 2,2-Dimethylpentane | 79.20 | 0.02 |
| Benzene | 80.10 | 0.15 |
| 2,4-Dimethylpentane | 80.50 | 0.08 |
| Cyclohexane | 80.74 | 0.71 |
| 1,1-Dimethylcyclopentane | 87.85 | 0.16 |
| 2,3-Dimethylpentane | 89.78 | 0.15 |
| 2-Methylhexane | 90.05 | 0.73 |
| 1-cis-3-Dimethylcyclopentane | 90.77 | 0.87 |
| 1-trans-3-Dimethylcyclopentane | 91.72 | 0.21 |
| 3-Methylhexane | 91.85 | 0.51 |
| 1-trans-2-Dimethylcyclopentane | 91.87 | 0.48 |
| 3-Ethylpentane | 93.48 | 0.06 |
| n-Heptane | 98.43 | 2.30 |
| Methylcyclohexane | 100.93 | 1.60 |
| Ethylcyclopentane | 103.47 | 0.16 |
| 1,1,3-Trimethylcyclopentane | 104.89 | 0.30 |
| 2,2-Dimethylhexane | 106.84 | 0.01 |
| 2,5-Dimethylhexane | 109.10 | 0.06 |
| 1-trans-2-cis-4-Trimethylcyclopentane | 109.29 | 0.22 |
| 2,4-Dimethylhexane | 109.43 | 0.06 |
| 2,2,3-Trimethylpentane | 109.84 | 0.004 |
| 1-trans-2-cis-3-Trimethylcyclopentane | 110.20 | 0.26 |
| Toluene | 110.62 | 0.51 |
| Total | | 16.47 |

## 1. Paraffins

Paraffins have the general formula $C_nH_{2n+2}$, with straight or branched chains but without any closed-ring structure. Paraffins are very stable, but the stability decreases when the molecular weight and the number of branched chains increase. Branched paraffins are more desirable than normal paraffins as components of gasoline because of their better antiknock properties. At atmospheric pressure, paraffins having from 1 to 4 carbon atoms in the molecule are gaseous, those having 5 to 15 carbon atoms are liquid, and those having 16 or more carbon atoms are solid.

## 2. Naphthenes

Naphthenes are also called alicyclic hydrocarbons or cycloparaffins and are expressed by the general formula $C_nH_{2n}$. They are saturated hydrocarbons with a closed ring structure that can have one or more paraffinic side chains. Naphthenic compounds constitute a substantial amount of the whole crude and the proportion varies with the type of crude. Naphthenes of the petroleum fractions have rings of five or six carbon atoms (cyclopentanes or cyclohexanes).

## 3. Aromatics

Aromatics are nonsaturated hydrocarbons containing one or more aromatic nuclei (benzene, naphthalene, phenanthrene), which may be linked up with naphthene rings and/or paraffinic side chains. The general formula is $C_nH_{2n-6}$. The percentage of aromatics in petroleum is lower than that of paraffins or naphthenes. For the same petroleum, the heaviest fractions are richer in aromatics than the lightest ones. Benzene, toluene, xylenes, trimethylbenzenes, naphthalene, alkylnaphthalenes, alkylphenanthrenes, and chrysenes are the most usual compounds. Aromatics are important components of gasoline because of their antiknock properties, and some of them are raw materials for the petrochemical industry.

## B. Other Organic Compounds

### 1. Oxygen Compounds

The oxygen content of petroleum increases with the boiling point of the fraction. The nonvolatile residue may have an oxygen content of up to 8% by weight. The main oxygen compounds found in petroleum are naphthenic acids, carboxylic acids, and phenolic and asphaltic compounds.

### 2. Sulfur Compounds

Sulfur compounds are among the most important heteroatomic constituents of petroleum. In general, the higher the density of the crude oil, the higher the sulfur content. Organic sulfur compounds are found in all crude oils, in amounts varying from as little as 0.05 wt % sulfur to as high as 5 wt %. The main sulfur compounds identified in petroleum are hydrogen sulfide, sulfuric acid, thiophene, alkylsulfurs, alkylsulfates, alkyl mercaptans, and free sulfur. The sulfur compounds are harmful because during petroleum processing they are frequently poisons for metallic catalysts, and their presence in fuels also promotes engine corrosion and causes environmental problems.

3. Nitrogen Compounds

Nitrogen compounds in petroleum may be arbitrarily classified as basic and nonbasic. Basic compounds are mainly pyridine homologues. Even though they are present throughout the whole boiling range, they are more concentrated in the highest-boiling fractions and in the residua. Nonbasic nitrogen compounds (pyrrole, indole, and carbazole derivatives) also occur in the highest-boiling fraction and in the residua. Most crude oils contain 0.1 wt % nitrogen or less, depending on the type of crude. The presence of nitrogen in petroleum is of much greater significance in refinery operations than might be expected from the small amounts present. Nitrogen compounds are responsible for the poisoning of cracking and reforming catalysts and for gum formation in products such as domestic fuel oils.

## C. Water

Normally, petroleum contains water in variable amounts. The presence of water is harmful because oil distillation becomes difficult: water produces foam and has a high heat of vaporization compared with petroleum.

## D. Metallic Constituents

These constituents are partially present as inorganic water-soluble salts, and some metals are present in the form of oil-soluble salts or oil-soluble organometallic compounds. Common metal complexes are those formed by the inclusion of a cation (such as vanadium or nickel) in porphyrins.

## III. TYPES OF CRUDE OILS

Crude oils vary because of different proportions of the various molecular types and molecular weights of hydrocarbons previously described. One crude oil may contain mostly paraffins, another mostly naphthenes; one may contain a large quantity of lower hydrocarbons, another may consist mainly of higher hydrocarbons and be highly viscous. Crude oils are usually characterized as belonging to one of three types depending on the relative amounts of waxes (higher-molecular-weight paraffins that are solid at room temperature) and asphalts present. The wax content correlates with the degree to which the crude is paraffinic. The presence of asphalts indicates an aromatic crude.

## A. Asphaltic or Naphthenic Crude Oils

These crude oils have a low paraffin concentration and the residue is mainly asphaltic (essentially condensed aromatics). Sulfur, nitrogen, and oxygen contents are usually high, and light and intermediate fractions have high percentages of naphthenes. These crudes are particularly suitable for the production of gasoline and asphalts. They are not adequate for kerosene because of the high content of cyclic hydrocarbons, which produce smoke during burning. Naphthenes and aromatics have a high variation of viscosity with temperature; for this reason these crudes are not good for the production of lubricating oils.

## B. Paraffinic Crude Oils

The paraffinic crude oils consist mainly of paraffinic hydrocarbons, usually giving good yields of paraffinic wax, high-grade lubricating oils for motors, and high-grade kerosene.

## C. Mixed Crude Oils

Mixed crude oils contain substantial proportions of both paraffin wax and asphaltic matter. Paraffins and naphthenes are present, together with a certain proportion of aromatic hydrocarbons. Practically all products can be obtained from these crudes, but with lower yields.

Table 4 shows the composition of different fractions obtained by distillation of both types of crude oils. Most of the crude oil components have boiling points higher than 300°C at atmospheric pressure. In the case of the crude oils shown in Table 4, only 45% of the paraffinic and 34% of the naphthenic crude oils are distilled before 300°C.

**Table 4** Composition of the Fractions of Crude Oils (% by Volume)[a]

| Fraction (°C) | Paraffinic crude oil A | N | P | Naphthenic crude oil A | N | P |
|---|---|---|---|---|---|---|
| 60–95 | 2 | 24 | 74 | 2 | 35 | 63 |
| 95–122 | 5 | 29 | 66 | 6 | 50 | 44 |
| 122–150 | 9 | 34 | 57 | 11 | 66 | 23 |
| 150–200 | 14 | 28 | 58 | 15 | 63 | 22 |
| 200–250 | 18 | 23 | 59 | 24 | 48 | 28 |
| 250–300 | 17 | 22 | 61 | 28 | 42 | 30 |

[a] A = aromatics, N = naphthenes, P = paraffins.

Simple physical and chemical data may indicate the quality of a crude oil. For example, the density is useful because it is influenced by the chemical composition, but quantitative correlations are difficult to establish. Nevertheless, it is generally admitted that a high concentration of aromatic-type compounds corresponds to high density, whereas an increase in saturated compounds results in a decrease in density. There are other properties that characterize a crude oil, such as viscosity, surface and interfacial tension, refractive index, and electrical and thermal properties, as detailed in Ref. (1).

## IV. PETROLEUM CONVERSION PROCESSES

Petroleum is processed in order to obtain fuels, lubricants, and petrochemical raw materials. Refinery processes are either simple, such as those used to separate crude oil into fractions, or more complicated when chemical reactions take place and the structure of the molecules is changed. The most important physical separation process is distillation (atmospheric and vacuum), based on differences in boiling points of the components of a mixture. Crude oil is primarily a complex mixture of hydrocarbons, some of which have the same or nearly the same boiling point. Consequently, except for the lowest-boiling hydrocarbons, it is not possible to separate crude oil into pure compounds by distillation. Crude oil is, therefore, separated into mixtures having a rather narrow boiling range. This distillation is carried out at atmospheric pressure and produces several cuts and a residue, or column bottom, boiling at temperatures higher than 300°C. Fractionation of the column bottom at atmospheric pressure is not possible because it is cracked before distilling. For this reason, it is fractionated by vacuum distillation. In order of their boiling ranges, the main cuts obtained in atmospheric distillation are liquefied petroleum gas (LPG), straight-run light and heavy naphtha, kerosene, gas oil, and fuel oil. In vacuum distillation of the atmospheric residue, the main products are gas oil, fractions of lubricant oils, and asphalt. The amount of gasoline precursor obtained by distillation is, in general, less than that required by the market; consequently, it is frequently necessary to transform heavier cuts into gasoline. This transformation is obtained by various cracking processes such as thermal cracking, catalytic cracking, and hydrocracking. The percentage of heavier cuts available to be cracked depends on the type of petroleum and on the required distribution of the other products.

Most of the products from a refinery are processed to improve their quality, with hydroprocessing and reforming as principal processes. Hydroprocessing removes undesirable components, mainly sulfur and nitrogen compounds, and catalytic reforming improves the quality of gasoline by increasing its octane number. The petroleum industry also provides hydrocarbons that are the raw materials for the petrochemical industry.

## V. OCTANE NUMBER

A very important property of an automobile gasoline is resistance to knock, detonation, or ping during service. The more the gasoline-air mixture is compressed in the cylinder before ignition, the greater the power that the engine can deliver. But this increase in performance goes together with an increase in the tendency for detonation or knocking. The detonation of a fuel depends on its composition and introduces a limit to the power and economy that an engine using that fuel can produce. The antiknock property of a gasoline is generally expressed as its octane number. This number is the percentage by volume of i-octane in a blend with $n$-heptane that matches the gasoline in knock characteristics in a standard engine run under standard conditions. Since the octane number of a gasoline varies with the type of engine and operation, it is desirable to perform the measurement under standardized conditions, which have been normalized by the American Society for Testing and Materials (ASTM). A set of conditions produces the research octane number (RON, indicative of normal road performance) and a more severe set of conditions gives the motor octane number (MON, indicative of high-speed performance). The arithmetic average of these two values, known as $(R + M)/2$, is being increasingly used. Octane numbers quoted in the literature are usually, unless stated otherwise, research octane numbers. In addition, it is important to measure the octane number of the proportion of fuel boiling below 100°C. The octane quality of this low-boiling fraction of gasoline is of particular importance under acceleration conditions. Under these conditions, fuel fractionation may occur; the lower-boiling fraction evaporates and the heavier components remain as liquid. If the octane quality of the low-boiling fraction is lower than that of the gasoline as a whole, the fraction may lead to engine knock.

Extensive studies of octane numbers of individual compounds have brought to light some general rules. Table 5 and Figure 2 show the octane numbers of several hydrocarbons. Normal paraffins have the least desirable knocking characteristics, and they become progressively worse as the molecular weight increases. $i$-Paraffins and naphthenes have higher octane numbers than the corresponding normal paraffins, and the octane number of $i$-paraffins increases as the degree of branching of the chain is increased. Olefins have markedly higher octane numbers than the corresponding paraffins and aromatics usually have very high octane numbers.

Comparing the different hydrocarbon series, aromatics (with the exception of benzene) are the hydrocarbons with the highest octane number for the number of carbon atoms in the molecule. It is clear that in order to have a large increase in octane number it is necessary to transform paraffins and naphthenes into aromatics.

In 1922, it was discovered that tetraethyl lead is an excellent antiknock material when added to gasoline in small quantities. Since that time, tetraethyl lead has been added to gasoline to increase its octane number. However, in recent years, in many countries the amount of lead has been decreased by legislation with the target of eliminating it because of possible health problems in the population. The elimination of lead additives has been compensated by increasing the severity of the reforming process (increase in aromatic concentration) as well as adding oxygenated compounds (ethers and alcohols). In addition, to decrease harmful vehicle emissions, the maximum concentration of benzene has been limited in several countries by government regulations. Some regulations establish the addition of oxygenated compounds, which promote more complete combustion of gasoline because of the presence of oxygen in their molecules. Methyl tertiary butyl ether (MTBE) and tertiary amyl methyl ether (TAME) are increasingly used, although they cost more per octane number than lead components. Alcohols such as methanol, ethanol, tertiary butyl alcohol, and alcohol blends can also be used as oxygenated compounds. Their octane numbers are shown in Table 5.

**Table 5** Research Octane Numbers of Pure Hydrocarbons[a]

| Pure hydrocarbons | RON |
|---|---|
| Paraffins | |
|   $n$-Butane | 113 |
|   Isobutane | 122 |
|   $n$-Pentane | 62 |
|   2-Methylbutane | 99 |
|   2,2-Dimethylpropane | 100 |
|   $n$-Hexane | 19 |
|   2-Methylpentane | 83 |
|   3-Methylpentane | 86 |
|   2,2-Dimethylbutane | 89 |
|   2,3-Dimethylbutane | 96 |
|   $n$-Heptane | 0 |
|   2-Methylhexane | 41 |
|   3-Methylhexane | 56 |
|   3-Ethylpentane | 64 |
|   2,2-Dimethylpentane | 89 |
|   2,3-Dimethylpentane | 87 |
|   2,2,3-Trimethylbutane | 113 |
|   $n$-Octane | −19 |
|   $n$-Onane | −17 |

(*continued*)

In 1922, it was discovered that tetraethyl lead is an excellent antiknock material when added to gasoline in small quantities. Since that time, tetraethyl lead has been added to gasoline to increase its octane number. However, in recent years, in many countries the amount of lead has been decreased by legislation with the target of eliminating it because of possible health problems in the population. The elimination of lead additives has been compensated by increasing the severity of the reforming process (increase in aromatic concentration) as well as adding oxygenated compounds (ethers and alcohols). In addition, to decrease harmful vehicle emissions, the maximum concentration of benzene has been limited in several countries by government regulations. Some regulations establish the addition of oxygenated compounds, which promote more complete combustion of gasoline because of the presence of oxygen in their molecules. Methyl tertiary butyl ether (MTBE) and tertiary amyl methyl ether (TAME) are increasingly used, although they cost more per octane number than lead components. Alcohols such as methanol, ethanol, tertiary butyl alcohol, and alcohol blends can also be used as oxygenated compounds. Their octane numbers are shown in Table 5.

**Table 5** Research Octane Numbers of Pure Hydrocarbons[a]

| Pure hydrocarbons | RON |
| --- | --- |
| Paraffins | |
| $n$-Butane | 113 |
| Isobutane | 122 |
| $n$-Pentane | 62 |
| 2-Methylbutane | 99 |
| 2,2-Dimethylpropane | 100 |
| $n$-Hexane | 19 |
| 2-Methylpentane | 83 |
| 3-Methylpentane | 86 |
| 2,2-Dimethylbutane | 89 |
| 2,3-Dimethylbutane | 96 |
| $n$-Heptane | 0 |
| 2-Methylhexane | 41 |
| 3-Methylhexane | 56 |
| 3-Ethylpentane | 64 |
| 2,2-Dimethylpentane | 89 |
| 2,3-Dimethylpentane | 87 |
| 2,2,3-Trimethylbutane | 113 |
| $n$-Octane | −19 |
| $n$-Onane | −17 |

(*continued*)

## Chemistry and Processing of Petroleum

**Table 5** Continued

| Pure hydrocarbons | RON |
|---|---|
| Olefins | |
|   1-Pentene | 91 |
|   1-Octene | 29 |
|   3-Octene | 73 |
|   4-Methyl-1-pentene | 96 |
|   Methylcyclopentane | 107 |
|   Ethylcyclopentane | 75 |
|   1,1-Dimethylcyclopentane | 96 |
|   1,3-Dimethylcyclopentane (*cis*) | 98 |
|   1,3-Diemthylcyclopentane (*trans*) | 91 |
|   1,1,3-Trimethylcyclopentane | 94 |
|   Cyclohexane | 110 |
|   Methylcyclohexane | 104 |
|   Ethylcyclohexane | 43 |
|   1,1-Dimethylcyclohexane | 95 |
|   1,1,3-Trimethylcyclohexane | 85 |
|   1,3,5-Trimethylcyclohexane | 60 |
|   Isopropylcyclohexane | 62 |
| Aromatics | |
|   Benzene | 99 |
|   Toluene | 124 |
|   *o*-Xylene | 120 |
|   *m*-Xylene | 145 |
|   *p*-Xylene | 146 |
|   Ethylbenzene | 124 |
|   *n*-Propylbenzene | 127 |
|   Isopropylbenzene | 132 |
|   1-Methyl-2-ethylbenzene | 125 |
|   1-Methyl-3-ethylbenzene | 162 |
|   1-Methyl-4-ethylbenzene | 155 |
|   1,2,3-Trimethylbenzene | 118 |
|   1,2,4-Trimethylbenzene | 148 |
|   1,3,5-Trimethylbenzene | 171 |
| Oxygenates | |
|   Methanol | 106 |
|   Ethanol | 99 |
|   2-Propanol | 90 |
|   Methyl *tert*-butyl ether (MTBE) | 117 |

[a] Calculated value of pure hydrocarbon from research method rating of clear mixture of 20% hydrocarbon and 80% primary reference fuel (60% isooctane + 40% *n*-heptane).
*Source*: From Refs. 1, 6, and 7.

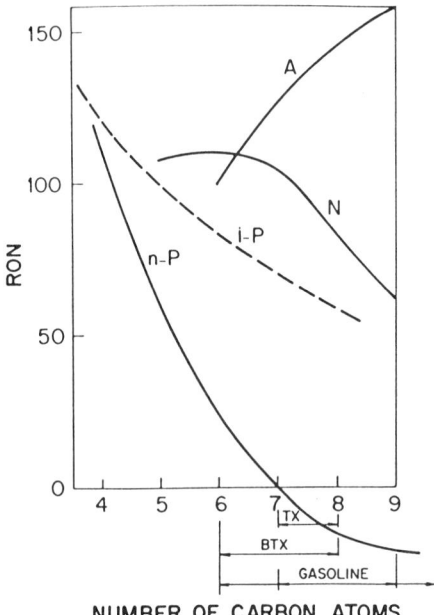

**Figure 2** Research octane number of pure hydrocarbons. $n$-P, normal paraffins; $i$-P, isoparaffins; N, naphthenes; A, aromatics. BTX, Main components of a cut to be reformed to obtain benzene, toluene, and xylenes; TX, to obtain toluene and xylenes. Gasoline, main components of a cut to be reformed to obtain high-octane gasoline.

## VI. CATALYTIC REFORMING

The main objective of the catalytic reforming process is the transformation of low-octane virgin naphthas into high-octane gasolines by increasing concentrations of aromatics and $i$-paraffins. The octane number of a gasoline increases linearly with the increment in aromatic concentration.

An appropriate gasoline for automobiles must have:

1. A maximum octane number to allow a high compression ratio, which increases the efficiency of the motor
2. A minimum capacity to form gums, which are produced by polymerization and oxidation of olefins and which can foul the motor
3. A minimum vapor pressure compatible with room temperature in order to avoid loss of vapors
4. A minimum capacity to produce smoke or smog, which are produced by heavy olefins and aromatics

Catalytic reforming produces a product that fulfills these requirements and produces, with good yields, high-octane gasolines (reformates) with research octane numbers in the order of 100. Catalytic reforming is also the main source of benzene, toluene, and xylenes. This is the reason for the universal application of catalytic reforming, which is the focus of this book.

During catalytic reforming, hydrogen and LPG (propane-butane) are produced in addition to the reformate. Gaseous products ($C_1$–$C_4$) have a low price, consequently, the selectivity of the catalyst for the production of low amounts of gases and high liquid yields is very important. The main purpose of this process is to rebuild or reform hydrocarbon molecules, producing molecules with a higher octane number without changing the number of carbon atoms. By means of paraffin dehydrocyclization and naphthene dehydrogenation, aromatics of the same number of carbon atoms are produced. For example, dimethylcyclopentane, ethylcyclopentane, heptanes, and methylcyclohexane should give toluene. By means of isomerization the molecule is reorganized, increasing the number of branched chains. Paraffin isomerization is also a highly desired reaction because it increases the octane number without producing aromatics, the concentration of which in gasoline is likely to be limited by government regulations in the future. But the problem is that, under reforming conditions, limited amounts of isomers are produced because isomerization is a reaction controlled by thermodynamic equilibrium. In addition, some of the isomers produced are subsequently transformed into other products.

Table 6 shows properties of several typical hydrocarbons that are contained in naphthas. For each hydrocarbon type, the boiling point increases with the molecular weight; the boiling point of an $i$-paraffin is lower than that of the $n$-paraffin of the same molecular weight. After transformation of paraffins into aromatics, the product has a higher density and a lower volatility. Paraffins of $C_6$–$C_9$ have densities of 0.66 to 0.71 g/mL and boiling points of 68.7 to 150.6°C, and the corresponding aromatics have densities of 0.86–0.88 g/mL and boiling points of 80.1–159.2°C. This means that after paraffin dehydrocyclization, a decrease in liquid volume and in vapor pressure will be observed. The dehydrogenation of naphthenes into aromatics produces a decrease in liquid volume and the isomerization of paraffins produces an increase in vapor pressure.

A usual feed to a reforming unit contains 45–70% paraffins, 20–50% naphthenes, 4–14% aromatics, and 0–2% olefins. During the process, the content of aromatics increases to 60–75% (depending on the severity of operation), paraffins and naphthenes decrease to 20–45% and 1–8%, respectively, and olefins virtually disappear. Most of the naphthenes are converted into aromatics, but the aromatization of paraffins is more difficult. For this reason, naphthenic naphthas are more easily reformed. In each series, the conversion is

**Table 6** Properties of Several Paraffins, Naphthenes, and Aromatics

| Compound | Molecular weight | Density (g/mL, 25°C) | Boiling point (°C) |
| --- | --- | --- | --- |
| n-Hexane | 86 | 0.655 | 68.74 |
| n-Heptane | 100 | 0.680 | 98.43 |
| n-Octane | 114 | 0.698 | 125.66 |
| n-Nonane | 128 | 0.713 | 150.58 |
| n-Decane | 142 | 0.730 | 174.10 |
| 2-Methylpentane | 86 | 0.649 | 60.27 |
| 2-Methylhexane | 100 | 0.674 | 90.05 |
| 3-Methylheptane | 114 | 0.702 | 118.93 |
| 2-Methyloctane | 128 | 0.710 | 143.26 |
| 2-Methylnonane | 142 | 0.728 | 166.80 |
| Methylcyclopentane | 84 | 0.744 | 71.81 |
| Cyclohexane | 84 | 0.774 | 80.74 |
| Methylcyclohexane | 98 | 0.765 | 100.93 |
| Ethylcyclohexane | 112 | 0.784 | 131.78 |
| Benzene | 78 | 0.874 | 80.10 |
| Toluene | 92 | 0.862 | 110.63 |
| o-Xylene | 106 | 0.876 | 144.41 |
| m-Xylene | 106 | 0.860 | 139.10 |
| p-Xylene | 106 | 0.857 | 128.35 |
| Ethylbenzene | 106 | 0.863 | 136.19 |
| n-Propylbenzene | 120 | 0.858 | 159.22 |
| i-Propylbenzene | 120 | 0.861 | 152.40 |

easier for hydrocarbons of higher molecular weight. During the process, the percentage of light paraffins is increased because they are produced by hydrocracking or hydrogenolysis.

Most of the reforming reactions are highly endothermic, producing a decrease in the temperature of the reacting stream and catalyst along the bed; therefore, a corresponding decrease in the rate of reactions is produced. To avoid this problem, the total mass of catalyst is distributed in three or four adiabatic reactors with heating before each reactor.

Reforming units are operated under severe conditions (high temperature and low hydrogen partial pressure) to achieve the greatest production of aromatics. However, catalyst stability is a limiting factor for the severity of the operation. Although hydrogen is harmful to the thermodynamics and kinetics of the desired reactions, the process is carried out in the presence of hydrogen to decrease the catalyst deactivation produced by coverage of the active surface with carbonaceous deposits.

The desired reforming reactions require two types of functions on the catalyst: a hydrogenation-dehydrogenation function (metals from Group VIII of the Periodic Table) and a cyclization or isomerization function. The last function is provided by oxides with acid properties. The first bifunctional metal-acid catalyst was introduced in this process in 1949 (8) and had platinum as metallic function. In 1968 (9), the addition of rhenium to promote platinum activity was patented, opening up the era of bimetallic catalysts. Bimetallic catalysts are more stable than monometallic ones, allowing operation at lower pressure with better selectivity to aromatics and a better liquid yield. Some of the first catalysts had silica-alumina or fluorine-promoted alumina as the acid function, but these strong-acid materials produced too much cracking. The acid function of the present reforming catalysts is provided by chloride-promoted alumina, which is more selective for isomerization reactions and its acidity can easily be regulated by changing the chloride concentration. These catalysts contain about 0.2–0.4% by weight platinum as metal function with one or more metals as modifiers, such as rhenium, tin, germanium, and iridium. This metal function is supported on the acid function: alumina with acidity promoted by the addition of nearly 1% chloride. Catalysts may be presulfided to decrease the great initial hydrogenolytic activity of the metal function. This reaction is very exothermic and may produce large amounts of methane. Catalysts with tin do not require presulfidation.

Platinum-rhenium catalysts are of two types: those with equal amounts of Pt and Re, with typical weight percentages of Pt(0.3%)-Re(0.3%)-S(0.04%)/$Al_2O_3$-Cl(0.9%), and the so-called skewed metal catalyst (10) with more rhenium than platinum—typically 0.2% Pt and 0.4% Re.

All naphthas contain sulfur, a poison of the catalyst metal function that must be eliminated before the reforming process. Elimination is accomplished by the hydrodesulfurization process: sulfur compounds are hydrogenolyzed, producing $H_2S$, which is easily separated. The hydrodesulfurization catalyst has to be a low-activity hydrogenating agent to avoid the hydrogenation of aromatics but active enough to hydrogenolyze the sulfur compounds. Usual catalysts are Co-Mo, Ni-Mo, or Ni-W supported on alumina. These hydrotreating catalysts also eliminate nitrogen (from nitrogen compounds) as $NH_3$, hydrogenate olefins, and withhold arsenic and metals.

## A. Reformer Feed

The most common feed to the catalytic reforming unit is the virgin (straight-run) naphtha distilled to obtain a narrower cut to be reformed. The light fraction is eliminated because it will mainly be cracked to gases. The heavy fraction is eliminated because it promotes catalyst deactivation (11). The feed may also contain gasoline produced in thermal or catalytic cracking units that contain olefins. These olefins are hydrogenated during the hydrodesulfurization of the feed, performed before reforming. The boiling range of the feed varies according to the process and products needed (gasoline or aromatics). If aromatics are to be produced, the reforming unit is run at high severity to produce a highly concentrated aromatic product.

The reforming of hydrocarbons of six carbon atoms differs greatly from the reforming of heavier hydrocarbons. Six-carbon naphthenes are unique naphthenes with a larger octane number than the corresponding aromatic, as shown in Figure 2 and Table 5 (methylcyclopentane 107 and cyclohexane 110, compared with benzene 99). The increase in octane number in the transformation of a hexane into benzene is not as large as in the aromatization of paraffins of a larger number of carbon atoms. Moreover, the transformation of hexanes into benzene is slower and less selective than the transformation of heavier paraffins. For this reason, heavier cuts are reformed (for example, a 105–160°C cut) in many plants. This is accomplished by separation of the $C_6$ straight-run naphtha prior to reforming, followed by isomerization of the $C_5/C_6$ $n$-paraffins to high-octane $i$-paraffins. The isomerate is added to the gasoline pool, increasing the octane number of the volatile fraction. The requirement of a low benzene concentration in gasoline is also met in this way. The lower part of Figure 2 shows the most convenient feed compositions to obtain gasoline or aromatics.

When benzene, toluene, and xylenes are the desired reforming products (BTX operation), an 80–120°C cut is more suitable as feed. Some petrochemical plants reform the $C_7$–$C_8$ fraction, thus obtaining toluene and xylenes (TX) as in Figure 2. In such a case, benzene is obtained by toluene or heavier aromatic dealkylation or extracted from other sources (i.e., pyrolysis naphthas and coke oven light oil).

Table 7 shows the composition of a typical naphtha feed used to obtain BTX. The composition of a naphtha cut to be reformed for high-octane gasoline production could be 1.5% $C_6$, 6.5 $C_7$, 39% $C_9$, 1% $C_{10}$. Requirements for benzene reduction will produce a reduction in reformer severity, changes in feedstock, and additional downstream processing. For example, instead of separating the light fraction before reforming, another possibility for decreas-

**Table 7** Composition of a Naphtha Feed for BTX Production, wt %[a]

|       | P  | N  | A | Total |
|-------|----|----|---|-------|
| $C_6$ | 20 | 6  | 2 | 28    |
| $C_7$ | 33 | 12 | 3 | 48    |
| $C_8$ | 13 | 8  | 3 | 24    |
| Total | 66 | 26 | 8 | 100   |

[a] P = paraffins, N = naphthenes, A = aromatics.

ing benzene concentration is to reform the whole and then split out the light reformate with benzene extracted for petrochemical usage or for alkylation with light olefins. The last reaction produces an increase in volumetric yield and octane number and the product can be added to the gasoline pool. New processes claim the possibility of alkylation of benzene diluted in the reformate.

## REFERENCES

1. J. G. Speight, *The Chemistry and Technology of Petroleum*, Chemical Industries 3. Marcel Dekker, New York, 1980.
2. *The Petroleum Handbook*, Royal Dutch/Shell Group of Companies, 6th ed. Elsevier, Amsterdam, 1983.
3. L. F. Hatch, and S. Matar, *From Hydrocarbons to Petrochemicals*, Gulf Publishing Co., Houston, 1981.
4. R. E. Kirk and D. F. Othmer, *Encyclopedia of Chemical Technology*, 3rd ed. Wiley, New York, 1982.
5. W. L. Nelson, *Petroleum Refinery Engineering*, 4th ed. McGraw-Hill, New York, 1958.
6. American Petroleum Institute Research Project 45, Sixteenth Annual Report (1954).
7. F. G. Clapetta, R. M. Dobres, and R. W. Baker, Catalytic reforming of pure hydrocarbons and petroleum naphthas. In *Catalysts* (P. H. Emmett, ed.), Reinhold, New York, 1958, Vol. VI, p. 495.
8. V. Haensel, U.S. Patents 2,479,109 and 2,479,110 (1949).
9. H. E. Kluksdahl, U.S. Patent 3,415,737 (1968).
10. P. A. Larsen. In *Question and Answer Session—Hydroprocsesing, Ketjen Catalysts Symposium* (H. Th. Rijnten and H. J. Lovink, eds.). AKZO Chemie, Ketjen Catalysts, The Netherlands, 1986, p. 75.
11. J. N. Beltramini, R. A. Cabrol, E. J. Churin, N. S. Figoli, E. E. Martinelli, and J. M. Parera. *Appl. Catal. 17*, 65 (1985).

# 2
# Basic Reactions of Reforming on Metal Catalysts

## Zoltán Paál

*Institute of Isotopes of the Hungarian Academy of Sciences, Budapest, Hungary*

## I. INTRODUCTION

Since the first application of industrial reforming for fuel upgrading using supported Pt catalysts, this large-scale commercial process has proved to be a driving force for research on metal-catalyzed hydrocarbon reactions. Laboratory studies, which frequently employed conditions vastly different from industrial ones, provided a scientific background for catalytic reforming, and these apparently remote investigations prepared the ground for several industrially important innovations in the past and will do so in the future. This chapter concentrates on a few points of laboratory-scale studies that will be of value for industry.

Several catalytic reactions in reforming involve rearrangement of the hydrocarbon skeleton; hence they can be termed "skeletal reactions." These include aromatization, isomerization, $C_5$ cyclization, and hydrogenolysis. The first three reactions are economically useful; the last one is disadvantageous for operation of a reforming unit.

This chapter reviews present-day knowledge on the possible metal catalysts and mechanisms of reactions catalyzed by them. This will be done by pointing out relevant problems and the numerous hypotheses suggested for their solution rather than by presenting ready and apparently finalized theories. The diversity of ideas, methods, approaches, and so forth truly reflects the present situation, in which the experimental results as a function of several parameters lack well-established and generally valid interpretations. For this reason, a relatively large number of references (to both classic and recent papers) have been included; still, the literature covered is far from being comprehensive.

## II. POSSIBLE MECHANISMS OF THE REACTIONS

The chemistry of chemical processes in catalytic reforming is discussed in Chapter 3 of this book by Parera and Fígoli. The main body of knowledge of reforming has been incorporated in an excellent review on the chemistry and engineering of reforming (1) summarizing much valuable information obtained in the 1960s and 1970s, which is not repeated here. A more concise review dealing with both chemistry and industrial aspects has been published recently (2). It was pointed out there that *all* the reactions important in reforming (aromatization, $C_5$ cyclization, isomerization, and fragmentation) can also proceed over catalysts having metallic activity only; this point will be stressed throughout the present chapter. The majority of the measurements to be discussed have been carried out in the temperature range 500–650 K and pressures up to 1 bar, which are much lower than conditions typical of industrial reforming. Insight into these reactions is derived from the useful information provided by these studies.

*Aromatization* (or $C_6$ dehydrocyclization) was first observed by a Russian group as the formation of a second aromatic ring from an alkylbenzene over monofunctional Pt/C catalyst; the same group also reported the formation of an aromatic $C_6$ ring from alkanes (3). Later they described the metal catalyzed $C_5$ *cyclization* of alkanes to alkylcyclopentanes (4). The aromatic ring is very stable under these conditions but $C_5$-cyclization is reversible: opening of the $C_5$ ring to alkanes also takes place (5). Metal-catalyzed *isomerization* (6) may occur (i) by the formation and splitting of the $C_5$ ring (7) and (ii) in the case of hydrocarbons whose structure does not allow the formation of a $C_5$-cyclic intermediate by a so-called bond shift mechanism (8, 9). The former isomerization route is often termed the *cyclic mechanism* (10). I prefer the name $C_5$-cyclic mechanism (11) because bond shift may also involve a $C_3$-cyclic intermediate (12). *Hydrogenolysis* of alkanes has also been a well-known and widely studied reaction (13). The reaction mechanisms of these reactions and their relative importance over various catalysts have been comprehensively reviewed (11–17). Here only a brief summary will be given.

Early ideas about *aromatization* (18) assumed dehydrogenation of an open-chain hydrocarbon and subsequent ring closure of the olefin directly to give a six-membered ring. Aromatization on carbon-supported metals was interpreted in terms of a direct 1,6 ring closure of the alkane molecule without its preliminary dehydrogenation (3). The concept of this 1,6 ring closure has fallen temporarily into the background in favor of the two-dimensional mechanism (1), which described very satisfactorily the reactions observed on bifunctional catalysts under industrial conditions. Still, the possibility of 1,6 ring closure has again surfaced because of the new evidence. The stepwise dehydrogenation of heptanes to heptenes, heptadienes, and heptatriene followed by cyclization has been shown over oxidic (19) and also unsupported metal catalysts (20, 21). This triene mechanism has been regarded as one of the possible reaction pathways over Pt/Al$_2$O$_3$, together with another, direct C$_6$ ring closure with no intermediates appearing in the gas phase (22). Tracer studies using $^{14}$C-labeled n-heptane indicated that, over Pt supported on a nonacidic Al$_2$O$_3$, 1,6 ring closure was the main reaction of aromatization (the pathway of which has not been established); 1,5 ring closure, opening, and repeated cyclization might also occur (23).

The assumption of dienes and trienes does not mean that these intermediates should appear in the gas phase. Instead, a "hydrocarbon pool" is produced on the catalyst surface upon reactive chemisorption of the reactant(s). These surface hydrocarbon species may undergo dehydrogenation, rehydrogenation, and, if they have reached the stage of surface olefins, double bond or *cis-trans* isomerization (11, 21). Their desorption is possible in either stage (hence hexenes, hexadienes, etc. may appear as intermediates). The true intermediates of aromatization are *surface* unsaturated species: those appearing in the gas phase are the products of surface dehydrogenation and desorption processes. Desorption should be less and less likely with increasing unsaturation of the surface intermediates (24). The loss of hydrogen produces either *cis* or *trans* isomers. The *cis* isomer of hexatriene is expected to aromatize rapidly, the chance of its desorption being practically zero. It is obvious, however, that the probability of dehydrogenation producing such structures is fairly low. The *trans* isomer, on the other hand, has to isomerize prior to cyclization, and during this process it also has a minor chance of desorbing to the gas phase. Alternatively, it may be a coke precursor (11, 21). It is also a misunderstanding to suggest that thermal cyclization of triene intermediates would have any importance in heterogeneous reactions (24) just because a gas-phase hexatriene molecule would cyclize spontaneously and very rapidly at or above about 400 K (21). The temperatures in any catalytic reaction exceed this value.

The role of hydrogen in this scheme is twofold. First, the metal-catalyzed *trans→cis* isomerization through half-hydrogenated surface intermediates (25) requires hydrogen. Second, the degree of dehydrogenation of the surface entities is often too deep; thus their removal to the gas phase is a hydrogenative process, e.g.,

$$C_6H_4(ads) + 2H(ads) \rightarrow C_6H_6(gas)$$

The reality of such processes has been confirmed experimentally by at least two independent methods, temperature-programmed reaction (26) and the transient response method (27) (see also Section V.B.5).

The formation of aromatics from alkanes with less than six C atoms in their main chain (e.g., methylpentanes) is also possible but proceeds much less readily than direct $C_6$ dehydrocyclization (11, 28). This mechanism should involve a sort of dehydrogenative bond shift isomerization. The same is valid for the $C_5 \rightarrow C_6$ ring enlargement, which can take place over metal catalysts but is not a favored process (28).

$C_5$-cyclization and $C_5$-cyclic isomerization are closely related reactions and they probably have a common surface intermediate (7, 29). An "alkene-alkyl insertion" (like that mentioned for $C_6$ cyclization) (30) and a "dicarbene" and also a "dicarbyne" mechanism involving surface intermediates attached by two double or two triple bonds to the surface (10) were proposed. An alternative pathway would involve a much less dissociated intermediate, where the position of the cycle would be roughly parallel to the catalyst surface (31). The relatively low extent of dehydrogenation of this latter intermediate has been shown by deuterium tracer studies (32). The forming or splitting of this type of intermediate of the $C_5$ cycle and its preferred 1,3 attachment to two sites of the catalyst have been suggested (11), with experimental support obtained by comparing several open-chain (33) and $C_5$-cyclic hydrocarbons (34).

The parallel occurrence of $C_5$-cyclic and bond shift isomerization has been studied extensively by the use of $^{13}C$ tracer (10, 29). Whenever the $C_5$-cyclic mechanism was possible, its importance in the overall conversion was usually found to be predominant; at the same time, strong sensitivity to the structure of the reactant and to the catalyst has been observed (16) (see also later). Bond shift means a transfer of a $C-C$ bond to a next carbon atom (1,2 bond shift); this is the route of this reaction on Pt and Pd. The interaction of these metals with the reactant is, however, claimed to be different (11); for example, the product composition pointed to the interaction of two methyl groups of 2,2-dimethylbutane with Pt, whereas one methyl group and the secondary C atom formed the preferred active surface intermediate on Pd (35). A 1,3 bond shift of methylpentanes has also been detected by $^{13}C$ tracer studies over iridium at 493 K (36).

Hydrogenolysis has often been found to be related to isomerization reactions (10–12, 35, 37). The Anderson-Avery mechanism (17) is the most widely accepted one for this reaction. This involves a rather deeply dehydrogenated surface intermediate. If the starting molecule has at least three carbon atoms, a 1,1,3 interaction with the surface is preferred.

For a more detailed discussion of the suggested surface intermediates and proposed mechanisms of metal-catalyzed reactions, the reader is referred to the literature (2, 10–16, 29, 37).

## III. CATALYSTS AND THEIR ACTIVE SITES

Platinum is the most important catalyst that is able to catalyze all types of hydrocarbon reactions mentioned. Metals of Group 8–10 of the Periodic Table (except for Fe and Os) as well as Re and Cu have been found to catalyze aromatization to at least a slight extent (11). Hydrogenolysis and bond shift isomerization seem to be an inherent property of several metals (38, 39). On the other hand, $C_5$-cyclization and related isomerization of 3-methylpentane were found (39) to be restricted to a few metals: Pt, Pd, and Ir (to some extent Rh and, perhaps, Ru). Pt, Pd, Ir, and Rh have another common property: they split a hydrocarbon molecule predominantly into two fragments (39). Multiple hydrogenolysis prevails over other metals. A fragmentation factor ($\zeta$) has been defined as the average number of fragments per decomposed $C_n$ hydrocarbon molecule. Its value is around 2 in the case of single hydrogenolysis and can go up to $n$ with multiple hydrogenolysis. Hardly any change is seen in the value of the zeta factor as a function of the degree of conversion with single hydrogenolyzing metals, whereas it usually increases up to $n$ in the case of multiple hydrogenolysis (Figure 1) (40). The intermediate position of Ru between singly and multiply hydrogenolyzing metals can be seen. Remember that this classification has nothing to do with the hydrogenolysis activity (13), which can

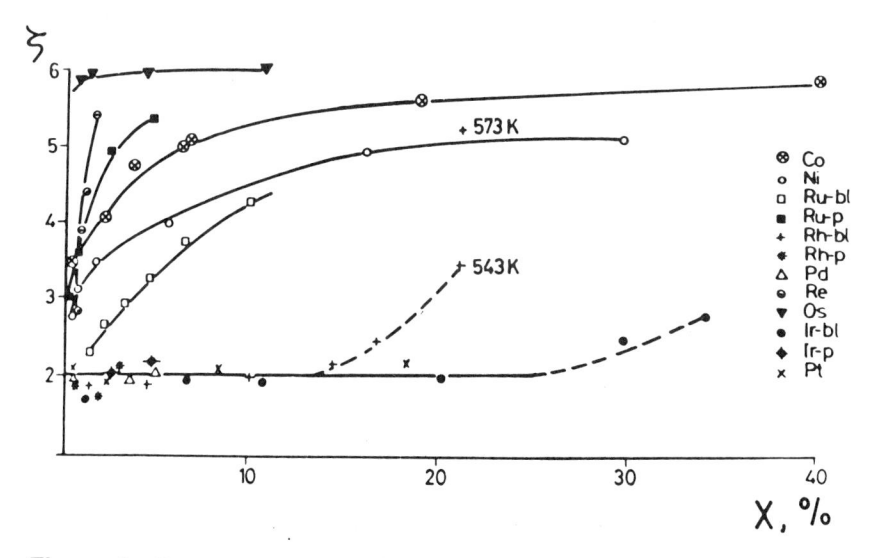

**Figure 1** Number of fragments per decomposed molecule ($\zeta$) in the hydrogenolysis of 3-methylpentane over various metal blacks, as a function of the percent conversion to hydrogenolysis products ($X$). Abbreviations: bl = high-area metal black; p = low-area metal powder. From Ref. 40.

be quite pronounced in the case of single hydrogenolysis, too (e.g., with Rh or Ir).

The probability of the rupture of individual C−C bonds of an alkane molecule is not equal, either. An ω factor has been defined as the ratio of actual rupture and random rupture at a given C−C bond (41). In fact, this calculation is valid for the case of single hydrogenolysis only; still, ω values provide useful information when the value of $\zeta$ is between 2 and 3 by considering the amounts of the larger fragments (e.g., the products $C_5$, $C_4$, and $C_3$ from n-hexane and disregarding the slight $C_1$ and $C_2$ excess).

Multiple fragmentation can occur by true disruption of the molecule during one sojourn on the catalyst (e.g., Os in Figure 1) or, alternatively, by subsequent end demethylation (e.g., Ni). A way to distinguish between these cases is to calculate the relative amounts of fragments $>C_1$ and methane. Their ratio, the so-called fission parameter ($M_f$), permits one to distinguish between terminal, multiple, or random hydrogenolysis (42). Both factors are strongly dependent on the nature of the metal ($M_f$ was determined for Pt, Ir, Ni-Cu, and Pd (35); ω for Pt, Pd, Ir, and Rh (43)).

Even though on a given catalyst various skeletal reactions are catalyzed, they are catalyzed most likely by different sites. The reactions whose active intermediate forms more than one chemical bond with the surface require active centers consisting of more than one atom. One Pt atom was found to be sufficient to deydrogenate propane to propene (44), whereas aromatization required ensembles with up to three Pt atoms (45). Active sites for $C_5$-cyclic isomerization should consist of more surface metal atoms than those for bond shift isomerization (46). Under certain conditions (e.g., lower temperature, higher hydrogen excess), however, single-atom active sites may also be operative (47). The parallel occurrence of multisite and single site reactions has been reviewed (12, 15).

Single-crystal studies permit one to elucidate the possible role of surfaces of various geometries by artificially creating various crystal surfaces (48). These results agree well with the conclusions mentioned in the preceding paragraph: platinum crystal faces with sixfold (111) symmetry—where active ensembles of three atoms are abundant—exhibited enhanced aromatization activity as compared with those with fourfold symmetry (100) planes. Single crystals with steps and corner sites were, as a rule, more active than flat planes (49). Much lower crystal plane sensitivity was observed in methylcyclopentane ring opening (50): two-atom ensembles can be present on any crystal plane configuration. Iridium single-crystal surfaces were less active than corresponding Pt surfaces in cyclohexane dehydrogenation and n-heptane dehydrocyclization at low pressures (51).

The catalytic properties of various single crystal planes have been compared directly with those of $Pt/Al_2O_3$ catalysts with various metal loadings and different crystallite sizes (52, 53). The importance of the so-called $B_5$ sites (containing five atoms along a step, sometimes termed a ledge structure) has been pointed out on both single-crystal and 10% $Pt/Al_2O_3$ catalysts for bond shift isomerization and related hydrogenolysis. A sample containing 0.2% $Pt/Al_2O_3$ and very small crystallites was active in catalyzing $C_5$-cyclic reactions only. No single-crystal analogy could be found for the catalytic behavior of this very disperse catalyst.

Ledge structures were found to be indispensable for alkane or cycloalkane hydrogenolysis on Pt single crystals (54). Rounded crystallites (like a field emission tip (55)) must contain a large number of these high-Miller-index sites; hence rounded crystallites should promote hydrogenolysis. Indeed, a supported Pt catalyst exhibiting large rounded crystallites showed higher hydrogenolysis selectivity than 6% $Pt/SiO_2$ (EUROPT-1) (56). This latter catalyst has smaller but more perfect crystallites approximated well with 55-atom cubooctahedra, as indicated by x-ray diffraction investigations (57). Aromatization has been attributed to the six-atom (111) facets of the cubooctahedra, while other skeletal reactions have been ascribed to (100)-type facets (58). The hydrogenolysis activity of single-crystal and disperse Ni catalysts also correlated well with the abundance of their ledge structures (59).

Not only the atoms constituting the active sites but also the adjacent atoms may be important for catalysis. All hydrocarbon reactions mentioned involve the dissociation of at least one C−H bond, which can occur with or without participation of hydrogen adsorbed on the catalyst (60). The kinetics of skeletal rearrangements of alkanes have been discussed in these terms, where the initial step of all reactions of, e.g., a heptane isomer would be a reactive chemisorption (61):

$$C_7H_{14} + H_a \rightarrow (C_7H_{15})_a \rightarrow (C_7H_{13})_a + H_2$$

as opposed to a dissociative chemisorption

$$C_7H_{14} \rightarrow (C_7H_{13})_a + H_a$$

Not all sites of the metal surface proper are active catalytically: chemisorption sites (or "landing sites" (62)) and reactive sites have been distinguished. One approach attributes the former to flat planes, while reaction would, accordingly, proceed on ledge structures or $B_5$ sites (53) (Figure 2). The underlying ideas are valid not only for large single-crystal planes but also for areas comprising terraces a few atoms wide and adjacent steps in the case of disperse catalysts.

The environment of the active metal atom can modify its catalytic properties, especially in the case of single-atom active sites (47). One of these latter factors may be the interaction of a support atom in the vicinity of the one active atom (63). This may be especially marked with $TiO_2$ supported metals, where

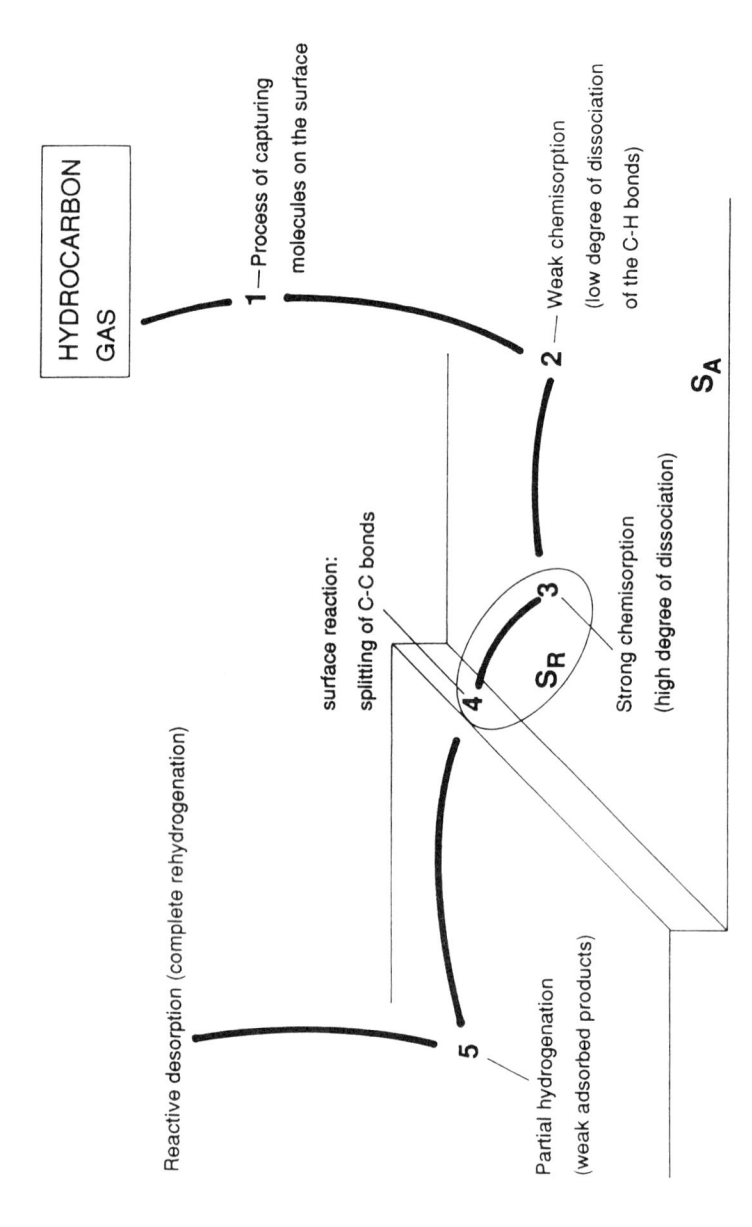

**Figure 2** Possible adsorption sites and reactives sites and migration of surface species on a stepped single-crystal surface. Reproduced by permission from Ref. 53. (Copyright by Academic Press, Inc., 1987.)

## Basic Reactions of Reforming on Metal Catalysts

so-called strong metal-support interactions may occur (64). This phenomenon has minor importance in most hydrocarbon reactions discussed. One important exception is the ring opening of cyclopentanes (a very important step of metal-catalyzed isomerization, too). "Nonselective" $C_5$ ring opening—which may occur in any position of the ring producing $n$-hexane and methylpentanes from methylcyclopentane—has been attributed to so-called adlineation sites at the metal-support borderline. The "selective" reaction occurring far from the alkyl substituent(s)—producing only methylpentanes from methylcyclopentane—in turn was attributed to sites consisting of metal atoms only (65). This suggestion obtained experimental support when artificial metal-support borderlines were created by vacuum depositing $Al_2O_3$ on to small Pt particles and observing that the abundance of $n$-hexane in the ring opening products increased (66).

One of the most important support effects in reforming reactions is the contribution of its acidic sites to the final product composition. The classical picture of reforming on so-called bifunctional catalysts containing both metallic and acidic sites attributes dehydrogenation reactions to the metallic sites, ring closure and $C_5 \rightarrow C_6$ ring enlargement reactions to acidic active centers (67). These are not necessarily situated on the same catalyst particles: a gas-phase transfer of reactive intermediates (which are stable alkene or cycloalkene molecules) in physical mixtures of nonacidic supported metal and acidic catalyst grains would also result in almost the same product composition (1). The fact that a nonnegligible fraction of the products can be produced over the metallic sites, "makes it necessary to alter the reaction scheme"—that is, the original two-dimensional scheme of Ref. 67—"to include the additional reactions (particularly cyclization and isomerization) which occur on the metal surface alone, without involving acidic centers" (1, pp. 286–287).

Attempts to construct such alternative, more complete reaction schemes have been put forward by Parera et al. (138) and Paál (68) as shown in Figure 3. The scheme includes stepwise aromatization pathway on the metallic sites; the metal-catalyzed $C_5$ ring closure, ring opening, and isomerization are also shown. It is believed that the reactions included in this scheme should be kept in mind, even if the cooperation of metallic and acidic centers (according to Mills et al. (67)) describes fairly well the processes occurring in the commercial reformer at steady state (1, 67).

A family of rather active catalysts for aromatization of hexane has been developed (69). These contain Pt on alkali or alkali-earth modified zeolites and have no or negligible acidity, and hence are monofunctional. Another type of support action may be effective here: reactive chemisorption of the reactant over metal particles within the zeolite channel, which involves a geometric constraint. If the reactant is forced within the narrow zeolite channel to form a six-ring pseudocycle and this comes in contact with an adjacent small Pt particle, $C_6$ dehydrocyclization will be facilitated (confinement model (70)). This hypothesis, which proposed $C_6$ ring closure over single Pt atoms of small Pt

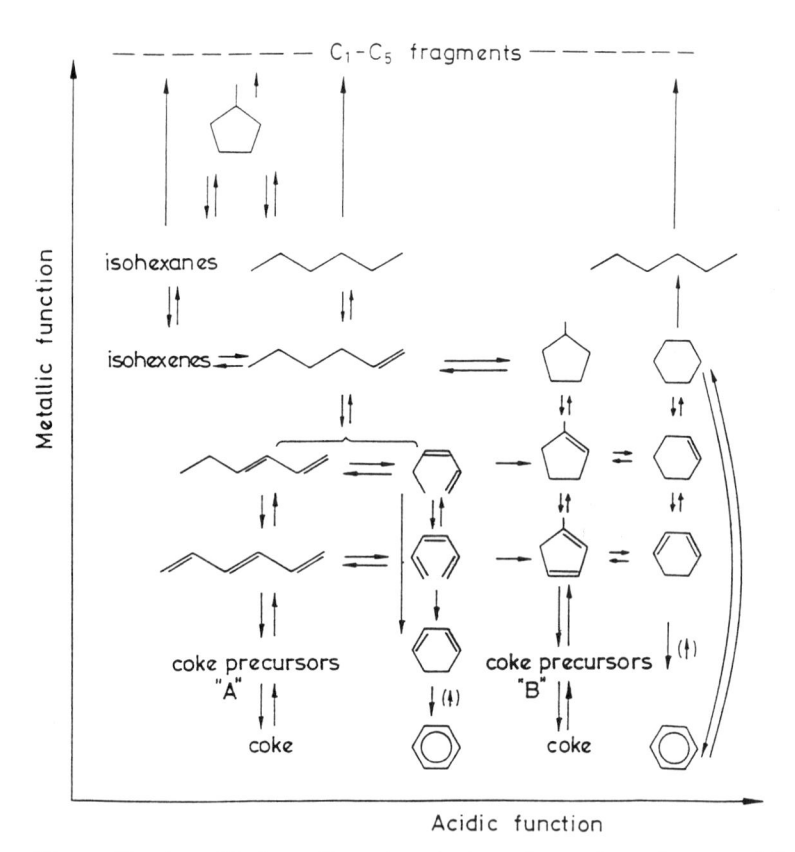

**Figure 3** A modified two-dimension reaction scheme of reactions over bifunctional catalysts. Adapted from Ref. 68. (Copyright by Academic Press, Inc., 1987.)

clusters exposed in the windows of zeolite cages of Pt/KL (70), represents an interesting example of single-atom active sites. A direct six-ring closure of *n*-hexane over nonacidic Pt/KL between 600 and 690 K was suggested and the confinement model was found to be valid by comparing this catalyst with Pt/KY (71). However, the confinement model has been claimed to be unsatisfactory because high benzene selectivity from *n*-hexane was observed with various zeolite-supported Pt catalysts (72). The role of the L-zeolite framework was claimed to be the stabilization of extremely small Pt particles in a nonacidic environment. A further possible reason for the enhanced dehydrocyclization selectivity over Pt/KL catalysts may be the hindered deactivation of metal particles within the zeolite channels (73). The course of action over these catalysts as well as that of other monofunctional supported Pt catalysts has been found to be in agreement with the stepwise aromatization pathway (74).

Counterclaims have appeared regarding the importance of basic sites in a zeolite framework: additional basic sites in the zeolite support did not enhance benzene formation from *n*-hexane at 733–783 K (72), whereas increasing aromatization was observed at 600 K over 0.8% Pt catalysts supported on Y-zeolites of increasing basicity (75). The interaction of *basic* sites with Pt may be another reason why Pt/MgO catalysts (with no confinement geometry) have also exhibited high aromatization selectivity (76). This interaction may manifest itself in electron donation from the alkaline support to the metal, resulting in (i) enhanced alkane activation and (ii) hindered hydrogenolysis and coke formation (77). In terms of the polyene mechanism, good aromatization catalysts might expose active ensembles favorable for the *cis*-triene precursor structure; their deactivation by the formation and/or polymerization of *trans*-polyenes in turn, is hindered. Both geometric and electronic effects have been proposed as the underlying reason. A third factor may be the better hydrogen utilization facilitating isomerization of unsaturated intermediates and hydrogenative desorption of aromatic end products.

## IV. RELATIONSHIP BETWEEN REACTANT STRUCTURE AND REACTIVITY

It can easily be understood that not all hydrocarbons have the same reactivity: the reactant structure permits or excludes certain catalytic processes (e.g., alkylbutanes are not able to participate in $C_5$-cyclic reactions). Furthermore, even if the reactions are theoretically possible, some structures must be more favorable for certain reactions.

The species in the partly dehydrogenated hydrocarbon pool responsible for stepwise aromatization seems to "remember" the structure of the hydrocarbon feed. For example, the structure of *trans*-2-hexene is more favorable for $C_5$ cyclization than for aromatization; more hexadienes appear from 1-hexene than from both 2-hexene isomers (78). Hardly any hexatriene was formed from 2,4-hexadiene, whereas 1,4-hexadiene produced rather high amounts of *trans*-1,3,5-hexatriene (21). It seems likely that there is an aromatization route involving the reaction 2-hexene $\rightarrow$ 2,4-hexadiene followed by cyclization of the latter, bypassing the stage of hexatrienes (11, 21). This route has also been confirmed by temperature-programmed reaction (TPR) studies in which a reactant (such as 1-hexene or 2-hexene) was first adsorbed on Pt black and Pt/$Al_2O_3$ catalysts and the removal of a product(s) of a surface reaction(s) followed during programmed heating (26). These measurements also indicate that a cyclization step of hexatriene $\rightarrow$ cyclohexadiene belongs to the rapid elementary reactions as compared with dehydrogenation and/or *trans*$\rightarrow$*cis* isomerization processes.

As far as skeletal rearrangement reactions are concerned, hydrocarbons were assumed (79) to react according to their structure as "$C_2$ units" (consisting of a two-atom molecular fraction containing primary and/or secondary C atoms only) or "iso-units" (in which one of the reacting C atoms is tertiary or quaternary). These structures react in a different way in skeletal isomerization and their hydrogenolysis pattern is also different. Garin and Maire (80) define an "agostic entity," which is the part of the molecule interacting with the catalyst surface during reaction. They consider the electronic state of the catalyst as the main factor determining the prevailing pathway of the reaction (80). This suggestion is based on the variation of the ratio of iso-unit mode to $C_2$ unit mode reactions. This ratio is between 0.85 and 1.2 on Pt single-crystal planes and as well as over $Al_2O_3$-supported Pt with dispersions between 0.04 and 1. If the predominating surface interaction is $\sigma$-alkyl, iso-unit mode reactions will prevail, while multiple carbon-metal bonds favor $C_2$-unit reactions. This is claimed to be the main reason why the catalytic properties of different metals are different. If the electronic factor has already decided the pathway of the primary reactive absorption, geometric effects control whether bond shift or $C_5$-cyclic mechanisms (with or without hydrogenolysis) will follow and in what ratio. Neohexane (2,2-dimethylbutane), consisting of a $C_2$-unit and an iso-unit, is an excellent probe molecule for studying the reactions of these structures (with or without fragmentation) (80a).

There is experimental evidence that a hydrocarbon with a $C_5$ main chain will undergo predominantly $C_5$ cyclization (and $C_5$-cyclic isomerization). At the same time, the reactivities of 3-ethylpentane and 3-methylhexane (= "2-ethylpentane") are different over Pt-black (33) and 6.3% Pt/$SiO_2$ (EUROPT-1) (81), simply because the attachment of a flat-lying intermediate is not favored by the steric structure of the latter molecule. This was one of the author's arguments for the flat-lying intermediate of $C_5$-cyclic reactions (33). The same conclusion can be drawn from the different reactivity of *cis*- and *trans*-dimethylcyclopentanes (34). The presence of a quaternary C atom in 1,1-dimethylcyclopentane, in turn, promotes its ring enlargement and subsequent aromatization (11, 34, 82). Toluene formation from 3,3-dimethylpentane is also large in comparison with other alkylpentanes with no quaternary C atom (81).

Bearing in mind the different energies of activation for different bond ruptures (11, 29), totally random rupture is an exception rather than a rule. The position of C−C bond rupture also showed a reactant structure sensitivity on supported Pt (83) as well as on unsupported Pt, Pd, Ir, and Rh catalysts (43). The preferential position of hydrogenolysis of *n*-hexane, heptane, and octane isomers indicated that the splitting of a $C_5$ unit is facilitated from all of those molecules (to give $C_1 + C_5$, $C_2 + C_5$, and $C_3 + C_5$ fragments, respectively) (84).

## V. EFFECT OF SECOND COMPONENTS ON THE ACTIVE SITES OF THE CATALYSTS

Modern reforming catalysts often contain at least one second component in addition to Pt (see also Chapter 3 and Chapter 7). These additives are, as a rule, metals. Nonmetallic components are either deliberately added (e.g., sulfur or hydrogen) or just cannot be avoided (e.g., carbon, oxygen, or nitrogen). Their effect is briefly reviewed in the following section.

## A. Metallic Components

Some of the added metals have catalytic properties on their own (Ir, Re), while others are catalytically inactive (Sn). Bimetallic or multimetallic catalysts are often regarded as alloys, although alloying would mean an intimate mixing of all components which would create new phases. For example, the shifts in the infrared band of chemisorbed CO indicate an electronic interaction between Sn and Pt (85). Even if the formation of such phases is possible, it should not be assumed that they are really formed under catalytic conditions.

Another assumption—especially with catalytically inactive additives such as Cu or Au—is that the second metal is totally inert and its effect is purely geometric (to dilute multiatomic active ensembles). This possibility has been discussed using the results of electron spectroscopy (12). Although several conclusions about ensemble effects and sizes of active centers have been drawn on the basis of this method (35), in reality this assumption may be far from true, since it is likely that there is no geometric effect without electronic interaction (86). Examples for bimetallic systems are mentioned throughout this chapter but a detailed discussion of bimetallic catalysts is beyond the scope of this chapter. The reader is referred to Chapter 7 of this book and other excellent review papers (87, 88).

## B. Nonmetallic Components

### 1. Sulfur and Nitrogen

It is generally accepted that under industrial conditions, Pt-Re/Al$_2$O$_3$ operates in the sulfided state (89). These catalysts were found to have a much lower sulfur tolerance than Pt/Al$_2$O$_3$ (90). The primarily adsorbed, irreversibly held sulfur alone was responsible for suppression of fragmentation reactions and consequent enhancement of aromatization over Pt-Re/Al$_2$O$_3$, while an additional, reversibly held sulfur was necessary to achieve this with Pt/Al$_2$O$_3$ (90).

A comparative study (91) of irreversibly chemisorbed sulfur using $^{35}$S-labeled H$_2$S for catalyst sulfidation shows that its amount increases in the order: Pt/Al$_2$O$_3$ < Pt-Re/Al$_2$O$_3$ < Re/Al$_2$O$_3$. The behavior of Pt-Re/Al$_2$O$_3$ in comparison with Pt/Al$_2$O$_3$ and Re/Al$_2$O$_3$ is nonadditive. At 773 K, pronounced desorption in H$_2$ is observed over Pt/Al$_2$O$_3$ followed by a minor amount of exchange with a mixture of H$_2$ and nonradioactive H$_2$S; small desorption and small exchange are seen with Re/Al$_2$O$_3$, whereas the mixed catalyst exhibits "Pt-like" desorption and "Re-like" exchange properties. This nonadditivity indicates the formation of mixed active sites in Pt-Re/Al$_2$O$_3$.

Following earlier investigations (e.g., 68), the effect of sulfur on supported Pt, Ir, Pt-Ir, Ru, and Pt-Re catalysts was compared (92, 93). It was reported that sulfur suppressed selectively the hydrogenolysis activity; the remaining activity in the sulfided state was 1 to 6% of that observed in the sulfur-free state. The activity for skeletal isomerization, in turn, decreased to a much lesser extent (residual activity, 25 to 75%); in two cases (with Ir and Pt-Re catalysts) the relative isomerization rates were even *higher* after sulfidation (Figure 4). On the other hand, the basic additive pyridine must have been selectively attached to acidic sites as it suppressed mainly isomerization rates (Figure 4). Hence, bifunctional activity could manifest itself even at a temperature as low as 620 K.

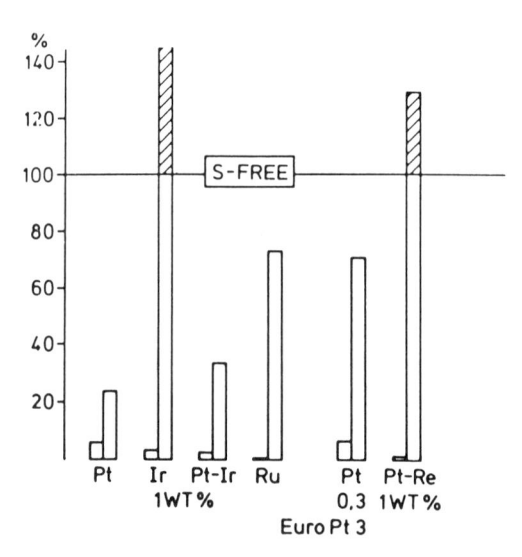

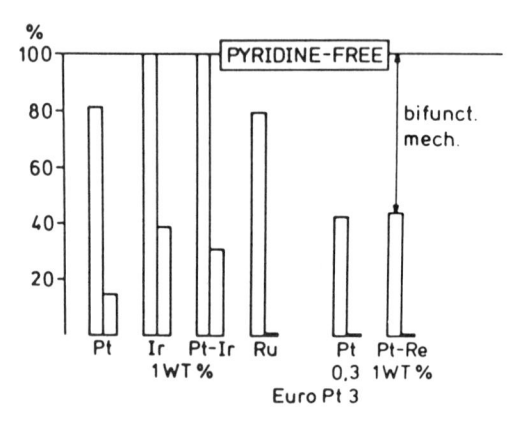

**Figure 4** Relative rates of hydrogenolysis (left bar of each pair) and skeletal isomerization (right bar of each pair) of *n*-hexane at 620 K over various $Al_2O_3$-supported catalysts after sulfidation (top) and pyridine addition (bottom). The sulfur-free and nitrogen-free states are taken as reference (100%). Reproduced by permission from Ref. 93. (Copyright by Elsevier Science Publishers, 1991.)

A Pt-Re single crystal showed highly enhanced hydrogenolysis activity with *n*-hexane model reactant as compared to pure Pt. Adding about one monolayer of sulfur suppressed this reaction and enhanced both aromatization and the formation of methylcyclopentane (94). Adding S even in very small quantities (about 1% of a monolayer) to a stepped Pt(557) single-crystal surface induces its reconstruction (95). The selectivity of hydrogenolysis of 2-methylpentane decreased gradually by a factor of 2 when the S content increased from 1 to 18% of a monolayer. At the same time, the overall rate of conversion per unit surface increased by a factor of 2.

## 2. Carbon

Carbonaceous deposits cause deactivation of industrial reforming catalysts. The reader is referred to Chapter 9 of this book. An excellent summary has been published by Parera and Fígoli (96). The results obtained under laboratory conditions and with model catalytic systems are described below.

The main problem concerning carbon on reforming catalysts is whether carbon-containing catalysts are active and, if so, to what extent and in which reactions. Another interesting problem concerns the way carbonaceous deposits form and their location on the catalysts.

Rather marked carbon coverage has been claimed to be the normal state of a Pt catalyst during its active period in hydrocarbon reactions (97). The surface analysis of single-crystal catalyst necessitates the use of an apparatus that can be evacuated to ensure ultrahigh vacuum (UHV) conditions. Hydrogen flushing of the catalyst before evacuation may be necessary to remove adsorbed hydrocarbons; if they lose sufficient hydrogen, they transform into carbonaceous deposits (97). The total carbon content of a Pt single crystal or polycrystalline Pt foil as measured by Auger electron spectroscopy (AES) was divided into three groups (98): (i) a constant amount of "residual carbon" (up to 4%), (ii) "reversible carbon" formed instantaneously during working conditions, and (iii) "irreversibly adsorbed carbon" accumulating during reaction (length of run, up to 10 h) and ultimately forming coke. Carbon coverages calculated from the C and Pt AES intensity ratio were 0.21 to 0.34 when the hydrocarbon reactant was evacuated after the run; the maximum coverage was, however, as low as 0.09 when the run was concluded by hydrogen flushing. This study regarded any carbon as a catalyst poison. Menon (99), on the other hand, classified various types of surface carbon as "beneficial," "harmful," "harmless," or "invisible." He also defined a "coke sensitivity" of catalytic reactions.

Sárkány (100) determined the clean and carbon-covered fraction of various Pt catalysts between 453 and 593 K. The carbon coverage was controlled by the hydrogen pressure in the $H_2$-hydrocarbon reactant mixture. The carbon coverage of Pt sites could reach the value of two C atoms per surface Pt (C/$Pt_s$) with Pt/$Al_2O_3$ and Pt black; 6.3% Pt/$SiO_2$ (EUROPT-1), an exception, retaining 0.3 to 0.5 C/$Pt_s$ under analogous conditions (101). These results have been interpreted in terms of three stages of the catalyst's life (100). Freshly regenerated (reduced) catalyst has been called a "Pt-H" system; this transforms into "Pt-C-H" after a rather short reaction time as a certain amount of firmly held hydrocarbon residue is accumulated. This catalyst exhibits almost steady-state activity. The surface hydrocarbons in the Pt-C-H state are probably not too deeply dehydrogenated hydrocarbon entities of the feed and may correspond to the "reversible" carbon as specified in Ref. 98. Their gradual dehydrogenation results in a fully or partly deactivated Pt-C catalyst state with irreversibly held carbon.

All these results indicate that Pt plus C catalysts possess some catalytic activity (87, 97). Reactions of hydrogen transfer have been attributed to mixed metal-carbon ensembles (102). One of the reasons for deactivation may be that carbonaceous deposits could hinder the migration of chemisorbed hydrocarbon species to reactive sites and also deactivate metallic active sites (97). The changes of carbon coverages with $p(H_2)$ were accompanied by marked selectivity changes in skeletal reactions of $n$-hexane: with the development of the Pt-C-H stage, isomerization activity dropped and that of $C_5$-cyclization increased; as the Pt-C state was approached, more and more olefins were produced at 600 K. Keeping the hydrogen pressure constant, a more rapid activity change due to carbon accumulation was observed at 693 K. This was true for both 6.3% $Pt/SiO_2$ (EUROPT-1) (103, 104) and Pt/KL zeolite (105). Nonuniform deactivation was reported at higher temperatures (773 K) with an industrial reforming catalyst (106) as well. Hydrogenolysis activity was deactivated first, followed by loss of isomerization and aromatization ability.

Single C atoms on valley positions of close-packed metal surfaces have been regarded as equivalent to inert metal activities (such as Cu or Au): both additives shift the selectivity toward processes requiring 1,3 interaction with the surface (35).

Accumulation of carbon also depends on the structure of the reactant. Less surface carbon was formed from $C_4$–$C_5$ alkanes than from $n$-hexane on any Pt single-crystal plane. Carbon deposition from light alkanes was more or less independent of crystal place structure, whereas rough surface structures favored coke formation from $n$-hexane (107).

At least two types of coke formation routes have been reported, one involving $C_1$ units and the other the polymerization of surface polyenes (108). The polyene route may result in three-dimensional carbon islands observed by electron spectroscopy over Pt black catalysts (109). The structure of their ultraviolet photoelectron spectra (UPS) resembled that of amorphous hydrogenated carbon (diamondlike) overlayers and it also contained strongly graphitized fractions. It is not excluded that these entities correspond to "ordered" or "disordered" carbonaceous deposits, the interconversion of which has been reported on Pt single crystals (48). These deposits may include "residual" and "irreversible" carbon (98); the detection of "reversible" carbon under ex situ conditions was not possible.

Sulfur may compete with carbon deposition on the active (or on the chemisorptive) sites. Sulfidation increased the yields of aromatics over $Al_2O_3$-supported Pt, Pt-Ir, Pt-Re, and Pt-Re-Cr catalysts from $n$-heptane and methylcyclopentane; however, in the latter case, the extent of coking increased (110). Obviously, the reactant structure is also important here: sulfur obviously cannot prevent polymerization of methylcyclopentadiene, which may represent a peculiar type of polyene route of coking (111).

Precursors of carbonaceous deposits may migrate from metallic sites to support sites and back in the case of supported catalysts. Coke situated on metallic and support sites can easily be distinguished by its temperature-programmed oxidation (111). One of the secrets of preparing high-performance catalysts is how to facilitate this migration to liberate more valuable metallic sites.

### 3. Silicon

Hexanes were reacted on an oxidized $Pt_{97}Si_3$ alloy at 5 torr pressure with 755 torr $H_2$ present at 623 K (112). Silicon was present as $SiO_2$ and $SiO_x$. This catalyst represents small Pt particles dispersed in an $SiO_2$ matrix, hence its behavior is similar to that of a surface exposing steps, edges, and corners instead of large terraces. It favors isomerization according to the $C_5$-cyclic mechanism. The relatively large amounts of labeled hexanes produced from labeled methylpentanes indicate a favored nonselective opening of the $C_5$-cyclic intermediate; this would also correspond to the abundance of adlineation sites at the borderlines between disperse Pt and $SiO_2$ (65). The presence of Si efficiently prevents carbon accumulation. A 6.3% $Pt/SiO_2$ catalyst pretreated with tetramethylsilane lost most of its activity, but the selectivity for $n$-heptane aromatization increased at the expense of hydrogenolysis (113). A high-temperature reduction of $Pd/SiO_2$ catalysts resulted in enhanced skeletal isomerization selectivity, most likely due to a Pd-Si interaction rather than changes in metal topography (114).

### 4. Oxygen

It is a customary industrial practice to pretreat Pt reforming catalysts by a "selective, controlled poisoning procedure with sulfur to reduce its initial hydrogenolysis activity" after their regeneration with oxygen (1, pp. 291–293). This high hydrogenolysis activity may be caused by the presence of surface oxygen. A Pt black pretreated with oxygen showed enhanced fragmentation selectivity from $n$-hexane (115) and so did Pt single-crystal catalysts (48). Also, the selectivity of methane formation from $n$-hexane on freshly regenerated $Pt/SiO_2$ approached 100% with a few seconds of contact time (104). The rate of ethane, propane, and $n$-butane hydrogenolysis was reported to be higher on $SiO_2$-supported Ni, Ru, Rh, Pd, Ir, and Pt catalysts when they were pretreated in $O_2$ at 773 K (116). Reconstruction upon $H_2/O_2$ cycling was demonstrated by electron microscopy and was considered a reason for these phenomena. Another possible reason may be the actual presence of surface oxygen. UPS of Pt black after its $O_2$ and subsequent $H_2$ treatment (even up to 700–800 K) showed the presence of intensive peaks attributed to surface OH and $H_2O$ species. Surface Pt oxide was present at most, in minor amounts, and an exposure to $H_2$ at a higher temperature transformed surface oxides to adsorbed OH and/or $H_2O$ (107, 117). Spectra recorded at 600 K show the same features. Therefore, it can be assumed that hydrogen treatment before a catalytic run does not remove all surface oxygen, in spite of the general belief.

Indeed, the reaction of $n$-hexane over Pt black containing some surface oxygen gave predominantly hydrogenolysis products with very short contact times. A few minutes were sufficient to produce a stabilized catalytic surface where active oxygenated species were obviously replaced by "C−H" deposits (see Section IV.B.2) (56, 116). Sintering or particle size redistribution of this Pt black sample of rather high crystallite size [20–50 nm (118)] is not likely; it is not excluded that the reconstruction observed by electron microscopy (116) and the activity changes have a common cause in the change of the amount of chemical state of surface oxygen, which is believed to consist of adsorbed species rather than metal oxide(s).

## 5. Hydrogen

Industrial and laboratory experiments are usually carried out in the presence of hydrogen. Hydrogen may induce surface reconstruction manifesting itself as sintering, particle migration, growth or coalescence, and particle shape changes with various supported and unsupported metals (116, 118). However, single crystal reconstruction has been attributed to sulfur impurities of hydrogen in the parts-per-million range (95).

Once a catalyst has a likely stable morphology, hydrogen effects in the catalytic reactions manifest themselves. Maximum yields of aromatization as a function of hydrogen pressure were reported as early as 1961 (119). These were observed for other skeletal reactions in laboratory experiments (120) and under the conditions of industrial reforming (121). Hydrogen effects have been summarized in detail elsewhere (56, 122, 123).

Hydrogen effects can be caused by the participation of hydrogen in the active ensemble for catalytic reaction. The idea of reactive chemisorption assumes that dissociative absorption of hydrocarbons occurs on metal-hydrogen sites, thus including the hydrogen in the adsorption equation. This hypothesis applies a single rate equation over a range of several orders of magnitude of hydrogen pressure assuming that adsorption is the rate-determining step (60).

Temperature-programmed desorption studies point to the presence of more than one type of surface hydrogen (for reviews, see Refs. 122 and 123). Following $H_2$ adsorption at 773 K, a $Pt/Al_2O_3$ catalyst contained almost three times as much high-temperature desorbable hydrogen as after adsorption at 300 K (124). This high-temperature hydrogen seemed to facilitate benzene formation from 2-hexene (26), possibly because hydrogen plays an active part in the *trans*→*cis* isomerization step(s) essential for the cyclization of a randomly dehydrogenated pool of surface unsaturated entities (25). The temperature-programmed dehydrocyclization of fully deuterated *n*-heptane ($C_7D_{16}$) gave three peaks of toluene (124). Apart from small $D_2$ peaks at lower temperatures (one of which corresponded to desorption of unreacted *n*-heptane), there was only one major desorption peak of $D_2$, coinciding with that of the highest-temperature toluene peak. This means that hydrogen split off during stepwise toluene formation at lower temperatures most likely joined the hydrogen pool of the surface [perhaps, according to reactive chemisorption theory (60–62)]. The highest temperature toluene peak concomitant with $D_2$ desorption might correspond to the direct aromatization pathway (22). The apparent changeover of the aromatization pathway from a stepwise to a direct route has been reported to occur over $Pt/SiO_2$ with increasing hydrogen concentration in the gas phase (104, 125). It must be noted that aromatization in TPR studies occurred up to 500 K, which is a much lower temperature than that of industrial reforming but the conditions are also different: the reaction took place in the adsorbed phase surrounded with vacuum instead of constant hydrogen and/or hydrocarbon pressure.

Ideas about the participation of surface hydrogen in the active site for the formation of saturated products (isomers, $C_5$ cyclics) have also been put forward (11, 31, 33) but never proved conclusively. Even if hydrogen does not participate directly in the surface reaction, maximum yields as a function of hydrogen pressure must mean that an optimum hydrogen pressure exists. The optimum is, at a rule, different for different reactions. Two explanations of these hydrogen effects have been proposed. A direct hydrogen effect can be interpreted in terms of competition between hydrogen and hydrocarbon reactant. Displacement of the reactive hydrocarbon species by hydrogen explains adequately the hydrogen response of overall rates at and above hydrogen pressures for maximum reaction rates. To interpret the section of the bell-shaped curve at hydrogen pressures lower than at maximum rate, two types of surface hydrocarbon entities are assumed: one deeply dissociated (as precursor of coke) and one less dissociated serving as a reaction intermediate (126). Increasing amounts of hydrogen would suppress deep dissociation first. The rate equation for this process contains a larger hydrogen exponent; at lower hydrogen pressures this will result in a positive hydrogen order. Of course, various reactions require intermediates with various degrees of dehydrogenation (128). That is why the maxima for different processes lie at different hydrogen pressures (56, 123).

An alternative explanation assumes an indirect hydrogen effect and proposes that "the surface-held hydrocarbons whose coverage is a function of the partial pressure of hydrogen contribute to the observed hydrogen sensitivity of the product selectivity" (129). In other words, actual hydrogen concentration governs the rate of transformation of Pt-H to Pt-C-H and Pt-C surfaces as discussed in Section V.B.4. The selectivity changes at different hydrogen/$n$-hexane pressure ratios are in agreement with this hypothesis (58, 104): at the lowest $H_2$ excess, hexenes are the only primary products and this is characteristic of a Pt-C state. The same conclusion has been drawn from transient response curves of the reactions of $n$-hexane on hydrogen-precovered Pt/$Al_2O_3$ at 793 K (27). Displacement of hydrogen by hydrocarbon was accompanied by marked changes in relative yields of hydrogenolysis, isomerization, $C_5$ cyclization, and aromatization. The response curves for 1-hexene formation had a different shape and were hardly affected by the presence or absence of hydrogen precoverage. The curves for all other reactions were essentially different over the hydrogen-free catalyst.

The two explanations are complementary rather than contradictory. The essence of the two approaches can be summarized in terms of whether surface hydrogen ultimately determines catalytic selectivity directly or by controlling the amount and/or the state of surface carbon. This latter interpretation could explain why higher hydrogen pressures would increase the selectivity of reactions that require larger clean metallic ensembles. Transient response studies do not exclude either possibility: they indicate that "the selectivity of $n$-hexane conversion is controlled by (i) the extent of dehydrogenation and geometry of the primary surface intermediates and (ii) the fraction of sites covered by hydrogen and hydrocarbon" (27). Recent (still unpublished) results of the author indicate that similar amounts of hydrocarbonaceous deposits can be removed from Pt catalysts after catalytic runs carried out at different hydrogen pressures. Therefore, various hydrogen pressures may influence the degree of dehydrogenation rather than the amount of sorbed hydrocarbons.

It is apparent that if a reaction needs optimum hydrogen surface coverage, higher pressures are required for this coverage at higher temperatures. Maximum rates are found at even higher hydrogen pressures as temperature increases (126). This can give rise to phenomena like inverse Arrhenius plots as reported for Pt (49) and Ni (59) single crystals, which were explained by shifting from the negative to the zero and then to the positive hydrogen order branch of the bell-shaped curve as the temperature increases at the same hydrogen pressure (129). Surface hydrogen depletion may be the reason why metal-catalyzed isomerization (which is favored by a high hydrogen content of the catalyst) almost ceases at the same (lower than atmospheric) hydrogen pressure when the temperature is increased from 603 to 693 K (104). At the lower temperature, reversible carbonaceous deposits have been claimed to determine the catalytic properties, while massive carbonization was assumed to occur above 703 K (131).

## Basic Reactions of Reforming on Metal Catalysts

The necessity of optimum surface hydrogen concentration as a function of temperature and pressure is clearly seen when the ratio of $C_5$ cyclics to aromatics is measured over both single crystals and disperse Pt catalysts (49, 113, 121, 124). More hydrogen promotes the formation of skeletal isomers rather than to $C_5$ cyclics from their common surface intermediate (103, 104, 124, 132). The ratio of fragments to nondegradative products behaves similarly, with lower $H_2$ pressure favoring hydrogenolysis. The favored position of hydrogenolytic splitting of an alkane chain (expressed by the $\omega$ value (41)) is also strongly hydrogen pressure dependent (43, 84). Terminal rupture prevails in the presence of less hydrogen, in agreement with Pt/C being the active ensemble for this process (88). Hydrogen effects seem to be superimposed on particle size effects (133). Hydrogen should also be considered as one of the factors controlling deactivation of reforming catalysts (97) (see also Chapter 3 of this book by Parera and Fígoli).

The maximum yields as a function of hydrogen pressure found at lower absolute pressures in low-temperature laboratory studies and at higher pressures with industrial catalysts (121, 123) may represent different sections of the same batch of curves in an $n$-dimensional space. Direct comparison of the same catalysts in a much wider pressure and temperature range would be necessary to elucidate all subtleties of the very complex phenomena occurring during reforming.

## VI. CONCLUDING REMARKS ON THE POSSIBLE ACTIVE SITES IN METAL-CATALYZED REACTIONS

The nature of the catalyst metal is the most important factor determining its properties, irrespective of defining this factor as an "electronic factor" (80) or "bond strength requirement" (134). One of the earliest catalytic theories, by Balandin (135), attributed the catalytic effect to atoms of large, stable, low-index clean metal areas. More recent investigations (136) point just to the opposite: (i) the surface of working catalysts is never clean; (ii) the catalytic effect is attributed to rough, high-Miller-index surfaces that (iii) reconstruct readily during catalysis.

Statement (i) is supported by surface analysis of Pt black catalysts (109, 117) as well as Pt single crystals (48, 49, 97, 134), where large amounts of metallic Pt are available for catalytic reactions in spite of their relative large carbon (and oxygen) content; the catalysts are, indeed, rather active in hydrocarbon reactions. The second statement is in agreement with the importance of ledge structures in skeletal reactions (52–54); this was formulated in the terms that "rough surfaces do chemistry" (136). As for (iii), the concept of active sites and active centers as entities undergoing constant changes during reaction has been discussed by Burch (86), who also put forward the concept of reaction-sensitive structures (137). On inhomogeneous surfaces (i) the reactions select optimum active sites themselves and (ii) additional active sites are created by surface reconstruction induced by the reactant or by other components present (hydrogen or oxygen) (95, 116, 118). One must remember that the strength of a chemical bond between a chemisorbed entity and the metal is close to the energy observed between atoms of a crystal and the formation and rupture of such bonds must involve energy transfer that is also important for the structure of the metal crystal itself.

A catalyst must not be regarded as a constant and immobile entity during reaction. The frequency of the changes on its active surfaces may be on the same time scale as that of the molecular processes (136). That may be the reason why the relevance of results of ex situ catalyst characterization methods to working catalysts—however valuable and informative they are for the conditions under which they were obtained—should be regarded with some skepticism. The catalytic reaction itself—using appropriate probe molecules—may be one of the best methods for catalyst characterization (80a).

## ACKNOWLEDGMENT

The author is pleased to express his thanks to Dr. Elek Bodor for the successful surgical intervention that restored his health, the immediate consequence of which was the completion of this chapter.

## REFERENCES

1. B. C. Gates, J. R. Katzer, and G. C. A. Schuit, *Chemistry of Catalytic Processes.* McGraw-Hill, New York, 1979, Chapter 3, pp. 184–324.
2. C. N. Satterfield, *Heterogeneous Catalysis in Industrial Practice.* McGraw-Hill, New York, 1991.
3. B. A. Kazansky and A. F. Plate, *Ber. Dtsch. Chem. Ges. B 69,* 1862 (1936); B. A. Kazanskii, A. L. Liberman and M. I. Batuev, *Dokl. Akad. Nauk SSSR 61,* 67 (1948).
4. B. A. Kazanskii, A. L. Liberman, T. F. Bulanova, V. T. Aleksanyan and Kh. E. Sterin, *Dokl. Akad. Nauk SSSR 95,* 77 (1954); B. A. Kazanskii, A. L. Liberman, V. T. Aleksanyan and Kh. E. Sterin, *Dokl. Akad. Nauk SSSR 95,* 281 (1954).
5. N. D. Zelinsky, B. A. Kazansky and A. F. Plate, *Ber. Dtsch. Chem. Ges. B 68,* 1869 (1935).
6. Yu. K. Yuryev and P. Ya. Pavlov, *Zh. Obshch. Khim. 7,* 97 (1937).
7. Y. Barron, G. Maire, D. Cornet, J. M. Muller and F. G. Gault, *J. Catal. 2,* 152 (1963); Y. Barron, G. Maire, J. M. Muller and F. G. Gault, *J. Catal. 5,* 428 (1966).
8. J. R. Anderson and B. G. Baker, *Nature 187,* 937 (1960); J. R. Anderson and N. R. Avery, *J. Catal. 5,* 446 (1966).
9. M. A. McKervey, J. J. Rooney and N. G. Sammann, *J. Catal. 30,* 330 (1973).
10. F. G. Gault, *Adv. Catal. 30,* 1 (1981).

11. Z. Paál, *Adv. Catal. 29*, 273 (1980).
12. H. C. de Jongste and V. Ponec, *Bull. Soc. Chim. Belg. 88*, 453 (1979).
13. J. H. Sinfelt, *Adv. Catal. 23*, 91 (1973).
14. Z. Paál and P. Télényi, in *Catalysis Specialists Periodical Reports* (G. C. Bond and G. Webb, eds.), Royal Soc. Chem., London, 1982, Vol. 5, p. 80.
15. E. van Broekhoven and V. Ponec., *Progr. Surf. Sci., 19*, 351 (1985).
16. J. K. A. Clarke and J. J. Rooney, *Adv. Catal. 25*, 125 (1976).
17. J. R. Anderson, *Adv. Catal. 23*, 1 (1973).
18. G. H. Twigg, *Trans. Faraday Soc. 35*, 979 (1939).
19. M. I. Rozengart, E. S. Mortikov, and B. A. Kazansky, *Dokl. Akad. Nauk SSSR 158*, 911 (1966); *166*, 619 (1966).
20. Z. Paál and P. Tétényi, *Acta Chim. Acad. Sci. Hung. 53*, 193 (1968); *54*, 175 (1967); *58*, 105 (1968).
21. Z. Paál and P. Tétényi, *J. Catal. 30*, 350 (1973).
22. F. M. Dautzenberg and J. C. Platteeuw, *J. Catal. 19*, 41 (1970).
23. L. Nogueira and H. Pines, *J. Catal. 70*, 404 (1981).
24. Z. Paál, Discussion remark, *Proceedings of the 10th International Congress on Catalysis, Budapest*, 1992, Part A, p. 903.
25. G. H. Twigg, *Proc. R. Soc. A178*, 106 (1941).
26. H. Zimmer, V. V. Rozanov, A. V. Sklyarov, and Z. Paál, *Appl. Catal. 2*, 51 (1982).
27. J. L. Margitfalvi, P. Szedlacsek, M. Hegedüs, E. Tálas, and F. Nagy, *Proceedings of the 9th International Congress on Catalysis, Calgary*, 1988, Vol. 3, p. 1283.
28. Z. Paál and P. Tétényi, *J. Catal. 29*, 176 (1973).
29. G. L. C. Maire and F. G. Garin, in *Catalysis, Science and Technology* (J. R. Anderson and M. Boudart, eds.), Springer, Berlin, 1984, Vol. 6, p. 161.
30. F. E. Shephard and J. J. Rooney, *J. Catal. 3*, 129 (1964).
31. A. L. Liberman, *Kinet. Katal. 4*, 128 (1964).
32. O. E. Finlayson, J. K. A. Clarke, and J. J. Rooney, *J. Chem. Soc. Faraday Trans. 1 80*, 345 (1980).
33. H. Zimmer and Z. Paál, *Proceedings of the 8th International Congress on Catalysis, Berlin*, 1984, Vol. 3, p. 417.
34. H. Zimmer and Z. Paál, *J. Mol. Catal. 51*, 261 (1989).
35. M. W. Vogelzang, M. J. P. Botman, and V. Ponec, *Faraday Discuss. Chem. Soc. 72*, 33 (1981).
36. F. Garin, P. Girard, F. Weisang, and G. Maire, *J. Catal. 70*, 215 (1981).
37. G. Maire and F. Garin, *J. Mol. Catal. 48*, 99 (1988).
38. J. F. Taylor and J. K. A. Clarke, *Z. Phys. Chem. Neue Folge 103*, 216 (1976).
39. Z. Paál and P. Tétényi, *Nature 267*, 234 (1977).
40. Z. Paál, P. Tétényi, and M. Dobrovolszky, *React. Kinet. Catal. Lett. 37*, 163 (1988).
41. G. Leclercq, L. Leclercq, and R. Maurel, *J. Catal. 50*, 87 (1977).
42. V. Ponec and W. M. H. Sachtler, *Proceedings of the 5th International Congress on Catalysis, Palm Beach*, 1972, Vol. 1, p. 645.
43. Z. Paál and P. Tétényi, *React. Kinet. Catal. Lett. 12*, 131 (1979).

44. P. Biloen, F. M. Dautzenberg, and W. M. H. Sachtler, *J. Catal. 50,* 77 (1977).
45. P. Biloen, J. N. Helle, H. Verbeek, F. M. Dautzenberg, and W. M. H. Sachtler, *J. Catal. 63,* 112 (1980).
46. J. R. H. Van Schaik, R. P. Dessing, and V. Ponec, *J. Catal. 38,* 273 (1975).
47. J. B. F. Anderson, R. Burch, and J. A. Cairns, *J. Catal. 107,* 364 (1987).
48. G. A. Somorjai, *Chemistry in Two Dimensions.* Cornell University Press, Ithaca, 1981; S. M. Davis and G. A. Somorjai, in *The Chemical Physics of Solid Surfaces and Heterogeneous Catalysis* (D. A. King and D. P. Woodruff, eds.). Elsevier, Amsterdam, 1982, Vol. 4, p. 217.
49. S. M. Davis, F. Zaera, and G. A. Somorjai, *J. Catal. 85,* 206 (1984).
50. F. Zaera, D. Godbey, and G. A. Somorjai, *J. Catal. 101,* 73 (1986).
51. B. E. Nieuwenhuys and G. A. Somorjai, *J. Catal. 46,* 259 (1977).
52. F. Garin, S. Aeiyach, P. Légaré, and G. Marie, *J. Catal. 77,* 323 (1982).
53. A. Dauscher, F. Garin, and G. Marie, *J. Catal. 105,* 233 (1987).
54. D. W. Blakely and G. A. Somorjai, *J. Catal. 42,* 181 (1976).
55. G. Ehrlich, *Adv. Catal. 14,* 255 (1963); a structure is depicted on p. 314.
56. Z. Paál, *Catal. Today 12,* 297 (1992).
57. V. Gnutzman and W. Vogel, *J. Phys. Chem. 94,* 4991 (1990).
58. G. C. Bond and Z. Paál, *Appl. Catal. 86,* 1 (1992).
59. D. W. Goodman, *Catal. Today 12,* 189 (1992).
60. A. Frennet, in *Hydrogen Effects in Catalysis* (Z. Paál and P. G. Menon, eds.), Marcel Dekker, New York, 1988, p. 399.
61. P. Parayre, V. Amir-Ebrahimi, F. G. Gault, and A. Frennet, *J. Chem. Soc. Faraday Trans. 1 76,* 1704 (1980).
62. A. Frennet, G. Liénard, A. Crucq, and L. Degols, *Surf. Sci. 80,* 412 (1979).
63. J. B. F. Anderson, R. Burch, and J. A. Cairns, *J. Catal. 107,* 351 (1987).
64. R. Burch, in *Hydrogen Effects in Catalysis* (Z. Paál and P. G. Menon, eds.), Marcel Dekker, New York, 1988, p. 347.
65. H. Glassl, K. Hayek, R. Kramer, *J. Catal., 68*: 397. (1981); R. Kramer and H. Zuegg, *J. Catal., 80,* 446 (1983); *85,* 530 (1984).
66. Z. Paál, *Catal. Today, 2,* 595 (1988).
67. G. A. Mills, H. Heinemann, T. H. Milliken, and A. G. Oblad, *Ind. Eng. Chem., 45,* 134 (1953).
68. Z. Paál, *J. Catal. 105,* 540 (1987).
69. J. R. Bernard, in *Proceedings of the 5th International Congress on Zeolites,* Naples, 1980, p. 686; T. R. Hughes, W. C. Buss, P. W. Tamm, and R. L. Jacobson, in *New Developments in Zeolite Science and Technology* (Y. Murakami, A. Iijima and J. W. Ward, eds.). Kodansha-Elsevier, Tokyo, 1986, p. 725.
70. E. G. Derouane and D. J. Vanderveken, *Appl. Catal. 45,* L15 (1988).
71. G. S. Lane, F. S. Modica, and J. T. Miller, *J. Catal. 129,* 145 (1991).
72. E. Mielczarski, S. B. Hong, R. J. Davis, and M. E. Davis, *J. Catal. 134,* 359 (1992).
73. E. Iglesia and J. E. Baumgartner, *Proceedings of the 10th International Congress on Catalysis, Budapest,* 1992, Part B, p. 993.
74. R. J. Davis and E. G. Derouane, *J. Catal. 132,* 269 (1991).
75. Z. Paál, Zh. Zhan, I. Manninger, and D. Barthomeuf, J. Catal. *146,* in press (1994).

76. E. G. Derouane, V. Jullien-Lardot, R. J. Davis, N. Blom, and P. E. Hojlund-Nielsen, *Proceedings of the 10th International Congress on Catalysis, Budapest,* 1992, Part B, p. 1031.
77. T. Baird, E. J. Kelly, W. R. Patterson, and J. J. Rooney, *J. Chem. Soc. Chem. Commun.* 1431 (1992).
78. Z. Paál, B. Brose, M. Räth, and W. Gombler, *J. Mol. Catal. 75,* L13 (1992).
79. K. Foger and J. R. Anderson, *J. Catal. 54,* 318 (1978).
80. F. Garin and G. Maire, *Acc. Chem. Res. 20,* 100 (1989).
80a. R. Burch and Z. Paal, *Appl. Catal.,* accepted (1994).
81. Z. Paál, K. Matusek, and H. Zimmer, *J. Catal. 141,* 648 (1993).
82. B. A. Kazansky, A. L. Liberman, G. V. Loza, and T. V. Vasina, *Dokl. Akad. Nauk SSSR 128,* 1188 (1959).
83. G. Leclercq, L. Leclercq, and R. Maurel, *Bull. Soc. Chim. Belg. 88,* 599 (1979).
84. H. Zimmer, M. Dobrovolszky, P. Tétényi, and Z. Paál, *J. Phys. Chem. 90,* 4758 (1986).
85. A. Palazov, Ch. Bonev, D. Shopov, G. Lietz, A. Sárkány, and J. Völter, *J. Catal. 103,* 249 (1987).
86. R. Burch, *Catal. Today 10,* 233 (1991).
87. J. K. A. Clarke, *Chem. Rev. 75,* 291 (1975).
88. V. Ponec, *Adv. Catal. 32,* 149 (1983).
89. J. Biswas, G. M. Bickle, P. G. Gray, D. D. Do, and J. Barbier, *Catal. Rev. Sci. Eng. 30,* 161 (1988).
90. P. G. Menon and J. Prasad, *Proceedings of the 6th International Congress on Catalysis, London,* 1976, Vol. 2, p. 1061.
91. L. Pönitzsch, M. Wilde, P. Tétényi, M. Dobrovolszky, and Z. Paál, *Appl. Catal. 86,* 115 (1992).
92. M. J. Dees, A. J. den Hartog, and V. Ponec, *Appl. Catal. 72,* 343 (1991).
93. V. Ponec, *Catal. Today 10,* 251 (1991).
94. Ch. Kim and G. A. Somorjai, *J. Catal. 134,* 179 (1992).
95. G. Maire, G. Lindauer, F. Garin, P. Légaré, M. Cheval, and M. Vayer, *J. Chem. Soc. Faraday Trans. 86,* 2719 (1990).
96. J. M. Parera and N. S. Figoli, in *Catalysis Specialists Periodical Reports,* Vol. 9 (J. J. Spivey, ed.), Royal Soc. Chem., London, 1992, p. 65.
97. G. A. Somorjai and F. Zaera, *J. Phys. Chem. 86,* 3070 (1982).
98. F. Garin, G. Maire, S. Zyade, M. Zauwen, A. Frennet, and P. Zielinski, *J. Mol. Catal. 58,* 185 (1990).
99. P. G. Menon, *J. Mol. Catal. 59,* 207 (1990).
100. A. Sárkány, *J. Chem. Soc. Faraday Trans. 1 84,* 2267 (1988).
101. A. Sárkány, *Catal. Today 5,* 173 (1989).
102. S. J. Thomson and G. Webb, *J. Chem. Soc. Chem. Commun.* 526 (1976); G. Webb, *Catal. Today 7,* 139 (1990).
103. Z. Paál, H. Groeneweg, and J. Paál-Lukács, *J. Chem. Soc. Faraday Trans. 86,* 3159 (1990).
104. Z. Paál, I. Manninger, Zh. Zhan, and M. Muhler, *Appl. Catal. 66,* 305 (1990).
105. I. Manninger, Z. Zhan, X. L. Xu, and Z. Paál, *J. Mol. Catal. 65,* 223 (1991).
106. C. A. Querini, N. S. Figoli, and J. M. Parera, *Appl. Catal. 53,* 53 (1989).

107. S. M. Davis, F. Zaera, and G. A. Somorjai, *J. Am. Chem. Soc. 104*, 7453 (1982).
108. A. Sárkány, H. Lieske, T. Szilágyi, and L. Tóth, *Proceedings of the 8th International Congress on Catalysis*, Berlin, 1984, Vol. 2, p. 613.
109. Z. Paál, R. Schlögl, and G. Ertl, *J. Chem. Soc. Faraday Trans. 88*, 1179 (1992).
110. M. Wilde, R. Stolz, R. Feldhaus, and K. Anders, *Appl. Catal. 31*, 99 (1987).
111. J. Barbier, in *Catalyst Deactivation 1987* (B. Delmon and G. F. Froment, eds.). Elsevier, Amsterdam, 1987, p. 1.
112. G. Maire, P. Légaré, S. Aeiyach, and F. Garin, *Catal. Today 12*, 201 (1992).
113. Z. Paál, H. Zimmer, and P. Tétényi, *J. Mol. Catal. 25*, 99 (1984).
114. Z. Karpinski, *Adv. Catal., 37*, 45 (1990).
115. E. Santacesaria, D. Gelosa, and S. Carrà, *J. Catal., 38*, 403 (1975).
116. S. Gao and L. D. Schmidt, *J. Catal. 115*, 473 (1989); L. D. Schmidt and K. R. Krause, *Catal. Today 12*, 269 (1992).
117. Z. Paál, R. Schlögl, and G. Ertl, *Catal. Lett. 12*, 331 (1992).
118. E. Ruckenstein and I. Sushumna, in *Hydrogen Effects in Catalysis* (Z. Paál and P. G. Menon, eds.), Marcel Dekker, New York, 1988, p. 259
119. Z. Paál, in *Hydrogen Effects in Catalysis* (Z. Paál and P. G. Menon, eds.), Marcel Dekker, New York, 1988, p. 293.
120. J. C. Rohrer, H. Hurvitz, and J. H. Sinfelt, *J. Phys. Chem. 65*, 1458 (1961).
121. Z. Paál and P. Tétényi, *Dokl. Akad. Nauk SSSR 201*, 1119 (1971).
122. J.-P. Bournonville and J.-P. Franck, in *Hydrogen Effects in Catalysis* (Z. Paál and P. G. Menon, eds.), Marcel Dekker, New York, 1988, p. 653.
123. Z. Paál and P. G. Menon, *Catal. Rev. Sci. Eng. 25*, 273 (1983).
124. Z. Paál, in *Hydrogen Effects in Catalysis* (Z. Paál and P. G. Menon, eds.), Marcel Dekker, New York, 1988, p. 449.
125. V. V. Rozanov, J. Gland, and A. V. Sklyarov, *Kinet. Katal. 20*, 1249 (1979).
126. Z. Paál and X. L. Xu, *Appl. Catal. 43*, L1 (1988).
127. Z. Paál, K. Matusek, and P. Tétényi, *Acta Chim. Acad. Sci. Hung. 94*, 119 (1977).
128. Z. Paál, G. Székely, and P. Tétényi, *J. Catal., 58*, 108 (1979).
129. A. Sárkány, *J. Chem. Soc. Faraday Trans. 1 85*: 1523 (1989).
130. Z. Paál, *J. Catal. 91*, 181 (1985).
131. E. G. Christoffel and Z. Paál, *J. Catal. 73*, 30 (1982).
132. Z. Paál, H. Zimmer, J. R. Günter, R. Schlögl, and M. Muhler, *J. Catal. 119*, 146 (1989).
133. O. V. Bragin, Z. Karpinski, K. Matusek, Z. Paál, and P. Tétényi, *J. Catal. 56*, 219 (1979).
134. W. M. H. Sachtler, *Faraday Disc. Chem. Soc., 72*, 7 (1981).
135. A. A. Balandin, *Z. Phys. Chem., B 2*, 289 (1929); *Adv. Catal. 19*, 1 (1969).
136. G. A. Somorjai, *Catal. Lett., 9*, 311 (1991); *Surf. Interf. Anal. 19*, 493 (1992).
137. R. Burch, in *Catalysis Specialists Periodical Reports*, Vol. 7 (G. C. Bond and G. Webb, eds.), Royal Soc. Chem., London, (1985) p. 149.
138. J. M. Parera, J. N. Beltramini, C. A. Querini, E. E. Martinelli, E. J. Churin, P. E. Aloe, and N. S. Fígoli, *J. Catal. 99*, 39 (1986).

# 3
# Reactions in the Commercial Reformer

**José M. Parera and Nora S. Fígoli**

*Instituto de Investigaciones en Catálisis y Petroquímica (INCAPE), Santa Fe, Argentina*

## I. INTRODUCTION

In this chapter the thermodynamic concepts of chemical equilibrium and catalyst selectivity applied to hydrocarbon reactions occurring during naphtha reforming are reviewed. The most important reactions (dehydrogenation of cyclohexanes, dehydroisomerization of alkylcyclopentanes, dehydrogenation of paraffins, isomerization of paraffins, dehydrocyclization of paraffins, hydrocracking, hydrogenolysis, and coke deposition) are analyzed individually, from the most rapid to the slowest one. For each reaction the thermodynamics, kinetics, and mechanisms are studied. The influence of operational parameters on the reactions is used to conclude which operational conditions are most convenient. Finally, kinetic models for naphtha reforming are briefly described.

## II. THERMODYNAMICS

The chemical thermodynamic equations that can be applied to hydrocarbon reactions are the following:

1. Change of standard free energy of reaction:

$$\Delta F^\circ_R = \sum \gamma_i \, \Delta F^\circ_{Fi} \tag{1}$$

where $\Delta F^\circ_{Fi}$ is the standard free energy of formation for each of the $i$ reacting species and $\gamma_i$ is its stoichiometric coefficient in the reaction, which is negative for reactants and positive for products.

2. Equilibrium constant:

$$K = \prod a_i^{\gamma_i} \tag{2}$$

where $\prod$ means the mathematical product of all the values of the equilibrium activity of reactants and products, $a_i$. $K$ is related to the change in free energy of reaction by

$$\Delta F^\circ_R = -RT \ln K \tag{3}$$

$\Delta F^\circ_R$ must be negative to have values of $K$ higher than one. For gaseous reactants, fugacities are used instead of activities. If the system behaves as an ideal gas, fugacities could be substituted by partial pressures, and the equilibrium constant is expressed as

$$K = K_p = \prod P_i^{\gamma_i} \tag{4}$$

3. Influence of temperature on the equilibrium constant:

$$\frac{d(\ln K)}{dT} = \frac{\Delta H_R}{RT^2} \tag{5}$$

where $\Delta H_R$ is the heat of reaction at temperature $T$.

4. Influence of total pressure ($P_T$) and composition:

$$K_p = K_y P_T^{\Sigma \gamma_i} \tag{6}$$

$$K_y = \prod y_i^{\gamma_i} \tag{7}$$

where $y_i$ is the mole fraction of component $i$.

Figure 1 shows the free energy of formation per carbon atom of several hydrocarbons as a function of temperature. For the elements, in this case carbon and hydrogen, $\Delta F^\circ_F$ is conventionally taken as zero at any temperature. A reaction is thermodynamically feasible when a decrease in free energy is produced; i.e., $\Delta F^\circ_R$ is negative. This means that in Figure 1 the product free energy level must be below the free energy level of the reactant.

Figure 1 shows that the greater the number of carbon atoms of a paraffin, the higher its free energy of formation. Therefore, cracking reactions producing hydrocarbons with a smaller number of carbon atoms are feasible. Methane is the hydrocarbon showing the lowest free energy of formation, and it should be the main component when total reaction equilibrium is reached. This means that total hydrogenolysis to methane is the most feasible reaction from the thermodynamic point of view and that oligomerization of methane is not thermo-

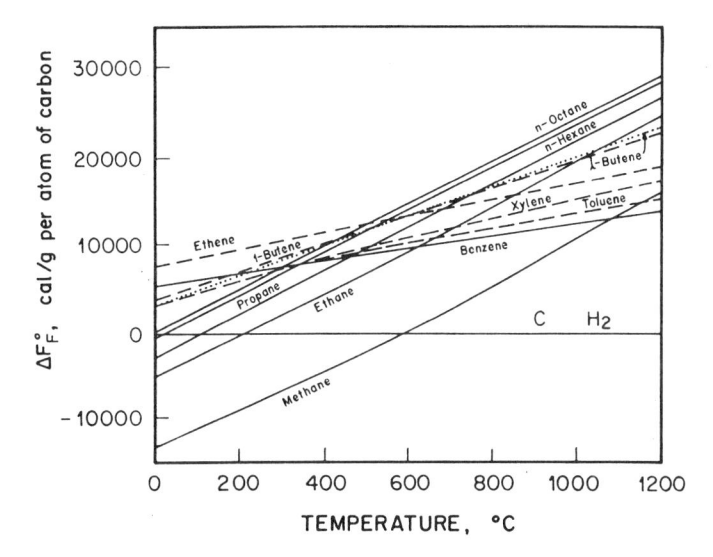

**Figure 1** Free energy of formation of hydrocarbons, per carbon atom, as a function of temperature. (From Ref. 1.)

dynamically possible. To make this reaction possible, it must be coupled with another reaction that is thermodynamically very feasible, as in the case of the oxidative dimerization of methane.

At low temperatures, paraffins have $\Delta F_F^\circ$ values lower than those of the corresponding olefins, and thus the hydrogenation of olefins is very feasible, whereas at high temperatures dehydrogenation of paraffins is thermodynamically favored. This can be seen in Figure 1 for ethane-ethene. At temperatures lower than 820°C, ethene hydrogenation shows a negative change in free energy, so that the reaction is thermodynamically feasible. At 820°C, lines for ethane and ethene intersect, no change of free energy in the reaction is observed, and the equilibrium is at an intermediate position. At higher temperatures, ethane has a higher free energy of formation than ethene, and dehydrogenation is more favored. The temperature of inversion of the hydrogenation-dehydrogenation equilibrium is lower when the number of carbon atoms in the molecule is larger.

Figure 1 shows that, at high temperatures, aromatic hydrocarbons are more stable than the corresponding paraffins. For this reason, paraffin dehydrocyclization is thermodynamically very feasible under these conditions. On the other hand, opening of the aromatic ring and hydrogenation to paraffins (hydrogenolysis) are favored at low temperatures.

Linear paraffins and olefins show a small difference in $\Delta F_F^\circ$ from their corresponding isomers, as shown in Figure 1 for 1-butene and *i*-butene. For this reason, the isomerization reaction generally has an intermediate equilibrium over a wide range of temperatures.

Figure 2 shows the change of standard free energy in reactions between hydrocarbons with six carbon atoms as a function of temperature. Exothermic reactions, such as isomerization (reaction A) and hydrogenation (reaction E) decrease in feasibility when temperature increases, whereas endothermic reactions, such as dehydrogenations (C and E), cracking (D), and dehydrocyclization (B), are favored when temperature increases. The slope is larger when the heat of reaction is higher.

Table 1 shows thermodynamic data for several reactions occurring at 500°C, a normal temperature for the naphtha reforming process, and considering $K = K_p$.

## III. KINETICS: CATALYTIC ACTIVITY AND SELECTIVITY

Each hydrocarbon has several reaction paths that are thermodynamically feasible. For example, a normal paraffin—at a high temperature—can be cracked, producing a mixture of paraffins and olefins of a smaller number of carbon

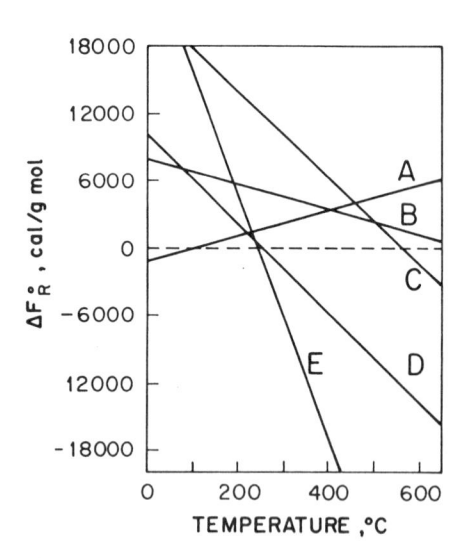

**Figure 2** Change in the standard free energy for several reactions of six-carbon-atom hydrocarbons as a function of temperature. A, Methylcyclopentane → cyclohexane; B, *n*-hexane → cyclohexane + $H_2$; C, *n*-hexane → 1-hexene + $H_2$; D, *n*-hexane → 1-butene + ethane; E, cyclohexane → benzene + $3H_2$. (Data from Refs. 1 and 2.)

Reactions in the Commercial Reformer

**Table 1** Thermodynamic Data, at 500°C, of Some Reactions That Occur in Naphtha Reforming

| Reaction | | $K_p$ ($p$ in atm) | $\Delta H_R$ (cal/mol) |
|---|---|---|---|
| Cyclohexane | ⇔ benzene + 3H$_2$ | 6 × 10$^5$ | 52,800 |
| Methylcyclohexane | ⇔ toluene + 3H$_2$ | 2 × 10$^6$ | 51,500 |
| Methylcyclopentane | ⇔ cyclohexane | 0.086 | −3,800 |
| Methylcyclopentane | ⇔ benzene + 3H$_2$ | 5.2 × 10$^4$ | 49,000 |
| $n$-Hexane | ⇔ 1-hexene + H$_2$ | 0.037 | 31,000 |
| $n$-Hexane | ⇔ 2-methylpentane | 1.14 | −1,400 |
| $n$-Hexane | ⇔ 3-methylpentane | 0.76 | −1,100 |
| $n$-Hexane | ⇔ benzene + 4H$_2$ | 0.78 × 10$^5$ | 63,600 |
| $n$-Heptane + H$_2$ | ⇔ butane + propane | 3.1 × 10$^3$ | −12,300 |
| $n$-Heptane + H$_2$ | ⇔ $n$-hexane + methane | 1.2 × 10$^4$ | −14,800 |

*Source*: Calculated from Ref. 2.

atoms; it can be isomerized to a branched paraffin, dehydrogenated to an olefin with the same number of carbon atoms, or dehydrocyclized producing a naphthenic or an aromatic hydrocarbon. At a certain contact time, the product distribution depends on the relative rate of these reactions, on the successive reactions that can occur, and on the thermodynamic equilibrium of the reactions which are so rapid that they virtually reach equilibrium. Rates of the individual reactions depend on the catalyst selectivity. An important point in hydrocarbon processing is to find a very active catalyst and, above all, one selective enough for the desired reactions.

Naphthas to be reformed contain paraffins, naphthenes with five or six carbon atom cycles, and aromatics. Since one of the main goals of the process is to increase the aromatic concentration, the desired reactions are dehydrocyclization of paraffins, dehydrogenation of cyclohexanes, and dehydroisomerization of alkyl cyclopentanes to aromatics. Even the best catalysts are not completely selective for these reactions. Some undesired reactions that decrease the liquid yield (gas formation by cracking) and deactivate the catalyst (formation of carbonaceous deposits on the catalyst) always occur to some extent. Paraffins, mostly linear, are the main components of virgin naphthas. On the classical bifunctional naphtha reforming catalyst, the main reactions of paraffins are the following: dehydrogenation to olefins, isomerization, dehydrocyclization to aromatics or naphthenes, and hydrocracking or hydrogenolysis to lighter paraffins. Naphthenic hydrocarbons are alkylcyclopentanes and alkylcyclohexanes, and the possible reactions are dehydroisomerization of alkylcyclopentanes and dehydrogenation of alkylcyclohexanes to produce aromatics, isomerization, and ring opening to produce paraffins. Paraffinic chains of alkylaromatics can be partially or totally hydrocracked, and the ring can also be hydrocracked to produce a paraffin.

Considering normal heptane as a typical component of a naphtha, a simple scheme for its reactions on the reforming catalyst is as follows:

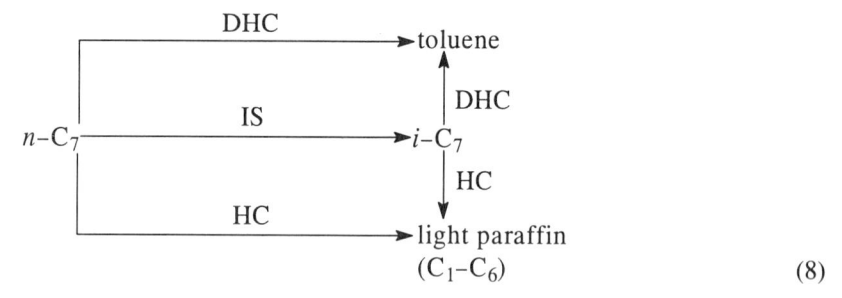

```
                    DHC
         ┌─────────────────────→ toluene
         │                         ▲
         │            IS           │ DHC
   n-C₇──┼─────────────────────→ i-C₇
         │                         │
         │                         │ HC
         │            HC           ▼
         └─────────────────────→ light paraffin
                                  (C₁–C₆)                    (8)
```

The desired reaction in this scheme is n-heptane dehydrocyclization to toluene, but i-heptanes and light paraffins are produced simultaneously. iso-Heptanes are intermediate products that can be dehydrocyclized or hydrocracked. The catalyst selectivity and the contact time determine the product distribution. At the same time as these reactions are taking place, a carbonaceous deposit is slowly produced on the catalyst surface due to condensation reactions of small amounts of unsaturated hydrocarbons present as reaction intermediates.

The most important reactions of catalytic naphtha reforming are analyzed separately in this chapter. There are many papers in the literature dealing with reforming reactions; some of the first papers related to Pt/Al₂O₃ catalysts were published together in 1953 (3–6). There are several reviews on reforming also including sections about individual reaction types (7–13).

## IV. REFORMING REACTIONS

### A. Dehydrogenation of Cyclohexanes

As shown in Table 5 and Figure 2 of Chapter 1, the formation of aromatics by dehydrogenation of cyclohexanes produces an increase in octane number, except for benzene. According to Table 6 of Chapter 1, this reaction leads to a decrease in liquid volume (aromatics density is approximately 0.1 g/mL greater) as well as in vapor pressure (higher boiling point of aromatics) and in weight yield due to the loss of hydrogen.

1. Thermodynamics

The thermodynamic equilibrium constant for dehydrogenation of cyclohexane (CH) to benzene (B) at 500°C, according to Eq. (4) and data from Table 1, is

$$K_p = \frac{P_B P_{H_2}^3}{P_{CH}} = 6 \times 10^5 \tag{9}$$

The reaction is completely displaced to benzene formation. Since this is an endothermic reaction, the equilibrium constant increases with temperature, Eq. (5). An increase in total pressure shifts the equilibrium toward the cycloparaffin according to Eq. (6). Since $K_p$ is independent of pressure and $P_T^3$ greatly increases with increasing $P_T$, $K_y$ must decrease in the same proportion. Hydrogen pressure is also unfavorable for this reaction, Eq. (7). Increasing the hydrogen mole fraction, $y_{H_2}^3$ increases and, in the same proportion, $y_B/y_{CH}$ must decrease to keep the $K_y$ value constant.

An increase in the molecular weight of the cycloparaffin favors the thermodynamic feasibility of its transformation to an aromatic. Table 1 shows that the transformation of methylcyclohexane to toluene is more feasible than that of cyclohexane to benzene.

## 2. Kinetics

The dehydrogenation of cyclohexanes is catalyzed by the metal function of the catalyst and is the most rapid of all the reactions occurring in the reforming process.

Sinfelt et al. (14) studied the kinetics of methylcyclohexane dehydrogenation on a $Pt/Al_2O_3$ catalyst and found that the rate of reaction increases with increasing methylcyclohexane pressure at low pressures and does not change at higher pressures. The hydrogen pressure has no influence on the rate. The near-zero order of reaction indicates that the active surface platinum sites are completely covered by molecules or radicals formed when the hydrocarbon is adsorbed. The authors suppose that the adsorption occurs on a surface site, S, with very rapid and irreversible elimination of hydrogen atoms from methylcyclohexane, and they consider the desorption of the toluene produced as the rate-limiting step:

$$\text{MCH} + \text{S} \xrightarrow[-H_2]{k_1} \text{Tol} - \text{S} \xrightarrow{k_2} \text{Tol} + \text{S} \tag{10}$$

The rate of adsorption of methylcyclohexane is equal to the rate of desorption of toluene at steady state:

$$k_1 P_M (1 - \theta_T) = k_2 \theta_T \tag{11}$$

$P_M$ is the methylcyclohexane pressure and $\theta_T$ the fraction of platinum surface covered by toluene, which from Eq. (11) is

$$\theta_T = \frac{bP_M}{1 + bP_M}, \qquad b = \frac{k_1}{k_2} \tag{12}$$

The experimental rate of reaction corresponds to the slowest step, toluene desorption:

$$r = k_2 \theta_T = \frac{k_2 b P_M}{1 + bP_M} \tag{13}$$

The value of $b$ is very high; then when $P_M$ is not very small, the rate is independent of $P_M$, $r = k_2$, which agrees with experimental results.

Shipikin et al. (15) studied the influence of the deactivation and regeneration of a Pt/Al$_2$O$_3$ reforming catalyst on the cyclohexane dehydrogenation reaction using catalysts that were fresh, spent, chlorided, and chlorided with elimination of impurities. They found that at 470°C and 40 atm the catalytic activity is affected only by the deactivation and regeneration of the metal function. The decrease or increase in chloride content, or an increase in sodium content or its elimination, does not affect the dehydrogenation reaction.

Ritchie and Nixon (16) studied the dehydrogenation of cyclohexane and methyl, dimethyl, ethyl, diethyl, and trimethyl cyclohexanes on Pt/Al$_2$O$_3$. They observed that, when increasing the number of methyl groups and considering first-order reactions, the value of the kinetic constant increases whereas the activation energy decreases. Thus, for alkylcyclohexane dehydrogenation, the larger the number of alkyl groups, the greater the thermodynamic feasibility as well as the rate of reaction. When the reaction takes place in an adiabatic reactor (commercial operation), the reaction temperature decreases along the catalyst bed due to the endothermicity of the reaction. In this case, and because of its lower activation energy, the rate of dehydrogenation for the higher homologues is decreased less than that of cyclohexane.

The dehydrogenation of cyclohexane on Pt is very rapid and selective. At 5 atm, 505°C, H$_2$/cyclohexane mole ratio = 4, and weight hourly space velocity = 4 (17), 100% of cyclohexane is converted on Pt/Al$_2$O$_3$ with a selectivity to benzene of 99.2%. When this catalyst is sulfided (Pt-S/Al$_2$O$_3$), the selectivity decreases to 93.7% because the rate of dehydrogenation decreases and some of the intermediates (cyclohexane and cyclohexadiene) are desorbed and cracked (ring opening), thus producing $n$-hexane and its cracking products. Sulfur changes the selectivity because it partially poisons the dehydrogenation on the metal function while the activity of the acid function of the catalyst is not affected. With unsulfided bimetallic Pt-Re/Al$_2$O$_3$, the selectivity to benzene is only 56.1%; the main by-product is methane produced by successive hydrogenolysis. When this catalyst is sulfided, as in the case of the commercial catalyst (Pt-Re-S/Al$_2$O$_3$), the hydrogenolysis reaction is selectively poisoned (0.4% selectivity to methane) and the selectivity to benzene increases to 97.3%. Under the conditions of these experiments, the total conversion of cyclohexane was always 100%.

## 3. Mechanism

The mechanism of cyclohexane dehydrogenation involves its adsorption on platinum and the subsequent elimination of six hydrogen atoms. According to Germain (18), the steps are the same but in a reverse order, as for the hydrogenation of benzene to cyclohexane. Figure 3 shows the mechanism proposed by this author. Cyclohexane loses H atoms, forming an $\alpha$, $\beta$, or $\pi$ adsorbed

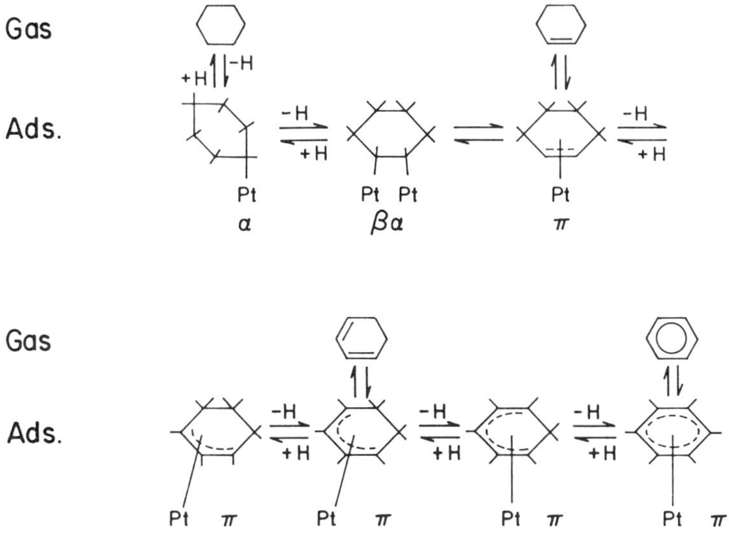

Reactions in the Commercial Reformer 53

**Figure 3** Mechanism of cyclohexane dehydrogenation.

cyclohexane. By successive removals of hydrogen atoms, a benzene bonded through π-electron interaction with metal $d$ orbitals is formed, which is finally desorbed. The intermediate olefin and diolefin can be desorbed, but their dehydrogenation is so rapid that it is completed without appreciable desorption. When cyclohexane or cyclohexadiene is fed instead of cyclohexane, it is adsorbed on the metal and dehydrogenated to benzene following the same steps.

## B. Dehydroisomerization of Alkylcyclopentanes

All naphthas contain alkylcyclopentanes which are transformed into aromatics through dehydrogenation steps, catalyzed by the metal function, and an isomerization step (ring enlargement) catalyzed by the acid function, as shown in Figure 4 for methylcyclopentane. Methylcyclopentane is first dehydrogenated to methylcyclopentene on a metal site, and dehydrogenation continues to methylcyclopentadiene on the same type of site. The adsorbed cyclodiolefin migrates

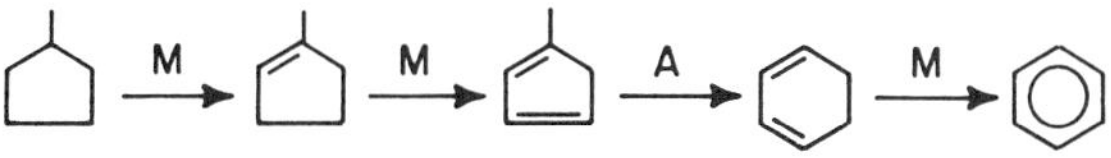

**Figure 4** Mechanism of methylcyclopentane dehydroisomerization. M, metal site; A, acid site.

to an acid site, where it is isomerized to cyclohexadiene, which then migrates to a metal site where it is finally dehydrogenated to benzene.

From data in Table 2, Weisz (19) clearly showed that the transformation of methylcyclopentane into benzene requires a bifunctional catalyst in order to give a good yield. $SiO_2$-$Al_2O_3$, an acid catalyst, is not able to start the reaction sequence because it has no dehydrogenation activity, with only some cracking being produced. Platinum is a good dehydrogenating catalyst, but it has no activity for isomerization, proceeding only up to methylcyclopentadiene. Only the presence of both catalysts produces the transformation into benzene.

Isomerization of methylcyclopentane to cyclohexane followed by dehydrogenation to benzene could be another possible mechanism. In this mechanism, the first step has very low thermodynamic feasibility ($K_p = 0.086$, Table 1 and Figure 2) because the five-carbon-atom ring is more stable than the six-carbon-atom ring. This mechanism has little possibility from the kinetic point of view because the isomerization of saturated hydrocarbons is very difficult. Expansion of a five-carbon-atom ring into a six-carbon-atom ring is easier in the case of unsaturated hydrocarbons: methylcyclopentene or methylcyclopentadiene.

Sinfelt et al. (20) studied the influence of the increase in platinum concentration on a Pt/$Al_2O_3$ catalyst in several reactions. Table 3 shows the results of experiments using $n$-heptane or methylcyclopentane as feed. It can be seen that the dehydroisomerization of methylcyclopentane is little affected by the platinum content, but it does not occur without platinum. A small amount of platinum is enough to produce the unsaturated hydrocarbons required by the acid function, which controls the whole reaction.

The dehydroisomerization of alkylcyclopentenes is less feasible thermodynamically (Table 1); it is also less rapid and has a higher activation energy than the dehydrogenation of alkylcyclohexanes. The value of the activation energy for dehydroisomerization of methylcyclopentane is 32,800 cal/mol, and that for dehydrogenation of cyclohexane is 18,100 cal/mol; both reactions are

**Table 2** Dehydroisomerization of Methylcyclopentane Catalyzed by Acid, Metal, and Mixed Catalysts. Concentrations in the Liquid Product, mol %[a]

| Catalyst | Methyl cyclopentane ⟶ | Methyl cyclopentene ⟶ | Methyl cyclopentadiene ⟶ | Benzene |
|---|---|---|---|---|
| 10 cm³ $SiO_2$ − $Al_2O_3$ | 98 | 0 | 0 | 0.1 |
| 10 cm³ Pt/$SiO_2$ | 62 | 19 | 18 | 0.8 |
| $SiO_2$ − $Al_2O_3$ + Pt/$SiO_2$ | 65 | 14 | 10 | 10.0 |

[a] Reaction conditions: 500°C, 0.8 atm hydrogen partial pressure, 0.2 atm methylcyclopentane partial pressure, and 2.5 s residence time. Catalysts: 0.3 wt % Pt on $SiO_2$, $SiO_2$ − $Al_2O_3$ with $S_g = 420 \, m^2/g$.
Source: From Ref. 19.

**Table 3** Effect of Pt Content on Reactivities of $n$-Heptane ($nC_7$) and Methylcyclopentane (MCP) over Pt − $Al_2O_3$

| Reaction rates[a] | Pt content (wt %) | | |
|---|---|---|---|
| | 0.10 | 0.30 | 0.60 |
| Isomerization of $nC_7$ | | | |
| 744 K | 0.035 | 0.035 | 0.038 |
| 800 K | 0.12 | 0.13 | 0.12 |
| Dehydrocyclization of $nC_7$ | | | |
| 744 K | 0.0022 | 0.0027 | 0.0045 |
| 800 K | 0.020 | 0.025 | 0.035 |
| Isomerization-dehydroisomerization of MCP | | | |
| 744 K | — | 0.019 | 0.021 |
| 772 K | — | 0.039 | 0.043 |

[a] Gram moles per hour per gram of catalyst at 21 atm and $H_2$/hydrocarbon mole ratio = 5.
*Source*: From Ref. 20.

endothermic. This means that if a mixture of both naphthenes is fed into an adiabatic reactor, due to the descending temperature profile along the catalytic bed, there will be a larger decrease in the rate of transformation of methylcyclopentane than that of cyclohexane.

The main reaction of methylcyclopentane on the reforming catalyst, under normal operating conditions for this process, is dehydroisomerization, and, in the second place, hydrogenolysis producing $n$- and $i$-hexane (ring opening) and $C_1$–$C_5$ paraffins. In a study of $Pt/Al_2O_3$, $Pt$-$Re/Al_2O_3$, and their sulfided forms (17), it was shown that $Pt/Al_2O_3$ produces more benzene than $Pt$-$Re/Al_2O_3$, but when the catalysts are sulfided, the increase in selectivity toward benzene is larger on $Pt$-$Re/Al_2O_3$. This large increase corresponds to higher poisoning of hydrogenolysis by sulfur.

## C. Dehydrogenation of Paraffins

Paraffins of six to nine carbon atoms are the main components of the stream to be reformed. These paraffins can be dehydrogenated to olefins on the metal function of the catalyst. Such olefins, in the presence of the acid function, can react according to the bifunctional mechanism, the main mechanism operating in the conditions of naphtha reforming.

### 1. Thermodynamics

The dehydrogenation of paraffins has low thermodynamic feasibility; the feasibility increases with an increase in temperature, because it is an endothermic reaction (Table 1):

$$R-CH_2-CH_3 \Leftrightarrow R-CH=CH_2 + H_2 \tag{14}$$

$$K_p = \frac{P_{\text{olef.}} \cdot P_{H_2}}{P_{\text{paraf.}}} \ll 1, \quad \Delta H_R > 0 \tag{15}$$

As cited above, at a certain temperature, an inversion of the hydrogenation-dehydrogenation equilibrium is observed. At lower temperatures the hydrogenation of the olefin is favored; at higher temperatures, the dehydrogenation of the paraffin is more feasible.

The thermodynamic equilibrium conversion for dehydrogenation decreases with increasing total and hydrogen pressures. As the naphtha reforming process is typically performed at a high hydrogen pressure, there are minor amounts of olefins in the products, although they are formed as reaction intermediates. When the pressure and/or the hydrogen-hydrocarbon ratio is decreased, olefins may appear more significantly in the reformate.

### 2. Kinetics

The dehydrogenation of naphtha paraffins on reforming catalysts is very fast, and thermodynamic equilibrium concentrations are rapidly reached. The equilibrium concentration of olefins is very small and remains virtually constant along the catalyst bed. As olefins are consumed by reaction, the rapid paraffin dehydrogenation restores their concentration on the catalyst.

A very small amount of platinum is enough to produce the olefins required by the acid function for the bifunctional mechanism, as shown in Table 3. In several cases, the metallic function of the reforming catalyst is partially deactivated but still is active enough to maintain the equilibrium concentration of olefins. Experiments performed by Silvestri et al. (21), using a catalyst with a metal function only, Pt/C, showed that when feeding $n$-heptane completely free of sulfur at 482°C, 12 atm $H_2$ partial pressure, and 1.13 atm $n$-heptane partial pressure, cyclic products and olefins were produced, with the olefins in thermodynamic equilibrium with $n$-heptane. By using $n$-heptane doped with 100 ppm thiophene, cyclization was completely suppressed, while olefins were still produced at equilibrium concentrations. This is due to the very small amount of Pt required to reach dehydrogenation equilibrium, enough even when the metal function activity is greatly diminished by sulfur. Dehydrogenation has a lower sensitivity to sulfur poisoning than cyclization, because it is a nondemanding reaction in the sense of Boudart et al. (22). For dehydrogenation, all metal sites are active and a particular structure of the metal surface is not required, whereas cyclization demands a particular arrangement of metal surface atoms that can easily be blocked by sulfur.

### 3. Mechanism

The mechanism of paraffin dehydrogenation is similar to the one for naphthenes shown in Figure 3 and is the reverse path of olefin hydrogenation:

$$R-CH_2-CH_3 \underset{+H}{\overset{-H}{\rightleftarrows}} \underset{Pt}{R-CH_2-CH_2} \underset{+H}{\overset{-H}{\rightleftarrows}} \underset{Pt\pi}{R-CH}-\underset{Pt}{CH_2} \rightleftarrows R-CH=CH_2 \quad (16)$$

The olefin is strongly adsorbed on the metal through a $\pi$ bond. When the adsorption is very strong there is a great possibility of continuing the dehydrogenation, particularly at a low hydrogen pressure. This leads to the formation of a polymeric hydrocarbonaceous deposit that covers the catalytic surface and produces its deactivation.

## D. Isomerization of Paraffins

The paraffin isomerization reaction is very important because naphthas contain a high percentage of normal paraffins, which, after isomerization, yield products with a higher octane number.

### 1. Thermodynamics

The isomerization of paraffins is limited by the thermodynamic equilibrium and temperature has little influence on it because the heat of reaction is low. Neither total nor hydrogen pressure influences the equilibrium. Data for n-hexane isomerization to 2- or 3-methylpentane at 500°C are shown in Table 1. By this reaction, the vapor pressure increases (lower boiling point; see Table 6 of Chapter 1) and density remains almost constant.

### 2. Kinetics

Paraffin isomerization is a moderately fast reaction on the reforming catalyst, not so rapid as dehydrogenation of cyclohexanes but more rapid than paraffin dehydrocyclization and hydrocracking. In the reforming process, the isomerization of paraffins reaches thermodynamic equilibrium and the i-paraffins are then consumed by dehydrocyclization and hydrocracking, as shown in scheme (8). Hettinger et al. (23), in a very exhaustive study, examined the product distribution in the reaction of n-heptane on Pt(0.6%)/Al$_2$O$_3$. Selected results are shown in Table 4. Extrapolating to zero conversion, a large isomerization rate can be observed; when the conversion reaches 95.3%, very small amounts of i-heptanes remain because they disappear through successive reactions. The authors found a maximum in the formation of i-heptane as a function of contact time, and Heinemann et al. (3) found a maximum as a function of temperature. At higher contact times or temperatures, C$_1$-C$_6$ paraffins predominate due to hydrocracking. Querini et al. (24) studied n-hexane and n-heptane reforming on samples of commercial Pt(0.3%)-Re(0.3%)-S(0.04%)/Al$_2$O$_3$ catalysts that were coked in a commercial operation up to 208 days on stream. Samples had 0 to 14% coke, according to the time elapsed in operation. In the case of n-hexane reforming, using a catalyst sample without coke, conversions were 38.9% to i-hexanes, 30.0% to hydrocracking products, and 4.2% to benzene. The total

**Table 4** Product Distribution in $n$-Heptane Reforming, in moles per 100 moles of $n$-Heptane Converted[a]

|  | At approaching zero conversion | At 95.3% conversion |
|---|---|---|
| $i$-Heptanes | 52 | 3.3 |
| Naphthenes plus aromatics | 10 | 20 |
| Methane plus hexane | 30 | 32.1 |

[a] Catalyst: Pt(0.6%)/Al$_2$O$_3$; reaction conditions: 34 atm, 496°C, H$_2$/$n$ – C$_7$ = 4.7.
*Source*: From Ref. 23.

conversion of $n$-hexane decreased, but the conversion to $i$-hexanes increased with increasing amount of coke up to 6%. At this coke concentration, there is a maximum in isomerization. The increase in isomer concentration is due to the large decrease in the successive reactions of hydrocracking and dehydrocyclization. For coke concentrations larger than 6%, hydrocracking and dehydrocyclization were already decreased and further coke deposition decreased isomerization. A similar analysis was made for $i$-heptanes. The authors (24) showed that the $i$- to $n$-butane ratio is higher than the equilibrium value when the feed is either $n$-hexane or $n$-heptane, indicating that formation of $i$-butane occurs by successive cracking of $i$-hexane or $i$-heptane and not by isomerization of $n$-butane produced by cracking.

### 3. Mechanism

Paraffin isomerization is a reaction typically catalyzed by very strong acid sites. More efficient than acid catalysts are metal-acid bifunctional catalysts. These catalysts are very active because they allow a dual site mechanism: $n$-paraffins are dehydrogenated to olefins on the metal, olefins are isomerized on the acid, and the $i$-olefins produced are hydrogenated to $i$-paraffins on the metal. The bifunctional catalyst is very convenient because olefins are more easily isomerized on the acid function than paraffins. Hydrogenation-dehydrogenation steps catalyzed by the metal function are very rapid, and olefin isomerization is the slowest, or rate-controlling, step.

Several authors found that isomerization rates are controlled by the acid function. Shipikin et al. (15) studied $n$-hexane isomerization with the same catalysts and under experimental conditions cited in Section IV.A, Dehydrogenation of Cyclohexanes. The authors found that the transformation of $n$-hexane into $i$-hexane increases with the recovery of acidity produced by chloride addition and with the elimination of impurities, such as Na$_2$O, that destroy acidity. Maslyanskii et al. (25) increased the acidity of Pt/Al$_2$O$_3$ by adding fluorine. They found a large increase in the rate of isomerization of $n$-hexane with an increase in fluorine content, working at 380–400°C, 40 atm, and H$_2$/$n$-hexane

= 6. By impregnating the catalyst with NaOH solutions, both acidity and catalytic activity were decreased.

Querini et al. (24), using industrially coked catalyst samples of Pt(0.3)-Re(0.3%)-S(0.04%)/Al$_2$O$_3$ with different amounts of coke and working at 5 atm, 500°C, weight hourly space velocity = 3, and H$_2$/$n$-pentane = 3, found that the catalytic activity for $n$-pentane isomerization decreased in the same pattern as the increase of coke deposition on the acid function. For further verification of the fact that the acid function of the catalyst controls the isomerization reaction, the authors doped $n$-hexane, $n$-pentane, and cyclohexane with $n$-butylamine and found that isomerization reactions were greatly deactivated by the base, while a metal-catalyzed reaction, cyclohexane dehydrogenation, remained unaffected.

An increase in the rate of isomerization is achieved by increasing the acid strength of the catalyst, but in this way hydrocracking is also promoted, producing undesirable light gas formation. A catalyst with an intermediate acid strength shows a good balance between isomerization and cracking. A suitable acid function is Al$_2$O$_3$ promoted by nearly 1% chloride, typical of commercial naphtha reforming catalysts. This chloride concentration can be changed during the reforming operation, producing modifications in the acid catalytic activity. According to Berteau and Delmon (26), the promotion of Al$_2$O$_3$ with chloride produces an increase in acid sites of intermediate strength.

As shown above, the dehydrogenation of paraffins to olefins has a small thermodynamic feasibility, thus giving a very small concentration of olefins at equilibrium. Although this dehydrogenation is the first step in paraffin isomerization, the reaction proceeds to a great extent because, on the bifunctional catalyst, the olefins formed are subsequently isomerized on an acid site. This allows the continuation of paraffin dehydrogenation that is always in thermodynamic equilibrium. All steps proceed until the isomerization equilibrium is reached:

$$n\text{-Hexane} \underset{\text{metal}}{\overset{-H_2}{\rightleftharpoons}} n\text{-hexene} \underset{\substack{\text{acid} \\ \text{rate-controlling} \\ \text{step}}}{\longrightarrow} i\text{-hexene} \underset{\text{metal}}{\overset{+H_2}{\rightleftharpoons}} i\text{-hexane} \quad (17)$$

Table 3 shows the effect of Pt concentration on the isomerization of $n$-heptane. There is no isomerization without Pt because the acid function is not strong enough to isomerize paraffins through a monofunctional acid mechanism. A small amount of Pt is enough to produce the olefins to be isomerized on the acid function. If the amount of Pt is very small, the isomerization rate is a function of Pt concentration because the metal function would be controlling it. However, for 0.1% Pt, the acid function becomes the controlling step. The Pt concentration on the commercial reforming catalysts is larger than this value, and the acid function always controls the bifunctional mechanism.

Considering the olefin isomerization as the controlling step, the rate of reaction would be proportional to the concentration of $n$-olefin adsorbed on the acid sites, $\theta_{n\text{-}o}$. Sinfelt (8) considered that this concentration is related to the pressure of the $n$-olefin in the gas phase by the simple expression

$$\theta_{n\text{-}o} = bP_{n\text{-}o}^n \tag{18}$$

where $b$ is a constant, $n = 0.5$ over the range of conditions studied, and n−o means $n$-olefin.

Since the dehydrogenation of $n$-pentane (n-p) is in thermodynamic equilibrium, from Eq. (15):

$$P_{n\text{-}o} = K_p \frac{P_{n\text{-}p}}{P_{H_2}} \tag{19}$$

and the rate expression becomes

$$r = k\theta_{n\text{-}o} = kbK_p^n \left[\frac{P_{n\text{-}p}}{P_{H_2}}\right]^n = k' \left[\frac{P_{n\text{-}p}}{P_{H_2}}\right]^n \tag{20}$$

This equation agrees with the experimental data of Sinfelt et al. (27) obtained with Pt(0.3%)/Al$_2$O$_3$ at 372°C and 7.7 to 27.7 atm total pressure. These authors also plotted the isomerization rate on Pt/Al$_2$O$_3$ versus $n$-pentane partial pressure calculated according to Eq. (19). In the same plot, the isomerization of 1-pentene over platinum-free alumina (acid function only) falls along the same line, showing that the reaction is controlled by the $n$-olefin isomerization on the acid sites, independent of whether the olefin is produced by in situ dehydrogenation of $n$-pentane on Pt or the olefin is already present in the feed.

Skeletal isomerization of paraffins can occur on the metal function through a monofunctional mechanism, without participation of the acid function. Evidence for this mechanism derives from studies using unsupported metals (evaporated films) or Pt on inert supports (Pt/charcoal, Pt/pumice, Pt/neutralized Al$_2$O$_3$). The monofunctional metal-catalyzed skeletal isomerization of paraffins is very important from the theoretical point of view, but under reforming conditions it shows a low rate compared with the bifunctional metal-acid isomerization. Maire and Garin (28) stated that, under reforming conditions, the platinum-only catalyzed skeletal reactions probably make just a minor contribution to the overall reaction occurring on bifunctional catalyst systems, because the presence of the acid function is essential to obtain a good activity and selectivity. Sinfelt (13) pointed out that, under commercial reforming conditions, dehydrocyclization and isomerization on the metal sites are negligible compared with the bifunctional mechanism. Vestiges of sulfur are enough to poison reaction mechanisms involving only the metal site. For these reasons, the bifunctional mechanism is considered in more detail.

The mechanism of metal-catalyzed isomerization of paraffins was reviewed in several papers (12, 28, 29) and is reviewed from a fundamental point of view in Chapter 2. According to those papers, skeletal isomerization on metal surfaces can occur by means of two different mechanisms, bond shift and cyclic. The bond shift mechanism consists of simple displacement of a carbon-carbon bond and the cyclic mechanism involves consecutive 1-5 ring closure and opening. The bond shift is the only possible mechanism for small molecules. In the case of larger molecules, with at least five carbons in the chain, the cyclic mechanism predominates for very high metal dispersion. Both mechanisms compete at low metal dispersion, and the contribution of each is evaluated by means of $^{13}C$ tracer experiments (28).

Figure 5 shows, on the right side, the isomerization of $n$-hexane through the cyclic mechanism. Carbons 1 and 5 of $n$-hexane lose one H atom and are bonded to neighboring metal sites, producing a five-membered ring. A C−C bond can be formed between these two C atoms, producing an adsorbed methylcyclopentane molecule that can be desorbed (dehydrocyclization of $n$-hexane to a five-membered ring species). Another possibility is change of the carbon atoms that are bound to metal atoms, without desorption. In this way, a C−C bond is formed in position 1-5 and another C−C bond between two adsorbed atoms is broken. After the addition of hydrogen to this surface species, 2- or 3-methylpentane is desorbed. The same scheme explains the isomerization of 2- or 3-methylpentane.

## E. Dehydrocyclization of Paraffins

$n$-Paraffin dehydrocyclization is the most important reaction in naphtha reforming because it is the one producing the largest increase in octane number and

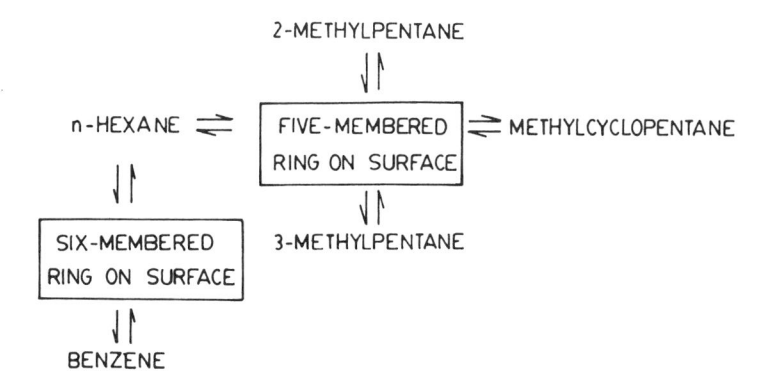

**Figure 5** $n$-Hexane reactions (isomerization and dehydrocyclization) on a metal surface through cyclic mechanisms.

also because *n*-paraffins are usually the main component of naphthas. *n*-Hexane (RON = 19) produces benzene (RON = 99) and *n*-heptane (RON = 0) produces toluene (RON = 124). The increment in octane number is larger when the molecular weight of the paraffin increases, as shown in Figure 2 of Chapter 1. The conversion of a *n*-paraffin into an aromatic produces a decrease in vapor pressure (increase of 9–12°C in boiling points) and in liquid yield (the density of the liquid increases about 30% and the mass of the liquid product decreases because of the hydrogen loss).

### 1. Thermodynamics

Table 1 shows that dehydrocyclization of *n*-hexane at 500°C is displaced toward the formation of benzene. The dehydrocyclization equilibrium is unfavored with an increase in pressure, because the reaction produces an increase in the number of molecules; it is favored with an increase in temperature due to the important endothermicity of the reaction. The thermodynamic feasibility of the reaction increases with increasing the length of the paraffin chain. This can be seen in Figure 6, where the aromatic content at equilibrium was calculated for three different feed mixtures; the influence of temperature and pressure is also shown.

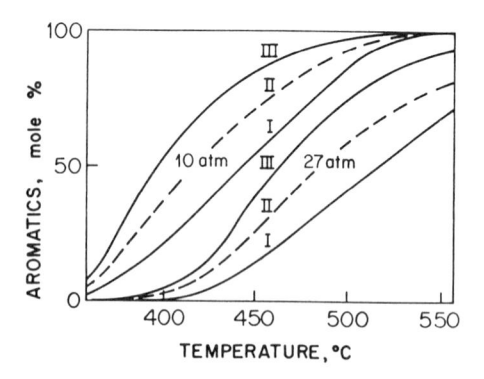

**Figure 6** Concentration of aromatics in thermodynamic equilibrium with paraffins for three mixtures of paraffins used as feed, as a function of the reforming temperature and pressure. Feed composition, paraffin mole %:

|     | $C_6$ | $C_7$ | $C_8$ | $C_9$ |
|-----|-------|-------|-------|-------|
| I   | 40    | 30    | 20    | 10    |
| II  | 25    | 25    | 25    | 25    |
| III | 10    | 20    | 30    | 40    |

(From Ref. 12, p. 383.)

## 2. Kinetics

Paraffin dehydrocyclization is a relatively slow reaction. It is slower than dehydrogenation of cyclohexanes, dehydroisomerization of cyclopentanes, and isomerization of paraffins, and its rate is comparable to that of hydrocracking. Because of such a small rate, thermodynamic equilibrium of paraffin dehydrocyclization is not reached in reforming units, whereas more rapid reactions virtually reach equilibrium in the first or second reactor. Dehydrocyclization occurs together with hydrocracking mainly in the last reactor, where the largest fraction of the total weight of catalyst is loaded in order to allow the maximum degree of dehydrocyclization.

The dehydrocyclization rate increases with an increase in the paraffin molecular weight. Under normal reforming conditions, little hexane is dehydrocyclized, the conversion of heptanes is larger, and the conversion of octanes and decanes is two and five times that of heptane, respectively.

Dehydrocyclization rate is favored by an increase in temperature and is decreased by an increase in pressure, as shown in the pioneering investigations of Hettinger et al. (23). Therefore, a high temperature and a low pressure are the desirable conditions for this reaction. For this reason in several commercial processes, such as Magnaforming (7), more complete dehydrocyclization of paraffins is obtained working at a lower pressure and at a higher temperature in the last reactor or reactors. These conditions are more severe and require more stable catalysts.

Figure 7 shows the detrimental effect of the increase in pressure on the dehydrocyclization of $n$-heptane. At a high pressure, not only is the conversion to aromatics small but also, when increasing the contact time, cyclic products are destroyed by the pressure-favored hydrogenolysis or hydrocracking.

## 3. Mechanism

Similar to paraffin isomerization, the reaction of dehydrocyclization can be produced by a mechanism involving only the catalyst metal function or by a bifunctional mechanism involving dehydrogenation-hydrogenation steps on the metal and isomerization on the acid function. Figure 8 shows the bifunctional mechanism for dehydrocyclization of $n$-hexane. This scheme is similar to the classical reaction network of Mills et al. (4), with the addition of the possibility of passage from a five- to a six-carbon-atom ring by isomerization of methyl cyclopentadiene. The possible formation of cyclohexane by cyclohexene hydrogenation is not included because, as stated above, cyclohexane is completely dehydrogenated to benzene under normal operational conditions. Reaction steps drawn parallel to the abscissa occur on the acid function of the catalyst and reactions drawn parallel to the ordinate axis occur on the metal function. $n$-Hexane is first dehydrogenated on the metal to give $n$-hexene, which migrates to a neighboring acid site, where it is protonated, producing a

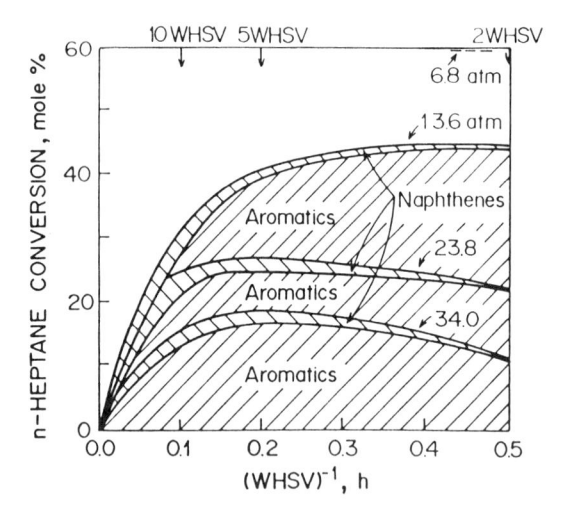

**Figure 7** Conversion of $n$-heptane to aromatics and naphthenes as a function of the inverse of weight hourly space velocity (contact time) at various pressures. $T = 496°C$. (From Ref. 23.)

secondary carbenium ion. This ion is cyclized on the acid function, producing methylcyclopentane that migrates to the metal function, where it is dehydrogenated to methylcyclopentene and methylcyclopentadiene. These five-carbon-atom ring olefins are then isomerized on the acid sites, enlarging the ring to six carbons. In this way, cyclohexene and cyclohexadiene are produced and are finally dehydrogenated to benzene.

As cited in Section C, dehydrogenation-hydrogenation reactions on the metal function are so rapid that they can be considered in thermodynamic equilbrium, whereas isomerization reactions on the acid function are slower. Table 5 gives thermodynamic equilibrium constants of the reactions involved in the bifunctional dehydrocyclization scheme. Reaction 1 has little thermodynamic feasibility because the process is usually carried out at a high hydrogen pressure. The concentration of $n$-hexene, which is in thermodynamic equilibrium, is very small and undetectable in the gas phase. The amount of $n$-hexene that is produced and reacted is controlled by reaction 2. Methylcyclopentane is in equilibrium with its dehydrogenated products, methycyclopentene and methylcyclopentadiene, because the dehydrogenations are very rapid and the acid cycle expansion is very slow. According to equilibrium constants, the concentrations of unsaturates are very small compared to that of methycyclopentane, which is the only five-carbon-atom ring compound that can be detected. Cyclohexene and cyclohexadiene are immediately dehydrogenated to benzene and cannot be detected. Therefore, in the dehydrocyclization of $n$-hex-

```
                    n-HEXANE
                      1 ↓↑   2
                    n-HEXENE ⇌ METHYLCYCLOPENTANE
                               3 ↓↑           5
                        METHYLCYCLOPENTENE ⇌ CYCLOHEXENE
                               4 ↓↑      6    7 ↓↑
                     METHYLCYCLOPENTADIENE ⇌ CYCLOHEXADIENE
                                               8 ↓↑
                                              BENZENE
```

(Left side labels: REACTIONS ON THE METAL SITES — DEHYDROGENATION / HYDROGENATION)

(Bottom label: ISOMERIZATION — REACTIONS ON THE ACID SITES)

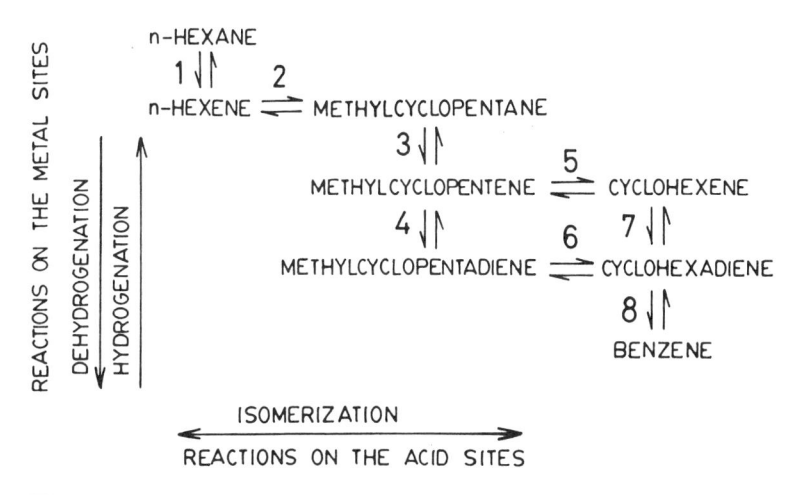

**Figure 8** Reaction network for *n*-hexane dehydrocyclization through the bifunctional mechanism.

ane into benzene through the bifunctional mechanism, methylcyclopentane is the only intermediate existing in measurable amounts. It seems that acid sites required for reactions 5 and 6 are different from those required for reaction 2 because, when the acid function is poisoned with *n*-butylamine, there is a large accumulation of methylcyclopentane (24). The increase in methylcyclopentane, when the acid function is deactivated, is used in some commercial units to

**Table 5** Thermodynamic Equilibrium Constants, at 500°C, for Reactions of the Bifunctional Scheme

| Reaction | $K_p$ |
| --- | --- |
| 1 | $8.40 \times 10^{-2}$ |
| 2 | 11.87 |
| 3 | $1.27 \times 10^{-1}$ |
| 4 | $4.20 \times 10^{-2}$ |
| 5 | $3.33 \times 10^{-1}$ |
| 6 | $3.31 \times 10^{-1}$ |
| 7 | $4.20 \times 10^{-2}$ |
| 8 | $6.36 \times 10^{7}$ |

*Source*: Calculated according to the method of van Krevelen (30).

follow catalyst deactivation and to indicate the need for an increase in chlorine addition and/or temperature.

The reaction network in Figure 8 was proposed for Pt/Al$_2$O$_3$ and extended to Pt-Re/Al$_2$O$_3$ by Selman and Voorhies (31). Some modifications of the scheme were published. For instance, Sinfelt (8) proposed the dehydrogenation of *n*-hexene to *n*-hexadiene on the metal site, which can then produce methylcyclopentene on the acid sites.

There is no catalyst capable of producing only the reactions shown in the scheme of Figure 8, and many other reactions occur simultaneously, depending on the catalyst and on the operating conditions. *n*-Hexene, produced by reaction 1, can be isomerized on the acid sites to *i*-hexene, which is a step of *n*-hexane bifunctional isomerization similar to Eq. (17) for *n*-hexane isomerization. Methylcyclopentane and benzene rings can be opened by hydrogenolysis, producing *n*- and *i*-hexanes. The hexanes can be transformed into C$_1$–C$_5$ paraffins by successive hydrogenolysis and hydrocracking.

Dehydrocyclization is similar to isomerization because both reactions occur through the bifunctional mechanism or by a mechanism involving only the metal function of the catalyst. The metal mechanism is similar for both reactions, as shown in Figure 5 for *n*-hexane. An *n*-hexane molecule is adsorbed through carbon-metal bonds on two neighboring metal atoms, producing a cyclic compound. If the cycle has five carbon atoms, on reorganization and desorption, 2- or 3-methylpentane or methylcyclopentane is produced. If the surface cycle has six carbon atoms, on dehydrogenation and desorption, benzene will be produced. Both five- and six-membered ring closures occur as parallel reactions of *n*-hexane. However, in the case of 2- or 3-methylpentane transformation into benzene, these compounds have only five carbon atoms in the chain which can be absorbed on the metal, producing only a five-membered ring species. From this surface species *n*-hexane can be produced. Then this *n*-hexane can be adsorbed as a surface six-membered ring that can lead to benzene. These mechanisms were verified by Dautzenberg and Platteeuw (32) using 0.5% Pt on nonacidic alumina as a catalyst at 440°C, 9.5 atm, and H$_2$/HC = 4. Feeding *n*-hexane, all the species (2- and 3-methylpentane, methylcyclopentane, and benzene) showed a nonzero rate of formation at a very small contact time, indicating that all are primary products. Feeding 2- or 3-methylpentane, the rate of formation of benzene is zero at a very small contact time, indicating that benzene is a secondary product.

Davis and Venuto (33) studied the dehydrocyclization of hydrocarbons of eight and nine carbon atoms on Pt(0.6%)/Al$_2$O$_3$ and on the same catalyst after acidity elimination by washing with aqueous ammonia. With the nonacid catalyst, at 482°C, aromatics that can be predicted from direct six-membered ring closure were produced. For instance, from *n*-octane, ethylbenzene and *o*-xylene can be obtained:

```
           ┌─────────────────────┐
           │                     │                ethylbenzene
       C—C—C—C—C—C—C—C
                       │         │
                       └─────────┘                o-xylene                (21)
```

When the acid Pt/Al$_2$O$_3$ catalyst was used, more than 50% of the products were *m*- and *p*-xylene. These results show the capacity of the acid function for isomerization of paraffins and aromatics. In the case of *n*-octane, after isomerization on the acid function to its three methylheptane isomers, the three xylenes can be produced by cyclization on the metal function:

```
                C
                │
       C—C—C—C—C—C—C           m-xylene
       │           │
       └───────────┘                                                     (22)
```

```
                C
                │
       C—C—C—C—C—C—C
       │   │       │
       │   └ p-xylene ┘
       │               │
       └── o-xylene ───┘                                                 (23)
```

```
                    C
                    │
       C—C—C—C—C—C—C           m-xylene
       │   │       │
       │   └───────┘
       └───────────┘                                                     (24)
```

Paál and Tetényi [(34) and papers cited therein] studied the dehydrocyclization of *n*-hexane on platinum and nickel in the temperature range 240–400°C. They found that the mechanism includes metal-catalyzed dehydrogenation via the intermediates hexene-hexadiene-hexatriene. Hexatriene rapidly undergoes a six-ring closure producing cyclohexadiene, which immediately dehydrogenates to benzene. A similar mechanism was proposed by Dautzenberg and Platteeuw (32). Because of thermodynamic limitations of these extensive dehydrogenation reactions, this dehydrocyclization route becomes important only at a low hydrogen pressure. More details on the mechanism of this metal-catalyzed dehydrocyclization are given in Chapter 2.

Results of Silvestri et al. (21), cited in Section C, show that cyclic products and olefins were produced when feeding pure *n*-heptane over Pt/C at 12 atm H$_2$ pressure. When *n*-heptane doped with 100 ppm thiophene was fed, cyclization was suppressed but dehydrogenation to olefins was little affected, with the concentration of 3-heptenes close to the value corresponding to thermodynamic equilibrium. This fact indicates that the metal mechanism of cyclization was poisoned but, since olefins were produced, the bifunctional mechanism might be possible in cases where an acid function was present. The same authors verified this supposition with data shown in Table 6. It can be seen that when feeding $n$-heptane doped with thiophene, no dehydrocyclization occurs on Pt/C or on $Al_2O_3$, but when both are mixed, the reaction occurs through the bifunctional mechanism.

In the commercial process of naphtha reforming, sulfur is always present. Even when the feed is hydrogenated, the catalyst is usually presulfided. So, according to Sinfelt (13), dehydrocyclization on the metal function is partially or totally poisoned, and the reaction is mainly or completely produced through the bifunctional mechanism.

Table 3, cited in Section B, shows that an increase in $n$-heptane dehydrocyclization is produced when the concentration of Pt on the catalyst increases and that such increase is less than proportional to the increase of Pt. Even though there is a contribution of the metal function to dehydrocyclization, the contribution is smaller than the dehydrocyclization through the bifunctional mechanism. Very pure $n$-heptane, free of sulfur, was used in these experiments; hence the metal function was not poisoned and contributed to the reaction.

## F. Hydrocracking

Under the conditions of naphtha reforming, hydrocarbon cracking is produced on the acid function of the catalyst. The reaction is called hydrocracking

**Table 6** Comparison of Ring-Compound Production from $n$-Heptane Doped with Thiophene with and without Mixing of Catalyst Components[a]

|  | Catalyst | | |
| --- | --- | --- | --- |
|  | 0.25 g of Pt/C | 0.50 g of $Al_2O_3$ | 0.75 g of mixed catalyst containing 33% Pt/C and 67% $Al_2O_3$ |
| Percentage of $n$-heptane converted to: | | | |
| Ethylcyclopentane | — | 0.02 | 0.28 |
| 1,1-Dimethylcyclopentane | — | — | 0.09 |
| 1,2-Dimethylcyclopentane | 0.11 | 0.06 | 1.03 |
| 1,3-Dimethylcyclopentane | — | 0.07 | 1.04 |
| Methylcyclohexane | — | — | 0.12 |
| Toluene | 0.01 | 0.01 | 0.77 |
| 3-Heptenes | 0.76 | — | 0.65 |

[a] Reaction conditions: 482°C, 12 atm $H_2$ partial pressure, 1.13 atm $n$-heptane partial pressure, and $n$-heptane liquid flow rate = 16 cm$^3$/h; sulfur added to $n$-heptane = 100 ppm.
*Source*: From Ref. 21.

because it is produced under hydrogen pressure. Olefins produced by cracking are immediately hydrogenated because of the presence of Pt. Hydrogenation is a very rapid and practically irreversible reaction, and hydrocracking is controlled by the acid function of the catalyst. The presence of the metal favors hydrocracking on the acid function because it produces olefins, which are more easily cracked than paraffins. This reaction produces light paraffins, either liquid or gaseous. A low degree of hydrocracking is beneficial. The produced liquid paraffins have a higher octane number and vapor pressure than the starting paraffins because of the lower number of carbon atoms. Under reforming conditions, naphthenes and aromatics can also be cracked; the ring opening is followed by hydrogenation and a paraffin is produced. Cyclopentanes are more susceptible to cracking than cyclohexanes and, within each hydrocarbon series, the compounds of larger molecular weight are more easily cracked.

Hydrocracking has the following characteristics: all products are saturated; both catalyst functions are involved; it is very exothermic; it is the slowest reaction in reforming; the catalyst deactivation by coke formation is very much slower than when cracking is produced on acid catalyst without metal sites; the reaction is favored by an increase in hydrogen pressure; it is the reaction with the greatest activation energy in reforming, hence it shows the largest increase with increasing temperature; and, finally, it is a hydrogen-consuming reaction. Under normal operational conditions, hydrocracking is slower than dehydrocyclization, but at high pressure and temperature it can be faster.

Only a moderate level of hydrocracking is convenient in the reforming process; therefore, the catalyst acidity must be the lowest possible consistent with good isomerization activity. An excessive degree of hydrocracking produces a reduced endotherm (or even an increase in temperature) through the last reactor, a reduction in hydrogen production, an increase in $C_1$–$C_4$ hydrocarbons, and a decrease in liquid yield. Figure 7 shows how the hydrocracking of cyclic products increases with increasing pressure. At a high pressure, aromatic production has a maximum beyond which the aromatic content decreases due to hydrocracking. At low pressures, aromatic production always increases with an increase in contact time.

Hydrocracking catalyzed by the acid function cracks the paraffin molecule approximately at a middle position, resulting in low methane production. Haensel and Donaldson (35) found that the products of hydrocracking, in general, are not further cracked within the reaction contact time range used in normal plants. Their results for hydrocracking in the catalytic reforming of $n$-heptane are shown in Table 7. Most of the hydrocracking products are $C_3$ and $C_4$ paraffins, and there is a correspondence between the yield of $C_1$ and $C_6$, $C_2$ and $C_5$, and $C_3$ and $C_4$, which suggests that the products come from primary hydrocracking. Other authors obtained different results; for instance, Hettinger et al. (23) found the product distribution shown in Table 8 when $n$-heptane conversion reaches 95.3%. For this high conversion, successive or secondary

**Table 7** Distribution of Light Products from the Catalytic Reforming of n-Heptane, Moles per 100 Moles of Feed

| Methane | 3.1 |
| --- | --- |
| Ethane | 5.1 |
| Propane | 25.1 |
| Butanes | 26.7 |
| Pentanes | 6.1 |
| Hexanes | 3.0 |

Source: From Ref. 35.

hydrocracking of $C_2$–$C_6$ paraffins increases the yield of lower paraffins. Cracking on the metal, called hydrogenolysis, produces an increase in the production of lower paraffins, mainly of methane.

### G. Hydrogenolysis

Hydrogenolysis of hydrocarbons involves the rupture of C–C bonds and the formation of C–H bonds, all produced on the catalyst metal function. Even though hydrogenolysis and hydrocracking give the same type of products (lower paraffins), active sites are different and the product distribution is different.

As with hydrocracking, hydrogenolysis is very feasible thermodynamically, and it is very exothermic (thermodynamic data for n-heptane hydrogenolysis to

**Table 8** Product Distribution in n-Heptane Reforming at 95.3% Conversion in Moles per 100 Moles of n-Heptane Converted[a]

| Aromatics $C_7$–$C_{10}$ | 17.4 |
| --- | --- |
| Cycloparaffins $C_5$–$C_7$ | 2.7 |
| Olefins | 0.6 |
|  | (near equilibrium) |
| Dimethylpentane | 2.6 |
| Ethylpentane | 0.6 |
| Methylhexane | 14.1 |
| n-Heptane | 4.7 |
| Hexanes | 9.1 |
| Pentanes | 15.7 |
| Butanes | 22 |
| Propane | 39 |
| Ethane | 29 |
| Methane | 24 |

[a] Catalyst: Pt(0.6)/$Al_2O_3$; reaction conditions: 34 atm, 496°C, $H_2/nC_7 = 4.7$.
Source: From Ref. 23.

$C_6 + C_1$ are shown in Table 1). Its rate increases with an increase in temperature and hydrogen pressure. Hydrogenolysis is an undesired reaction because it produces low-value gases and consumes hydrogen.

The mechanism for cyclopentane hydrogenolysis is that mentioned in Section D (Figure 5). The size of the Pt crystallite influences the product distribution of this hydrogenolysis. For a very high metal dispersion (crystallites $\leqslant 2$ nm), the ring rupture is not selective. All C−C bonds of the ring have the same probability of being broken, and products correspond to a statistical distribution as shown in Figure 9a. The C−C bond rupture is selective for Pt crystallites larger than 2 nm, and there is no formation of n-hexane, as shown in Figure 9b.

Benzene hydrogenolysis to n-hexane occurs through the six-membered ring species adsorbed on the metal, as shown in Figure 5. The same mechanism operates for alkylbenzene and cyclohexane hydrogenolysis. The elimination of the alkyl chain in alkylbenzenes occurs prior to ring opening.

Larger metal ensembles are required for hydrogenolysis than for isomerization and dehydrocyclization (36). Barbier and Marecot (37) studied cyclopentane hydrogenolysis on Pt/Al$_2$O$_3$ and found that the activity per Pt atom decreased with an increase in metal dispersion. Small crystallite are less active because they do not have the required number of Pt atoms. An effect similar to the decrease in the metal particle size is obtained by sulfurization of Pt or by alloying it with inactive atoms (Au, Sn, Pb) (38). In all these cases, the number of large ensembles of adsorbing surface atoms is drastically reduced (geometric effect), thereby decreasing the hydrogenolysis activity.

Unsulfided Pt-Re/Al$_2$O$_3$ shows a high selectivity for hydrogenolysis, which Shum et al. (36) considered a characteristic of Pt-Re alloy formation. This harmful property is selectively eliminated by presulfiding the catalyst. Sulfur adsorbed on Re reduces the number of contiguous Pt atoms available for the formation of the chemisorption complex required for hydrogenolysis. The Pt-Re/Al$_2$O$_3$ commercial catalyst is always sulfided before the operation to avoid an initial hyperhydrogenolytic activity that could produce excessive methane formation and a dangerous increase in temperature (runaway). Pt-Ir/Al$_2$O$_3$ also

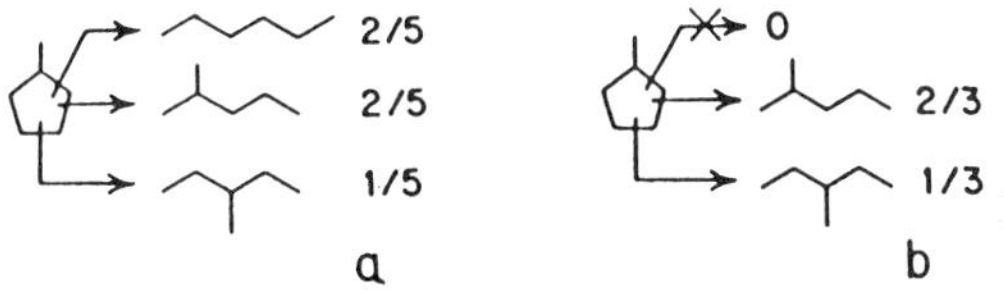

**Figure 9** Hydrogenolysis of methylcyclopentane. (a) Nonselective hydrogenolysis on highly dispersed metal catalysts. (b) Selective hydrogenolysis on low-dispersion metal catalysts.

has intense hydrogenolytic activity and requires presulfiding, but Pt-Sn/Al$_2$O$_3$ and Pt-Ge/Al$_2$O$_3$ do not require presulfiding because Sn and Ge are inert diluents that produce a decrease in the number of Pt ensembles suitable for hydrogenolysis.

In Chapter 2, Z. Paál presents a comprehensive study of the mechanism of hydrogenolysis, the active sites, and the varying selectivity for hydrogenolysis on different metals.

### H. Coke Deposition

A carbonaceous deposit, generally known as coke, is produced on the catalyst surface simultaneously with the reforming reactions. This deposit blocks the active surface, producing a decrease in catalytic activity and modifications in selectivity. A detailed review of the deactivation of reforming catalysts by coke deposition is presented in Chapter 9 of this book, where it is stated that only one carbon atom out of 200,000 activated by the reforming catalyst remains on its surface as coke. The problem is that this deposition is cumulative. In Chapter 9, the authors describe the characterization of coke, how coke is formed as a function of time, the effect of the metallic function and of the support, and the effects of the reaction conditions and of sulfurization on coke deposition on bifunctional reforming catalyst.

Coke deposition produces catalyst deactivation and is the cause of the periodic interruption of the reforming operation to regenerate the catalytic activity. This regeneration is done by coke elimination through oxidation with diluted air. Most of the improvements in the reforming of naphtha are due to the development of catalysts that produce less coke, allowing operation at lower pressure with great improvement in selectivity.

## V. COMPARISON OF REFORMING REACTIONS

The catalytic reforming of naphtha is a process strongly influenced by reaction kinetics and thermodynamics. For comparative purposes, Table 9 summarizes the properties of reforming reactions reviewed in Section IV. Reactions are ordered according to reaction rates, starting with the most rapid: the dehydrogenation of cyclohexanes. All the reactions produce an increase in octane number. Coking is not included because its product is not present in the reformate.

The most rapid reactions reach thermodynamic equilibrium, and the others are kinetically controlled. A high process temperature and a low pressure favor the thermodynamic feasibility as well as the reaction rate in the three most important reactions: dehydrogenation of cyclohexanes, dehydroisomerization of cyclopentanes, and dehydrocyclization of paraffins. For the other reactions—isomerization, hydrocracking, and hydrogenolysis—temperature and

**Table 9** Thermodynamic and Kinetic Comparison of the Main Naphtha Reforming Reactions[a]

| Reaction | Reaction rate | Thermal effect | Reach thermodynamic equilibrium | Thermodynamic P | Thermodynamic T | Reaction P | Reaction T | H$_2$ | Vapor pressure | Density | Liquid yield |
|---|---|---|---|---|---|---|---|---|---|---|---|
| Cyclohexane dehydrogenation | Very rapid | Very endothermic | Yes | − | + | − | + | Produce | d | i | d |
| Cyclopentane dehydroisomerization | Rapid | Very endothermic | Yes | − | + | − | + | Produce | d | i | d |
| Paraffin isomerization | Rapid | Slightly exothermic | Yes | None | Sl − | + | + | None | i | sl d | sl i |
| Paraffin dehydrocyclization | Slow | Very endothermic | No | − | + | − | + | Produce | d | i | d |
| Hydrocracking hydrogenolysis | Very slow | Exothermic | No | None | − | ++ | ++ | Consume | Great i | d | Great d |

[a] +, an increase in either pressure or temperature produces an increase in equilibrium conversion or reaction rate; ++, produces a great increase; −, produces a decrease; i, increase; d, decrease; sl, slight.

pressure have only a slight influence on equilibrium. Overall, high temperatures and low pressures would seem most desirable, but the same conditions favor deactivation of the catalyst and the reactions of hydrocracking and hydrogenolysis. For this reason the process conditions are a compromise.

The three most important reactions are endothermic and dominate the heat balance, producing a decrease in temperature along the catalyst bed. Only in the last reactor, where the slowest reactions ocur, is the endothermicity of dehydrocyclization of paraffins partially neutralized by the exothermicity of hydrocracking and hydrogenolysis. Only these last two molecule-rupture reactions consume hydrogen, and the process is a net hydrogen producer.

## VI. KINETIC MODELS

A detailed reaction model for naphtha reforming would be very complicated because there are many naphtha components, each one undergoing several different reactions. For this reason, naphtha components are lumped in groups with similar properties and kinetic behaviors. Krane et al. (39) studied the reforming of individual hydrocarbons having up to 10 carbon atoms in the molecule. They tabulated the first-order rate constants of 53 individual reactions on Pt(0.3%)/Al$_2$O$_3$ at 6.8–30 atm, H$_2$/hydrocarbon = 5, and 496°C. Experimental values for the constants showed the general principles cited above: with an increase in the molecular weight of the reactant, the kinetic constant increases, and in hydrocracking the highest kinetic constant values are for rupture in the middle of the molecule. Twenty components are included in this model: 10 paraffins ($C_1$ to $C_{10}$), 5 naphthenes ($C_6$ to $C_{10}$), and 5 aromatics ($C_6$ to $C_{10}$). One ordinary differential equation for each component describes its formation and disappearance due to the reactions in which it is involved. Experimental values under isothermal conditions were compared with values calculated at the same temperature, and they agreed acceptably. To reduce the number of equations to be solved, all the reactions can be reduced to eight and all components lumped in average concentration values of paraffins, $C_P$; naphthenes, $C_N$; and aromatics, $C_A$. The eight reactions are (1) dehydrocyclization of paraffins to naphthenes, (2) dehydrogenation of naphthenes to aromatics, (3) hydrocracking of paraffins, (4) hydrogenolysis of aromatics (ring opening), (5) hydrocracking of naphthenes, (6) hydrogenation of aromatics, (7) hydrocracking of aromatics, and (8) hydrogenolysis of naphthenes (ring opening). Equations to be solved are as follows:

$$\frac{dC_P}{d(1/SV)} = -k_1 C_P - k_3 C_P + k_4 C_A + k_8 C_N \tag{25}$$

$$\frac{dC_N}{d(1/SV)} = k_1 C_P - k_2 C_N - k_5 C_N + k_6 C_A - k_8 C_N \tag{26}$$

$$\frac{dC_A}{d(1/SV)} = k_2 C_N - k_4 C_A - k_6 C_A - k_7 C_A \tag{27}$$

where $k_i$ is the first-order rate constant of reaction $i$ (1 to 8) and SV the weight hourly space velocity. The solution of these equations gives the concentrations of paraffins, naphthenes, and aromatics as a function of the inverse of space velocity, or contact time.

Smith (40) used naphtha as feed, which was idealized as a mixture of three average components: a paraffin, a naphthene, and an aromatic. Four reactions were considered: (1) naphthenes to aromatics, (2) naphthenes to paraffins, (3) hydrocracking of paraffins, and (4) hydrocracking of naphthenes. The operating pressure was very high, 40 atm, and under these conditions hydrocracking was more important than dehydrocyclization. A system of four adiabatic reactors was studied, and energy balance must be added to the kinetic equations. The solution of these differential equations gives the temperature profiles through the four reactors.

Kmak and Stuckey (41) studied the kinetic simulation of the Power-forming process, working over a wide range of operating conditions and using pure compounds, mixtures of pure compounds, and naphthas as feeds. By solving the kinetic and heat balance equations, they were able to predict pilot plant and commercial unit behavior for a wide range of feeds and operating conditions. They calculated the concentration profiles for the 22 components of the kinetic model through the four adiabatic reactors for a target octane number of 100, 103, or 108.

Ramage et al. (42) from Mobil Research and Development Corporation published a very complete kinetic reforming model, which was extended to include catalyst deactivation. The model includes hydrocarbon conversion kinetics on fresh catalysts (start-of-cycle kinetics) and the modification of the kinetics with time on stream due to catalyst deactivation. Due to the large number of reacting species, the inherent complexity of a kinetic model for naphtha feeds was reduced by proper lumping, leading to a reasonable number of chemical reaction pathways.

The start-of-cycle reaction network defines the interconversion between 13 kinetic lumps, including reactions of cracking, ring closure and isomerization, and dehydrogenation. Reactions at the same carbon atom number level are assumed to be reversible, and those coupling species with different carbon atom numbers are considered irreversible. Additional assumptions and approximations of this model are the following:

1. Rate expressions are of nonlinear pseudomononuclear form.
2. Kinetics with Langmuir-Hinshelwood type adsorption is assumed.
3. Microscopic reversibility is satisfied with equilibrium constraints calculated from free energy data.

4. Possible reaction steps not consistent with a known pure component behavior are not allowed (i.e., benzene cracking).
5. Internal diffusional effects were included in the apparent kinetic constants.
6. Catalyst deactivation is due only to coke formation.
7. First order irreversible aging rates are used.
8. Deactivation obeys a single-site, nondissociative Langmuir-Hinshelwood poisoning model.

A system of three reactors was studied, with 20% of the total mass of catalyst in the first reactor and 40% in each of the following ones. Figure 10a shows temperature profiles in the reactors, and Figure 10b shows the reactor composition profiles by molecular type. The feed enters the first bed at 510°C and, because of the very rapid endothermic cyclohexane dehydrogenation, the temperature decreases rapidly. In this reactor a large part of the cyclopentanes is also dehydroisomerized, with the rapid transformation of both classes of naphthenes responsible for the large increase in aromatics. The remarkable

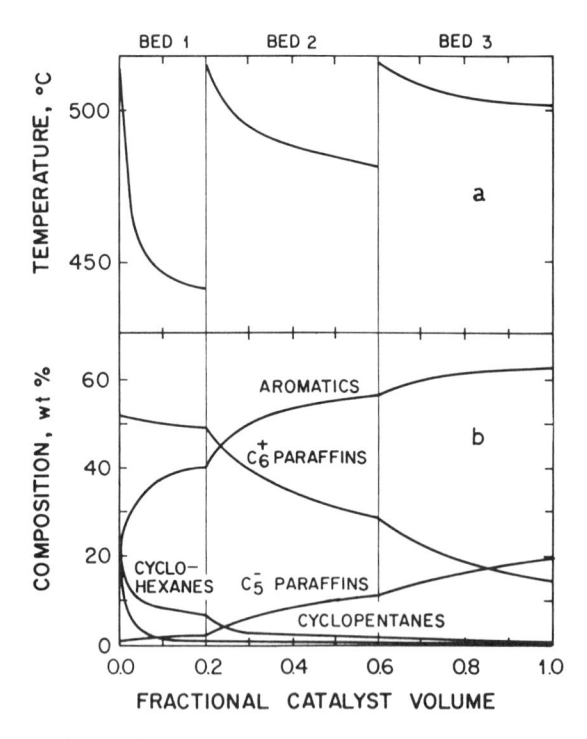

**Figure 10** (a) Temperature profiles and (b) composition profiles by molecular type, as a function of the fractional catalyst volume through the three reforming reactors. (From Ref. 42.)

temperature decrease, 70°C, almost quenches all reactions and more catalyst in the first reactor would not provide additional conversion. The product of the first reactor is reheated to 510°C before entering the second reactor. In this reactor, cyclopentanes complete their dehydroisomerization to benzene, reaching thermodynamic equilibrium, as cyclohexanes did in the first reactor, and paraffin isomerization reaches its equilibrium. Temperature in the second reactor is higher and the amount of catalyst is larger than in the first reactor; therefore, there is a certain degree of transformation of paraffins ($C_6^+$) into aromatics by the slow reaction of dehydrocyclization and into light paraffins ($C_5^-$) by the slow reaction of hydrocracking. After reheating to 510°C, paraffin dehydrocyclization and hydrocracking occur in the third reactor. Since dehydrocyclization is endothermic and hydrocracking exothermic, only a small drop in temperature in the last reactor is observed. The two reactions are very slow and, even though the temperature is high, concentrations do not reach thermodynamic equilibrium.

The Mobil kinetic model considers catalyst deactivation and can predict the increase in the inlet reactor temperature as a function of time on stream in order to keep the target octane number. This increase in temperature produces a large increase in hydrocracking together with the corresponding decrease in liquid yield.

Chapter 14 of this book, by L. E. Turpin, is devoted to modeling of commercial reformers. The process simulation model allows business decisions regarding catalyst, yields, process changes, and control and optimization of the process with profit as the objective function. Examples of the economic benefit of the method applied to typical refinery problems are shown using a kinetic model.

## REFERENCES

1. G. S. Parks and H. M. Huffman, *Free Energy of Some Organic Compounds.* Reinhold, New York, 1932.
2. F. D. Rossini, K. S. Pitzer, R. L. Arnett, R. M. Braum, and G. C. Pimentel, *Selected Values of Physical and Thermodynamic Properties of Hydrocarbons and Related Compounds.* API Res. Project 44. Carnegie Press, Pittsburgh, 1953.
3. H. Heinemann, G. A. Mills, J. B. Hattman, and F. W. Kirsch, *Ind. Eng. Chem.* 45, 130 (1953).
4. G. A. Mills, H. Heinemann, T. H. Milliken, and A. G. Oblad, *Ind. Eng. Chem.* 45, 134 (1953).
5. F. G. Ciapetta and J. B. Hunter, *Ind. Eng. Chem.* 45, 147 (1953).
6. F. G. Ciapetta and J. B. Hunter, *Ind. Eng. Chem.* 45, 155 (1953).
7. F. G. Ciapetta, R. M. Dobres, and R. W. Baker, Catalytic reforming of pure hydrocarbons and petroleum naphthas. In *Catalysis* (P. H. Emmet, ed.). Reinhold, New York, 1958, Vol. VI, p. 495.
8. J. H. Sinfelt, *Adv. Chem. Eng.* 5, 37 (1964).

9. F. G. Ciapetta and D. N. Wallace, *Catal. Rev. 5*, 67 (1971).
10. M. J. Sterba and V. Haensel, *Ind. Eng. Chem. Prod. Res. Dev. 15*, 2 (1976).
11. B. C. Gates, J. R. Katzer, and G. C. A. Schuit, *Chemistry of Catalytic Processes*. McGraw-Hill, New York, 1979.
12. R. Prins and G. C. A. Schuit, eds., *Chemistry and Chemical Engineering of Catalytic Processes*. Sijthoff & Noordhoff, Alphen aan den Rijn, 1980.
13. J. H. Sinfelt, *Catal. Sci. Technol. 1*, 257 (1981).
14. J. H. Sinfelt, H. Hurwitz, and R. A. Shulman, *J. Phys. Chem. 64*, 1559 (1960).
15. V. V. Shipikin, G. N. Maslyanskii, B. B. Zharkov, and N. R. Bursian, *Neftekhimiya 6*, 401 (1966).
16. A. Q. Ritchie and A. C. Nixon, *Am. Chem. Soc. Div. Petrol. Chem. Preprint 11, No. 3*, 93 (1966).
17. J. M. Parera, J. N. Beltramini, C. A. Querini, E. E. Martinelli, E. J. Churin, P. E. Aloé, and N. S. Figoli, *J. Catal. 99*, 39 (1986).
18. J. E. Germain, *Catalytic Conversion of Hydrocarbons*. Academic Press, New York, 1969.
19. P. B. Weisz, 2nd Cong. Int. Catal., Technip, Paris, 1961, p. 937.
20. J. H. Sinfelt, H. Hurwitz, and J. C. Rohrer, *J. Catal. 1*, 481 (1962).
21. A. J. Silvestri, P. A. Naro, and R. L. Smith, *J. Catal. 14*, 386 (1969).
22. M. Boudart, A. Aldag, J. E. Benson, N. A. Dougharty, and C. G. Harkins, *J. Catal. 6*, 92 (1966).
23. W. P. Hettinger, C. D. Keith, J. L. Grings, and J. W. Teter, *Ind. Eng. Chem. 47*, 719 (1955).
24. C. A. Querini, N.S. Figoli, and J. M. Parera, *Appl. Catal. 52*, 249 (1989).
25. G. N. Maslyanskii, N. R. Bursian, and S. A. Barkan, *Zhur. Prikl. Khim. 39*, 3, 650 (1966).
26. P. Berteau and B. Delmon, *Catal. Today 5*, 121 (1989).
27. J. H. Sinfelt, H. Hurwitz, and J. C. Rohrer, *J. Phys. Chem. 64*, 892 (1960).
28. G. L. C. Maire and F. G. Garin, *Catal. Sci. Technol. 6*, 161 (1984).
29. F. G. Gault, *Adv. Catal. 30*, 1 (1981).
30. R. C. Reid, J. M. Prausnitz, and T. K. Sherwood, *The Properties of Gases and Liquids*, 3rd ed. McGraw-Hill, New York, 1977, p. 278.
31. D. Selman and A. Voorhies, *Ind. Eng. Chem. Prod. Res. Dev. 14*, 12 (1975).
32. F. M. Dautzenberg and J. C. Platteeuw, *J. Catal. 19*, 41 (1970).
33. B. H. Davis and P. B. Venuto, *J. Catal. 15*, 363 (1969).
34. Z. Paál and P. Tétényi, *Acta Chim. Acad. Sci. Hung. 58*, 105 (1968).
35. V. Haensel and G. R. Donaldson, *Ind. Eng. Chem. 43*, 2102 (1951).
36. V. K. Shum, J. B. Butt, and W. M. H. Sachtler, *J. Catal. 99*, 126 (1986).
37. J. Barbier and P. Marecot, *Nouv. J. Chim. 5*, 393 (1981).
38. W. M. H. Sachtler and R. A. van Santen, *Adv. Catal. 26*, 69 (1977).
39. H. G. Krane, A. B. Gruh, B. L. Schulman, and J. H. Sinfelt, *Proceedings of the Fifth World Petroleum Congress*, New York, 1959, Sect. III, p. 39.
40. R. B. Smith, *Chem. Eng. Prog. 55(6)*, 76 (1959).
41. W. S. Kmak and A. N. Stuckey, Paper No. 56a, AIChE National Meeting, New Orleans, March 1973.
42. M. P. Ramage, K. R. Graziani, P. H. Schipper, F. J. Krambeck, and B. C. Choi, *Adv. Chem. Eng. 13*, 193 (1987).

# 4
## Catalyst Preparation

**J. P. Boitiaux, J. M. Devès, B. Didillon, and C. R. Marcilly**
*Institut Français du Pétrole, Rueil-Malmaison, France*

## I. INTRODUCTION

Platinum-based reforming catalysts have evolved considerably since their discovery and use in the late 1940s. The changes have essentially concerned their chemical formulation, which became significantly more sophisticated by the addition to platinum, the principal metal, of other metals and, to a lesser degree, the methods of preparation which were adapted to these new formulations. In the late 1960s the introduction of Ge, Sn, Re, and Ir conferred better selectivity, better stability of performance over time, and the possibility of operating at lower pressure.

Most conventional reforming catalysts are based on well-dispersed platinum on an alumina promoted by a halogen, usually chlorine. Since the early 1970s, however, many attempts have been and are still being made to try to find better-performing systems. Zeolite X containing tellurium is very selective for the dehydrocyclization of $n$-hexane to benzene (1) but is rapidly deactivated. Systems consisting of platinum deposited on zeolites exchanged with alkaline ions (2), especially zeolites K-L (2, 3), K-Ba-L, and Ba-L (4, 5), are highly selective for the dehydrocyclization of $n$-hexane and $n$-heptane but are unfortunately extremely sensitive to poisoning by sulfur compounds (6). Activity and selectivity performances similar to those of platinum on zeolite L were obtained by replacing the latter zeolite with a support based on mixed aluminum and magnesium oxides obtained by the decomposition of hydrotalcite (7–10). Zeolite H-MFI containing gallium helps to convert light paraffins with three or four carbon atoms to aromatics (11–20). Other supports are also mentioned in various patents (21–23). None of these catalyst systems has taken a dominant position or appears today capable of replacing conventional reforming catalysts.

This chapter deals with the preparation of bifunctional reforming catalysts. The first part is devoted to monometallic Pt/Al$_2$O$_3$-Cl catalysts and more precisely to the description of physicochemical phenomena that take place during the impregnation of alumina with aqueous acid solutions of H$_2$PtCl$_6$. The second part presents the various preparation techniques mentioned in the literature for the most common bimetallic catalysts, i.e., Pt-Re, Pt-Sn, Pt-Ir, and Pt-Ge formulations.

## II. REFORMING CATALYST SUPPORTS

Prior to the successful use of platinum in 1949, reforming catalysts consisted of chromium or molybdenum oxides deposited on alumina (24). The first patents claiming the use of platinum, due to Haensel (25–27), were published a few years after the end of World War II. One of the patents (25) describes the deposition of chloroplatinic acid on an acidic support of silica/alumina/zirconia (Pt contents about 0.5 to 3 wt %). In other patents, the platinum is deposited from H$_2$PtCl$_6$ on a fluorinated alumina (26) or chlorinated alumina (27).

All modern industrial reforming catalysts consist of platinum, usually combined with one or more metals, deposited on a chlorinated alumina. Given its vital importance in the process, alumina is the only support considered in this chapter. The first part of this section describes the two main forms of alumina used, their synthesis, structural and surface characteristics, and their shaping. The second part describes the various steps of impregnation, drying, and calcination of the alumina support.

### A. γ- and η-Al$_2$O$_3$

Aluminas represent a large group including aluminum hydrates, transition aluminas, and α-Al$_2$O$_3$. Aluminum hydrates include amorphous hydroxide, crystallized trihydrates, gibbsite, bayerite, and norstrandite and the monohydrates boehmite and diaspore. Several methods are available for preparing these hydrates (28): acidification of sodium aluminate, neutralization of an aluminum salt, and hydrolysis of an aluminum alcoholate.

The hydrate conversions are complex and serve to obtain several low-temperature transition aluminas: $\rho$, $\chi$, $\eta$, and $\gamma$. These can in turn be converted to one or more of the following high-temperature transition aluminas: δ, κ, and θ. The final stage in these transformations is α-$Al_2O_3$, which is thermodynamically the most stable. Figure 1, based on references 28–41, gives an idea of the complexity of these relationships between the different aluminas.

The two main alumina supports used in catalytic reforming are the two transition aluminas η- and above all γ-$Al_2O_3$. γ-$Al_2O_3$ is usually obtained by the calcination in air of boehmite. Boehmite, which is generally described as an aluminum monohydrate, actually represents a broad continuum of products $Al_2O_3 \cdot nH_2O$ with varying degrees of hydration and organization without any clear borderlines: gelatinous boehmite is the most hydrated with $n$ possibly even exceeding 2, pseudoboehmite or microcrystalline boehmite, crystalline boehmite, and finally highly crystallized boehmite in large rhombohedral crystals with $n$ close to 1. Furthermore, depending on the conditions used, boehmite crystals with different shapes can be obtained—for example, acicular or platelets (28, 39). The transformation of an alumina hydrate is a topotactic reaction in which the morphology and size of the particles are preserved. The properties of γ-$Al_2O_3$ and especially the shape of the crystals and its specific surface area hence depend on the initial boehmite. Pseudoboehmite or microcrystalline boehmite leads to γ-$Al_2O_3$ with a large area ($> 350 \, m^2 g^{-1}$ at 500°C) which is very poorly organized (28, 30). Calcination around 500°C of the crystalline boehmite helps to obtain a γ-$Al_2O_3$ with a surface area close to 250 $m^2 g^{-1}$. Highly crystallized boehmite in the form of large rhombohedral crystals yields an alumina similar to γ-$Al_2O_3$ with a low area ($< 100 \, m^2 g^{-1}$ at 500°C), which is not used as a support for reforming catalysts.

η-$Al_2O_3$ is obtained by the calcination above 250 to 300°C in air or in vacuum of the alumina trihydrate bayerite, or in vacuum of gibbsite. The usual precursor, bayerite, is always obtained in the form of large crystals bigger than 0.1 μm (28). But the liberation of the water caused by calcination generates fine micropores in the large particles obtained. The specific surface area of the η-$Al_2O_3$ varies widely with the final calcination temperature: from 500 to about 400 $m^2 g^{-1}$ between 250 and 450°C and from 400 to about 250 $m^2 g^{-1}$ between 450 and 550°C (29).

The structures of γ-$Al_2O_3$ and η-$Al_2O_3$ are similar (29, 42). They are based on the compact cubic arrangement of the oxygen of the spinel $MgAl_2O_4$ structure (29, 41), but with a slight tetragonal deformation. This deformation caused by a disorder in the stacking of the oxygens is more pronounced in η-$Al_2O_3$ than in γ-$Al_2O_3$. The $Al^{3+}$ cations are distributed over the 32 octahedral sites and the 64 tetrahedral sites. Tetrahedral sites are slightly more occupied in η-$Al_2O_3$ than in γ-$Al_2O_3$ (29, 42). The nature of the exposed crystal faces at the surface of the two aluminas also appears to be different. Based on spectroscopic data, Knözinger and Ratnasamy (43) concluded that the surface consists of the three faces (111), (110), and (100) in varying proportions in the two aluminas, with a predominance of the (111) face in η-$Al_2O_3$ and the (110) face

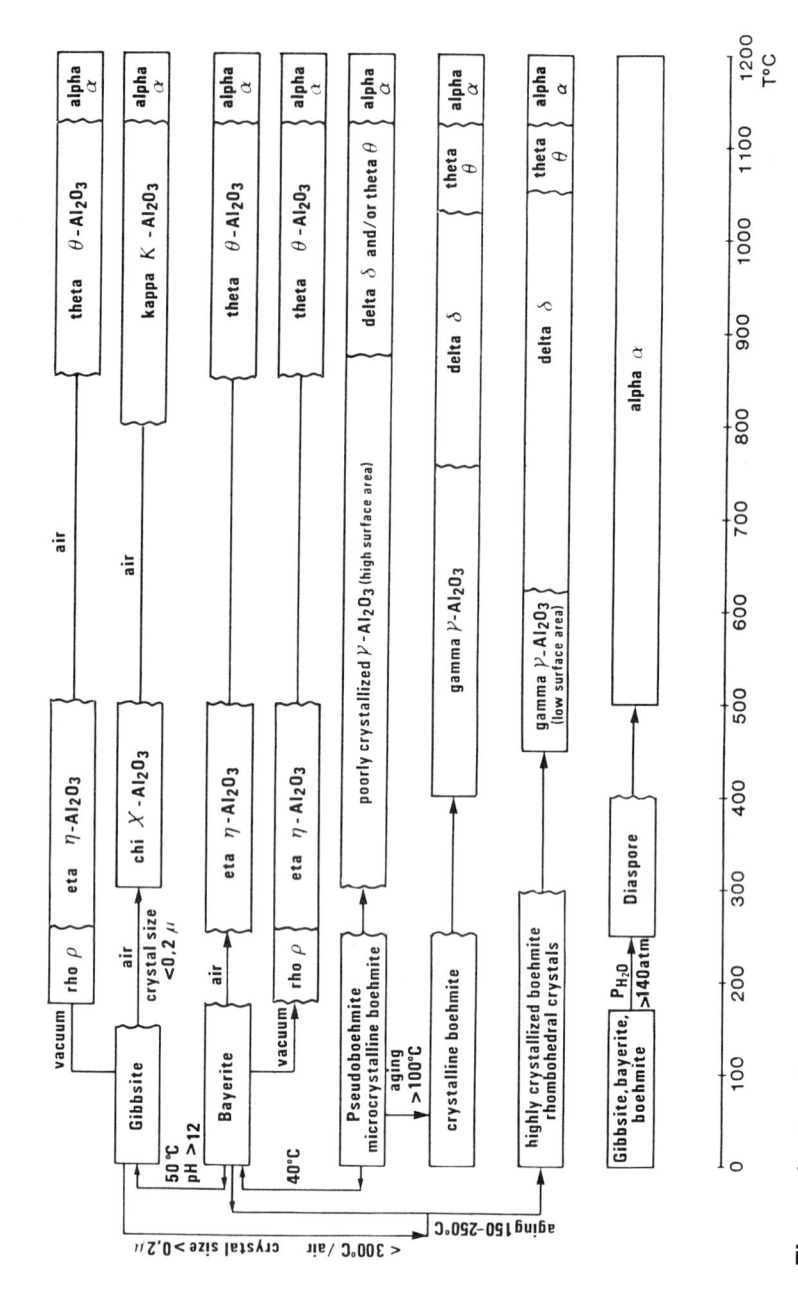

**Figure 1** Transformation sequence of aluminas.

in $\gamma$-Al$_2$O$_3$. This assumption is confirmed by Nortier et al. (39), who moreover observed that the ratio of the proportions of the (110) and (111) faces exposed varies according to the morphology and the size of the crystals of $\gamma$-Al$_2$O$_3$.

The two main models of the surface acidity-basicity that have been proposed for $\gamma$- and $\eta$-Al$_2$O$_3$ are those of Peri (44) and of Knözinger and Ratnasamy (43). The latter model considers the environment of the Al$^{3+}$ cations connected to the hydroxyls for the (111), (110), and (100) faces of the two aluminas. The number of different OH sites is five on the (111) face, three on the (110) face, and only one on the (100) face. The authors attributed the catalytic properties of the alumina to the combination of one of these OH sites with neighboring surface defects which have an exceptional configuration and hence a very low probability of existence. Those catalytic sites are formed only above 60% dehydroxylation. The degree of dehydroxylation of the alumina, which governs the number of defects, appears as an essential parameter for acidity (43, 44–46).

### B. Influence of Halogens on Alumina Acidity

Whereas the increased acidity caused by the fixation of a halogen on the surface of an alumina has been known for many years, the nature of the halogenated surface complexes has not yet been fully clarified. Contradictory results were obtained with fluorinated aluminas (47–55). Acidity seems to be promoted in different ways according to the halogen used: F or Cl (47–49). Chlorination of alumina was studied using CCl$_4$ or HCl (56–63). The type of acidity, Brönsted or Lewis, seems to depend closely on the fixed chloride content and the chloride precursor used. The maximum level of chloride fixed on $\gamma$-Al$_2$O$_3$ using gaseous HCl appears limited to around 2 wt % (i.e., around 2.5 atoms of Cl per nm$^2$) (58). At such low chloride contents, Gates et al. (41) proposed the following mechanism for the promotion of Brönsted acidity, which seems to play an important role in part of the numerous reforming reactions:

$$\begin{array}{cccccc} H & H & H^{\delta-} & H^{\delta-} & H & H \\ | & | & | & \| & | & | \\ O & O & O \searrow Cl \swarrow O & O & O \\ & Al & & Al & & Al \end{array}$$

The acidity of an OH group is strengthened by the inductive effect exerted by a Cl$^-$ ion adjacent to the OH group. This model agrees with the observations of Tanaka and Ogasawara (57).

For high chloride contents, more than a few wt % and up to more than 10 wt %, strong acid sites of the Lewis type are formed (58–63). They correspond to the formation of a gem-dichlorinated aluminum complex at the surface (58, 61–63). These sites were obtained only with compounds having at least two chlorine atoms on the same carbon atom: $CH_2Cl_2$, $CHCl_3$, $CCl_4$, etc. (63). The acidity corresponding to these high chloride contents is unsuitable for reforming catalysts, for which the usual chloride content is about 1% by weight.

## C. Alumina Shaping

The use of $\gamma$- or $\eta$-$Al_2O_3$ as a reforming catalyst support requires a shaping operation adapted to the type of process: moving bed or fixed bed. For a moving bed, it is necessary to prepare 1- to 2-mm-diameter beads to facilitate circulation and to limit the mechanical abrasion of the catalyst. For the fixed bed, the support can be in the form of either beads or cylindrical extrudates 1 to 2 mm in diameter. The catalyst support is shaped essentially by three methods: granulation, drop coagulation, and extrusion (28, 64).

In the granulation process, a powder is agglomerated in the form of spherical beads by progressive humidification in a large bowl called a pan granulator with granulation seeds and water to which a peptizing agent may be added. As the bead grows, gravity and the centrifugal force tend to push its trajectory to the side of the granulator, where it is ultimately ejected upon reaching a certain size.

Drop coagulation, especially "oil drop," is a particularly important technique for the formation of alumina beads used as reforming catalyst supports. The first patents issued on this technique date from the 1950s (65–67). In general, the oil drop technique consists of dropping an alumina hydrosol in a water-immiscible liquid in a vertical column. The alumina hydrosol can be prepared by the hydrolysis of an acid salt of aluminum, such as aluminum chloride (65), nitrate (65), or sulfate (68), or by the digestion of aluminum metal under heat by an aqueous aluminum salt solution (65, 69). The water-immiscible liquid is preferably an oil (66, 68, 70), for example, a light gas oil (69) or a paraffinic cut (71) with a high interfacial tension with respect to water. The higher the interfacial tension, the greater the sphere-forming tendency of the hydrosol in the liquid. The liquid is then heated between 50 and 105°C for a given time to permit the hydrosol spheres to transform progressively into a hydrogel. To promote gelification of the hydrosol droplets, ammonia additives such as hexamethylenetetramine (65, 66, 69, 72–74), urea (69, 72, 73), or a solution of ammonium acetate and ammonium hydroxide (66) can be added to the hydrosol, or the water-immiscible organic liquid can be saturated with ammonia (62).

In practice, the hydrogel spheres are retained and aged in the oil medium for an extended period to avoid deformation of the spheres during removal. In most cases, a subsequent complete coagulation is achieved in an aqueous alkaline media, usually an ammonium hydroxide solution, for a further extended period (62, 69). This latter procedure imparts a high mechanical resistance to the spheres. Various modifications of the general oil drop technique have been mentioned in other patents (70, 72, 75–78).

The oil drop technique previously described applies easily to the synthesis of $\gamma$-$Al_2O_3$ beads. It can also be used to produce $\eta$-$Al_2O_3$ beads provided that bayerite alumina spheres are formed through a final aging step (73).

Extrudates are formed in two main steps. In the first mixing step, a peptizing agent is added to a mixture of water and alumina powder and the mixture is thoroughly stirred to form a plastic paste that may typically contain 40% by weight of alumina. In the second step, the paste is forced through dies of a shape and diameter selected in accordance with the desired end product. The support beads or extrudates thus obtained are dried and then calcined in air, generally between 400 and 600°C.

The shaping operation normally has a negligible effect on the microporosity (pore diameter $d_p < 5$ nm) and hence on the specific surface area of the aluminas, because it does not alter the size of the precursor crystallites. It may nevertheless have a significant influence on the macroporosity ($d_p > 50$ nm) and hence on the pore volume of the support beads (28). This macroporosity stems essentially from the free voids between the more or less large agglomerates of crystallites and is conditioned by the shaping method. Moreover, it can be adjusted, before shaping, by the addition of a pore-forming substance that is combustible or decomposable, or through the addition of a nondispersive inorganic filler (28, 64, 79).

## III. MONOMETALLIC CATALYSTS

When preparing the bifunctional monometallic $Pt/Al_2O_3$-Cl catalyst, the following main features must be achieved to guarantee optimal performance:

- Low Pt content (usually < 0.5 wt %) with uniform macroscopic distribution in the catalyst particle and maximum accessibility, i.e., maximum atomic dispersion. This high dispersion also prevents metal sintering during high-temperature treatments.
- Acid sites in the close vicinity of the atoms or small particles of Pt.
- Adequate mechanical properties ideally undiminished from those of the support.

To deposit platinum on an alumina support, two types of impregnation techniques can be used: impregnation with or without interaction. In the first case, the platinum precursor forms an electrostatic or chemical bond with the surface of the support. In the second, the precursor displays no affinity for the surface and remains localized in the solution. It has been demonstrated that impregnation techniques with interaction are substantially superior to the other in terms of metal dispersion and catalyst performance (80, 81). The following discussion of the preparation of a monometallic catalyst is therefore limited to this category.

## A. Industrial Preparation

Modern techniques for the preparation of reforming catalysts use solutions of chloroplatinic acid as a platinum precursor with the addition of hydrochloric acid as a competing agent (Section B.3). These solutions are placed in contact with the alumina support in different techniques.

1. The dried alumina is impregnated with a volume of solution corresponding exactly to the quantity required to fill the pore volume $V_P$ ("capillary" or "dry" or "without excess solution" impregnation).
2. The dried alumina is immersed in a volume of solution substantially larger than $V_P$ (impregnation with excess solution).
3. The alumina, which is previously saturated with water or with a solution of hydrochloric acid, is immersed in the aqueous solution containing $H_2PtCl_6$ (diffusional impregnation).

Strong interaction develops in an acidic medium between the support and $H_2PtCl_6$, slowing down the diffusion of the latter toward the center of the beads. The rate of diffusion determines the impregnation time. In technique 1, capillary aspiration of the solution by the support enables the rapid penetration (a few dozen seconds to a few minutes) of the solution into the pores (82–84). On completion of the first impregnation phase, the solution which reaches the center of the bead will be depleted of the Pt precursor, owing to the strong interaction of the latter with the support. A homogeneous Pt profile from the outside to the core of support grains is attained in the second phase of this technique, in which platinum diffusion from the solution in the pore volume occurs. In technique 3, the $H_2PtCl_6$ acid must diffuse in the aqueous phase from the external solution to the centers of the beads, and impregnation is purely diffusional. The diffusion is slow and may require many hours to ensure good distribution of the Pt in the bead (85–87). Technique 2 is intermediate between the other two techniques.

In technique 1, which may also be carried out by spraying the support with the solution, the considerable liberation of heat that occurs may raise the temperature by a few dozen degrees (82). If uncontrolled, this could lead to insufficient penetration of the metallic precursor. In technique 2, the excess solution favors rapid elimination of liberated heat. If the very dilute solution contains only the quantity of metal to be introduced onto the support in a single operation, a long immersion of a few hours is required to exhaust the metallic precursor from the solution. If, on the other hand, the solution is concentrated to enable the introduction of the desired quantity of platinum by simply filling the pores of the support, the latter must be dipped and then withdrawn very

# Catalyst Preparation

quickly to prevent rapid depletion of the precursor from the solution. In this case, the removal of the heat is obviously less effective than in the other methods.

During capillary impregnation carried out in techniques 1 and 2, air bubbles are trapped in the pores, and especially in the micropores, where they are highly compressed (82). The Young-Laplace law $P = 2\gamma/r$ expresses the overpressure applied to the air bubbles trapped by the solution with a surface tension $\gamma$ in pores with radius $r$. If $r$ is very small, very high pressures may be reached (for example, $P = 14$ MPa approximately for $r = 10$ nm), which may be detrimental to the mechanical properties of the beads (82). Under the effect of such pressures, the air is progressively dissolved and migrates toward the macropores where less compressed and hence larger bubbles tend to appear. Removal of most of the imprisoned air to the exterior of the beads is generally complete after a few dozen minutes (82).

More practical information can also be obtained in Ref. 74.

## B. Impregnation

### 1. Interactions and Chemical Processes Involved

The hydroxyl groups present at the surface of a transition alumina behave like bases with respect to the protons of an HX acid in an aqueous solution (88–100). According to various authors, the reaction is represented by two different mechanisms, A and B:

$$\text{—Al—OH} + H_3O^+ X^- \underset{B}{\overset{A}{\rightleftarrows}} \begin{cases} \text{—Al—O}^{\oplus}(H)(H)\ X^- + H_2O \\ \text{—Al—X} + 2H_2O \end{cases} \qquad (1)$$

In mechanism A, the OH group is protonated and its interaction with the anion is electrostatic (88–90, 92, 94, 98). In mechanism B (93–97, 99), the $X^-$ anion replaces the $OH^-$ group in the complexation sphere of the surface aluminum. Mechanism A is expected to predominate because the surface of the transition aluminas is mainly covered with OH groups (43–46).

Protonation of an oxygen atom of the surface can also occur and lead to the scission of an Al-O bond according to reaction (2):

$$\text{Al—O—Al} + H_3O^+ X^- \rightleftarrows \text{—Al—O}^{\oplus}(H)(H)\ X^- + \text{—Al—OH} \qquad (2)$$

The interactions of halogenic acids, especially hydrochloric, and chloroplatinic acids with alumina can be tentatively rationalized more precisely by confronting both reactions (1A) and (2) with the numerous results from the literature.

At pH values lower than that of the point of zero charge (ZPC), which is approximately 8 for alumina, the surface is positively charged by protonation of hydroxyl groups and can therefore fix, by electrostatic bonding, the $X^-$ anion of the acid. We can reasonably imagine that only reaction (1A) occurs as long as nonprotonated hydroxyl is still present on the alumina surface. The maximum quantity of anions that the surface can fix—i.e., the anion exchange capacity—is directly related to the number of protonated OH groups (88, 89, 92–94, 101). Logically, it depends on the equilibrium constant of reaction (1A) and, as shown by Figure 2 (28), it rises as the pH of the aqueous solution decreases (28, 88, 89, 92, 94, 101).

In addition to the pH, the other essential parameter governing the quantity of anions fixed at equilibrium is the charge of the anions (90): the higher the charge, the stronger the interaction. The influence of the anion concentration of the medium on the quantity fixed is significant only if the available sites are not fully saturated. The ionic strength parameter of the medium plays only a minor role (90). The process of anion fixation on the surface is kinetically very fast (90) and rarely limits the overall kinetics. The kinetically limiting step is

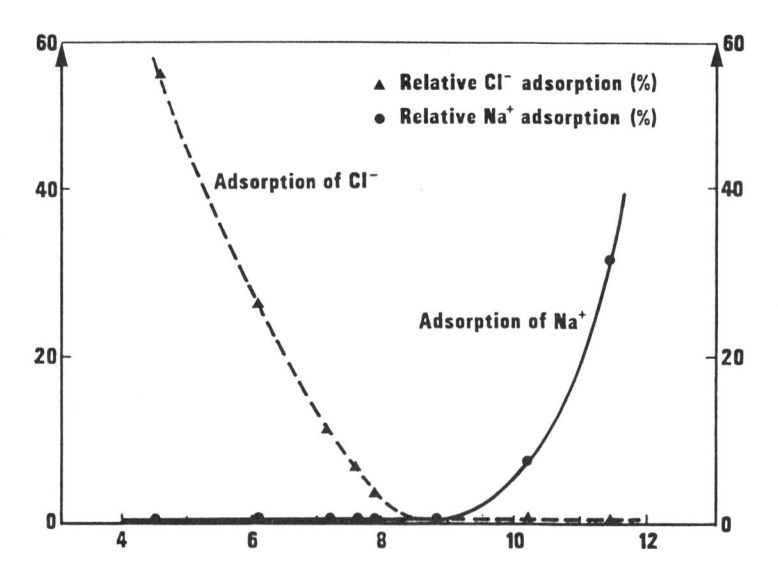

**Figure 2** Adsorption of $Na^+$ and $Cl^-$ on gamma alumina versus equilibrium pH. (From Ref. 28.)

expected to be the diffusion of the ionic species in the pores of the alumina particle. The diffusion rate will depend on the type of porosity (tortuosity, pore size) and on the superficial chemical reactivity of the support, especially the sensitivity to hydrothermal alteration, as discussed later. In other words, the diffusion rate depends on the type of alumina. This was observed by Sivasanker et al. (99) who showed that, at comparable surface areas (about 200 $m^2 g^{-1}$), $\eta$-$Al_2O_3$ fixes $Cl^-$ anion faster and in larger amounts than $\gamma$-$Al_2O_3$. The fixation capacities of the two aluminas range between $10^{13}$ and $10^{14}$ ions per $cm^2$ (0.1 to 1 ion/$nm^2$), i.e., one or two orders of magnitude below the density of the superficial OH.

In an acidic medium, simultaneously with the protonation reaction (1A) of the OH groups, a process of attack and dissolution of the support occurs (88, 98, 101–104). This process, illustrated in Figure 3 (88), occurs below about

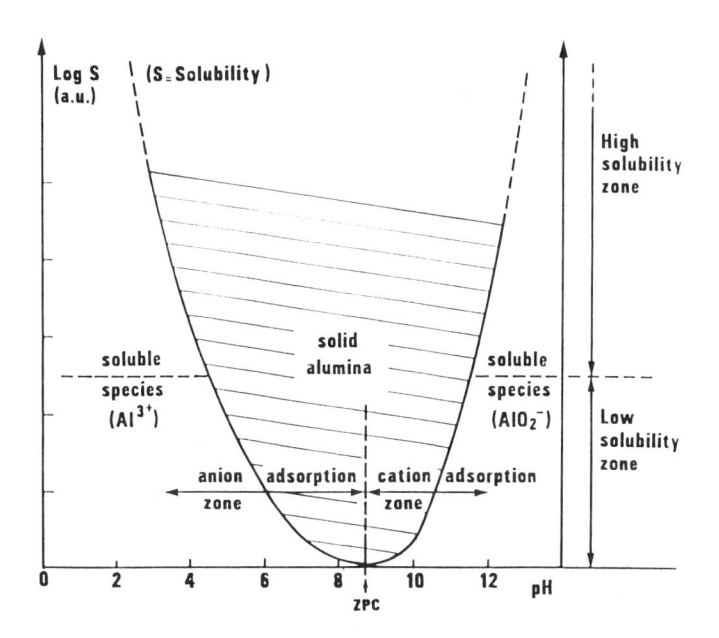

**Figure 3** At acidic pH, soluble alumina species formed are positively charged and are of the type $[Al(H_2O)_y(OH)_x]^{(3-x)+}$. At basic pH, soluble alumina species are negatively charged and written as $AlO_2^-$. When the pH differs substantially from the pH corresponding to ZPC (zero point charge: the net global charge of the support is zero), the solubility increases both in the acidic and the basic domain. The acidic domain, where the surface of the alumina support is positively charged, corresponds to an anion adsorption zone. In the basic domains, the alumina surface is negatively charged, corresponding to a cation adsorption zone. It can be arbitrarily defined that solubility is negligible above a pH of about 4.5 and under pH 11.

pH 4, and the solubility of the support rises rapidly with decreasing pH. Reaction (2) probably represents the first step of the formation of soluble compounds of the type $[Al(H_2O)_y(OH)_x]^{(3-x)+}$ with $x + y = 6$. The acid chemical attack of the alumina should be more sensitive to the pH than reaction (1A), because detachment of an aluminum atom from the alumina surface requires the breaking of at least two or three Al−O bonds, i.e., at least two or three successive reactions (2). Reaction (2) is thus expected to occur only after reaction (1A) is complete. Several experimental observations, such as that illustrated in Figure 4 (82), support these considerations. If alumina beads are placed in contact with an HCl solution (Figure 4), the following sequence of events occurs: a rapid fixation of the $Cl^-$ anion probably due to reaction (1A), a subsequent slowdown of this fixation, then a decrease in the quantity of chloride fixed probably due to the progressive dissolution of the external part of the alumina particle (reaction 2) where the pH is the lowest, and finally a slow fixation of the $Cl^-$ anion. If the initial quantity of acid used is small (less than 4–5 wt % Cl), the final event (the slow chloride fixation) resumes after several minutes (82). This happens because the situation at the particle scale is still far from equilibrium: the pH in the external solution is low (see Figure 4) but higher in the internal solution near the center of the particle. The pH progressively evolves toward the equilibrium where the pH is slightly higher than 3. At such a pH, chlorinated aluminum species which were solubilized at the lower pH during the beginning of the reaction will redeposit on the support. On the other hand, for high chloride contents, the pH in the solution remains notably lower for a few hours. Under these conditions, the redeposition of chlorinated aluminum species is not visible, although it probably still occurs. After reaching a maximum chloride fixation capacity, which occurs within the first 15 minutes, a continuous decrease in the quantity of fixed chloride is observed (88). In other words, any excess acid in the solution continues to react with the alumina support to form a soluble aluminum salt (80, 88, 105). This leads to a slow and progressive neutralization of the mother solution up to a pH limit close to 3. Irrespective of the HCl acid concentration of the initial solution, the solubility of $\eta$-$Al_2O_3$ is always higher than that of $\gamma$-$Al_2O_3$ (99), which means that the hydrothermal alteration of the $\eta$-$Al_2O_3$ surface is faster and more pronounced. Hence, at low and identical chloride concentrations in the initial solution, chloride should react and the pH should increase faster with $\eta$-$Al_2O_3$, resulting in more rapid and greater redeposition of chlorinated aluminum species and thus faster and higher fixation of chloride than for $\gamma$-$Al_2O_3$ (99). Note that at the concentrations used for the preparation of monometallic reforming catalysts, the solubilities of $\eta$- or $\gamma$-$Al_2O_3$ are very low (no more than a few hundred ppm).

The process of attack and dissolution of the support also occurs if the HCl acid is replaced by chloroplatinic acid (101, 103). The maximum amount of

*Catalyst Preparation* 91

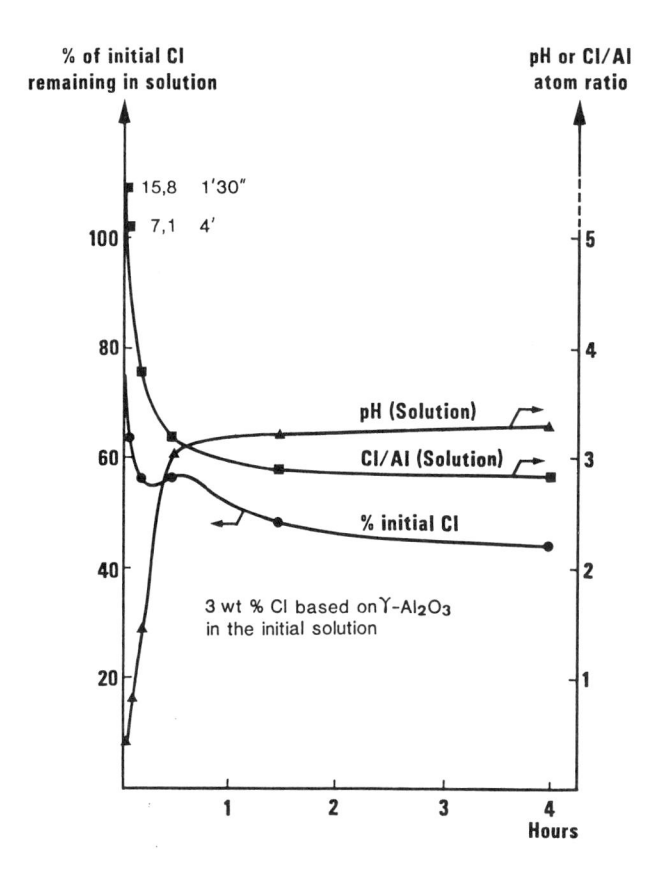

**Figure 4** Dried γ-alumina beads were immersed in an aqueous solution of HCl containing 3 wt % Cl based on alumina. The evolutions of the following characteristics of the solution were recorded and/or calculated as a function of time: pH, Cl/Al atomic ratio, percentage of the initial Cl remaining in the solution. The first measurements were made after 1 minute 30 seconds and 4 minutes.

platinum fixed on the alumina from $H_2PtCl_6$ acid, without any significant attack of the support, differs widely according to various authors (74, 83, 88, 101, 103, 106, 107), varying between 2 and about 8% by weight. It appears to be strongly related to the type and essentially the specific surface area of the transition alumina concerned (88, 103). The influence of alumina pretreatment has not been clearly determined (101, 105). Most of the values obtained at saturation of γ-$Al_2O_3$ with surface areas ranging between 150 and 250 $m^2 g^{-1}$ are normally found to be between 2 and about 4% by weight (74, 88, 106, 107).

At this stage of impregnation, two major questions arise:

1. On the atomic scale, in what form is the adsorbed platinum complex?
2. How many minutes or hours after impregnation is the platinum precursor uniformly distributed in a catalyst bead, if a small amount of platinum is involved initially?

## 2. Type of Chlorinated Platinum Precursor Fixed on the Alumina

$H_2PtCl_6$ is a strong diacid (108), stable in acidic aqueous solution. In a basic solution, the $PtCl_6^{2-}$ anion may undergo varying degrees of hydrolysis to yield an intermediate series $[PtCl_x(OH)_{6-x}]^{2-}$ called the Miolati series. In a neutral or weak acid medium, the $PtCl_6^{2-}$ anion may undergo a number of aquation and hydrolysis reactions (or even polymerization) (88, 108–111), such as:

$$[PtCl_6]^{2-} + H_2O \rightleftharpoons [PtCl_5(H_2O)]^- + Cl^- \qquad (3)$$

$$[PtCl_5(H_2O)]^- + H_2O \rightleftharpoons [PtCl_4(H_2O)_2] + Cl^- \qquad (4)$$

The partially aquated species like $[PtCl_5(H_2O)]^-$ behave like weak acids:

$$[PtCl_5(H_2O)]^- \rightleftharpoons [PtCl_5(OH)]^{2-} + H^+ \qquad (5)$$

$$[PtCl_4(H_2O)_2]^- \rightleftharpoons [PtCl_4(OH)(H_2O)]^- + H^+ \qquad (6)$$

$$[PtCl_4(OH)(H_2O)]^- \rightleftharpoons [PtCl_4(OH)_2]^{2-} + H^+ \qquad (7)$$

Replacement of the Cl ligand of the complexation sphere by the OH ligand, during the fixation of the $PtCl_6^{2-}$ anion on alumina, has been proposed by various authors (96, 105, 112, 113).

According to Summers and Ausen (105), if a small amount of $H_2PtCl_6$ acid (0.25% by weight with respect to the support), present at a low concentration ($5 \times 10^{-3}$ mol/L) in an aqueous solution, is placed in contact with $\gamma$-$Al_2O_3$ in the absence of hydrochloric acid, the anion is fixed with the elimination of two chloride ligands from the coordination sphere after about 1 h. The elimination of four ligands has been observed after 24 h in the absence of hydrochloric acid (113). The aquated and/or hydrolyzed species thus formed display only slight interaction with the support (105).

Characterization studies in the visible ultraviolet (113) or by EXAFS (114) show that, if the impregnation solution contains hydrochloric acid, platinum is fixed in the form of anions, most of which preserve their octahedral environment of six chloride ligands. However, the EXAFS study (114) has indicated the presence of small quantities of platinum complexes whose coordination sphere contains a small amount of oxygen, which could be the $[PtCl_5OH]^{2-}$ or $[PtCl_5H_2O]^-$ complexes reported elsewhere (113). The overall results can be explained on the basis of reactions (3) to (7), which are favored in the neighborhood of the hydrated surface of alumina and at slightly acidic pH corresponding to the end of impregnation and inhibited in the presence of an excess of Cl⁻ and H⁺ ions, hence of hydrochloric acid.

A Chinese study (115) concluded that the $[PtCl_6]^{2-}$ complex is preferentially fixed on high-energy sites: first the kink sites, followed by the step sites, and, when the first two are saturated, the terrace sites.

## 3. Macroscopic Distribution of the Chlorinated Platinum Complex

Many authors have observed a strong interaction between the chlorinated complex of platinum, $H_2PtCl_6$, and $\gamma$-$Al_2O_3$ (74, 83, 88, 97, 102, 107). At impregnation conditions approaching those used for the preparation of reforming catalysts (i.e., small amounts of platinum, generally less than 0.6% by weight of platinum with respect to the support, and low pH, hence high anion exchange capacity), platinum tends to be fixed rapidly and strongly on the first sites encountered. Maatman and Prater (83) characterized the interaction by an adsorption equilibrium constant:

$$K = \frac{1}{C_s} \frac{\Theta}{1 - \Theta}$$

where $\Theta$ is the fraction of adsorption sites covered by the anion and $C_s$ is the anion concentration at equilibrium. $K$ is high, about $10^4$ for a chi-type alumina. When the solution is exhausted, the complex is located on the peripheral layer of the catalyst grains (extrudates or beads) (83, 88, 97, 107, 110). A relatively long impregnation time (from a few hours to a few days) is needed to achieve a uniform distribution on the macroscopic scale, i.e., an identical superficial concentration from the periphery to the center of the bead (83, 102). This results from the rapid and virtually total fixation of the anions from the solution on the support, due to an excessively high adsorption coefficient. On the contrary, if the adsorption coefficient of the anionic complex is low, an anion distribution equilibrium is established between the surface of the support and the solution, and in these conditions a uniform distribution of the metal in the grain is observed (105). It is therefore clear that the diffusion of platinum toward the center of the grain is favored by the liquid phase and stops when this phase is exhausted.

To ensure rapid migration of platinum, it is necessary to add a competitor ion to the solution, whose role is to act on the equilibrium of the reaction between the platinum precursor and the support surface. The usual competitor is HCl. After adding a sufficient quantity of HCl, an exchange equilibrium rapidly occurs between the Cl⁻ and $PtCl_6^{2-}$ ions of the surface according to the reaction

$$[PtCl_6]^{2-}_{Al} + 2Cl^-_{aq} \rightleftharpoons [PtCl_6]^{2-}_{aq} + 2Cl^-_{Al} \qquad (8)$$

The quantity of Cl⁻ ions added to the system is usually adjusted to leave a sufficient quantity of $[PtCl_6]^{2-}$ in the aqueous solution to ensure easy migration to the center of the grain, so that, at the end of the operation, the residual nonadsorbed amount is only a small fraction of the total quantity of platinum involved. In contrast to $\gamma$-$Al_2O_3$, for which the $Cl^-$ anion in the form of HCl is an effective competitor that ensures uniform distribution of the $[PtCl_6]^{2-}$ in the grains, the presence of chloride in variable amounts of $\eta$-$Al_2O_3$ appears barely to affect its capacity to fix $[PtCl_6]^{2-}$ (99).

Competitors other than HCl can be used (97, 102, 107, 112, 116–122, 123). Acids such as $HNO_3$ and $H_2SO_4$ are usually more effective than the corresponding salts, such as ammonium, aluminum, and sodium nitrates (102). Nitric acid is about five times more effective than acetic acid (117). Hydrofluoric acid has a much stronger interaction with alumina than chloroplatinic acid (118): HF blocks the alumina sites, inhibiting the adsorption of the $[PtCl_6]^{2-}$ anions and forcing these ions to migrate farther toward the center of the grain. Among the other competitors used are $CO_2$ (123); phosphoric acid (116, 122); organic acids such as formic, oxalic, propionic, lactic, salicyclic, tartaric, and citric acids (97, 107, 112, 116, 119–122); miscellaneous salts such as alkaline phosphate (107), alkaline halogenides (107, 122), and alkaline benzoate (107); and amines such as monoethanolamine (97).

Some of these "competitors," such as monoethanolamine, really act not by a competition effect on adsorption but rather by a pH effect (97). In fact, a rise in the pH tends to lower the exchange capacity of alumina, causing more uniform occupation of the surface by the metallic complex. Competitors help in obtaining a large number of metal distribution profiles on the support surface between the periphery and the grain center (84, 107, 124, 125), depending on the strength of their interaction with the support. Four main types of profiles (84, 124, 125) serve to describe all the others: uniform, film or eggshell, internal ring or egg white, central or egg yolk. HF (118) and lactic, tartaric, and citric acids (119), for example, help to obtain an egg yolk distribution of chloroplatinic acid. A number of theoretical models of migration and deposition of metallic ions over time in porous supports, with or without the presence of competitor compounds, based on competing liquid-phase diffusion and adsorption mechanisms, have been proposed to account for the experimental results (85–87, 118, 126–131).

## C. Drying, Oxidation, and Reduction of the Impregnated Support

Drying is intended to eliminate most of the aqueous solution from the pores. It is well known that this operation can cause substantial movements of the solution and hence a significant redistribution of the precursors still present in the dissolved state in the particles of the support. For reforming catalysts in which the uniformly impregnated precursor in the presence of HCl acid is virtually entirely fixed, such a redistribution can be ignored.

However, drying modifies the form of the adsorbed platinum complex. For preparations made under industrial or similar conditions ($< 0.6\%$ by weight of platinum and $< 5\%$ by weight of chloride expressed with respect to the support, in the initial solution), heating slightly above 100°C converts the $[PtCl_6]^{2-}$ ionic species, in electrostatic interaction with the support, to a mononuclear species of platinum whose complexation sphere contains chloride and oxygen ligands. Visible ultraviolet spectroscopy indicates a species of the $[PtCl_yOH_x]^{2-}$ type (113). EXAFS indicates a species of the "$[PtCl_4O_2]$" type (114), where the complexation sphere contains about two oxygen atoms.

Oxidation in air causes progressive replacement of the chloride ligand by oxygen, with the formation of a complex close to "$[Pt^{IV}O_xCl_y]$" between 500 and 600°C, such as the four-ligand complex $[PtO_2Cl_2]^{2-}$ according to Lieske et al. (112) or the six-ligand complex "$[PtO_{4.5}Cl_{1.5}]$" according to Berdala et al. (114). This complex $[Pt^{IV}O_xCl_y]$ would preferentially occupy the high-energy sites of the alumina (56), first saturating the kink sites, then the step sites, and finally occupying the terrace sites.

Simultaneously, oxidation in air causes a decrease in the chloride content of the alumina, whether it is $\eta$- or $\gamma$-$Al_2O_3$ (99). The higher the chloride content after drying, the higher is the decrease. This decrease is also accentuated by the presence of small amounts of water in the air (99). Under identical conditions, chloride retention depends on the alumina employed: thus, after oxidation at 550°C, a $\gamma$-$Al_2O_3$ with a surface area of 190 $m^2 g^{-1}$ containing 0.9% by weight of chloride loses 40% of the chlorine, whereas a $\eta$-$Al_2O_3$ with 200 $m^2 g^{-1}$ containing 1% by weight of chloride loses only about 17% (99).

The reduction step, typically conducted between 500 and 550°C, converts the above platinum species to highly dispersed platinum metal (113). EXAFS indicates the formation of small metal particles in which the metal atoms have four to six immediately neighboring platinum atoms and also indicates the presence of some Pt-O bonds (114).

## IV. BIMETALLIC CATALYSTS

The deposition of chloride and platinum on alumina occurs through complex reactions as discussed in Section III. Both elements must be uniformly distributed in the grains of the support and platinum must be well dispersed at the end of the impregnation step. The good dispersion, obtained as a result of a strong metallic precursor–support interaction, and a uniform macroscopic distribution, obtained in spite of this strong interaction, must be preserved in the subsequent steps of drying, oxidation, and reduction in order to attain excellent catalytic performance and resistance to sintering. These considerations are still valid for multimetallic catalysts, although every additional metal brings its own unique chemistry, which may impose adaptation and optimization of the means of introducing the various metals. In this section the main preparation methods for Pt-Re, Pt-Sn, and Pt-Ir reforming catalysts are described.

## A. Platinum-Rhenium

Today, formulations based on Pt and Re (typically containing between 0.1 and 0.8 wt % of each metal) are among the most important industrial reforming catalysts. Such combinations, initially developed by Chevron in 1969 (132, 133), increase the stability of catalytic performance, allowing much longer operating periods of the unit between regenerations. This improved stability is explained by a double effect of rhenium: higher resistance to deactivation by coking and stabilization of the metallic phase on the support (24, 41, 134). However the presence of rhenium induces a high degree of hydrogenolysis, which necessitates a sulfurization step during the initial start-up of the catalyst.

It is generally accepted that the optimal Pt-Re phase consists of Pt and Re completely reduced to the metallic state (135–138). A fraction of rhenium is alloyed with platinum as bimetallic clusters, and this fraction may vary according to the conditions of preparation. The reduction of rhenium oxides is catalyzed by the presence of platinum. Moreover, water increases the mobility of Re oxide on the surface of alumina, hence its movement toward neighboring platinum. This favors its reduction and the alloy formation (139, 140).

Introduction of a metal into the alumina gel before shaping has been described in some of the literature. In such a case, the elaboration of the support becomes a specific step in the preparation of the bimetallic catalyst. As far as Re is concerned, this possibility is mentioned in only a few patents (141, 142) and does not seem to be applied industrially.

The widely used platinum precursor is the $PtCl_6^{2-}$ ion (similar to monometallic catalysts). Two types of rhenium precursors can be distinguished: mineral compounds and organometallic compounds.

### 1. Mineral Compounds

The main mineral compounds reported in the literature are the heptaoxide, perrhenates (especially $NH_4ReO_4$), and some halides ($ReCl_3$). The heptaoxide, which is very soluble in an aqueous acidic solution, easily gives $ReO_4^-$ anions, which can be fixed on the alumina surface (95, 143).

Usually, alumina is impregnated with an aqueous solution containing $H_2PtCl_6$, $HReO_4$, and HCl as a competitor. A mathematical model based on diffusion and adsorption phenomena and successfully predicting the radial profiles of Pt and Re has been proposed by Ardiles et al. (144). Since both precursors show different affinities for the support, it seems rather difficult to obtain similar distribution profiles of metals even with adjusted conditions (118, 144, 145). This is probably why successive impregnations of the two metals are claimed in some patents (141, 142).

After the deposition of precursors, impregnated catalysts are thermally treated. The reactions occurring during the thermal treatment are complicated and poorly understood. Besides the removal of impregnation solvent from the porosity of the support, precursors are transformed into oxidized species and then reduced to a highly dispersed bimetallic phase. During these treatments, it is important to control the mobility of Re species on the support surface.

An important feature of bulk rhenium precursors or oxides is their high volatility: the volatilization, which begins at a low temperature (near 100°C), is complete above 400–450°C. The volatility is expected to decrease considerably by interaction with the alumina support. Therefore, the oxidation temperature of the impregnated catalyst can be adjusted so as to decompose the precursor while preventing the volatilization of rhenium (146). Thermal oxidation provokes the dehydration of alumina, leading to a higher interaction between $Re_2O_7$ and the support (139). A high-stability Re precursor requires high temperature for its volatilization and/or decomposition and leads to an increased dehydroxylation level of the alumina as well as increased interaction with alumina. From this point of view and according to Reyes et al. (147), ammonium perrhenate, which exhibits the highest stability in oxygen, seems to be the best-suited precursor. A strong interaction between $Re_2O_7$ and the support limits the volatilization of rhenium oxides and the mobility of rhenium species on the support. As mentioned above, such mobility is necessary to favor the platinum-rhenium interaction (140). Moisture is thus expected to be an important parameter. Indeed, the reduction of a dried but not oxidized catalyst leads mainly to the formation of alloy, whereas after a very dry oxidation, reduction leads to a large proportion of free Re. The oxidation step for this kind of precursor has a slight effect on the macroscopic distribution, which is determined during impregnation, but has a strong effect on the "bimetallic nature."

## 2. Organometallic Compounds

The use of $Re_2(CO)_{10}$ dirhenium decacarbonyl as an organometallic precursor has been reported in numerous papers and patents (144, 148). Techniques using such a precursor require either impregnation with an organic solvent or sublimation in an inert gas in the 100–200°C range. This compound can interact either with the alumina support or with the supported platinum. The interaction with alumina is accompanied by a partial decarbonylation, and ligands exchange with the surface of the support (148, 149). However, the affinity of rhenium carbonyl is higher for reduced platinum. This limits the fixation of the precursor on the support sites and favors the formation of an alloy (150, 151). The total decomposition step is performed in an inert gas in order to avoid platinum oxidation (150). The decomposition can be complicated by the presence of hydroxyl groups, which can lead to oxidized surface complexes (152).

Other types of Re organometallic compounds have been decomposed on reduced platinum. Compared to perrhenate, some of these compounds, such as Re(OC$_2$H$_5$)$_3$, could be favorable in the formation of more dispersed and more stable Pt-Re bimetallic particles (153). From an industrial point of view, considering the problems of handling and maintaining reduced catalysts, this kind of precursor is less attractive than water-soluble mineral precursors.

## B. Platinum-Tin

The appearance of platinum-tin catalysts in reforming began in the late 1960s with the first patent claiming the use of this type of catalyst for dehydrocyclization reactions (154). Compared with monometallic systems, tin increases the selectivity and stability of the catalyst. Moreover, resistance to agglomeration of the bimetallic catalysts during coke combustion has increased compared with the corresponding monometallic systems (155).

From an industrial point of view, the platinum and tin salts used most frequently as precursors are chloro derivatives such as H$_2$PtCl$_6$, SnCl$_2$, and SnCl$_4$. The tin and platinum contents of industrial catalytic systems are always less than 0.8 wt %.

After reduction at temperatures higher than 400°C, it is generally accepted that platinum is in a metallic state. The oxidation state of tin is still subject to discussion. Generally, tin is found as Sn(II) together with a more or less important fraction of metallic tin. The ratio of these two tin species depends on the catalyst preparation conditions, analysis techniques used, and treatments undergone by the sample before analysis. This parameter is therefore not treated in this chapter.

In contrast to rhenium oxide, the high interaction between tin oxide and alumina does not favor mobility of tin on the surface during high-temperature treatments. This is why the interaction between platinum and tin in platinum-tin bimetallic catalysts must take place during the impregnation step.

### 1. Alumina-Supported Tin Oxide Preparation

*a. Sol-Gel Techniques.* Techniques using the sol-gel transition have been used to prepare reforming catalyst supports containing tin. These techniques consist of introducing a tin salt directly in the alumina sol. In a general way this preparation is performed in an aqueous environment by introducing stannic chloride in the sol obtained by hydroclorhic acid attack of aluminum (156). After calcination, tin is present only in its oxidized state (stannic oxide-like compound). No tin chloride species can be detected (157). This technique has been used with organometallic complexes (tetrabutyltin and tributoxide aluminum) in an alcoholic medium (158). With this technique, tin is uniformly distributed on the support, and tin oxide particles larger than 10 nm cannot be detected (159). Such a distribution of tin seems important to ensure good performance of the final catalyst (156, 159).

## Catalyst Preparation

*b. Impregnation of Tin Salts.* Although impregnation of tin salts (most often $SnCl_2$ or $SnCl_4$) is often the first step in preparing a reforming catalyst (160), the interaction between tin salts and an alumina surface has seldom been studied in the literature.

The surface complexes formed during the impregnation of stannic chloride on alumina have been studied by Li et al. (161). They showed that, whatever the impregnation medium of tin may be (water, alcohols, ketones), the surface species formed are the same. Spectroscopic data can be interpreted in terms of tin complexation by a surface hydroxyl. The mechanism visualized with water as the solvent is the following:

$$>Al-O^{H/} + H_2O \longrightarrow SnCl_4 \rightleftharpoons >Al-O^{H/} \longrightarrow SnCl_4 + H_2O \quad (9)$$

To fix tin(IV) chloride irreversibly on the alumina surface, one approach is to activate the hydroxyl functions with butyllithium (162):

$$>AlOH + n\text{-BuLi} \longrightarrow >AlOLi + n\text{-BuH} \quad (10)$$

$$n >AlO\text{-Li} + SnCl_4 \xrightarrow{\text{acetone}} (>AlO)_n SnCl_{(4-n)} + n\text{-LiCl} \quad (11)$$

However, for stannous chloride impregnation on nonchlorinated alumina, the mechanism proposed by Kuznetsov et al. (163), and confirmed by Homs et al. on silica (164), occurs via the substitution of a chloride ligand by the oxygen of an OH group. This reaction occurs without previous activation of the support. In acetone the mechanism would be the following (163):

$$>Al-O^{H/} + (CH_3)_2C=O \longrightarrow SnCl_2 \rightleftharpoons >Al-O-\overset{Cl}{\underset{|}{Sn}} \leftarrow O=C(CH_3)_2 + HCl \quad (12)$$

On chlorinated alumina, stannous chloride is only physisorbed on the support surface according to the reaction:

$$>Al-Cl + (CH_3)_2C=O \longrightarrow SnCl_2 \rightleftharpoons >Al-Cl \longrightarrow SnCl_2 + (CH_3)_2C=O \quad (13)$$

The tin-support interaction thus obtained should be weaker than the one obtained on nonchlorinated alumina in which a chloride ligand is lost by tin. During the drying that follows impregnation, oxidation of tin(II) to tin(IV) is only partial (157).

In all the cases described, calcination at a high temperature (500°C) results in the formation of supported tin(IV) ($SnO_2$-like compounds, $SnO_xCl_{4-x}$).

Contrary to the systems obtained with a sol-gel preparation, tin can keep chloride ligands even after calcination (157). Moreover, $SnO_2$ crystallites cannot be detected after calcination.

Reduction at around 500°C produces supported tin(II). The nature of this tin species has not yet been clearly described in the literature (SnO-like compounds, surface tin aluminate, etc.). Except for high tin contents (> 10 wt %), the reduction of alumina-supported tin oxide never gives metallic tin.

### 2. Impregnation of Platinum Derivatives on Alumina-Supported Tin Oxide

When tin on alumina is obtained by a first impregnation of tin chloride without subsequent oxidation, part of the tin, which is weakly linked on the support, can be dissolved during the subsequent impregnation in the chloroplatinic acid solution. In this case, the impregnation of $H_2PtCl_6$ can be described as an impregnation of a mixture of tin and platinum salts (see Section B.4).

If the impregnation of platinum salts is performed on an oxidized tin on alumina, two kinds of reaction can occur. Chloroplatinic acid can react with alumina as described in Section III.B or with supported tin oxide. The reaction of chloroplatinic acid with the tin oxide surface has been described (165). As far as Sn-Pt interaction is concerned, hydroxyl groups in the coordination sphere of the hydrated platinum salt would be displaced by an oxygen from a surface Sn-OH group:

$$\longrightarrow SnOH + Pt(OH)_n Cl_{6-n}^{2-} \longrightarrow \longrightarrow SnOPt(OH)_{n-1} Cl_{6-n}^{2-} + H_2O \quad (14)$$

The surface complexes can then be decomposed by hydroxylation and loss of chloride. It is obvious that in order to increase the interaction between tin and platinum during impregnation, the anchoring of the platinum complex must be promoted on tin oxide sites rather than on alumina.

### 3. Impregnation of Tin Derivatives on Alumina-Supported Platinum

*a. Tin Chloride Derivatives.* Formation of bimetallic complexes has been observed when an oxidized $Pt/Al_2O_3$ precursor is immersed in a water solution containing stannic chloride (166). During this step, part of the supported platinum oxide is extracted and reacts with tin chloride to form a bimetallic complex. This technique gives results almost similar to those with coimpregnation (see Section B.4).

If stannous chloride is used as a tin precursor, the redox potential values of Sn(IV)/Sn(II) and Pt(IV)/Pt(II) couples suggest the occurrence of an electrochemical process when tin(II) is impregnated on supported platinum(IV), leading to an interaction between both metals (167). Obviously, such a process cannot be envisaged if an alumina-supported tin oxide is impregnated by platinum(IV) salts. These remarks may explain why successive impregnations

*Catalyst Preparation* 101

of chloroplatinic acid and stannous chloride generally lead to higher interactions between the two metallic components than successive impregnations carried out in the reverse order (167, 168).

*b. Organometallic Tin Complexes.* Hydrogenolysis of tin(IV) organometallic complexes on supported metals has been studied during the past decade (169). This method leads to different kinds of bimetallic catalysts and especially alloys (170).

Impregnation of tetraethyltin, in argon, on a reduced alumina-supported platinum precursor gives a selective deposit of tin on the support (171). On the other hand, if the impregnation of the tin complex is performed in hydrogen, tetraethyltin reacts preferentially on platinum particles leading to a metal surface complex (169, 171):

$$Pt-H + Sn(C_2H_5)_4 \longrightarrow PtSn(C_2H_5)_3 + C_2H_6 \tag{15}$$

At high temperature the surface complex is decomposed, resulting in a supported platinum tin alloy (169, 171).

### 4. Coimpregnation

The coimpregnation technique offers the advantage of a reduced number of steps in comparison with successive impregnation techniques (172). During the dissolution of stannous chloride and chloroplatinic acid, the formation of bimetallic complexes has been observed (173). Baronetti et al. (167) have shown that the first step of the reaction is a reduction of platinum(IV) by tin(II):

$$PtCl_6^{2-} + SnCl_3^- + Cl^- \longrightarrow PtCl_4^{2-} + SnCl_6^{2-} \tag{16}$$

For tin/platinum ratios higher than one, the first reaction is followed by the formation of bimetallic complexes (167, 174–176):

$$PtCl_4^{2-} + SnCl_3^- \longrightarrow PtCl_3(SnCl_3)^{2-} + Cl^- \tag{17}$$

$$PtCl_3(SnCl_3)^{2-} + SnCl_3^- \longrightarrow PtCl_2(SnCl_3)_2^{2-} + Cl^- \tag{18}$$

These complexes are generally stable in an acidic environment in an inert atmosphere. During impregnation of such complexes on nonchlorinated alumina, precipitation of platinum can occur. This precipitation is due to adsorption of chloride on the support, leading to destruction of the bimetallic complex. Ensuring the stability of the bimetallic complex during all the impregnation steps is important in preserving the platinum-tin interaction. For example, increasing the chloride concentration of the impregnation solution or using a prechlorinated alumina can prevent platinum precipitation. After drying and oxidation, this preparation method leads to catalytic systems with a better platinum-tin interaction compared with successive impregnations of tin and platinum salts (155, 177–180).

## C. Platinum-Iridium

A bimetallic catalyst of some industrial importance is platinum-iridium on chlorinated alumina, which appeared in the early 1970s (181). It can be manufactured from chloroplatinic and chloroiridic acids, either by two successive impregnations (181), where the platinum is the first metal introduced, or by coimpregnation. The respective behaviors of platinum and iridium in reducing and oxidizing atmospheres are so different that there is doubt about easily maintaining intimate contact between both metals once it has been obtained. Foger and Jaeger (182) concluded that obtaining a single-phase Pt-Ir alloy is possible only if the concentrations of Pt and Ir are nearly equal. After oxidation below 300°C, iridium alone is transformed to oxide, and above this temperature iridium oxide crystals segregate. Above 550°C, $IrO_2$ is transported through the gas phase. Such results are completely in accordance with those obtained by Garten and Sinfelt (183). Highly disperse bimetallic clusters are obtained through coimpregnation of chloroiridic and chloroplatinic acids if the exposure to air is maintained below 375°C. Around 600°C, large crystallites of iridium oxide are formed, and after reduction the catalyst consists of highly disperse platinum or platinum-rich clusters and of large iridium or iridium-rich crystallites. Sequential impregnation of platinum and then iridium gives exactly the same results. These considerations emphasize the importance of the physical treatment subsequent to the catalyst impregnation.

Huang et al. (184) concluded that incorporation of platinum into iridium clusters retards the oxidative agglomeration of iridium. When bimetallic catalysts were oxidized at 320°C, the majority of the surface species were bimetallic Pt-Ir oxychlorides; no significant $IrO_2$ agglomerates were observed. A few patents claim improved results with the addition to the Pt-Ir couple of a third element capable of promoting a significant change in the metal-support interaction: chromium oxide (123, 185) and silicon, calcium, magnesium, barium, or strontium (187).

A platinum-iridium-chromium catalyst is prepared by coimpregnation of alumina with chloroplatinic, chloroiridic, and chromic acids. Before impregnation, the alumina beads are pretreated under $CO_2$ flow. Such a $CO_2$ treatment has been extensively studied by Kresge et al. (123). Introduction of either iridium, platinum, or mixed acids by incipient wetness impregnation leads to an eggshell profile.

Pretreatment of alumina with $CO_2$ at room temperature promotes a desirable uniform metal distribution. Infrared characterization demonstrates that, even after calcination, such a pretreated alumina is covered by hydrogenocarbonate species.

As explained in Section III.B, a uniform distribution profile is normally obtained by competitive impregnation of $H_2PtCl_6$ and a rather high concentration of HCl. Under such low-pH and high-chloride-concentration conditions, the exchange sites of alumina are numerous and the chloride ligands of platinum are well protected. By contrast, $CO_2$ is not acidic enough to maintain a low pH and does not protect the chloride environment of platinum. In these conditions, the lower acid site number of the support and the formation of hydrolyzed platinum species having a lower affinity for the support lead to a uniform distribution profile of platinum. This positive influence of $CO_2$ has also been used during the introduction of platinum and iridium on alumina modified by other elements introduced either before (Si, Mg) or after (Ca, Ba, Sr) shaping (187).

## D. Platinum-Germanium

Germanium compounds have chemistries very similar to those of tin. Similar methods of preparation as employed for platinum-tin bimetallic catalysts would be anticipated. The cost of germanium compounds is however substantially higher than that of tin. The preceding considerations on the preparation of bimetallic platinum-tin catalysts with organotin compounds are applicable to germanium. Among the organogermanium compounds that may be employed are tetrabutyl or tetramethylgermane. Some commercial catalysts are cited to contain platinum and germanium, but no precise information is obtainable from the suppliers on content or methods of preparation.

A patent issued to UOP cites coimpregnation of alumina with a mixed solution of chloroplatinic acid and germanium dioxide (189). Obtaining a homogeneous solution of both compounds requires selection of the hexagonal germanium dioxide, the only soluble form. To improve the solubility, the solution is prepared with hot water immediately before impregnation.

Other multimetallic catalysts containing platinum and germanium have been claimed to have substantially higher selectivities than platinum and platinum-rhenium catalysts. For example, Antos (187) has shown a drastic increase in the hydrogen purity and the gasoline yield with a catalyst containing platinum, rhenium, and germanium. The source of rhenium and germanium is a mixed carbonyl species: $[ClGeRe(CO)_5]_3$. A drawback of this preparation is that it requires the use of anhydrous organic solvents.

## E. Other Bimetallic and Multimetallic Catalysts

Some of the patent literature on bimetallic reforming catalysts gives extensive examples of metallic formulations, but, to our knowledge, none until now has really been applied. A brief extract of such claims is presented in Table 1.

Among all these catalysts, many formulas are lab curiosities, but it can be estimated that tungsten and indium have reasonable chances of future industrial application. The incorporation techniques for these various elements are industrially available and no problems of extrapolation are consequently foreseen. It

**Table 1** Examples of Bimetallic and Multimetallic Reforming Catalysts

| Composition | Method of preparation | Advantage | Reference |
|---|---|---|---|
| Pt-Ir-Ag | Successive impregnations | Higher activity, higher selectivity, and higher sulfur tolerance | 188 |
| Pt-Ir-Ni | Coimpregnation | | |
| Pt-Ir-Pd | Coimpregnation | | |
| Pt-Ge-Ni | Coimpregnation | Higher stability | 190 |
| Pt-Co, Ni, Fe, Cu, Sn, Pd | Successive ionic exchanges | Higher aromatization activity | 191 |
| Pt-Pb, Pt-V, Pt-Ni, Pt-Zr, Pt-Y, Pt-Tl, Pt-Ag, Pt-Hg, Pt-Th-Ta, Pt-Ti-Fe, Pt-U | Coimpregnation | Activity and selectivity improvement | 194 |
| Pt-Pd-Cr | Coimpregnation | Higher aromatic yields | 195 |
| Pt-Zr | Successive impregnation | Stability increase | 196 |
| Pt-W | Platinum impregnation on tungsten | Higher aromatic yield | 197, 198 |
| Pt-In-Sn | Sn in the alumina sol | Higher activity and selectivity | 199 |
| Pt-Fe | Fe in the alumina sol | Higher activity and selectivity | 200 |
| Pt-F | F in the alumina sol | Higher activity | 201 |

will, however, be necessary to see that the mechanical properties of the solids obtained are compatible with their use in present and future reforming units.

## V. CONCLUSION

The catalytic reforming process is considered to be a mature process that progresses through incremental changes. This is also true for reforming catalysts. Catalytic breakthroughs have been and will be achieved by the invention of multimetallic catalysts able to develop a positive synergy between the different metals. Obtaining an optimum synergy necessitates mastery of the final interactions between the metals and the support and between the metals themselves. This implies the control of all steps of catalyst preparation and the optimization of the preparation method. The information presented in this chapter clearly demonstrates that the different preparation steps are necessitated by a complex chemistry so that the preparation of a desired multimetallic catalyst is a difficult challenge.

Potential improvements in catalyst preparation procedures are no doubt possible. Fundamental studies continue to be needed in order to achieve better control of the individual preparation steps and to find new preparation routes, which will lead to highly tuned multimetallic catalysts or even entirely new formulations.

## REFERENCES

1. W. H. Lang, R. J. Mikousky, and A. J. Silvestri, *J. Catal.* 20, 293-298 (1971).
2. J. R. Bernard, Proc. 5th Int. Zeolite Conf., Heyden, London: 686-695 (1980).
3. C. Bezuhanova, J. Guidot, D. Barthomeuf, M. Breysse, and J. R. Bernard, *J. Chem. Soc. Faraday Trans. I*, 77, 1595-1604 (1981).
4. T. R. Hughes, W. C. Buss, P. W. Tamm, and R. L. Jacobson, Stud. Surf. Sci. Catal., New Developments in Zeolite Science and Technology 28, 725-732, Y. Murakami, A. Lijima, J. W. Ward (1986).
5. P. W. Tamm, D. H. Mohr, and C. R. Wilson, Stud. Surf. Sci. Catal., *Catalysis* 38, 335-353, J. W. Ward (1988).
6. J. L. Kao, G. B. McVicker, M. M. J. Treacy, S. B. Rice, J. L. Robbins, W. E. Gates, J. J. Ziemiak, V. R. Cross, and T. H. Vanderspurt, *Proc. 10th Int. Congr. Catal.*, Budapest, 1993, 1019-1028.
7. E. G. Derouane, V. Julien-Lardot, R. J. Davis, N. J. Blom, and P. E. Hojlund-Nielsen, *Proc. 10th Int. Congr. Catal.*, Budapest, 1993, 1031-1040.
8. R. J. Davis and E. G. Derouane, *Nature 349*, 313-315 (1991).
9. N. J. Blom, E. G. Derouane, Eur. Patent 475357A1, Topsoe (1991).
10. E. G. Derouane, R. J. Davis, and N. J. Blom, Eur. Patent 476489A1, Topsoe (1991).
11. A. W. Chester and Y. F. Chu, U.S. Patent 4,350,835, Mobil (1982).
12. D. Dave and A. H. P. Hall, Eur. Patent 50021, BP (1982).
13. C. D. Telford and D. Young, Eur. Patent 119023; B. R. Gane and P. Howard, Eur. Patent 119027, BP (1984).
14. J. Johnson and G. K. Hilder, NPRA Annu. Meet, San Antonio, 1984.
15. J. R. Mowry, R. F. Anderson, and J. A. Johnson, *Oil Gas 128*, 2 (1985).
16. R. F. Anderson, J. A. Johnson, and J. R. Mowry, AIChE Spring National Meeting, Houston (1985).
17. P. C. Doolan and P. R. Pujado, *Hydrocarbon Process.* 72 (1989).
18. G. N. Roosen, A. Orieux, and J. Andrews, Dewitt's Houston Conf., 1989.
19. L. Mank, A. Minkkinen, and R. Shaddick, *Hydr. Technol. Int.*, 69 (1992).
20. M. Guisnet, N. S. Gnep, and F. Alario, *Appl. Catal. (A) 89*, 1-30 (1992).
21. W. J. Porter, W. M. Smith, and R. E. Schexnailder, U.S. Patent 2,976,232, Esso (1961).
22. French Patent 2046355, B.P. (1971).
23. R. J. Houston and S. M. Csicsery, U.S. Patent 3,617,521, Chevron (1971).
24. D. M. Little, *Catalytic Reforming*. Pennwell Publishing Co., Oklahoma, 1985.

25. V. Haensel, U.S. Patent 2,478,916, UOP (1949).
26. V. Haensel, U.S. Patent 2,611,736, UOP (1952).
27. V. Haensel, U.S. Patent, 2,623,860, UOP (1952).
28. R. Poisson, J. P. Brunelle, and P. Nortier, *Catalyst Supports and Supported Catalysts*, A. B. Stiles, Butterworth, Boston, 1987, 11–55.
29. B. C. Lippens, Ph.D. thesis, University of Delft, Netherlands (1961).
30. B. C. Lippens and J. J. Steggerda, in *Physical and Chemical Aspects of Adsorbents and Catalysis* (B. G. Linsen, ed.). Academic Press, New York, 1970, chapter 4.
31. B. C. Lippens, *Chem. Week B1 6*, 336 (1966).
32. D. C. Cocke, E. D. Johnson, and R. P. Merrill, *Catal. Rev. Sci. Eng. 26*, 163–231 (1984).
33. V. J. Lostaglio and J. D. Carruthers, *Che. Eng. Prog. 82*, 46–51 (1986).
34. Rhône-Poulenc Documentations: Spheralite, Catalyst Carriers from Rhône-Poulenc and Activated Alumina.
35. R. Montarnal, I.F.P. Internal report 21074 (1973).
36. N. H. Brett, K. J. D. Mackenzie, and J. H. Sharp, *Quart. Rev. 24*, 185–207 (1970).
37. E. J. Rosinski, T. R. Stein, and R. H. Fischer, U.S. Patent 3,876,523, Mobil (1975).
38. G. P. Vishnyakova, V. A. Dzis'ko, L. M. Kefeli, L. F. Lokotko, I. P. Olen'kova, L. M. Plyasova, I. A. Ryzhak, and A. S. Tikhova, *Kinet. Katal. 11*, 1287–1292 (1970).
39. P. Nortier, P. Fourre, A. B. Mohammed Saad, O. Saur, and J. C. Lavalley, *Appl. Catal. 61*, 141–160 (1990).
40. A. J. Leonard, F. Van Cauwelaert, and J. J. Fripiat, *J. Phys. Chem. 71*, 695–708 (1967).
41. B. C. Gates, J. R. Katzer, and G. C. A. Schuit, *Chemistry of Catalytic Processes*. McGraw-Hill, New York, 1979.
42. D. Papee, and R. Tertian, *J. Chim. Phys.* 341 (1958).
43. H. Knözinger and P. Ratnasamy, *Catal. Rev. Sci. Eng. 17*, 31–70 (1978).
44. J. B. Peri, *J. Phys. Chem. 69*, 211–219 (1965); *J. Phys. Chem. 69*, 220–230 (1965); *J. Phys. 69*, 231–239 (1965).
45. H. Knözinger, *Adv. Catal. 25*, 184–271 (1976).
46. Z. Vit, J. Vala, and J. Malek, *J. Appl. Catal. 7*, 159–168 (1983).
47. Webb, A. N., *I.E.C. 49*, 261–263 (1957).
48. H. P. Boehm, *Adv. Catal. 16*, 179–274 (1966).
49. P. Berteau and B. Delmon, *Catal. Today 5*, 121–137 (1989).
50. I. D. Chapman and M. L. Hair, *J. Catal. 2*, 145–148 (1963).
51. T. V. Antipina, O. V. Bulgarov, and A. V. Uvarov, *Proc. 4th Int. Congr. Catal.* Moscow, 1968, 376–387.
52. E. V. Ballou, R. T. Barth, and R. T. Flint, *J. Phys. Chem. 65*, 1639–1641 (1961).
53. A. E. Mirschler, *J. Catal. 2*, 428–439 (1963).
54. V. A. Chernov and T. V. Antipina, *Kinet. Katal. 7*, 651–653 (1966).
55. V. C. F. Holm and A. Clark, *Ind. Eng. Chem. 2*, 38–39 (1963).

56. J. He, J. Ai, K. Wu, and X. Luo, *5th Nat. Conf. Petrol. Petrochem.*, Shandong, China, 1989.
57. N. Tanaka and S. Ogasawara, *J. Catal. 16*, 157–163 (1970); *J. Catal. 16*, 164–172 (1970).
58. J. M. Basset, Ph.D. thesis University of Lyon I, France, 1969.
59. E. Garbowski, J. P. Candy, and M. Primet, *J. Chem. Soc. Faraday Trans. I, 79*, 835–844 (1983).
60. A. Melchor, E. Garbowski, M. Mathieu, and M. Primet, *J. Chem. Soc. Faraday Trans I, 82*, 1893–1901 (1986).
61. A. Roumegous, Ph.D. thesis, University of Paris VI (1978).
62. R. G. McClung, J. S. Sopko, R. Kramer, and D. G. Casey, NPRA Annual Meeting, San Antonio, 1990.
63. A. Goble and P. A. Lawrence, *Proc. 3rd Int. Congr. Catal.*, Amsterdam, 1964, 320–324.
64. J. P. Brunelle and R. Poisson, *Matériaux de l'Avenir*, Rhône-Poulenc Publication, 1991.
65. J. Hoekstra, U.S. Patent 2,620,314, UOP (1952).
66. J. Hoekstra, U.S. Patent 2,666,749, UOP (1954).
67. C. Wankat, U.S. Patent 2,672,453, UOP (1954).
68. Hoekstra, J., U.S. Patent 2,774,743, UOP (1956).
69. M. W. Schoonevea, U.S. Patent 4,318,896, UOP (1982).
70. C. Wankat, U.S. Patent 2,733,220, UOP (1956).
71. E. Michalko, U.S. Patent 3,027,232, UOP (1962).
72. J. C. Hayes, U.S. Patent 3,887,492, UOP (1975).
73. J. C. Hayes, U.S. Patent 3,887,493, UOP (1975).
74. J. F. LePage, *Applied Heterogeneous Catalysis*, IFP Publications, Editions Technip, Paris, 1987.
75. R. W. Moehl, U.S. Patent 2,759,898, UOP (1956).
76. M. J. Murray, U.S. Patent 2,736,713, UOP (1956).
77. K. D. Vesely, U.S. Patent 3,496,115, UOP (1970).
78. E. Michalko, U.S. Patent 4,216,122, UOP (1980).
79. D. L. Trimm and A. Stanislaus, *Appl. Catal. 21*, 215–238 (1986).
80. J. P. Brunelle and A. Sugier, *Compt. Rend. Acad. Sci. Serie C 276*, 1545–1548 (1973).
81. T. A. Dorling, B. W. J. Lynch, and R. L. Moss, *J. Catal. 20*, 190 (1971).
82. C. Marcilly and J. P. Franck, *Rev. Inst. Fr. Petr. 39*, 337–364 (1984).
83. R. W. Maatman and C. D. Prater, *Ind. Eng. Chem. 49*, 253–257 (1957).
84. A. V. Neimark, L. I. Kheifez, and V. B. Fenelonov, *Ind. Eng. Chem. Process Des Dev. 20*, 439–450 (1981).
85. P. B. Weisz, *Trans. Faraday Soc. 63*, 1801–1806 (1967).
86. P. B. Weisz and J. S. Hicks, *Trans. Faraday Soc. 63*, 1807–1814 (1967).
87. P. B. Weisz and H. Zollinger, *Trans. Farady Soc. 63*, 1815–1823 (1967).
88. J. P. Brunelle, *Pure Appl. Chem. 50*, 1211–1229 (1978).
89. J. A. Schwarz, *Catal. Today 15*, 395–405 (1992).
90. M. J. D'Anielo, Jr., *J. Catal. 69*, 9–17 (1981).
91. C. P. Huang and Stumm, W., *J. Colloid Interf. Sci. 43*, 409–420 (1973).

92. J. A. Davis, R. O. James, and J. O. Leckie, *J. Colloid Interf. Sci. 63*, 480-499 (1978).
93. H. Hohl and W. Stumm, *J. Colloid Interf. Sci. 55*, 281-288 (1976).
94. A. M. Ahmed, *J. Phys. Chem. 73*, 3546-3555 (1969).
95. B. J. K. Acres, A. J. Bird, J. W. Jenkins, and F. King, The design and preparation of supported catalysts, *Specialist Periodical Reports Catalysis 4*, 1-30 (1981).
96. A. T. Bell, *Catalyst Design, Progress and Perspectives.* Wiley-Interscience, New York, 1987.
97. A. K. Aboul-Gheit, *J. Chem. Tech. Biotechnol. 29*, 480-486 (1979).
98. L. J. Jacimovic, J. Stevovic, and S. Veljkovic, *J. Phys. Chem. 76*, 3625-3632 (1972).
99. S. Sivasanker, A. V. Ramaswamy, and P. Ratnasamy, Stud. Surf. Sci. Catal., *Preparation of Catalysts II*, 3:185-196, B. Delmon, P. Grange, P. A. Jacobs, and G. Poncelet (1979).
100. E. Borello, G. Della Gatta, Bice Fubini, C. Morterra, and G. Venturello, *J. Catal. 35*, 1-10 (1974).
101. E. Santacesaria, D. Gelosa, and S. Carra, *Ind. Eng. Chem. Process Des. Dev. 16*, 45-47 (1977).
102. R. W. Maatman, *I.E.C. 51*, 913-914 (1959).
103. R. W. Maatman, P. Mahaffy, P. Hokestra, and C. Addink, *J. Catal. 23*, 105-117 (1971).
104. F. Umland, and W. Fischer, *Naturwisseneschaften 40*, 439-440 (1953).
105. J. C. Summers and S. A. Ausen, *J. Catal. 52*, 445-452 (1978).
106. E. I. Gil'Debrand, *Intern. Chem. Eng. 6*, 449-480 (1966).
107. Y. S. Shyr and W. R. Ernst, *J. Catal. 63*, 425-432 (1980).
108. J. R. Anderson, *Structure of Metallic Catalysts*, Academic Press, New York, 1975, 164-217.
109. W. P. Griffith, *The Chemistry of the Rarer Platinum Metals*, Interscience, London, 1967.
110. C. M. Davidson and R. F. Jameson, *Trans. Farady Soc. 61*, 2462-2467 (1965).
111. G. H. Van Den Berg and H. T. Rijnten, Stud. Surf. Sci. Catal., *Preparation of Catalysts II*, 3:265-277, B. Delmon, P. Grange, P. A. Jacobs, and G. Poncelet (1979).
112. H. Lieske, G. Lietz, H. Spindler, and J. Völter, *J. Catal. 81*, 8-16 (1983).
113. G. Lietz, H. Lieske, H. Spindler, W. Hanke, and J. Völter, *J. Catal. 81*, 17-25 (1983).
114. J. Berdala, E. Freund, and J. Lynch, *J. Phys. 47*, 269-272 (1986).
115. X. Luo, J. Ai, J. He, and J. Luo, *React. Kinet. Catal. Lett. 43*, 55-61 (1991).
116. V. Haensel, U.S. Patent 2,840,532, UOP (1958).
117. G. N. Maslyanskii, B. B. Zharkov, and A. Z. Rubinov, *Kinet. Katal. 12*, 699-701 (1971).
118. L. L. Hegedus, T. S. Chou, J. C. Summers, and N. M. Potter, Stud. Surf. Sci. Catal., *Preparation of Catlysts II*, 3: 171-183, B. Delmon, P. Grange, P. A. Jacobs, and G. Poncelet (1979).
119. W. Jianguo, Z. Jiayu, and P. Ll, *3rd Int. Symp. Scientific Bases for the Preparation of Heterog. Catal.*, Sept. 1982, paper A4, Louvain-la-Neuve, 1982.

120. T. A. Nuttal, CSIR Report CENG 182, CSIR, Pretoria, South Africa, 1977.
121. E. R. Becker and T. A. Nuttall, Stud. Surf. Sci. Catal., *Preparation of Catalysts II*, 3: 159-167, B. Delmon, P. Grange, P. A. Jacobs, and G. Poncelet (1979).
122. M. S. Heise and J. A. Schwarz, Stud. Surf. Sci. Catal., *Preparation of Catalysts IV*, 31: 1-13, B. Delmon, P. Grange, P. A. Jacobs, and G. Poncelet (1987).
123. C. T. Kresge, A. W. Chester, and S. M. Oleck, *Appl. Catal. 81*, 215-226 (1992).
124. M. Komiyama, *Catal. Rev. Sci. Eng. 27*, 341-372 (1985).
125. E. R. Becker and J. Wei, *J. Catal. 46*, 365-381 (1977).
126. P. Harriott, *J. Catal. 14*, 43-48 (1969).
127. R. C. Vincent and R. P. Merrill, *J. Catal. 35*, 206-217 (1974).
128. M. Komiyama, R. P. Merrill, and H. F. Harnsberger, *J. Catal. 63*, 35-52 (1980).
129. M. Komiyama and R. P. Merrill, *Bull. Chem. Soc. Jpn. 57*, 1169 (1984).
130. S. Y. Lee and R. Aris, Stud. Surf. Sci. Catal., *Preparation of Catalysts III*, 16: 35-45, G. Poncelet, P. Grange, and P. A. Jacobs (1983).
131. A. A. Castro, O. A. Scelza, E. R. Benvenuto, G. T. Baronetti, S. R. De Miguel, and J. M. Parera, Stud. Surf. Sci. Catal., *Preparation of Catalysts III*, 16: 47-56, G. Poncelet, P. Grange, and P. A. Jacobs (1983).
132. R. L. Jacobson, H. E. Kluksdahl, C. S. McCoy, and R. W. Davis, *Proc. Amer. Petrol. Inst. Div. Refining 49*, 504 (1969).
133. R. L. Jacobson, H. E. Kluksdahl, and B. Spurlock, U.S. Patent 3,434,960, Chevron (1969).
134. R. J. Bertolacini and R. J. Pellet, Stud. Surf. Sci. Catal., *Catalyst Deactivation 6*, 73-77, B. Delmon and G. F. Froment (1980).
135. M. F. Johnson and V. M. Leroy, *J. Catal. 35*, 434-440 (1974).
136. B. D. McNicol, *J. Catal. 46*, 438 (1977).
137. H. Charcosset, French-Venezuelian Congress, Caracas, 1983.
138. C. Betizeau, C. Bolivar, H. Charcosset, R. Frety, G. Leclercq, R. Maurel, and L. Tournayan, Stud. Surf. Sci. Catal., *Preparation of Catalysts I*, B. Delmon, P. A. Jacobs, and G. Poncelet 525-536 (1976).
139. S. B. Ziemecki, G. A. Jones, and J. B. Michel, *J. Catal. 99*, 207-217 (1986).
140. N. Wagstaff and R. Prins, *J. Catal. 59*, 434-445 (1979).
141. J. C. Hayes, U.S. Patent 3,775,301, UOP (1973).
142. H. E. Kluksdahl, U.S. Patent 3,558,477, Chevron (1968).
143. P. Pascal, *Chimie Minérale 10*, Masson, Paris, 1978.
144. D. R. Ardiles, S. R. De Miguel, A. A. Castro, and O. A. Scelza, *Appl. Catal. 24*, 175-186 (1986).
145. S. R. De Miguel, O. A. Scelza, A. A. Castro, G. T. Baronetti, D. R. Ardiles, and J. Parera, *Appl. Catal. 9*, 309-315 (1984).
146. C. Bolivar, H. Charcosset, R. Frety, M. Primet, L. Tournayan, C. Betizeau, G. Leclercq, and R. Maurel, *J. Catal. 39*, 249-259 (1975).
147. J. Reyes, G. Pecchi, and P. Reyes, *J. Chem. Res.*, 318-319 (1983).
148. A. K. Smith, A. Theolier, J. M. Basset, R. Ugo, D. Commereuc, and Y. Chauvin, *J. Am. Chem. Soc. 100*, 2590-2591 (1978).
149. A. F. Danilyuk, V. L. Kuznetsov, A. P. Shepelin, P. A. Zhdan, N. G. Maksimov, G. I. Magomedov, and Y. I. Ermakov, *Kinet. Katal. 24*, 919-925 (1983).

150. G. J. Antos, U.S. Patent 4,136,017 and 4,159,939, UOP (1979).
151. J. R. Bernard and M. Breysse, French Patent 2479707, Elf (1980).
152. A. Brenner and D. A. Hucul, *J. Catal. 61*, 216-222 (1980).
153. Y. I. Ermakov, B. N. Kuznetsov, and A. N. Startsev, *Kinet. Katal. 18*, 808-809 (1977).
154. Compagnie Française de Raffinage, French Patent 2031984 (1969).
155. F. Yining, Z. Jingling, and L. Liwu, *J. Catal.* (Cuihua. Xuebao) *10*, 111-117 (1989).
156. R. E. Rausch, U.S. Patent 3,745,112, UOP (1973).
157. Y. X. Li, K. J. Klabunde, and B. H. Davis, *J. Catal. 128*, 1-12 (1991).
158. R. Gomez, V. Bertin, M. Ramirez, T. Zamudio, P. Bosch, I. Schifter, and T. Lopez, *J. Non. Cryst. Solids 147*, 748 (1992).
159. T. Chee, W. M. Targes, and M. D. Moser, U.S. Patent 4,964,975, UOP (1990).
160. P. Engelhard, G. Szabo, and J. E. Weisang, U.S. Patent 4,039,477, CFR (1977).
161. Y. X. Li, Y. F. Zhang, and K. J. Klabunde, *Langmuir 4*, 385-391 (1988).
162. J. Margitfalvi, E. Tálas, M. Hegedüs, and S. Göbölös, *Proc. 6th Int. Symp. Heterogeneous Catalysis*, Sofia, 1987, 345-353.
163. V. I. Kuznetsov, A. S. Belyi, E. N. Yurchenko, M. D. Smolikov, M. T. Protasova, E. V. Zatolokina, and V. K. Duplyakin, *J. Catal. 99*, 159-170 (1986).
164. N. Homs, N. Clos, G. Muller, J. Sales, and P. Ramirez de la Piscina, *J. Mol. Catal. 74*, 401-408 (1992).
165. D. F. Cox, G. B. Hoflund, and H. A. Laitinen, *Langmuir 1*, 269-273 (1985).
166. R. Burch, *J. Catal. 71*, 348-359 (1981).
167. G. Baronetti, S. De Miguel, O. Scelza, A. Fritzler, and A. Castro, *Appl. Catal. 19*, 77-85 (1985).
168. F. M. Dautzenberger, J. H. Helle, P. Biloen, and W. M. H. Sachtler, *J. Catal. 63*, 119-128 (1980).
169. C. Travers, J. P. Bournonville, and G. Martino, *Proc. 6th Int. Congr. Catal.*, Berlin, 1984, 891-902.
170. C. Vértes, E. Talas, I. Czako-Nagy, J. Ryczkowski, S. Göbölös, A. Vertes, and J. Margitfalvi, *Appl. Catal. 68*, 149-159 (1991).
171. V. D. Stytsenko, O. V. Kovalenko, and A. Y. Rozovski, *Kinet. Katal. 32*, 163-169 (1989).
172. H. E. Kluksdahl and R. L. Jacobson, French Patent 2076937, Chevron (1971).
173. J. F. Young, R. D. Gillard, and G. Wilkinson, *J. Chem. Soc.*, 5176-5189 (1964).
174. V. H. Berndt, H. Mehner, J. Völter, and W. Meisel,*Z. Anorg. Allg. Chem. 429*, 47-58 (1978).
175. E. N. Yurchenko, V. I. Kuznetsov, V. P. Melnikova, and A. N. Sartsev, *React. Kinet. Catal. Lett 23*, 113-117 (1983).
176. Jin, L., *Appl. Catal. 72*, 33-38 (1991).
177. G. Baronetti, S. De Miguel, A. Castro, O. Scelza, and A. Castro, *Appl. Catal. 45*, 61-69 (1988).
178. B. H. Davis, *Proc. 10th Int. Congr. Catal.*, Budapest, 1992, 889-897.
179. A. Sachdev and J. Schwank, *Proc. 9th Int. Congr. Catal.*, Calgary, 1988, 1275-1283.
180. G. Baronetti, G., S. De Miguel, O. Scelza, and A. Castro, *Appl. Catal. 24*, 109-116 (1986).

181. W. C. Buss, U.S. Patent 3,554,902, Chevron (1971).
182. K. Foger, and H. Jaeger, *J. Catal.* 70, 53–71 (1981).
183. R. L. Garten and J. H. Sinfelt, *J. Catal.* 62, 127–139 (1980).
184. Y. J. Huang, S. C. Fung, W. E. Gates, and G. B. McVicker, *J. Catal.* 118, 192–202 (1989).
185. K. Anders, K. Becker, P. Birke, S. Engels, R. Feldhaus, W. Hager, H. Lausch, P. Mahlow, H. D. Neubauer, D. Sager, M. Wilde, and H. G. Vieweg, DDR Patent 212 192, Leuna Werke (1982).
186. W. C. Baird, U.S. Patent 4,966,879, Exxon (1990).
187. G. J. Antos, U.S. Patent 4,312,788, UOP (1992).
188. H. L. Mitchell and J. R. Hayes, French Patent 2249161, Exxon (1974).
189. K. R. McCallister and T. P. O'Neal, French Patent 2078056, UOP (1971).
190. F. C. Wilhelm, French Patent 2081634, UOP (1971).
191. E. E. Davies, J. S. Elkins, and R. C. Pitkethly, French Patent 2089516, BP (1971).
192. W. C. Buss, French Patent 2132676, Chevron (1972).
193. Asahi Kasei Kogyo Kabushiki Kaisha, French Patent 2030396 (1970).
194. Asahi Kasei Kogyo Kabushiki Kaisha, French Patent 2090058 (1971).
195. N. Kominani, T. Iwaisako, K. Tanaka, and K. Ohki, U.S. Patent 3,554,901, Asahi Kasei Kogyo Kabushiki Kaisha (1971).
196. W. J. Porter, W. M. Smith, and R. E. Schexnailder, U.S. Patent 3,002,920, Exxon (1961).
197. V. Haensel, U.S. Patent 2,957,819, UOP (1960).
198. J. L. Contreras, G. Del Toro, I. Schifter, and G. A. Fuentes, Stud. Surf. Sci. Catal., *Catalysis 1987 38,* 51–59 (1988).
199. F. C. Wilhelm, U.S. Patent 3,951,868, UOP (1976).
200. J. C. Hayes, U.S. Patent 3,379,641, UOP (1968).
201. E. Michalko, J. Hoekstra, and R. M. Smith, U.S. Patent 2,927,088, UOP (1960).

# 5
# Characterization of Naphtha Reforming Catalysts

**Burtron H. Davis**

*University of Kentucky, Lexington, Kentucky*

**George J. Antos**

*UOP Research Center, Des Plaines, Illinois*

## I. INTRODUCTION

Naphtha reforming involves heterogeneous catalysis, with the catalyst constituting a separate phase (1). Furthermore, naphtha reforming occurs by bifunctional catalysis (2). This means that for a $Pt-Al_2O_3$ reforming catalyst, some of the processes occur at the surface of platinum or other metal(s) and others at the acidic sites on the alumina or other support. For optimum performance, these two or more types of sites are intermixed on the same primary particles. Characterization of naphtha reforming catalysts therefore presents many obstacles. One must be aware of the assumptions that enable one to convert the experimental data into conclusions that define the catalyst structure. Unfortunately, all too often the assumptions are overlooked in developing models of reforming catalysts.

At the most elementary level, characterization of a reforming catalyst involves only two topics: (1) a measure of the amount together with the strength and distribution of the acid function and (2) a measure of the amount and activity of the metallic function. Perhaps needless to say, topic 2 becomes more difficult when the catalyst contains two or more metallic components.

Two levels of characterization data may be distinguished. One type of data is needed to address engineering applications. Here one is concerned with characterization of the features that will (1) permit the catalyst manufacturer to prepare repetitively a catalyst with the same properties and (2) provide the process operator with the ability (a) to bring the catalyst on-stream with the required activity and selectivity, (b) to monitor catalyst performance, and (c) to adjust the state of the working catalyst to maintain performance specifications for a long period of operation. The engineering approach requires only that the properties that are characterized can be related to the performance of the catalyst; it does not have to provide an accurate measure of the absolute value of a particular feature of the catalyst. For example, if it is found that 20 ppm of chlorine in the exit gas provides the optimum activity and selectivity for naphtha reforming with a particular catalyst, the engineering method requires only a measure of the chlorine in the exit gas. It is not necessary to know the amount of chloride incorporated or its chemical state in the working catalyst. Characterizations for engineering purposes are essential for the successful application of catalysts in commercial naphtha operations.

However, discovery of new catalyst formulations or improvement of existing formulations normally results from application of scientific models of the catalyst. For this purpose, the engineering characterizations seldom have value. What is needed for new catalyst design may be considered to be standard characterization procedures which allow an accurate measure of the absolute value and the chemical nature of a specific catalyst feature or catalytic site. Here one must define a specific feature of the catalyst and devise a method to make an accurate and exact measure of the feature. Of necessity, this involves definitions and terminology (1). At first glance, this requirement appears to be easy to meet; however, in practice it proves frequently to be an extremely demanding task.

It would be desirable to describe in detail the experimental techniques that are appropriate for the characterization of naphtha reforming catalysts, along with results from studies and their interpretation. However, a large volume would be required to do this. Thus, the present chapter will emphasize results obtained using many experimental techniques rather than an in-depth discussion of the techniques. Interpretations of these studies for a number of reforming catalysts are presented. The reader may see the diversity of experimental approaches and interpretations which are in the literature and may use this chapter as a guide for further investigations.

## II. ALUMINA SUPPORTS

### A. Surface Area and Porosity

The physical characteristics of naphtha reforming catalysts are determined primarily by the material which serves as a support for the metal or bimetallic function. Alumina is the support for nearly all reforming catalysts. The strength of the macroscopic catalyst particle is an important property. Measurement techniques are not straightforward however. For most fixed-bed operations, if a catalyst can survive the handling during manufacturing and loading, it has adequate strength. Moving-bed operations have their own strength requirements. The reader is best referred to catalyst suppliers or the American Society for Testing and Materials (ASTM) for further information.

The surface area is one of the most important physical properties of the catalyst. One of the first techniques introduced for the characterization of catalysts was the measurement of the surface area by the use of gas adsorption and the application of the Brunauer-Emmett-Teller (BET) equation (3). This method grew from a desire to learn whether it was the specific nature or the extent of the surface that controlled catalytic activity and selectivity (4). The BET method was introduced more than 50 years ago and today it is still the most widely utilized catalyst characterization technique. Detailed descriptions of this method are plentiful (5). Today the gas adsorption is accomplished by automated instruments that permit measurements to be made simultaneously on multiple samples. These instruments can be utilized for making measurements of both the surface area and the porosity.

The validity of the BET technique as an absolute method is still debated. The BET equation is based on a simple model, and the validity of some of the assumptions made for its deprivation is questionable (6). It is a very adequate technique for measuring the total surface area, and as an engineering method the technique is quite acceptable (7). In spite of numerous attempts to place the BET equation on a firmer basis or even to supplant it, for example, with the introduction of fractal theory (8), a realistic assessment would lead one to conclude that these attempts have led to more complex, and not more accurate, equations. For the current naphtha reforming catalysts, the BET equation therefore provides both the engineering and the absolute characterization technique for measuring the total surface area.

A complete assessment of porosity is usually obtained by a combination of gas adsorption and mercury penetration measurements. Gas adsorption is applicable for pore sizes falling within the range of about 0.5 to 40 nm; however, some of the newer instruments are claimed to be capable of making measurements that permit the upper limit to be extended to 100 nm. For pores in the range of about 10 to 5000 nm diameter, mercury porosimetry is applicable. For most materials, a direct comparison of the results from the two measurements can be made (9). Early calculations of the porosity followed the approach of Barrett et al. (10), and then the one formulated by Wheeler (11) in his classic treatment of the role of diffusion in catalysis. With the introduction of automated instrumentation, the isotherm of Dollomire and Heal (12) was utilized frequently. Other more complex approaches have been utilized for the calculation (13–23).

Nearly all physisorption isotherms may be grouped into the six types shown in Figure 1a (24,25). From the type of isotherm, a general idea of the structure

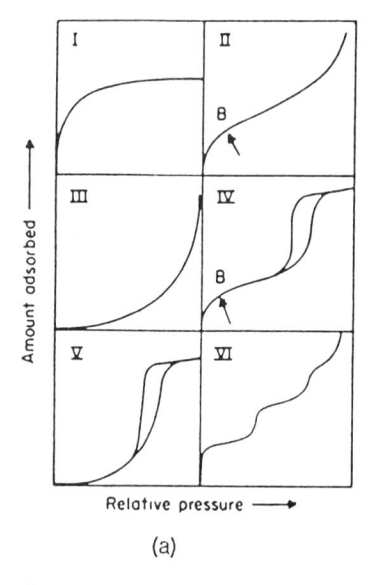

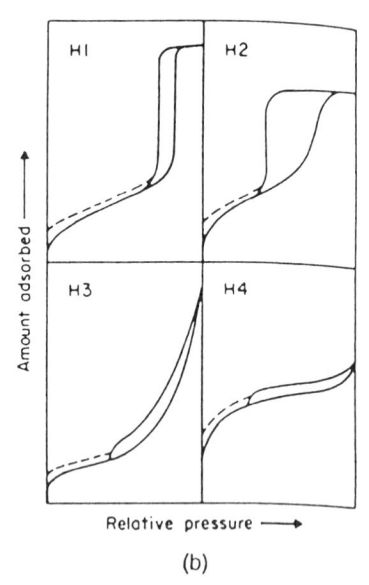

**Figure 1** (a) Types of physisorption isotherms. (b) Types of hysteresis loops. (From Ref. 24.)

of the material may be deduced. Four types of hysteresis are illustrated in Figure 1b. In most cases the naphtha reforming catalyst will produce either a type II or type IV isotherm with H1, H2, or H3 type of hysteresis. In these cases an analysis of the pore size can be obtained from either the adsorption or desorption isotherm, with the desorption isotherm being utilized more frequently (26-31). For comparative purposes, consistency of use of model and standard isotherm is probably more important than the actual choice of model.

The theory of mercury penetration used for measuring porosity was developed by Washburn in 1922 but the first measurements were made more than 20 years later when Ritter and Drake (32) introduced an experimental approach that was eventually developed into a commercial instrument. Today, commercial, computer-controlled instruments are available. For materials with pore sizes that permit a direct comparison between the results of the two techniques—gas desorption isotherm and mercury penetration—the two methods provide reasonable agreement. There are small differences in the distributions that are calculated using different models (26-31) with nitrogen adsorption or desorption isotherm data, and there is reasonable agreement with distributions calculated from mercury penetration data. The difference between the two methods becomes less than 20% when various correction procedures are used. The presence of metals on a support may cause a significant alteration of the contact angle needed to provide agreement of the two methods, and the angle needed may depend on the metal loading (33).

Other techniques are available for porosity measurements but they are not ordinarily utilized with reforming catalysts. For example, Ritter utilized the data from small-angle x-ray scattering measurements to calculate a pore size distribution. Although there have been significant advances in the theory needed to calculate a pore size distribution from small-angle x-ray scattering, currently it is not frequently used to characterize reforming catalysts. As noted below, this will probably change.

## B. Acidity

The other important feature of the support is a measure of its acidity. Measurements of acidity have assumed an increasingly important role in catalyst characterization. Benesi and Winquist (34) provided a concise and precise description of what is involved in terms of electron-pair acceptors for Lewis and Brönsted acid sites on surfaces of metal oxides. A characterization of acidity should provide a measure of at least three qualities: (1) the acid type (Brönsted or Lewis), (2) the acid site density, and (3) the acid strength distribution. A characterization of acidity therefore involves a definition of all three qualities. A number of approaches have been utilized for the characterization of these, and they include:

1. Hammett indicators—these are compounds that combine with the acid to form a color that differs from that of the uncombined molecule.
2. Probe molecules—these molecules are strong enough bases that they appear to react irreversibly with the acid site.
3. Probe reactions—these involve a simple reaction whose rate or selectivity depends on the acidity of the catalyst. Since reactions are covered in other chapters, this topic will not be discussed here.

Hammett acidity measurements involve the adsorption of indicators from suitable nonaqueous solvents that do not interact with the catalyst acid sites. Walling (35) defined the acid strength of a solid as its proton-donating ability, $H_0$. Benesi (36) utilized a number of indicators that could be related through their color change to the composition of a sulfuric acid solution that gave an equivalent color. Thus, the use of a series of indicators allows bracketing of the acidity of the catalyst between a high and low range.

Hammett indicators have several disadvantages when they are employed to measure the acidity of solids: (1) visually it is difficult to detect color changes, (2) many of the indicators are too large to penetrate any microporosity that is present, (3) the measurements are nearly always made far from the reaction conditions, and (4) they may not distinguish Lewis and Brönsted acid sites. Alternatively, arylmethanols react with strong protonic acids with the resulting conjugate acid being a colored carbenium ion. These arylmethanol–carbenium ion equilibria have been used to define an alternative acidity function (37–39) which has been previously designated as $C_O$, $J_O$, and $H_R$. The adsorption of these bases may provide a measure of the number and strength of the acid sites, but they reveal very little information about the structure of the catalyst site (40–42).

Alumina is not the ideal material to characterize using Hammett indicators. Following activation at moderately high temperatures (400–700°C) in either vacuum, air, or oxygen, alumina loses many, but not all, of the hydroxyls groups. Those that remain do not exhibit strong acidity (43) and exist in a variety of coordination states (44). One of the types of hydroxyls is considered to be basic because it will react with $CO_2$ to form the bicarbonate ion (45–47). The dominant portion of the activated alumina surface is comprised of several types of oxide ions; many are a result of the elimination of water from two hydroxyl groups during the activation process. The surface also contains coordinately unsaturated sites (cus): aluminum ions that impart Lewis acidity to the activated alumina (43,48). Hammett indicators cannot begin to characterize adequately the complete surface of aluminas.

## 1. Infrared Spectroscopy

In general, infrared is not very well suited to the direct examination of the reforming catalyst. Thus, the characterization nearly always involves the adsorption of one or more molecules on the reforming catalyst using infrared to distinguish the interaction of the probe molecule with a feature of the catalyst. Eischens and Pliskin (49) showed the utility of infrared spectroscopy for the measurement of acidity and metal-adsorbate interactions. These authors showed that the adsorption of a base such as ammonia could provide a quantitative measure of Brönsted and Lewis acid sites. Brönsted sites were identified with bands characteristic of the formation of the ammonium ion and Lewis sites with bands characteristic of covalent bonding. Since then there have been tremendous advances in the instrumentation, with computer-controlled instruments providing subtraction of background absorption and allowing detection of adsorbed species at much lower levels than was previously possible (50–53).

The spectrum of hydroxyl groups of alumina are shown in Figure 2 (54). Three bands with maxima at about 3800 (I), 3745 (II), and 3700 (III) cm$^{-1}$ are reported in most papers (55–58). Depending on the extent of dehydration, additional absorption bands at 3780, 3760, and 3733 cm$^{-1}$ may be obtained. The bands do not appear to depend on alumina crystal phase to a significant extent (54). Based on extensive study, Peri (44) proposed a detailed model of an alumina surface (Figure 3). In this model a surface hydroxyl group may have 0, 1, 2, 3, or 4 oxygen ions in their nearest environment, and this classification of the hydroxyl types made it possible to explain the position and relative intensities of the bands shown in the spectra in Figure 2 after different heat treatments. The concentration of a completely hydroxylated $\gamma$-$Al_2O_3$ was determined by deuterium exchange to be $1.3 \times 10^{15}$/cm$^2$. The concentration of a similar $\gamma$-$Al_2O_3$ after evacuation at 500°C was $3.6 \times 10^{14}$/cm$^2$ (59).

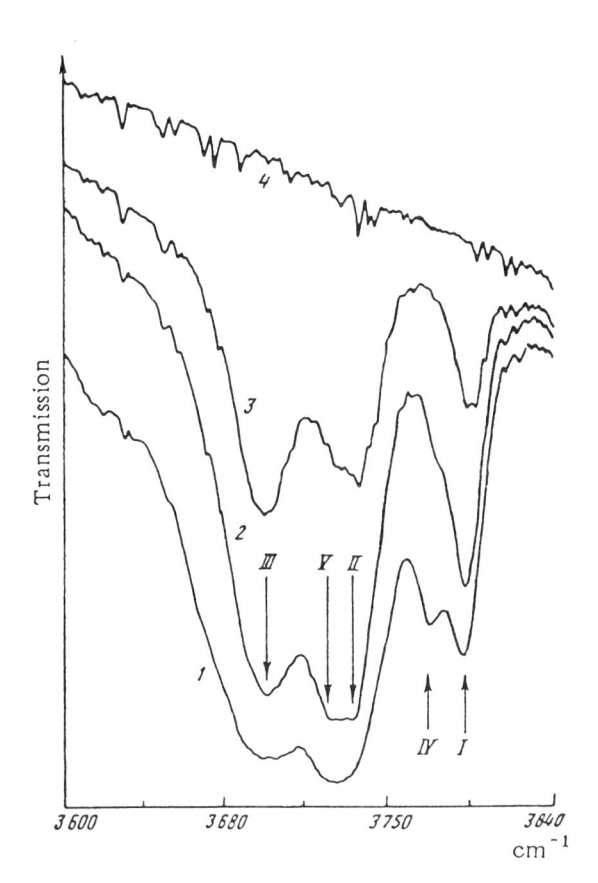

**Figure 2** Spectrum of hydroxyl groups of alumina: (1) after evacuation at 700°C; (2) the same, at 800°C; (3) the same, at 900°C; (4) emission background. The Roman numerals correspond to the interpretation of the bands given in Figure 3. (From Ref. 54.)

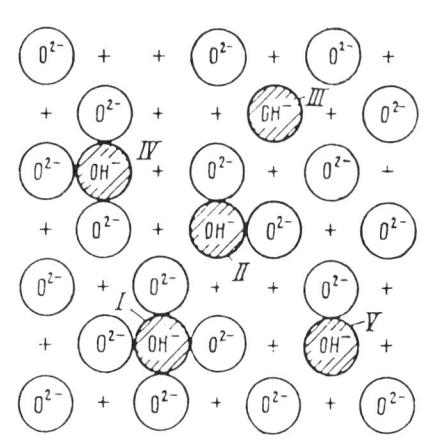

**Figure 3** Five types of hydroxyl ions (designated by Roman numerals) on the surface of partially dehydroxylated alumina. The + sign designates the $Al^{3+}$ ion in a deeper layer. (From Ref. 54.)

The adsorption of ammonia was followed by combined gravimetric measurements and infrared spectrometry (60). It was found that ammonia bonded with all five types of hydroxyl groups, and in addition to molecular adsorption of ammonia, surface reactions may occur at higher temperatures. The spectra of pyridine following adsorption on alumina and subsequent evacuation at increasing temperatures (Figure 4 [43] show molecularly adsorbed pyridine, which is removed by evacuation at 150°C, and also bands at 1632 and 1459 cm$^{-1}$ which are not removed even at 565°C. These bands are similar to those that result from the complex formed between pyridine and gas-phase Lewis acids, so it was concluded that these bands are characteristic of bonding due to Lewis acids. The spectrum of adsorbed pyridine on alumina did not show the absorption band at 1540 cm$^{-1}$ characteristic of the pyridinium ion formed by interaction with a Brönsted site.

The acid strength of the Lewis sites is significant and shows a broad distribution. The acid strength of the Lewis site depends on the degree of unsatura-

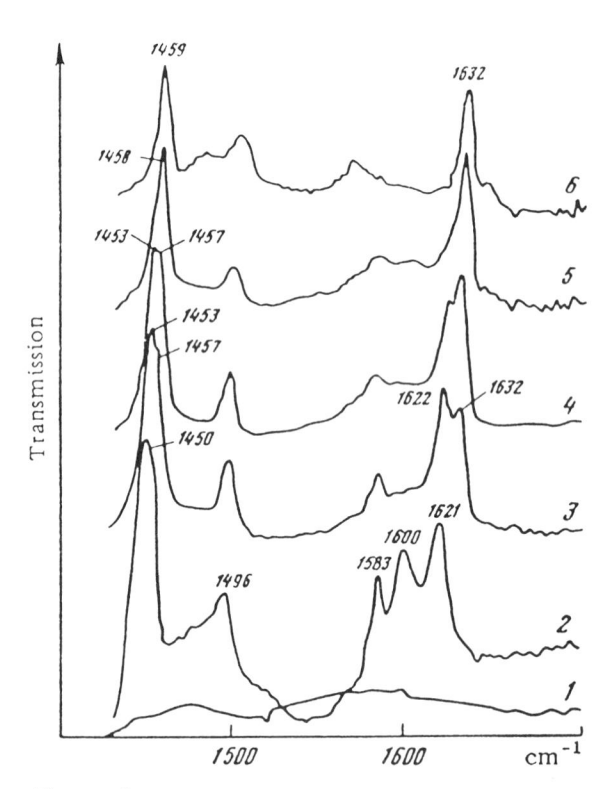

**Figure 4** Spectra of pyridine adsorbed on alumina: (1) spectrum of γ-Al$_2$O$_3$ after evacuation at 450°C for 3 h; (2) after adsorption of pyridine at 25°C; (3) after evacuation for 3 h at 150°C; (4) the same, at 230°C; (5) the same, at 325°C; (6) the same, at 565°C. (From Ref. 43.)

tion of the $Al^{3+}$ ion and the tetragonal $Al^{3+}$ ion exposed in a vacancy is a stronger site than an octahedral $Al^{3+}$ site (61). Heats of chemisorption for pyridine on γ-alumina after heating at 770 K range from 90 to over 120 kJ/mol, and chemisorbed pyridine cannot be quantitatively desorbed at temperatures below 750 K, where it begins to decompose (61). Steric hindrance is critical with bulky probe molecules for infrared studies. For this reason, work has been typically carried out with smaller molecules to probe acid sites.

The results obtained by Kazansky et al. (62) show that the low-temperature adsorption of dihydrogen is a promising approach for the characterization of Lewis acid sites. A frequency shift of 180 cm$^{-1}$ toward lower values relative to the H−H stretching frequency of the free molecule was observed for η-alumina pretreated at 870 K. This shift in frequency is taken to be a measure of the polarizing power of the $Al^{3+}$ (cus).

Carbon monoxide has also been utilized to probe the acidity of alumina (63). The spectra recorded for adsorption of CO on γ-alumina with increasing partial pressure at 77 K resulted in bands ascribed to CO σ-bonded to strong cationic Lewis acid sites (2238 cm$^{-1}$), CO σ-bonded to bulk tetrahedral $Al^{3+}$ ions on the surface (2210–2190 cm$^{-1}$), CO σ-bonded to octahedral $Al^{3+}$ ions on the surface (2165 cm$^{-1}$), and physically adsorbed CO (3135–2140 cm$^{-1}$). For adsorption of CO at room temperature, the peaks are not well resolved and depend on the alumina sample. Thus, for well-crystallized and pure-phase η-alumina two relatively well-resolved bands are found, whereas for microcrystalline specimens ex-boehmite (γ- and δ-alumina) there are probably more than two bands which are not well resolved (64). For the γ-alumina dehydrated at a low temperature the two CO bands are centered at about 2230 and 2200 cm$^{-1}$; dehydration at 1023 K causes the whole 2250–2190 cm$^{-1}$ spectral range to be occupied by a broad, asymmetric, and unresolved band. For aluminas calcined at a higher temperature to produce mixed δ,θ-alumina phases, the band becomes sharper with the elimination of a significant fraction of the stronger acid sites that produce the higher-frequency band.

## 2. Calorimetry

The reaction of an acid with a base generates heat, so another way to determine the acidity is by calorimetry; chemisorption on the metal function is also exothermic and calorimetry is applicable for this measurement as well. The equipment used for this measurement is illustrated in a publication using a commercially available instrument modified to make it applicable for catalyst studies (65). The instrument was capable of operation down to 200 K. In addition, a number of changes were made in the gas-handling system and the calorimeter that enhanced the sensitivity and accuracy by minimizing baseline perturbations after switching from the purge gas to a stream containing the adsorbate.

An investigation of the acidity and basicity of 20 metal oxides was conducted by Gervasini and Auroux (66) with microcalorimetry. Alumina was among the group of oxides showing amphoteric character by adsorbing both ammonia and carbon dioxide. Cardona-Martinez and Dumesic (67,68) included alumina in their study of the differential heat of pyridine adsorption. They found three regions of nearly constant heats of adsorption with increasing amine coverage; this implies that there are three sets of acid sites of different strength. It is surprising that more use has not been made of calorimetry in the characterization of naphtha reforming catalysts.

## C. Chlorided Catalysts

The regulation of the Cl concentration on the surface of the bifunctional catalyst is a key factor in optimization of the reforming process. An optimum chloride concentration allows enhancement of the acidic function of the catalyst (69) and an improvement in the self-regeneration capability of the "coked" Pt/ $\gamma$-Al$_2$O$_3$ system, presumably because of more effective H$_2$ spillover (70,71). In addition, oxidative treatment (400 $\leq T \leq$ 550°C) in the presence of chlorine compounds (CCl$_4$, CHCl$_3$, HCl, etc.) is a procedure that is claimed to provide redispersion of sintered Pt/$\gamma$Al$_2$O$_3$ reforming catalysts (72). Several studies have therefore been devoted to rationalizing the factors controlling the retention and the leaching of chloride either during activation treatments or under reaction conditions (73–77). The effects of chloride adsorption on the physicochemical properties and reactivity of $\gamma$-Al$_2$O$_3$ surfaces have also received attention (76–79).

Arena et al. (79) followed the loss of chloride upon heating in dry or wet conditions. They found that the initial rate of chloride loss is proportional to the initial chloride content of the sample and loss could be expressed by a first-order expression involving the actual surface Cl$^-$ concentration. The activation energy for loss was 6.2 kcal/mol, which is similar to the value of 6.0 ± 0.1 kcal/mol reported by Bishara et al. (80) for chloride loss from a Pt–Cl–Al$_2$O$_3$ during air calcination. It was found that steam increased the rate of chloride removal. The activation energy obtained for wet and dry conditions is the same; the increased rate results from a considerably higher preexpotential factor for the wet conditions. Ayame et al. (81) characterized the form of dehydrated alumina halogenated with chlorine at 773–1273K. They reported that the higher-temperature chlorine treatments resulted in the formation of adjacent strong Lewis acid sites, which were induced by the chloride ions bonding to aluminum cations. Treatment with HCl at 773 K produced materials with strong Brönsted acidity (82). It was shown by Garbowski and Primet (83) that aluminas chlorided by CCl$_4$ or HCl at 573 K strongly adsorbed benzene to form coke precursors and hexadienal cation, respectively. Arena et al. (79) compared the change in the zero-point charge with chloride addition to the catalytic activity for the isomerization of cyclohexene to methylcyclohexene and found a linear relationship between the two.

## D. Fluorided Alumina

Several authors have claimed that strong Brönsted acid sites are formed by fluorination of alumina (84–89). The extent of acidity, determined from ammonia adsorption, was shown to increase and then decrease as the concentration of fluoride increased (Figure 5) (90); infrared was used to follow the reaction of sterically hindered nitrogen bases (91,92) with fluorided aluminas. A model of the surface modifications has been proposed (55).

Hirschler (93) obtained a measure of the acidity of an Alcoa F-10 alumina as well as a sample of the fluorided alumina (both calcined at 500°C). The acidity titrations showed that the treatment with HF greatly increased the acid strength using $H_R$ indicators. Webb (84), on the basis of the effect of temperature of ammonia chemisorption, concluded that HF treatment of alumina did not increase the number of acid sites but did considerably increase their strength. Weber (94) showed that fluorided alumina impregnated with platinum salts has a high hydrocracking activity, whereas on a chlorided alumina the hydrocracking activity is low. Holm and Clark (95,96) reported that fluoriding alumina considerably increased its activity for $n$-octane cracking, $o$-xylene iso-

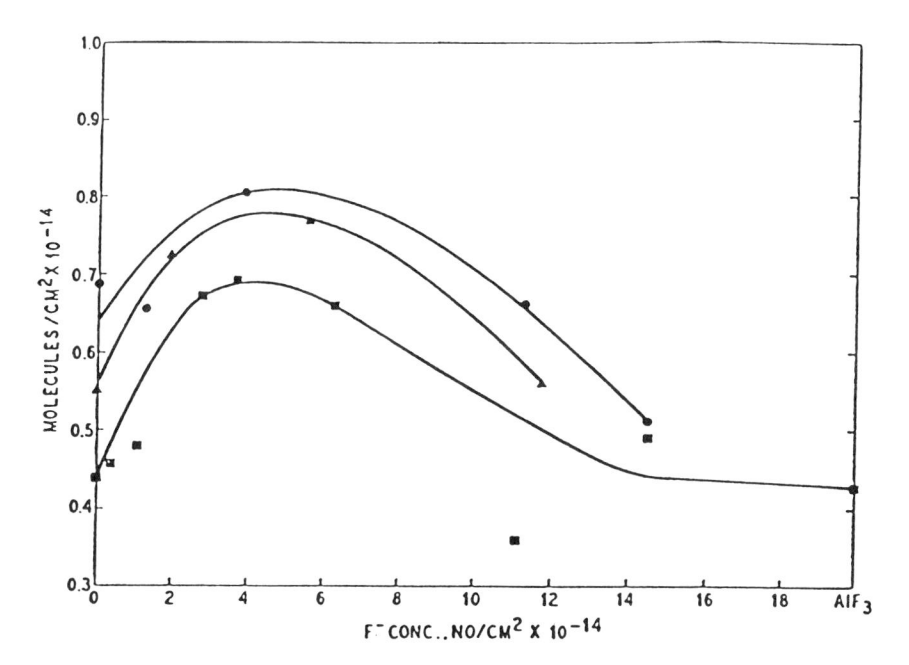

**Figure 5** Effect of fluoride concentration on the adsorption of ammonia. (●) Holm and Clark, 400°C, 10 torr, ammonium bifluoride impregnation; (▲) Holm and Clark, 400°C, 10 torr, HF impregnation; (■) Gerberich, Lutinski, and Hall, 500°C, 100 torr, HF impregnation. (From Ref. 90.)

merization, and propylene polymerization. The heats of ammonia adsorption and the observations with arylmethanol indicators could be reconciled if it was assumed that the acid centers on fluorided alumina are of a different type (e.g., protonic) from those on alumina. The fact that fluorided alumina converted 1,1-diphenylethylene to the carbenium ion and that it showed no increase in acid strength with the Hammett indicators supported this assumption, from which it followed that protonic acids rather than Lewis acids convert arylmethanols to their corresponding cations.

Typical IR spectra of pyridine adsorbed on samples with increasing F show bands for Lewis sites (1455, 1496, 1580, and 1620 cm$^{-1}$) (91). Brönsted acid sites (1545, 1562, and 1640 cm$^{-1}$) are not detected on alumina but are present in the fluorided alumina samples. The number of Brönsted sites was found to increase to about 0.6 sites/nm$^2$ as the F content increased; however, this site density was attained only after 10–20% F had been added to the alumina. Similar qualitative results were obtained for the Brönsted sites that react with 2,6-dimethylpyridine; however, the maximum number of sites is obtained at a much lower F content. Corma et al. (91) also determined the acid strength distribution by titration with butylamine and found that the maxima in total acidititity (pK $\leqslant$ 6.8) and in the strong acid sites (pK $\leqslant$ 1.5) are found for fluorine content between 2 and 4%. Furthermore, only a small fraction of the Brönsted sites created by fluorination exhibit strong acidity.

In the past, to resolve some of the contradictory data, researchers have studied fluorided aluminas using a variety of instrumental techniques (88, 97–101). Some of the inconsistencies can be attributed to variations in preparation methods; however, most of the inconsistencies involve data from X-ray diffraction (XRD). It is very possible that some preparations lead to well-dispersed phases that cannot be detected using XRD. In particular, $^{27}$Al nuclear magnetic resonance data reveal the presence of phases that are not detected by XRD (102). DeCanio et al. (102) carried out a multitechnique investigation of a series of F/Al$_2$O$_3$ samples in which the fluorine loading was varied from 2.0 to 20.0 wt %. The results showed that at low levels, fluoride served to block Lewis acid sites, but at higher levels its predominant role was to increase the Brönsted acidity of the alumina surface, and that fluoride strengthens the remaining Lewis acid sites.

## III. PLATINUM-ALUMINA CATALYSTS

For both platinum and the bimetallic systems there have been long-term conflicting views of the chemical state of the metal under typical reforming conditions. The early view that platinum was present as the metal was shattered by the report by McHenry et al. (103) that a significant fraction of the platinum in a reforming catalyst could be extracted with dilute HF or with acetylacetone. Acetylacetone extraction had been developed as a method for recovering the metallic platinum from reforming catalysts so that XRD could be utilized to obtain an average crystal size in the absence of the interfering peaks of the alumina support (104). Results (105-107) supporting the view of soluble platinum appeared following this observation. Others questioned the presence of soluble platinum, indicating that it was present only when the reduced catalyst had been exposed to air following extraction (108-111). The latter view is probably most widely accepted today. However, absence of platinum ions in the reduced reforming catalyst is not universally accepted (112,113).

The experimental observation that the best metallic function would be one of the expensive Group VIII metals dictated that the metallic function of the naphtha reforming catalyst be optimized. The cost of the noble metal required that the dispersion of the metal, platinum, be maximized. In the following we consider first the definition and a measurement of dispersion of a single function and then of the function dispersed on a high-surface-area support.

## A. Dispersion

Dispersion is easy to define but almost impossible to measure precisely. The dispersion may be defined as the number of atoms in the exposed surface ($N_S$) divided by the total number of atoms present in the catalyst ($N_{total}$). Although $N_{total}$ can be determined precisely, $N_S$ depends on the definition of surface as well as the experimental approach used for the measurement and the model used in the calculations. For example, some of the crystal faces of a metal are more densely packed than others; the openness of the outermost layer of metal atoms determines the extent to which the second layer will be exposed to the gaseous phase.

Except for the chemisorption techniques described below, dispersion is obtained from calculations based on the particle size. Thus, to make comparisons of the results from various techniques one needs to consider the relationship between dispersion and the particle size. The most common shape of particle is a sphere, or a hemisphere, especially for platinum. However, as the dispersion approaches unity, two-dimensional plates ("rafts") may be encountered. Thus, the dispersion is related to the crystal size. The relationship is not simple. The supported particles seldom, if ever, have a uniform size (monodispersed) but have a distribution of particle sizes (polydispersed). The "average size" may depend on the experimental technique used to make the measurement. Lemaitre et al. (114) considered these problems in some detail, only a brief outline is given.

## B. X-ray Diffraction (XRD)

The XRD and line broadening (XLBA) techniques are based on the fact that the breadth of the x-ray reflections, apart from an instrumental contribution, is related to the dimensions of the crystals giving rise to the reflections. The metal reflection must be intense enough to give a signal measurable above the background of the support. This requirement is easily met in the case of the $Pt-SiO_2$ catalyst but not with the $Pt-Al_2O_3$ catalyst. The two most intense peaks for Pt metal fall at angles of $2\theta$ values where intense peaks from the crystalline alumina support mask the Pt peaks. Thus, x-ray line broadening has limited utility for the characterization of $Pt-Al_2O_3$ reforming catalysts. This was the reason that Adams et al. (115) utilized a $Pt-SiO_2$ catalyst for a comparison of the metal sizes obtained by microscopy, XLBA, and chemisorption techniques.

The crystallite size is calculated from x-ray line broadening (116). The more common approach is to measure the line width at half-maximum (LWHM). Instrumental line broadening is usually taken into account by measuring the LWHM for a sample with very large crystallites and using this to correct the experimental LWHM for the sample. The XLBA technique returns "crystallite" size rather than "particle" size. For small particles, such as Pt in the naphtha reforming catalyst, the two are the same. However, as the particles become larger, they may comprise two or more crystals; in this case the crystallite size can be considerably smaller than the particle size (117).

### C. Transmission Electron Microscopy (TEM)

Adams et al. (115) measured the distribution of Pt particle sizes for a particular $Pt-SiO_2$ catalyst and found the number average diameter for the size distribution to be 2.85 nm, the surface average diameter 3.05 nm, and the volume average diameter 3.15 nm. These authors considered the probable error in these average diameters to be about 10%; this was based on the variation of the values obtained among eight observations.

Rhodes et al. (118) prepared a series of $Pt-Al_2O_3$ catalysts by sintering a 46% dispersion sample to produce lower dispersions of 26 and 15%. From the transmission electron microscopy (TEM) size distribution, the dispersions of the samples were estimated. The results of the calculation are largely independent of the shape of the crystal used (119). Evaluating the dispersions, they obtained 16, 22, and 61%, in reasonable agreement with the ones measured by chemisorption. They attributed the poor agreement for the sample with the highest dispersion to a breakdown of the model for small particles. Thus, the model would not be applicable for naphtha reforming catalysts that are of most interest, the highly dispersed materials. The measurement of the particle diameter will be subject to more error as the particles become smaller. This results from the inability to measure particles below some size that depends on the resolution of the electron microscope used for the measurement.

Overlapping contrast from the support, especially one that is crystalline, may affect the ability to observe metal particles. White et al. (120) found this to be the case for a high-resolution electron micrograph of a $Pt-Al_2O_3$ reforming catalyst. The contrast from the support tended to obscure the metal particles, making them harder to detect. The effects on resolution and contrast from the support on the determination of the sizes and shapes of small metal particles have been considered in theoretical papers (121). Based on image calculations, a 1.2-nm cubo-octahedron could be undetected when viewed through an amorphous support of 1.9-nm thickness in a microscope having a 0.2-nm point resolution. Particles larger than about 1 nm in diameter can usually be readily detected by bright-field and dark-field microscopy (122). For smaller particles, high-angle annular dark-field imaging is useful because the electrons scattered at high angles are more sensitive to atomic number; hence, the sensitivity for Pt with respect to the support increases at high angles. This technique, coupled with digital image processing, produced images of particles containing as few as three Pt atoms on alumina support (122) and three-atom clusters of Os on γ-alumina (123–125). Datye and Smith (122) contend that the difficulty in detecting the presence of highly dispersed metallic species may be caused as much by the mobility of the species as by the problem of obtaining sufficient contrast.

## D. Chemisorption Techniques

The chemisorption of hydrogen by metals had been related to the number of exposed metal atoms of platinum and other metals (126,127). However, it was the work of Emmett and Brunauer that showed the utility of the technique for the characterization of more complex catalytic materials that comprise two or more components (128–130). Chemisorption measurements are easily applied to Pt-alumina reforming catalysts. Several gases, including CO, $H_2$, $O_2$, and NO, have been used for this purpose. Hydrogen adsorption isotherms on silica-supported platinum catalysts are typical, the adsorption corresponding to chemisorption is completed at pressures of 0.1 mm Hg or less, and the adsorption at saturation is greater at lower temperatures. The amount of adsorption at saturation depends on the evacuation temperature following reduction and cooling in hydrogen. Adams et al. (115) found that chemisorption volume increased with increasing temperature of evacuation up to 250°C, remained constant up to 800°C, and then decreased at 900°C, presumably due to sintering of the platinum and/or support.

Although there was much background data available, Adams et al. (115) appear to be among the first to have made a detailed comparison of the dispersion calculated from the results of several techniques using hydrogen adsorption at −78 or 0°C. To calculate the dispersion, it is first necessary to define an approach to obtain the volume of hydrogen or other chemisorbed gas that corresponds to complete coverage of the metal. One can utilize two approaches to obtain the volume of gas corresponding to a monolayer of coverage. First, one can extrapolate the adsorption isotherm to zero pressure (Figure 6) and

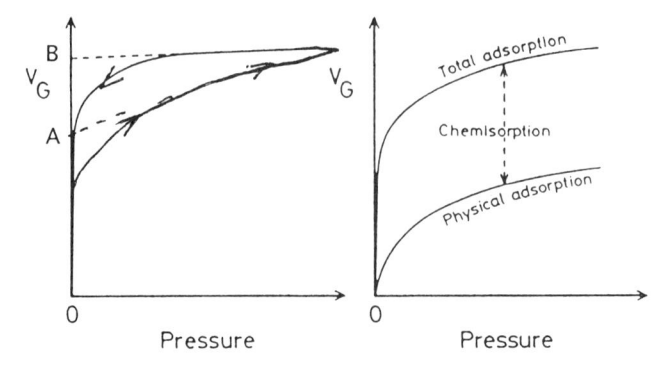

**Figure 6** Typical measurement of chemisorption. (Left) Extrapolation of Langmuir-type isotherm to zero pressure. (Right) Total adsorption at 90 K and physical adsorption at 90 K after evacuation at 195 K. The difference between the two gives the chemisorbed amount.

take this as the appropriate volume of gas. The second approach is to make a first measurement of the isotherm to obtain 4–10 data points up to pressures of about 40 cm Hg, then evacuate the sample at the adsorption temperature at a vacuum of $10^{-3}$ mmHg or better, and then measure a second adsorption isotherm. At any pressure the difference between the first and second isotherms should correspond to the amount of gas that is chemisorbed. For most Pt-Al$_2$O$_3$ catalysts, it is found that the two methods of obtaining the amount of hydrogen chemisorbed provide similar volumes.

To calculate the available platinum surface, one must know the area occupied by a chemisorbed hydrogen. The surface area for a platinum black sample calculated from the chemisorption data should be equal to the BET surface area calculated from the volume of nitrogen adsorbed. The area occupied by a hydrogen atom on platinum black is 1.12 nm$^2$. The particle size can then be calculated from the chemisorption data assuming the same density as bulk platinum.

Via et al. (131) have measured hydrogen adsorption at room temperature on a Pt-Al$_2$O$_3$ catalyst and on the support (Figure 7). The difference between curves A (adsorption on Pt-Al$_2$O$_3$) and B (adsorption on the alumina support) remains constant over the hydrogen pressures used for the measurement (5–15 cm Hg). However, extrapolation of the hydrogen adsorption on the support alone to zero pressure does not give the expected value of zero adsorption. For this particular catalyst, the difference between the two curves (A − B) returns a value of H/Pt = 0.9. In contrast, if the adsorption isotherm for the Pt-Al$_2$O$_3$ catalyst is extrapolated to zero hydrogen pressure (curve A), a value H/Pt = 1.2 is obtained. The "correct" procedure for obtaining a measure of H/M has

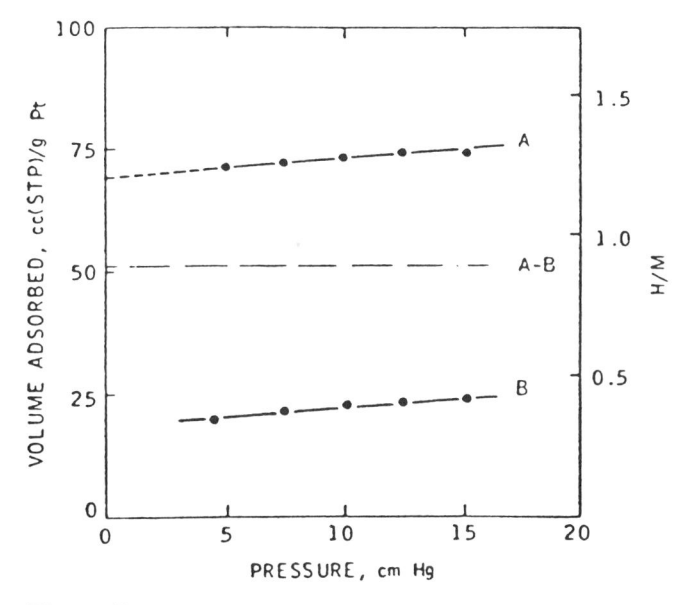

**Figure 7** Typical chemisorption isotherms at room temperature. The isotherms are for hydrogen chemisorption on the platinum on alumina catalyst. Isotherm A is the original isotherm, and B is a second isotherm determined after evacuation of the adsorption cell for 10 min subsequent to the completion of isotherm A. The difference isotherm A − B is obtained by substracting isotherm B from isotherm A. (From Ref. 131.)

been widely debated at ASTM Committee D.32 meetings and the ASTM Committee E.42 on surface analysis. The E.42 committee has approved surface analysis techniques that are now published in volume 3.06 of the *Annual Book of ASTM Standards*; several subcommittees are active in this area (132).

It is still not certain what H/Pt stoichiometry is correct. For most widely used platinum metal catalysts, an H/M stoichiometry of unity has been used and this assumption has been tested using XRD and TEM data (133–135). Surface science measurements also show that a maximum of one hydrogen atom per metal atom could be chemisorbed on the (111) faces of face-centered cubic (fcc) metal single crystals (136). It is generally assumed that metal particles larger than 1–2 nm consist for the most part of (111) faces, and the use of H/Pt = 1 seems reasonable. However, reports of stoichiometries greater than 1 date back over 30 years. For $Pt/Al_2O_3$ catalysts values of H/Pt = 1.2–2.5 have been obtained (137–143). Even higher values of near 3.0 have been obtained for supported Ir catalysts. McVicker et al. (144) found an upper limit of two adsorbed hydrogen atoms per Ir atom for $Ir/Al_2O_3$ if they based the calculation only on strongly adsorbed hydrogen; when the total adsorbed hydrogen was used for the calculation they obtained H/Ir values exceeding 2. Similarly,

Krishnamurthy et al. (145) found that a 0.48% Ir/Al$_2$O$_3$ catalyst adsorbed up to 2.72 hydrogen atoms per iridium atom, and about H/Ir = 0.28 was weakly adsorbed. A series of catalyst studies of Pt, Rh, and Ir metals supported on Al$_2$O$_3$, SiO$_2$, and TiO$_2$ have been made, and these show H/M values exceeding unity for both Rh and Pt catalysts (146-148). Values exceeding 2.0 were obtained for supported Ir catalysts (149,150).

A number of explanations have been given for obtaining values of H/M greater than 1. In most catalysts, reversibly adsorbed hydrogen should be used for the determination of the metal surface area (144,145,151-154). Others have explained that the high H/M ratio is due to hydrogen spillover (143,155-157). Some contend that the atoms located at the corners and/or edges of small metal particles may be responsible for adsorbing more than one hydrogen atom per surface metal (144,145,158). Another possible source of the high H/M ratio is that part of the hydrogen can be bonded to atoms under the outermost surface layer (159-161). Kip et al. (149) offered the explanation of adsorption beneath the metal surface or multiple adsorption on parts of the metal surface. In some instances, such as Pt/TiO$_2$, the high value of H/M may be due to partial reduction of the support itself.

## IV. TITRATION METHODS

A hydrogen-oxygen titration method was introduced by Benson and Boudart in 1965 (162). This technique was expected to reduce or eliminate errors introduced by hydrogen spillover and to provide an increase in the sensitivity over that of the adsorption of hydrogen. The titration should be the result of simple stoichiometries as shown by the following reactions:

Pt + ½H$_2$ = Pt−H (H chemisorption, HC)

Pt + ½O$_2$ = Pt−O (O chemisorption, OC)

Pt−O + 3/2H$_2$ = Pt−H + H$_2$O (H titration, HT)

2Pt−H + 3/2O$_2$ = Pt−O + H$_2$O (O titration, OT)

Menon (163) summarized the conflicting stoichiometries for HC:OC:HT as found in the literature. Prasad et al. (164) reported that much of this controversy was perhaps due to the fact that every research group used the very first hydrogen chemisorption on a fresh catalyst as the basis for all calculations, and if the surface was given an "annealing" treatment by a few H$_2$ − O$_2$ cycles at ambient temperature, it could behave normally in subsequent titrations. Furthermore, if the H$_2$-titer value was used as the basis for calculations after the H$_2$ − O$_2$ cycles, the stoichiometry was always found to be 1:1:3, independent of Pt crystallite size and of the pretreatment of the catalyst. Whether the freshly reduced catalyst or one that has been subjected to repeated H$_2$ − O$_2$ cycles is representative of the working reforming catalyst is not defined at this time. Isaacs and Petersen (165) also found that the dispersion from hydrogen chemisorption was greater than obtained for oxygen chemisorption or by the titration technique. The authors point out that the ratio (dispersion from hydrogen titration/dispersion from hydrogen chemisorption) of 0.82 is in very good agreement with the value of 0.81 obtained by Kobayashi et al. (166) and by Freel (167).

O'Rear et al. (168) addressed the stoichiometry of the titration technique using a clean surface platinum powder. They point out that several other authors have studied the chemisorption of oxygen on various types of platinum surfaces using ultrahigh vacuum techniques; these are reviewed by Gland (169). O'Rear et al. imply that the observed low values are due to inefficient clean-off of the gas used for the reduction or that the oxygen pressure was not sufficiently high to cause reconstruction of some Pt single crystal faces to a "complex" phase where high values of oxygen chemisorption are obtained. The authors (168) also compiled literature data to make comparisons of the platinum area average particle size calculated from hydrogen chemisorption and the number average particle size measured by TEM or the volume average crystallite size measured by XLBA. There was good agreement between the particle sizes measured by hydrogen chemisorption and the independent techniques (Figure 8) (117,137,170–172).

Measurements for a commercial reforming catalyst (CK-306; the same formulation but not the same sample) obtained in three laboratories are summarized in Table 1, and they show similar values even when different experimental procedures are utilized (165,173).

### A. Small Angle X-ray Scattering (SAXS)

X-rays experience scattering at the interface of materials with sufficient differences in density. An alumina support provides a two-phase system—the alumina particles and the porosity represented by the void space. When the alumina contains platinum there are now three interfaces. SAXS can be used to obtain interphase surface areas of a system such as alumina-supported platinum catalysts. Recent developments in the experimental techniques and the theory for interpreting the data promise to make this a useful technique for catalyst characterization. Summaries of the general scattering principles for such systems (175–180) and the experimental techniques (352) have been published.

One approach to obtaining information about the platinum particles is to reduce the three-phase system to a two-phase system by filling the pore structure with a liquid of the same electron density as the alumina support. Compounds such as $CH_2I_2$ are suitable for this purpose. It is difficult, however, to fill the pores completely because of wetting problems and the inaccessibility of some pores. Whyte et al. (181) utilized the masking technique to investigate the

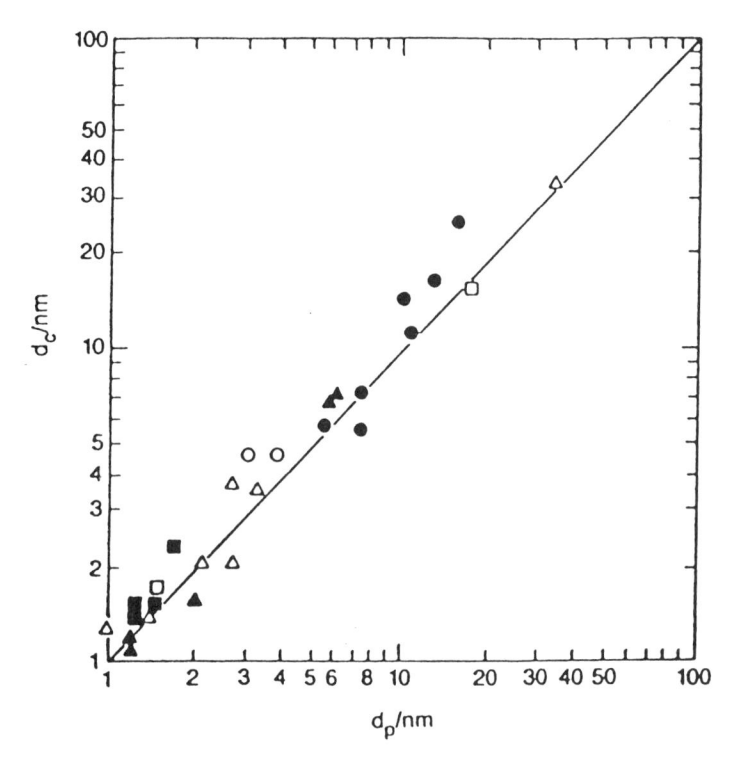

**Figure 8** Supported platinum particle size as measured by hydrogen chemisorption ($d^c$) assuming a stoichiometric coefficient y = 1 and measured by transmission electron microscopy, x-ray line broadening, and x-ray small-angle scattering ($d_p$). Symbols denote the sources of the data. (From Ref. 168.)

**Table 1** Comparisons of Pt Dispersion Values from Different Methods

| Method[a] | Gas Adsorbed mL STP/g | Dispersion, % |
|---|---|---|
| A | 0.282 | 82 |
| B | — | 73 |
| C | 0.278 | 81 |
| D | 0.271 | 79 |

[a] A, Volumetric chemisorption; B, hydrogen titration of Pt–O surface volumetrically; C, hydrogen titration of Pt–O surface gas chromatographically; D, oxygen titration of Pt–H surface gas chromatographically.
*Source*: From Ref. 174.

Pt−Al₂O₃ system. Most results were in reasonable agreement with their chemisorption data. Cocco et al. (182) made a detailed study of supported Pd and Pt materials. They excluded scattering from inaccessible voids by subtracting the scattering of the masked, metal-free support and used experimental techniques to make intensity measurements on an absolute basis. Calculated size distribution functions were found to be bimodal. Somorjai et al. (183) used extremely high pressure to reduce the pores present in the alumina to such a small size that they would not contribute significantly to the scattering. It is necessary to assume that the compaction at these very high pressures does not modify the platinum particles.

Brumberger and co-workers have made several experimental studies of catalysts using the SAXS technique (176,184–188). These workers used the support-subtraction method, assuming that addition of metal does not change the support morphology to an appreciable extent. This technique requires that both support and metal-support receive the same preparative and pretreatment procedures and that both respond in exactly the same way to these treatments. They generally obtained good agreement between BET and SAXS surface areas (186). Brumberger et al. (188) utilized SAXS to follow the sintering of Pt−Al₂O₃ catalysts with high Pt loadings (up to ≈11 wt %) in air. The data indicate an initial redispersion at 400–500°C and that the surface then decreases as the temperature is raised further, in agreement with chemisorption data. The SAXS data show that the sintering response of the catalyst to temperature changes is rapid. The authors emphasize that the SAXS method is capable of following changes in surface areas accurately, nondestructively, and continuously in situ, for a wide variety of temperature, pressure, and ambient atmosphere conditions.

Small-angle neutron scattering (SANS) can also be utilized. Hall and Williams (180) have made a comparison of the surface area and porosity of a range of silicas, aluminas, and carbons obtained from SANS and BET surface areas. They reported that the SANS areas of the nonporous and mesoporous aluminas and silicas were of the correct order of magnitude compared with the BET areas. In summary, SAXS and SANS measurements still have a long way to go before they become a reliable approach for naphtha reforming catalyst characterization. However, rapid advances are being made in both the experimental and theoretical approaches, so the methods must be considered to show promise.

Lemaitre et al. (114) contrasted the results obtained using the methods discussed so far and showed schematically how the dispersion is approached through the various experimental methods. They concluded that the dispersion values are usually most easily compared by expressing them all in terms of an average particle size.

## B. X-ray Absorption Fine Structure Analysis (XAFS)

XAFS is an experimental technique that is atom specific and can give structural information about supported metal catalysts. XAFS gives information on the atomic level but very limited and indirect information on the morphology of the catalyst particles (e.g., size, shape, crystalline imperfections). XAFS is, however, unique in providing a method for determining coordination numbers, interatomic distances and from this the "average" nearest neighbor atoms, and the vibrational motions of the metal atoms. Although the phenomenon was first observed over 60 years ago (189), it is only recently that XAFS has become a useful analytical tool. One reason is that the use of synchroton radiation increased the available flux by $10^5$–$10^6$, allowing faster, more accurate XAFS experiments. [For Experimental details see (190,191).]

### 1. Extended XAFS (EXAFS)

EXAFS spectra generally refer to the region 40–1000 eV above the absorption edge. Excellent reviews on EXAFS in general (192,193) and its application in catalysis (194–201) are available. The pre-edge region contains valuable bonding information, such as the energetics of virtual orbitals, the electronic configuration, and the site symmetry. The edge position also contains information about the charge of the absorber. In between the pre-edge and the EXAFS regions is the x-ray absorption near-edge structure (XANES).

Transmission is just one of several modes of EXAFS measurements. The fluorescence technique involves the measurement of the fluorescence radiation (over some solid angle) at a right angle to the incident beam. Other more specialized methods include (1) surface EXAFS (SEXAFS) studies, which involve measurements of either the Auger electrons or the inelastically scattered electrons (partial or total electron yield) produced during the relaxation of an atom following photoionization, and (2) electron energy loss (inelastic electron scattering) spectroscopy (EELS). The latter methods, which require high vacuum, are useful for light-atom EXAFS with edge energies up to a few keV. The accuracy of the structural parameters depends on many factors. Typical accuracies for the determination of parameters obtained in an unknown system have been quoted as 1% for interatomic distances, 15% for the coordination numbers, and 20% for the thermal mean-square displacements (202).

In the pioneering work in the application of EXAFS, Sinfelt, Lytle, and their co-workers (131) compared EXAFS and hydrogen chemisorption data for Os, Ir, and Pt catalysts. The hydrogen chemisorption data indicated high dispersion with H/M near 1 (Table 2). The data for the interatomic distances are the same as for the bulk metal, within experimental error, except possibly for Pt–$Al_2O_3$. This suggests that Pt interacts more strongly with alumina as the coordination number (7.2) for Pt–$Al_2O_3$ was the lowest of the three catalysts.

**Table 2** Comparison of Chemisorption and XAFS Data

| Catalyst[a] | H/M[b] | CO/M[b] | N[c] | R,A[d] |
|---|---|---|---|---|
| Os – SiO$_2$ | 1.2 | 1.0 | 8.3 | 2.702 (2.705) |
| Ir – SiO$_2$ | 1.5 | 0.8 | 9.9 | 2.712 (2.714) |
| Ir – Al$_2$O$_3$ | 1.3 | 0.9 | 9.9 | 2.704 (2.714) |
| Pt – SiO$_2$ | 0.7 | | 8.0 | 2.774 (2.775) |
| Pt – Al$_2$O$_3$ | 0.9 | 0.9 | 7.2 | 2.758 (2.775) |

[a] Catalysts contained 1 wt % metal.
[b] The number of H or CO chemisorbed per metal atom.
[c] Average coordination number (nearest metal atom neighbors).
[d] Interatomic distance (nearest metal atom neighbor); data in parentheses are for bulk metal.
*Source*: From Ref. 131.

The metal dispersion, as represented by the coordination number, is one of the more important EXAFS parameters for catalyst characterization. Although the experimental value of 7.2 is lower than for the bulk metal (12), it still appears high compared with the H/M value of 0.9. Zhao and Montano (203) point out that the small coordination numbers obtained by Sinfelt and coworkers are consistent with very small clusters (dimers, trimers, etc.) but that the interatomic distances correspond to those of the bulk metal. Based on detailed analysis, Zhao (204) contended that the metal particles in the 1% Pt catalyst contain more than 50 atoms and that the real first shell coordination number could be as large as 10.

Lytle et al. (205,206) reported additional data for in situ studies of 1% Pt on Cab-O-Sil in the presence of He, H$_2$, or benzene. In this instance the hydrogen-reduced sample has Pt present as a mixture of 10–15 Å disks and polyhedra. The authors included the phase shift in the transformation of the Pt EXAFS data that was shown by Marques et al. (207) to be very useful in sharpening and simplifying the peaks of the Fourier transform of elements, such as Pt, which have very nonlinear phase shifts. The latter results showed a clear demonstration of Pt – O bonds to the support at low temperatures. At temperatures above 600 K the Pt – O bonds break and the raftlike clusters curl up to be more spherelike. Concurrently with the bond breaking, electrons flow to the Pt $d$ band to provide a $d$ electron surplus in the Pt clusters relative to bulk Pt.

*a. In Situ EXAFS.* Although in situ high temperature (208) and high temperature (209) cells for EXAFS were known, Guyot-Sionnest et al. (201) claim the first in situ cell used to examine the activation of a Pt-alumina reforming catalyst. Dexpert (210) reported that on heating a catalyst prepared using chloroplatinic acid from room temperature to 200°C in air, there was destabilization of the Pt – Cl bond with formation of a Pt – O bond. With heating at

200°C in hydrogen the complex rapidly decomposed with a decrease in the Cl−Pt nearest-neighbor bonding and an increase in Pt−Pt nearest-neighbor bonding. The value of the coordination number of Pt−Pt bonds was consistent with a raft structure for Pt.

Guyot-Sionnest et al. (211) found that a sample yielded a well-formed oxide species after oxidation, with three coordination shells of the oxide visible (Figure 9). Following reduction at 460°C, Pt−Pt bonds having an average interactive distance of 2.67–2.68 Å became evident. This is a much shorter bond than in bulk Pt (2.75 Å); it was believed that the low Pt−Pt coordination numbers (4–5) could cause each Pt atom to share more electrons with neighbors, thereby making a shorter bond. Since the coordination of Pt was lower in the chlorided catalyst, it was concluded that Cl aids in maintaining Pt particle dispersion. Following reduction at 460°C, the pressure was increased to 5 atm and *n*-heptane was passed over the catalyst for 4 h. A peak at the appropriate Pt−C bond distance was observed. The authors conclude that, since the number of Pt−C bonds detected by EXAFS remained unchanged while dehydrocyclization activity decreased, deactivation of the catalyst was not due to changes in the number of Pt−C bonds.

## 2. XANES

At an absorption edge, the x-ray absorption coefficient increases abruptly with increasing energy. In an absorption spectrum for a given element, absorption edges are observed at certain energies characteristic of the element. The abrupt rise in absorption occurs when the energy of the x-ray photons is equal to that

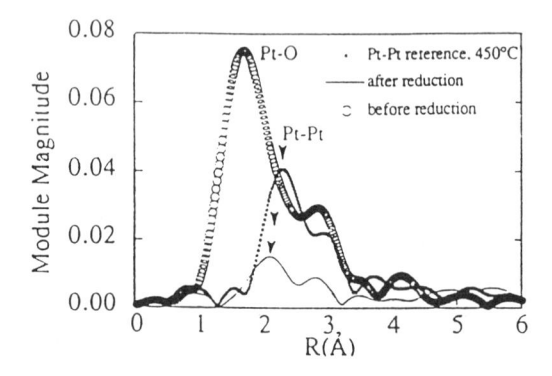

**Figure 9** Evolution of the Fourier transform modules for chlorinated 1.0 wt% Pt/Al$_2$O$_3$ before reduction (○), under flowing H$_2$ ($T = 25°C$), and after reduction (—) ($T = 260°C$), p(H$_2$ total) = 1 atm. Disappearance of Pt−O coordination and formation of Pt−Pt coordination is shown as temperature is raised to 450°C ( · ). (From Ref. 211.)

required to excite electrons from an inner level of the absorbing atom to the first empty electronic states. Edges are identified by the letters K, L, M, etc. to indicate the particular electronic shell from which the electrons are excited by the x-ray photons. By comparing with measurements on well-characterized systems, one can use $L_{II,III}$ edge studies to determine $d$-band occupancy in transition metal compounds and alloys (212).

Lytle (213) reported results for the $L_{III}$ absorption edges for supported catalysts. Meitzner et al. (199,214) more recently presented similar data for metals more frequently associated with naphtha reforming catalysts. The $L_{III}$ absorption edges of platinum, present in a Pt/Al$_2$O$_3$ catalyst with a dispersion approaching unity and with a metal dispersion of 0.2 corresponding to large crystallites, are shown in Figure 10. A careful inspection of the curves shows that the resonance may be slightly more intense for the small metal clusters but the differences are near the experimental uncertainty. Similar results were obtained for supported Ir and Os catalysts. The earlier reports (215,216) showing larger effects attributed to metal dispersion were apparently confounded by sample thickness effects. Thus, it appears that the metal atom in a 10-Å crystal (a size where the ratio of surface to total atoms approaches unity) of Pt, Os, or Ir, supported on either alumina or silica, exhibits electronic properties not very different from those of an atom present in a metal crystal representative of the bulk metal. Similar electronic effects are apparently experienced in metal crystals that are 10 Å and larger; likewise, the extent of electronic interaction of the metal clusters with either alumina or silica support is minimal. This conclusion appears to conflict with the EXAFS data described above (211).

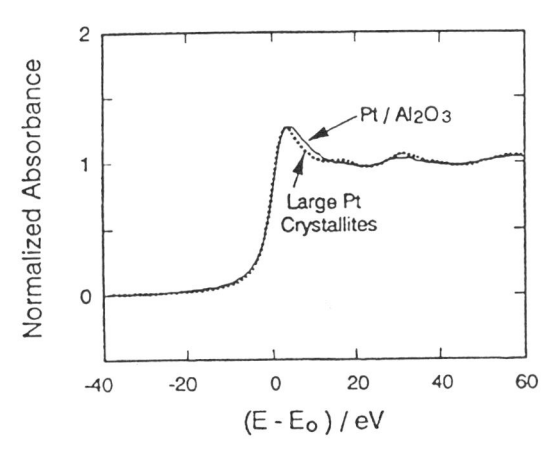

**Figure 10** Comparison of $L_{III}$ absorption edge for the platinum clusters in a Pt/Al$_2$O$_3$ catalyst (metal dispersion = 1.0) with that for large platinum crystallites (dispersion = 0.2). (From Ref. 214.)

The chemisorption of a monolayer of oxygen on highly dispersed clusters of Pt, Ir, or Os results in a substantial increase of the intensity of the $L_{III}$ edge. Data for an $Ir/Al_2O_3$ catalyst containing 1 wt % Ir, with clusters of Ir on the order of 10 Å, indicate that the number of unoccupied $d$ states of the metal increases as a result of interaction with the chemisorbed oxygen. While the metal becomes more electron deficient, chemisorption of oxygen at room temperature does not lead to a bulk oxide.

## C. Nuclear Magnetic Resonance (NMR)

Two types of NMR measurements are possible: a study of the nuclei of the catalyst itself and a study of nuclei of molecules adsorbed on the surface of the catalyst. The $^{195}Pt$ nuclei have sufficient natural isotopic abundance, a reasonably strong gyromagnetic ratio, and a spin of $I = \frac{1}{2}$ which eliminates the complexity due to electric quadrupole effects. Furthermore, $^{195}Pt$ has one of the largest Knight shifts of any metal ($-3.37\%$) and therefore offers the possibility of resolving the NMR peak of surface layers of atoms from that of the bulk.

Yu et al. (217) obtained $^{195}Pt$ NMR spectra for small unsupported Pt particles that corresponded to the bulk metal. Rhodes et al. (118), obtained spectra for a Pt supported on an alumina. The dispersion of these catalysts was measured by chemisorption and by TEM. The line shapes for six samples scaled to the same area are shown in Figure 11. They are broad and extend the full range of Knight shifts from that of nonmetallic compounds to that of bulk Pt metal. In going from samples of larger particles to those of progressively smaller particles, the intensity of the line shifts from the metallic end of the line shape to the nonmetallic end. The bulk peak at 1.13 kG/MHz is very prominent in the line shapes of samples with the larger particles (Pt-4-R, Pt-11-R, and Pt-15-R). The peak at 1.089 kG/MHz was taken to correspond to resonance of atoms present in the surface. The authors calculated a dispersion based on the relative area of the total line shape and the one due to surface atoms. These dispersions are presented together with those obtained from TEM and hydrogen chemisorption measurements in Table 3. The agreement among the data shown in Table 3 was taken as confirmation of the hypothesis that the low-field NMR peak measures surface atoms.

de Ménorval and Fraissard (218) showed that the NMR chemical shift of hydrogen adsorbed on $Pt-Al_2O_3$ varied with surface coverage and Pt particle size. The spectrum of hydrogen adsorbed on $Pt-Al_2O_3$ consisted of only one line whatever the coverage. The chemical shift δ was constant when the diameter was greater than 70 Å and then decreased when the hydrogen coverage increased beyond 0.5. Rouabah et al. (219) showed that for the $Pt-SiO_2$ EUROPT-1 catalyst, the chemical shift depended on the coverage, the size of the metal particles, and the temperature. From the variation of the chemical shift with the number of adsorbed hydrogen atoms, it was possible to determine

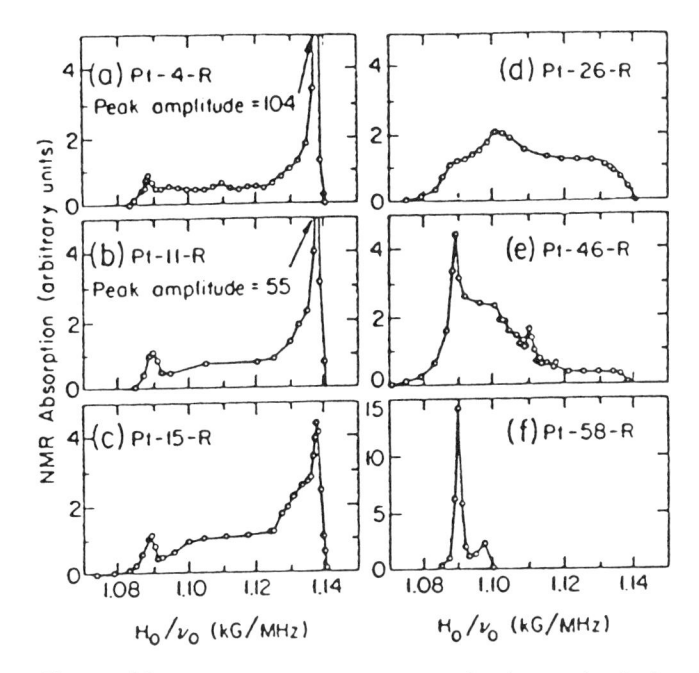

**Figure 11** NMR absorption line shapes for six samples (a–f) at 77 K and $v_o = 74$ MHz. (From Ref. 118.)

the metal dispersion. It is claimed that the NMR technique does not require the stoichiometry of the chemisorption on the metal to be known. The average particle size determined by $^1H$ NMR was smaller than the dispersion based on chemisorption or electron microscopy.

$^{129}Xe$ NMR has been used to estimate the average number of Pt atoms per cluster for samples Pt contained in an NaY zeolite (Pt/NaY) (221). Thus, an average of four to eight Pt atoms per cluster was estimated when the progres-

**Table 3** Dispersion (in %) of Pt Catalyst Samples Measured by Three Methods

| Sample | Chemisorption | Microscopy | NMR |
|---|---|---|---|
| Pt−4−R | 4 | — | 5 |
| Pt−11−R | 11 | — | 8 |
| Pt−15−R | 15 | 16 | 10 |
| Pt−26−R | 26 | 22 | 19 |
| Pt−46−R | 46 | 61 | 40 |
| Pt−58−R | 58 | — | 79 |

*Source*: From Ref. 118.

sive chemisorption of $H_2$ at room temperature in a bed of Pt/NaY powder was followed by Xe NMR. These results agreed with an average value of six Pt atoms per cluster which had been inferred earlier from the rapid exchange with $D_2$ of OH groups in a Pt/CaY zeolite as followed by infrared spectroscopy (222).

The determination of the number of Pt atoms per cluster attracted attention, as it was the first time the number of atoms in a supported metallic cluster less than $\approx$ 1 nm in size had been reported (222). However, the study of similar or identical samples of Pt/Y zeolites by TEM (223), SAXS (172), wide-angle X-ray scattering (WAXS) (224), and EXAFS (225) indicates Pt clusters of about 1 nm in size in the supercages of the Y-zeolite (1.3 nm in diameter), with about 16–40 Pt atoms per cluster (118). In particular, a combined investigation of Pt/Y samples by TEM, SAXS, WAXS, and EXAFS suggested that the Pt clusters were substantially larger than those probed by IR and Xe NMR. Detailed analysis of the experiments has reconciled both types of observations to the higher value (219,221,228–233).

Boudart et al. (226) utilized Pt supported on $SiO_2$, $\gamma$-$Al_2O_3$, or Y-zeolite for their NMR studies. The percent metal exposed for each sample was measured by titration of prechemisorbed O by $H_2$ (162), by irreversible H chemisorption, and by irreversible O chemisorption at room temperature in a standard volumetric system. The average size for the Pt clusters was obtained from the percent metal exposed, assuming a spherical shape and an average Pt surface number density of $1.10 \times 10^{19}$ m$^{-2}$ (226).

The Xe NMR spectra was then taken at various values of nominal surface coverage, H/Pt$_S$. All spectra show two peaks: one with $\delta \approx 0.5$ ppm, assigned to Xe in the gas phase (231), and another single peak with $\delta$ between 85 and 285 ppm, corresponding to the average interaction of adsorbed Xe with the Pt surface and the support (229). Boudart et al. (226) reported a puzzling observation in the discrepancy between the value of the chemical shift $\delta$ measured on prereduced Pt samples without any chemisorbed hydrogen. With all details of the reduction and evacuation procedures being the same (226,229) straight lines for $\delta$ versus H coverage extrapolate to the value measured for zero H coverage for all samples except Pt/$\gamma$-$Al_2O_3$. For this sample, the measured value of $\delta$ on the sample without H is considerably lower than that extrapolated from the $\delta$ versus H coverage line. For as little as 10% coverage by H, the measured value of $\delta$ is exactly on the $\delta$ versus H coverage line.

Boudart et al. (226) rationalized the anomaly by the following speculation. In vacuo, the shape of Pt clusters on $SiO_2$ or in Y-zeolite is approximately spherical, and the interface between metal and support is minimal. But with a $\gamma$-$Al_2O_3$ support, clusters of platinum about 1 or 2 nm in size probably have the shape of a pillbox or raft in vacuo as a result of the high interfacial energy. Indeed, high-resolution TEM reveals epitaxial growth of 2-nm clusters of Pd on $\gamma$-$Al_2O_3$ (232), and with comparable lattice parameters, both metals should behave similarly on $\gamma$-Al$_2$O$_3$. In the raftlike shape, xenon interacts with the free Pt surface but senses the surface of the cluster in contact with the Al$_2$O$_3$ platelets. At this contact, there are weak bonds between Pt and oxygen ions of the Al$_2$O$_3$. Hence the chemical shift of Xe is smaller than that corresponding to bare Pt, just as $\delta$ of Xe decreases as oxygen is added to bare Pt clusters (229). When hydrogen is introduced to the sample, the weak bonding between the metal and the support disappears as a result of chemisorption of hydrogen.

### D. Temperature-Programmed Desorption or Reduction (TPD or TPR)

TPR is one of the chemical methods for catalyst characterization; however, it does not provide a direct measure of chemical structure or chemical state. It provides a quantitative measure of the extent of reduction but does not provide direct information about what is being reduced. An analogous technique is temperature-programmed desorption (TPD). TPR suffers from a lack of knowledge of the species undergoing reduction and whether the weight loss is due to reduction or to desorption. It is possible to eliminate some of the uncertainty about the nature of the weight loss by monitoring the exit gas stream with a mass spectrometer. The TPR provides a "fingerprint" of the reducibility of the platinum group metal catalysts (234). For example, the reduction of Pt-, Ir-, and Pt–Ir–Al$_2$O$_3$ catalysts has been studied (235). The reduction of platinum in the Pt–Al$_2$O$_3$ catalyst occurs over a broad temperature range that peaks at about 400°C, whereas iridium oxide is reduced over a narrow but higher temperature range that is centered at about 550°C. In contrast to the two single metals, the reduction of Pt–Ir–Al$_2$O$_3$ is not a composite of the two curves representing the reduction of each metal on the support; rather, the reduction is effected at a lower temperature. These data are consistent with the formation of a Pt–Ir alloy on the support; however, the data do not provide direct evidence for the formation of an alloy.

### E. X-ray or Ultraviolet Photoelectron Spectroscopy (XPS or UPS); Electron Spectroscopy for Chemical Analysis (ESCA)

Photoelectron spectroscopy is based on the photoelectron effect, in which ionization occurs and an electron is expelled from the sample with a kinetic energy that depends on the energy of the incident photon, the element that is ionized, and the valence of that element in the sample. Depending on the energy of the incident photon, emission of the photoelectrons will be from the valence band only (UPS) or from both the valence and core levels (XPS). XPS provides a measure of the concentration and chemical valence of only the elements in the surface of the sample. However, knowledge of the mean free paths of electrons in solids is not completely defined even today. A schematic of the escape probabilities for an electron through surface layers to the vacuum of the instrument is illustrated in Figure 12.

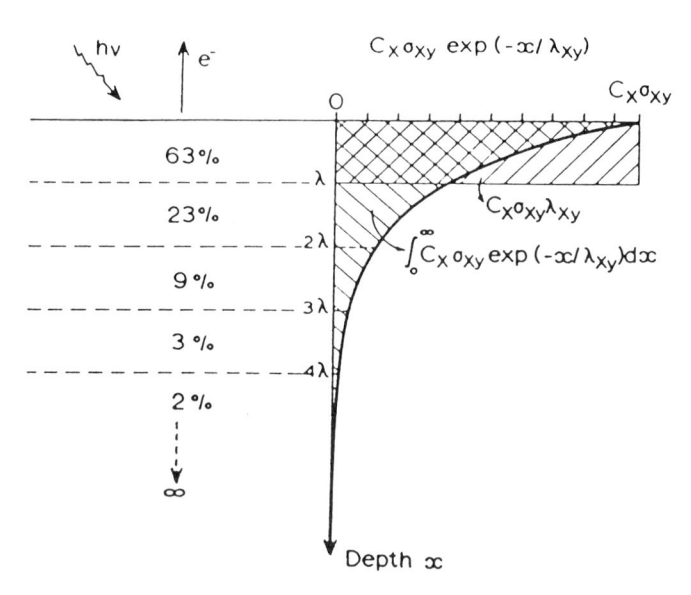

**Figure 12** Contribution of successive layers of thickness $\lambda$ to the total XPS intensity for $\theta = 90°$. (From Ref. 237.)

For the characterization of naphtha reforming catalysts, it is usually necessary to pretreat the sample before analysis by UPS or XPS. This is commonly accomplished in a sample cell, attached to the chamber used for the measurement. Following the pretreatment, the sample is transferred without exposure to the atmosphere to the high-vacuum chamber where the actual measurements are made. In nearly all of the instruments it is necessary to form the sample into a pellet so that a reasonably smooth surface is provided to the photon beam and to maintain sample integrity during the pretreatment operations.

Semiconductor or insulator samples will undergo charging because, in general, ions are formed more rapidly than the charge can be dissipated through conduction to the instrument. Thus, the measured binding energy must be corrected to account for the charging effect if the data are to have value for defining the valence of an atom. The common way to correct for the charging effect is to reference it to the C $1s$ line from the contamination overlayer. The C $1s$ line is normally taken to be 284.6 eV, and all elements are corrected by the same amount that is required to correct that of the C $1s$ line to 284.6 (237). Although there is some question of the validity of the implied assumption that the adventitious carbon layer is in good contact with the sample, several authors (238–240) showed that it is sufficiently reliable ± 0.1 to 0.2 eV) in most cases. Another option is to use one of the elements of the sample, usually the support, as the reference. For example, Ogilvie and Wolberg (241) showed that the standard deviation for the Al $2p$ binding energy in a series of alumina catalysts were reduced from 0.49 to 0.14 eV when $O$ $1s$ from the support is chosen as the reference rather than C $1s$. Round-robin studies have emphasized the difficulties in arriving at accurate binding energy values for catalysts (242).

## F. Auger Electron Spectroscopy (AES)

Auger electrons are ejected when an atom containing a core-level vacancy (e.g., as created in the XPS measurements when an x-ray photon interacted to eject an electron) de-excites by emitting an electron from an upper orbital. The energy needed to eject this electron comes from the nearly simultaneous decay of a third electron from another upper level into the inner core vacancy. Because XPS and AES are similar, spectrometers are frequently built that allow one to obtain either type of spectrum. AES and XPS have approximately the same sensitivity toward the surface. AES is usually more sensitive, with the ability to detect surface species at lower concentrations than XPS. Consequently, AES is utilized more frequently than XPS in depth profiling. To effect depth profiling, successive layers of the material are sputtered from the sample using a beam of ions and recording the surface concentration of the elements of interest between the periods of sputtering. One application of depth profiling would be to determine whether there was an enrichment of an element on the surface, for example, tin in a $Pt-Sn-Al_2O_3$ catalyst.

Recent advances suggest that AES may find wider applications in catalyst characterization. Angle-resolved AES (ARAES) provides a method whereby the primary beam incidence angle and the collection angle can be varied independently from 0 to 80° off sample normal (243). A surprisingly large increase in surface sensitivity is obtained by using grazing incidence and collection angles. Techniques have been developed for collecting high-energy-resolution AES (HRAES) spectra rapidly, making it useful for obtaining chemical state information from AES. It is possible to combine ARAES and HRAES to obtain depth-sensitive, chemical-state information, as illustrated in the study by Asbury and Hoflund (244).

## G. Ion-Scattering Spectroscopy (ISS)

ISS is extremely sensitive to the outermost layers of a surface, and this makes it a particularly effective tool for studying the adsorption of a species onto a surface. Sensitivity factors have been used for quantitations of the surface of binary alloys, but little has been done on real catalyst surfaces. A linear relationship exists between the ISS signal and the coverage at the beginning of the adsorption. The scattered ion yield is directly related to the number of surface atoms. Attenuation of the ion yield may result from a number of factors, including matrix effects, preferential sputtering, surface roughness, and a changing background. Determination of neutralization probabilities is difficult, and for this reason most catalyst ISS data are given in terms of ratios instead of absolute intensities. A number of factors prevent a calculation of neutralization probabilities from first principles.

Dwyer et al. (245,246) used ISS and TPD to show that the amount of chemisorption of hydrogen and CO was directly dependent on the number of exposed platinum atoms (Figure 13). The authors interpreted the data as showing that the deposited $TiO_2$ suppressed CO or $H_2$ chemisorption by blocking the adsorption sites of the platinum. The chemisorptive capacity of the system decreased on reduction but was recovered on reoxidation.

It is also possible to study the dispersion of a supported oxide or metal using this technique. Ewertowski et al. (247,248) impregnated γ-alumina with chloroplatinic acid to various extents and then calcined and reduced the materials. They observed that with increasing coverage by Pt metal a decrease in intensity of the narrow component of the momentum distribution curve was obtained. It was verified that the change was due to Pt and not the Cl added during catalyst preparation.

## H. Infrared Spectroscopy for Metallic Function

The adsorption of hydrogen onto platinum supported on γ-alumina produces bands at 2105 and 2055 cm$^{-1}$ (249). The high-frequency band is due to more weakly absorbed H. These IR bands gave the shift expected for the isotope effect. The structure associated with the more weakly bonded hydrogen was believed to be due to two hydrogen atoms adsorbed per platinum, and the more strongly bound hydrogen to a single adsorbed hydrogen atom per metal atom. Adsorption of CO on the metallic function likewise produces two bands (250). This has been interpreted as CO bonded to one Pt and to multiple Pt atoms through direct and bridging bonding, respectively. Bridging-type bonding becomes more pronounced at higher surface coverage. In some instances, especially at higher coverage, there may be multiple CO bonding to a single metal atom. This finding made investigators aware of the uncertainty associated with chemisorption measurements that utilize CO/Pt = 1 to calculate dispersion.

## I. Calorimetry

Lantz and Gonzalez (251) constructed a calorimeter and measured the heats of adsorption of hydrogen on platinum as a function of Pt crystal size. They found that the initial heat of adsorption decreased only slightly with increasing dispersion. A maximum in the heat of adsorption was obtained for all samples at a surface coverage, $\theta$, of 0.4. The heat of adsorption should not increase with coverage and several possibilities to account for this, including diffusion into the powder, were discussed. Sen and Vannice (252) later reported results that were similar.

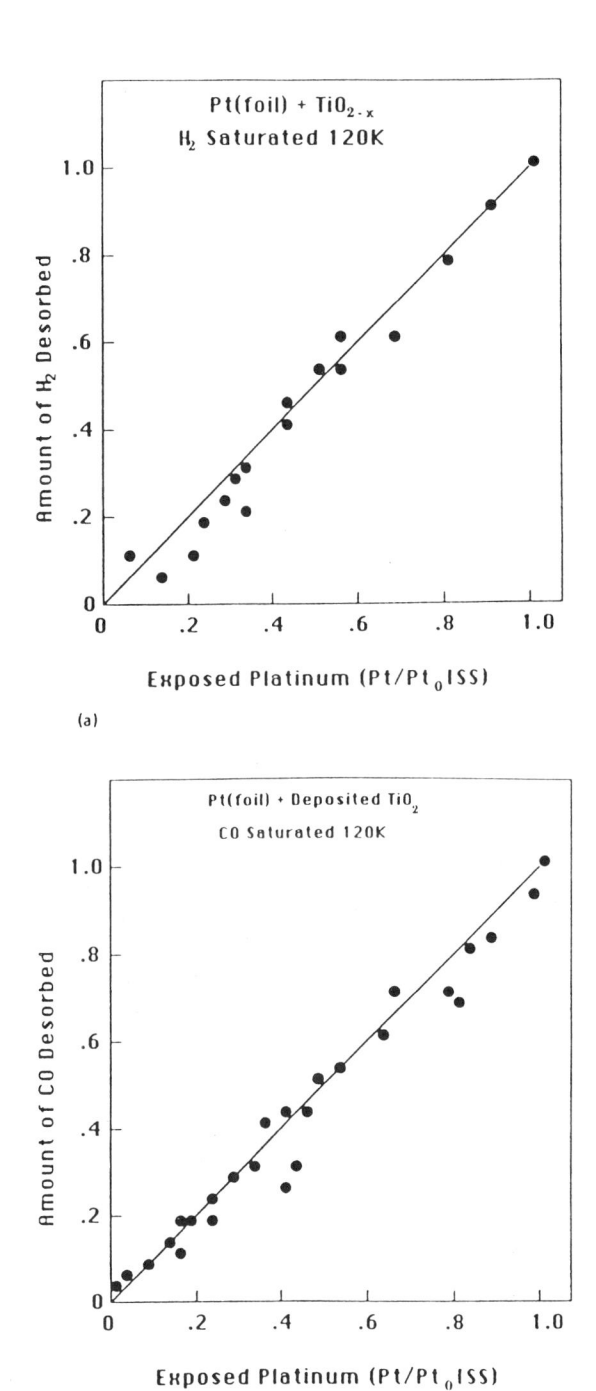

**Figure 13** Fractional coverage of (a) $H_2$ and (b) CO that can be chemisorbed at 120 K versus the fraction of exposed Pt as measured by ISS. (From Ref. 245.)

## V. CHARACTERIZATION OF PLATINUM-RHENIUM CATALYST

TPR data have been used more extensively for the Pt−Re system than for many of the other reforming catalyst formulations. Johnson and LeRoy (253) obtained evidence that the reduction stopped with rhenium in the oxidized state of $Re^{4+}$. Webb (254) obtained data using this technique that showed both platinum and rhenium were reduced to the metallic state. Johnson (255) pointed out that there was not necessarily any contradiction between these data because Webb used very dry conditions while Johnson employed a water partial pressure typical of reforming conditions. Water presumably would serve to maintain rhenium in an oxidized state.

Later TPR studies indicate that rhenium is reduced to the metal even under rather mild conditions. McNicol (256) carried out the reduction in TPD experiments as well as in fixed-bed reactors (Table 4). McNicol did not observe an influence of Pt on the reduction as had been reported in an earlier study by Bolivar et al. (257) and attributed the difference between the observations to the fact that the studies started with calcined and uncalcined samples, respectively.

Yao and Shelef (258) utilized chemisorption, TPR, and electron paramagnetic spectroscopy (EPR) techniques to investigate the reduction of Re/alumina catalysts. They reported that Re interacts strongly with $\gamma$-alumina and can be reduced to the metal in hydrogen only at temperatures greater than 500°C. Even when reduced to the metal, Re did not chemisorb hydrogen at room temperature. Reoxidation, even at 500°C, produces $Re^{4+}$, and the limiting coverage of alumina surface by this species is about 10%. This limitation in oxidation explains why Re is not volatilized as $Re_2O_7$ during catalyst regeneration. At high loadings, the three-dimensional phase can be reduced by hydrogen to

**Table 4** Isothermal and Dynamic TPR of Pt−Re-Alumina Catalysts[a]

| | Isothermal[a] | | | Dynamic TPA[c] | | |
|---|---|---|---|---|---|---|
| | Extent of reduction (%)[b] | | | Extent of reduction (%) | | |
| Temp. (°C) | 15 min | 30 min | 60 min | Temp. range (°C) | Pt | Re |
| 200 | 11 | 20 | 34 | 25–300 | 100 | 0 |
| 300 | 57 | 65 | 72 | 25–500 | 100 | 76 |
| 500 | 70 | 75 | 81 | 25–700 | 100 | 100 |
| 550 | 84 | 89 | 95 | | | |

[a] Catalyst containing 0.375 wt % Pt/0.2 wt % Re/alumina calcined at 525°C.
[b] Assuming that Pt and Re are in the 4+ and 7+ valency states, respectively, in the oxidized material.
[c] Heating rate of 1°C/min.
Source: From Ref. 256.

$Re^0$ at 350°C. The interaction of oxygen with the reduced, dispersed phase at 25°C produces species identified by EPR as surface nondissociated oxygen molecule-ions ($O_2^{2-}$) and $Re^{2+}$ ions.

Further studies (259–261) confirmed the reducibility of rhenium to $Re^0$ when present as Re/alumina or Pt—Re/alumina; moreover, these studies showed that the ease and extent of reduction are dependent on the sample history. Studies indicated that the reduction temperature decreases with increasing loading of Re and with increasing oxidation temperature (261); for a freshly oxidized sample about 90% reduction of $Re^{7+}$ to $Re^0$ took place up to 600°C (259). After reoxidation the associated TPR spectra exhibited a shift to higher temperatures with increasing oxidation temperature, and rhenium could easily reoxidize to the $Re^{7+}$ state. The Re(II) state was considered to reflect oxygen chemisorption. Wagstaff and Prins (259) find that adsorption of oxygen at temperatures up to 100°C leaves the bimetallic clusters largely intact, but subsequent high-temperature treatment in the absence of extra oxygen leads to segregation of Pt and Re species. These authors therefore take the view that "bimetallic clusters" are formed and these, in the presence of oxygen, are thermodynamically unstable, but under mild conditions the rate of segregation is slow.

The results shown in Figure 14 are typical of those obtained by most investigators for Pt—Re/alumina catalysts because they show that at least a portion of the Re is reduced at a lower temperature in the bimetallic catalyst than in a Re/alumina catalyst. Increasing oxidation temperature of the freshly prepared Pt—Re/alumina catalyst causes an increase in the fraction of Re that is not reduced at the lower temperature (Figure 15); however, for these bimetallic catalysts none of the time does the TPR of the Pt—Re/alumina catalyst become equal to one that is the sum of a mixture of the monometallics. Isaacs and Petersen (260) conclude that for drying temperatures of 500°C or less, a substantial fraction of the $Re_2O_7$ is at least partially hydrated and able to migrate to Pt reduction centers. Isaacs and Petersen conclude that hydrogen spillover is too slow (by at least two orders of magnitude) to allow the reduction rate observed up to 400°C. For this reason the authors conclude that some rhenium oxide species migrates to the platinum site, where it is reduced; the presence of water, as proposed by Bolivar et al. (257) and by Wagstaff and Prins (259), influences the mobility of $Re_2O_7$. However, Isaacs and Petersen (260) contend that the TPR curves indicate (Figure 15) the degree of hydration is set by the drying temperature; the higher the drying temperature, the lower the degree of hydration and thus the lower the mobility of the Re species. As the drying temperature is raised, the TPR temperature necessary for the Re species to migrate to the Pt reduction center, which results in a TPR peak due to Re reduction, increases.

There are many points of agreement between the data and conclusions of Mieville (261) and Isaacs and Petersen (260); however, there are some major

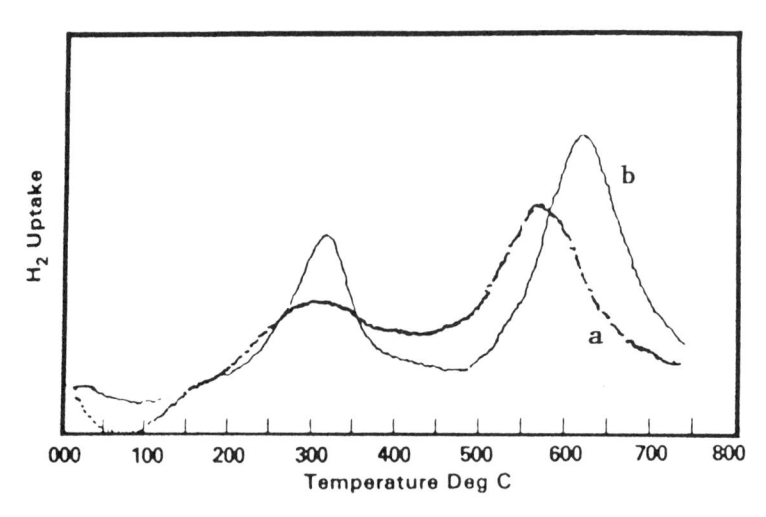

**Figure 14** TPR of Pt–Re vs. superimposed Pt + Re. (a) 0.35 wt% Pt–0.35 wt% Re/Al$_2$O$_3$. (b) 0.4 wt% Al$_2$O$_3$ and 0.35 wt% Re/Al$_2$O$_3$. Preoxidized at 500°C. (From Ref. 262.)

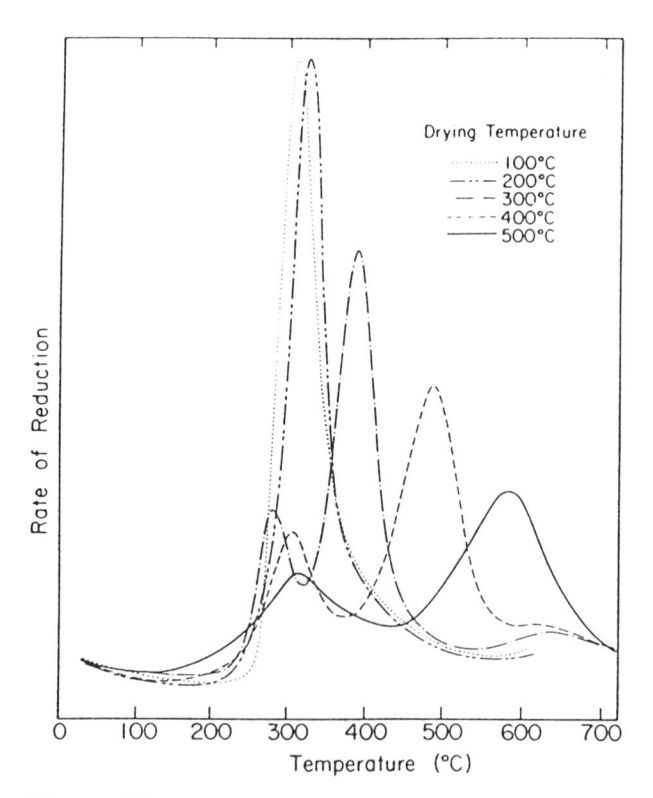

**Figure 15** TPR of Pt/Re/Al$_2$O$_3$. (From Ref. 165.)

differences. Mieville found that the superimposed Pt + Re TPD curve was essentially the same as that of a Pt−Re-alumina catalyst that had been preoxidized at the same temperature (500°). Mieville prefers the view that the catalyzed Re reduction may be due to a spillover mechanism and that only a small fraction of the total Re reduction needs to be by hydrogen that spills over because only a small fraction would be needed to generate sufficient Re metal which could then catalyze the further reduction of the rhenium oxide. In other words, the small amount of hydrogen that spills over could initiate what would be in effect an autocatalytic reduction of the rhenium oxide. Mieville points out that a catalyst containing 0.35 wt % Re and 0.01 wt % Pt will have a maximum reduction temperature of Re of 410°C; this compares to a temperature of 370°C for a catalyst containing 0.35 wt % Pt. With a value of 28.5 kcal/mol for the activation energy for the rate of diffusion for hydrogen spillover on an alumina surface (155), the rates should differ by about the factor of 4 that is observed for catalysts with the preceding Pt loadings. Ambs and Mitchell (262) observed a similar rate factor of 3 for catalysts containing 0.5 and 0.02 wt % Pt. One of the most important implications of the results obtained by Mieville (261) is that the temperature of maximum TPR rate increases as the Re loading decreases; the maximum for an Re/alumina catalyst containing 0.35 wt % Re is about 150 degrees higher than for one containing 1.1 wt % Re.

Menon et al. (264) took advantage of the fact that oxygen adsorbed on Re cannot be titrated by hydrogen at room temperature to effect a separate determination of the Pt and Re dispersion in Pt−Re-alumina catalysts. Using a pulse technique, they carried out the following room temperature measurements: (1) oxygen pulsing to measure chemisorption by Pt + Re; (2) reduction in hydrogen to covert Pt−O to Pt−H; (3) renewed oxygen pulsing to yield chemisorption by Pt only; (4) the difference between (1) and (3) for the chemisorption by Re only. These authors found that the dispersion of Pt is little altered in a composite Pt−Re catalyst compared to that in a Pt catalyst of the same metal content; however, the dispersion of Re seemed to be considerably enhanced in the composite catalyst. Eskinazi (264) compared hydrogen and oxygen chemisorption and hydrogen titration data for a series of catalysts with different Pt/(Pt + Re) ratios. The maximum metal surface areas were obtained with 55 mol % rhenium, alumina supports with surface areas greater than 150 $m^2/g$, and with a catalyst pretreatment involving an oxidation at 500°C prior to reduction.

Kirlin et al. (265) used XPS to follow the reduction of two Pt−Re-alumina catalysts as well as a Re-alumina catalyst. They reported that reduction of the bimetallic catalysts with hydrogen at 500°C and 1 atm gave predominantly $Re^{4+}$. However, addition of small amounts of water to the hydrogen feed effected the reduction to $Re^0$ and complete removal of chloride from the catalyst. In this case water assists in reduction, whereas Johnson (255) argued that water maintained a higher oxidation state. Reduction of the catalyst in butane + hydrogen at 500°C resulted in the reduction of $Re^0$ with retention of the chloride. Under identical conditions, the Re in an Re-alumina catalyst was reduced only to $Re^{4+}$, indicating that Pt is necessary to effect the reduction to $Re^0$. The XPS data are therefore consistent with the metals being present in the zero-valent state in the bimetallic reforming catalyst under reaction conditions. Adkins and Davis (266) confirmed the results of Kirlin in their XPS study.

EPR data indicated that the majority of Re was present as $Re^0$ in a Pt−Re-alumina catalyst (267). The surface of the reduced bimetallic Pt−Re-alumina catalyst contains both $Re^0$ and $Re^{4+}$, although the $Re^{4+}$ could account for only about 10% or less of total Re at the loadings investigated (0.2–0.3 wt %). The results showed that the $Re^{4+}$ surface sites do not chemisorb CO, whereas $Re^0$, either separately dispersed or in contact with the Pt, strongly chemisorbs CO. The authors also concluded that $Re^0$ was the valence state of Re which is decisive for the catalytic function of the Pt−Re-alumina reforming catalyst.

Onuferko et al. (268) utilized XPS, EXAFS, and XANES in their study of Pt−Re-alumina catalysts. Following in situ reduction, all of their results show that $Pt^0$ and $Re^{4+}$ are the dominant species. The Pt in the reduced catalyst was different from the bulk metal or the monometallic catalyst. The data also suggested that the Re was not significantly associated with Pt (268,269). Kelley et al. (270) utilized ISS and EDAX in TEM for analysis of the Pt−Re-alumina catalyst and found that Re was not significantly associated with Pt but was widely dispersed on the support surface. Sulfiding essentially covered the Re but left most of the Pt exposed.

Meitzner et al. (271) obtained a different structure for reduced one-to-one atomic ratio Pd−Re− and Pt−Re-alumina catalysts using EXAFS. In agreement with most of the TPR data, essentially all of their rhenium had an oxidation state of zero after reduction in hydrogen at 775 K. The EXAFS data also provided evidence for significant coordination of rhenium to platinum in bimetallic clusters. These bimetallic clusters were not characterized by a single interatomic distance. This finding could be an indication of variation from one cluster to another or intracluster variation of composition. This latter possibility is commonly associated with one of the components concentrating at the boundary or surface of a cluster as a consequence of differences in the surface energies of the components. If this were so, the average number of the nearest-neighbor metal atoms about one component will be larger than the number about the other. This was not observed, suggesting intercluster variation. For both catalysts, the interatomic distance for the unlike pair of atoms was significantly shorter than that for either pair of like atoms in the clusters or the pure metal. This indicates that the Pt−Re bonds in the cluster are stronger than would be anticipated.

Meitzner et al. (271) also examined a reduced Pt−Re-alumina catalyst after exposure to $H_2S$. The rhenium EXAFS results indicated that the sulfur had little disruptive influence on the clusters. There was no evidence for the formation of ReS$_2$. The Pt−Re and Re−Re bond distances do not change significantly when sulfur is adsorbed and the extent of coordination of rhenium to platinum was essentially unchanged. It was therefore concluded that the catalyst contained Pt−Re bimetallic clusters with sulfur strongly chemisorbed on the rhenium.

Sinfelt and Meitzner (199) report XANES studies of a Pt−Re-alumina (1% Pt, 1% Re). The data for the L$_{III}$ edges of rhenium show that the Re in the catalyst does not differ significantly from that in bulk rhenium, and the edge is much less intense than the edge for rhenium in the +7 oxidation state in ReO$_4^-$. Similar measurements were made for the reduced catalyst following sulfiding. These data show that the edge of rhenium after exposure to hydrogen sulfide does not differ significantly from that before exposure to sulfur but is much less intense than for ReS$_2$. These edge absorption data are consistent with the EXAFS data in showing that the Pt−Re clusters are not disrupted by sulfur. The sulfur appeared to be in the form of sulfur atoms chemisorbed on exposed rhenium atoms in the Pt−Re cluster.

Bazin et al. (272) obtained somewhat different results in their study of Pt−Re-alumina (240 m$^2$/g alumina). Both the white line and the EXAFS data for platinum indicated that it was present in the zero-valent state. However, whereas an average environment of one Pt atom in a monometallic catalyst was composed of 6 metal atoms at 2.75 Å and 0.7 oxygen atoms at 2.04 Å, in the bimetallic Pt−Re catalyst the analysis leads to 3 (reduction at 300°C) and 5 (reduction at 500°C) metal atoms around one Pt with hardly a detectable amount of oxygen. This shows that Pt is present in a more highly dispersed state in the bimetallic catalyst than in the monometallic one. The white line of the rhenium L$_{III}$ edge does not resemble the metallic state and the EXAFS indicates that one rhenium atom is surrounded by 6 to 8 metal atoms and 2 oxygens, and hence in an oxidized state. In their model, a rhenium oxide interface is between the platinum and the alumina, which they contend is in line with other experimental and theoretical results (271).

Hilbrig et al. (273) utilized coupled XANES and TPR techniques to study the reduction of Pt−Re-alumina catalysts. They measured the chemical shift of the XANES for the two metals in the catalyst relative to metallic Pt and Re. The authors then compared the derivative of the change in the chemical shift with that for hydrogen uptake (both as a function of temperature of reduction) on the reference monometallic catalysts and found that the two methods of following the reduction are quite similar in the case of each metal. Pt is completely reduced when the temperature has reached 380°C, as indicated by a chemical shift of 0.0 eV. A sharp peak at 460°C for Re corresponds to reduction to Re$^{4+}$; further reduction occurs so that the average Re oxidation state is +1.5 at 650°C. There is a considerable difference in the reduction profile for the Pt−Re-alumina catalyst compared with the monometallic catalysts. For a bimetallic catalyst with higher Re loading, the Re$^{7+}$ to Re$^{4+}$ reduction occurs around 275°C, and it occurs at 310°C for lower Re loading. Pt in both catalysts shows a reduction profile similar to that of the Pt monometallic catalyst. In both samples, however, the chemical shift is not zero even at 450°C; this is interpreted to be due to Pt−Re interaction rather than incomplete reduction of Pt (Figure 16). Pretreatment at higher temperatures decreases the reducibility of the Re. Even for catalysts that have been calcined at either 400 or 580°C there is a significant fraction of Re that is reduced at the lower temperature, as well as the higher temperature. This indicates that some of the Re is reduced to form bimetallic clusters, while some of the Re, reduced at the higher temperatures, is most likely present as the Re metal.

Vuurman et al. (274) utilized in situ Raman and IR spectroscopy and TPR data in developing a model of the state of Re on alumina as well as silica, zirconia, and titania. Their model should be applicable to the bimetallic catalyst for catalyst preparation and oxidation. Summarizing earlier data (275–277) and their data, the authors conclude that of the five different hydroxyl groups present on alumina, rhenium oxide reacts with specific hydroxyl groups as a function of coverage. At low loadings the so-called basic hydroxyl groups are titrated, and at higher loadings the more neutral and acidic groups are consumed. At the low loadings normally encountered in reforming catalysts only the first type of bonding would be expected. After dehydration, two surface rhenium oxide species are present. The concentration ratio of the two species is a function of the coverage, and their structures are similar, having three terminal Re=O bonds and one bridging Re−O support bond. The bridging Re−O−support bond strength decreases with increasing coverage. Hardcastle et al. (278) concluded, based on XANES and laser Raman spectroscopy (LRS) data, that the following reversible changes in the rhenium species on alumina take place as dehydration proceeds: perrhenate ion in solution → solvated surface species → unsolvated surface species → gaseous dimeric $Re_2O_7$.

Jothimurugesan et al. (279) utilized proton-induced x-ray emission (PIXE) and Rutherford backscattering spectrometry (RBS) together with SEM and chemisorption data to characterize Pt-, Re-, and Pt−Re-alumina catalysts. PIXE utilizes a beam of energetic protons to excite the inner shell electrons and the resulting x-rays are used for the identification and quantification of the elements of interest. PIXE can give in-depth concentration profiles of a metal. RBS is based on the elastic backscattering of ions from the catalyst sample. This technique has the possibility of high sensitivity, and surface concentrations on the order of a monolayer of heavy atoms present on a lighter atom support can easily be detected. RBS data present surprising results. For the monometallic Pt catalyst, the depth profile indicated that Pt was uniformly distributed from the surface to the interior of the pellet for samples representing the oxidized, reduced, or used catalyst. For the monometallic Re catalyst, the concentration increased slightly from the surface to the interior for the oxidized, reduced, and used catalyst. For the bimetallic catalyst, the Pt−Re depth

## Characterization of Naphtha Reforming Catalysts

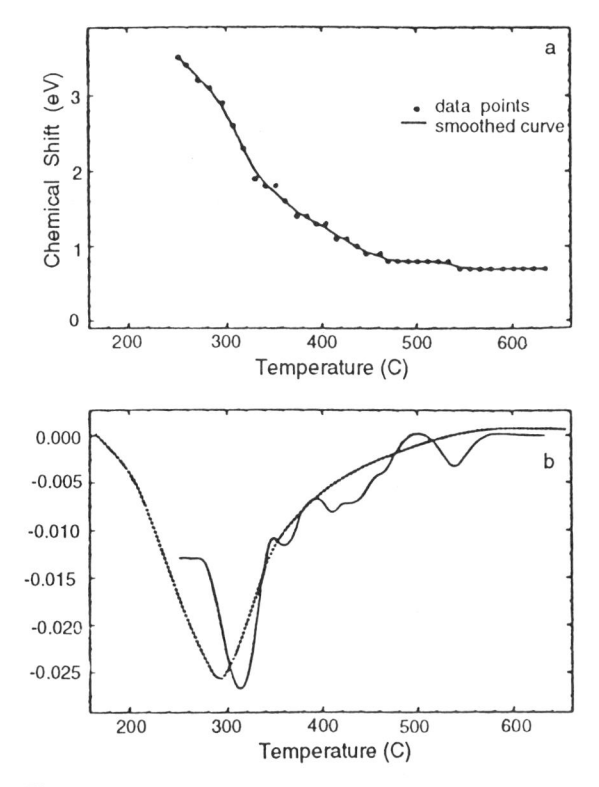

**Figure 16** TPR of 0.3 wt % Pt–0.3 wt % Re/Al$_2$O$_3$ pretreated at 250°C. (a) The chemical shift (relative to metallic Re) as a function of temperature of reduction. (b) The derivative of the chemical shift (—) and H$_2$ consumption (···) as a function of temperature of reduction. (From Ref. 273.)

profile for the oxidized catalyst resembles that of the Re-monometallic catalyst; however, there is about a fivefold increase in the metal concentration near the surface of the support pellet for the reduced and used catalyst samples.

Kelley et al. (270,280) used ISS to study Pt−Re−Al$_2$O$_3$. They found that Re was highly dispersed on the surface. Upon sulfiding of the catalyst, they found that rhenium was covered and platinum was exposed. The increase of the Pt + Re peak was strongly associated with a decrease in either the sulfur or chlorine peak.

## VI. CHARACTERIZATION OF PLATINUM-TIN-ALUMINA CATALYSTS

TPR studies of Pt−Sn−Al$_2$O$_3$ catalysts suggest that Sn is not reduced to the zero-valent state (281,282). Lieske and Völter (283) reported, based on TPR studies, that a minor part of the tin is reduced to the metal, which combined with Pt to form "alloy clusters," and the major portion of this tin is reduced to Sn(II) state. The amount of alloyed tin increased with increasing tin content. This disagrees with TPR data reported by Burch and Garla (284,285), who conclude that the tin is reduced only to Sn(II) and not to elemental tin. They believe that the altered Pt properties are due to an interaction with Sn(II) species, which is considered by Burch and Mitchell (286) to be the primary influence on the reforming properties. This agrees with the work of Müller et al. (287), who used hydrogen and oxygen adsorption measurements to show that neither metallic tin nor a Pt/Sn alloy was present in their catalysts.

In the following, data are reported from methods that make a more direct measure of the chemical or physical state of the elements present in Pt−Sn-alumina catalysts. XPS studies permit one to determine the chemical state of an element but the data do not permit one to define whether, for example, Sn(0), if present, is in the form of a Pt−Sn alloy. Furthermore, the major Pt XPS peak is masked by a large peak of the alumina support. Thus, XPS can provide only data to show that an alloy is possible; it cannot be used to prove the presence of a Pt−Sn alloy.

The early XPS studies revealed that the tin is present only in an oxidized state (282,288,289). However, some observations were due to the use of oxygen-containing pump oils to maintain the high vacuum. Li et al. (290,291) later reported that a portion of the tin in Pt−Sn-alumina catalysts was present in the zero-valent state; furthermore, the apparent composition of the Pt−Sn alloy contains increasing Pt/Sn ratios as the Pt/Sn ratio present on the catalyst is increased. However, it must be kept in mind that XPS provides a measure of the surface, and not the bulk, composition; Bouwman and Biloen (292) showed that the tin is concentrated in the surface of reduced unsupported Pt/Sn alloys, in general agreement with others (293-299).

$^{119}$Sn Mössbauer data have been utilized in a number of studies (300-315). Direct evidence for PtSn alloy formation was obtained from Mössbauer studies (300,310-313); however, many of these studies were at high metal loadings and even then such a complex spectrum was obtained that there was some uncertainty in assigning Sn(0) to the exclusion of tin oxide phases. Mössbauer results clearly show changes upon reduction in hydrogen but the width of the peaks for the reduced sample prevents a specific assignment for some states of tin. This was exemplified by the results of Kuznetsov et al. (315).

There are many complicated relations between the electronic state of the tin ion and the associated Mössbauer spectrum. The usual rules do not always apply. However, this is normally not a problem because each tin-bearing phase usually has a distinct characteristic set of parameters. Multiphase sample spectra can be resolved using standard curve fitting and statistical techniques so that the contributions of each phase may be separated. The relative amount of tin in each phase can be determine from the peak area associated with that phase.

Li et al. (314) utilized a series of 1% Pt/variable % Sn catalysts for Mössbauer studies. Tin was observed to be present in forms whose isomer shifts were similar to or the same as those of $SnO_2$, SnO, $SnCl_4$, $SnCl_2$, Sn(O), and PtSn alloy when alumina was the support. Representative Mössbauer spectra are shown in Figure 17. If it is assumed that the Pt/Sn alloy corresponds only to a 1:1 alloy, one finds that for lower Sn/Pt ratios (5 or less), little difference is observed in the extent of alloy formation and the distribution of the oxidized species for a low and a high surface area alumina support. The fraction of Pt present in an alloy phase increases with increasing tin composition and approaches complete alloy formation only at Sn/Pt > ≈ 5.

For Pt supported on a particular coprecipitated tin oxide-alumina catalyst, a smaller extent of alloy formation was found than for a material prepared by impregnation with the chloride complex (312,314). In another study, a catalyst was prepared by impregnating a nonporous alumina with a surface area of 110 $m^2/g$ with an acetone solution of $Pt_3(SnCl_3)_{20}^{2-}$ (316). A 5 wt % Pt catalyst was reduced in situ in the chamber of an XRD instrument (317); the diffraction pattern matches very well the pattern reported for PtSn alloy. In the pattern, it can be seen that with the 5 wt % Pt catalyst, a small fraction of the Pt is present as crystalline Pt. Similar results for the PtSn alloy are obtained for a catalyst that contains only 0.6 wt % Pt, with the same Sn/Pt ratio. These in situ XRD studies support alloy formation with a stoichiometry of Pt/Sn = 1:1. The Sn in excess of that needed to form this alloy is present in an X-ray "amorphous"

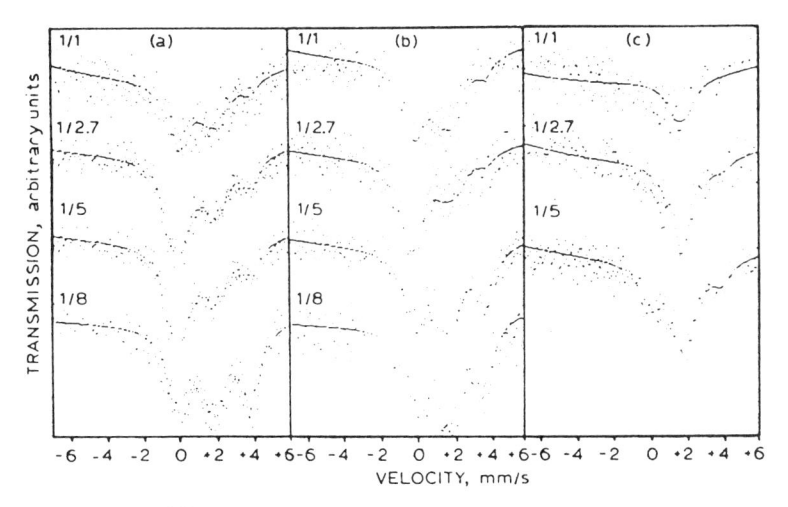

**Figure 17** $^{119}$Sn Mössbauer spectra for reduced Pt−Sn catalysts: (a) Pt−Sn/$Al_2O_3$ (250 $m^2/g$), (b) Pt−Sn/$Al_2O_3$ (110 $m^2/g$), and (c) Pt−Sn/$SiO_2$ (700 $m^2/g$). (Pt/Sn mole ratios are indicated by the numbers shown in the figure.) (From Ref. 314.)

form and is postulated to be present in a shell layer with a structure similar to tin aluminate.

XRD studies of other variations indicated that, irrespective of the Sn/Pt ratio, the only crystalline phase detected by XRD was SnPt (1:1). The XRD intensity of lines for the SnPt alloy phase increases with increasing Sn/Pt ratios, indicating the presence of unalloyed Pt in the samples containing low tin loadings. The integrated intensity measured for the most intense PtSn peak (102) clearly shows that the amount of Pt present as crystalline alloy increases with increasing Sn/Pt ratios. Likewise, the crystallite size calculated from the XRD data shows that the radius of the crystallite increases from about 10 to 16 nm with increasing tin content.

XANES and EXAFS were obtained for a series of dried, oxidized and reduced (773 K, 1 bar hydrogen) preparations of Pt/Sn-loaded silica and alumina supports (318,319). The Pt was maintained at 1 wt % and the Sn content was varied from 0.39 to 3.4%. In general, it was found that increasing the Sn loading on either support decreased the $d$ band vacancy for Sn. In contrast, alloying Pt with Sn leads to an increase in the $d$ band vacancies, in general agreement with Meitzner et al. (320). The Sn K-edge spectra indicate that the arrangement of atoms about the tin species was essentially the same for the high-area alumina and high-area silica supports. Sn-O atom pairs dominate even after reduction, but their contribution decreases somewhat with increasing Sn loading. The two supports led to profoundly different configurations of atoms about the platinum.

An electron microdiffraction technique was employed to identify crystal structures developed in two Pt–Sn-alumina catalysts. One catalyst was prepared by coprecipitating Sn and Al oxides and then impregnating the calcined materials with $H_2PtCl_6$ to give a Pt/Sn = 1:3 atomic ratio. The second catalyst was prepared by coimpregnating Degussa alumina with an acetone solution of chloroplatinic acid and stannic chloride to provide a Pt/Sn = 1:3. Pt–Sn alloy was not detected by x-ray diffraction for the coprecipitated catalyst, although evidence for PtSn alloy was found for the coimpregnated catalyst (321). Electron microdiffraction studies clearly showed evidence for Pt/Sn = 1:1 alloy phase in both catalysts. Evidence for minor amounts of the Pt/Sn = 1:2 phase was also found for the coprecipitated catalyst.

## VII. CHARACTERIZATION OF OTHER BIMETALLIC CATALYSTS

Following the successful introduction of the $Pt-Re-Al_2O_3$ bimetallic catalyst, a rash of metal combinations appeared in the open and patent literature. These catalyst combinations can be conveniently grouped into three general classifications of alumina supported catalysts: (1) a Group VIII metal, almost always Pt, together with a nonreducible metal; (2) a Group VIII metal, usually Pt, together with a metal at least partially reducible to the metal; and (3) a combination of Group VIII metals or a combination of a Group VIII and Group IB metal. In view of the large number of catalysts reported (for example, more than 500 papers concerning Pt and Ir catalysts appeared during the past 10 years), space permits only a brief outline of the results for these catalysts.

## A. Group VIII or Group VIII/Group IB Bimetallics

Of these catalyst types, the Pt−Ir combination appears to have been the most widely studied for naphtha reforming and is reported to have been used in commercial operations (322). Much of the early characterization work on Pt−Ir−$Al_2O_3$ catalysts have been reviewed by Sinfelt (322). Chan et al. (323) followed the steps in the preparation of Pt−Ir-alumina catalysts using LRS: impregnation with the appropriate chloroplatinic acid and chloroiridic acid solutions and heating in air at varying temperatures. The spectra indicated significant differences in the nature of the surface species remaining when Pt/$Al_2O_3$ and Ir/$Al_2O_3$ preparations were subjected to such thermal treatments. After drying at 110°C, the Pt/$Al_2O_3$ preparation exhibits Raman bands assigned to the $PtCl_6^{2-}$ ion, whereas Ir/$Al_2O_3$ yields a spectrum significantly different from that reported for the $IrCl_6^{2-}$ ion. With increasing temperature, both the platinum and the iridium species decompose with loss of chloride ligands. After being heated in air at 500°C, the Ir/$Al_2O_3$ sample exhibits a Raman spectrum characteristic of crystalline $IrO_2$. The presence of platinum in the bimetallic Pt−Ir/$Al_2O_3$ sample inhibits the formation of crystalline $IrO_2$ to some degree.

Subramanian and Schwarz (324) utilized TPR and thermogravimetry to show that $PtCl_4$, $IrCl_3$, and $PtCl_4$−$IrCl_3$ exist as supported entities when the catalyst, prepared from $H_2PtCl_6$ and $H_2IrCl_6$, is dried in air at 150°C. These authors reported that bimetal formation does not occur when the dried precursors are directly reduced. The TPR results indicate that there is no interaction between platinum and iridium (prior to reduction) when the precursors are dried at 423 K in air. Although they could not rule out bimetal formation during the reduction process, the spectrum for the bimetal precursor (Figure 18) resembled the profile generated by adding those observed for the corresponding monometal precursors (dashed line). However, Huang et al. (325) have reported that for samples dried at 383 K, the TPR spectrum of the bimetallic catalyst is not the superposition of the spectra of the individual components. Subramanian and Schwarz explained this apparent discrepancy on the basis of the methods used to prepare the catalysts. A wet impregnation procedure (50 $cm^3$ of impregnation solution/20 g of alumina) was used by Huang et al., rather than the incipient wetness technique using an amount of impregnant

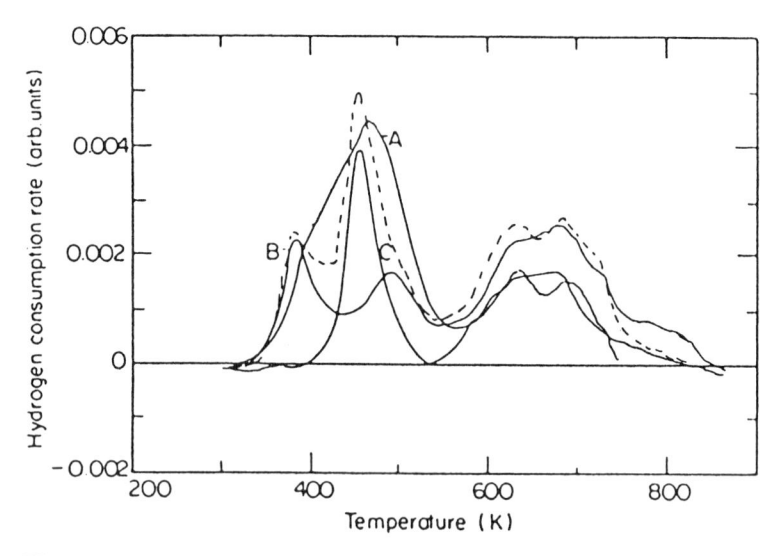

**Figure 18** Temperature-programmed reduction profile for A: Pt−Ir/Al$_2$O$_3$, B: Ir/Al$_2$O$_3$, and C: Pt/Al$_2$O$_3$ precursors. (From Ref. 324.)

corresponding to the pore filling volume of the support (0.5 cm$^3$/g). Recent studies (326) have shown that, at constant metal weight loading, a change in the amount of solution used to prepare the precursors results in a significant change in their reducibility when supported on porous alumina.

Huang et al. (325) found evidence that platinum and iridium in the hydrogen-reduced catalyst were in their metallic state, forming highly dispersed bimetallic Pt−Ir clusters. Huang and Fung (327) reported that the formation of IrO$_2$ during oxidation, and subsequent reduction to Ir metal, was inhibited by the presence of chloride. Furthermore, the presence of water vapor during the reduction removed chloride with increased formation of IrO$_2$. The degree of IrO$_2$ agglomeration of the 693 K oxidized Ir/Al$_2$O$_3$ catalyst was much larger than that of Pt−Ir/Al$_2$O$_3$ catalysts and indicated that the Pt−Ir interaction reduced the degree of IrO$_2$ agglomeration. Chloride promotes the interaction between Pt and Ir to form mixed oxychlorides and oxides in an oxidizing environment. The effect of Cl on IrO$_2$ agglomeration is, therefore, more significant for the Pt−Ir/Al$_2$O$_3$ catalyst than the Ir/Al$_2$O$_3$ catalyst. McVicker and Ziemiak (328) employed XRD and H$_2$ and CO chemisorption in their studies of the agglomeration behavior of a commercial Pt−Ir catalyst (0.3% Pt-0.3% Ir and 0.67% Cl) and showed that 4-h air oxidation at 773 K resulted in 58% IrO$_2$ agglomeration. The Cl/M ratio in this case is 1.1. A substantial increase in the Cl/M ratio is required to suppress IrO$_2$ agglomeration when the oxidation temperature is greater then 733 K. Thus, a subtle difference in water concentration during air oxidation could result in a large change in catalytic properties.

$^{193}$Ir Mössbauer studies have been conducted for Pt−Ir catalysts. von Brandis et al. (329) reported that silica supported materials prepared by impregnation with H$_3$IrCl$_6$ and H$_2$PtCl$_6$ contained Ir in a trivalent state, presumably [IrCl$_6$]$^{3-}$, following drying. Oxidation at 400°C converted the Ir to IrO$_2$. Following reduction at 200°C of the oxidized material, there was a strong tendency toward segregation of Ir and Pt. Similar results were obtained for coimpregnation using a solution containing Ir(NH$_3$)$_5$Cl$_3$ and Pt(NH$_3$)$_4$Cl$_2$.

Sinfelt et al. (330) utilized EXAFS to investigate their alumina- or silica-supported Pt, Pt−Ir and Ir catalysts. These studies extended their earlier investigations of the 5-20 wt % Pt + Ir catalysts with XRD, which indicated the presence of 2.5-5.0-nm bimetallic clusters. The EXAFS data for the lower loading of these metals showed that these catalysts with highly dispersed metals consisted of bimetallic clusters even as the ratio of surface to total atoms approached unity. The data obtained to date are typical of much of the work reported on catalysts that are utilized commercially. Exxon workers appear to be able to prepare catalysts which consist of bimetallic clusters even at high dispersion. These workers point out the sensitivity of the Pt−Ir catalyst formulations to the drying, oxidizing, and activating procedures. Most of the other workers obtain data that suggest little interaction of the supported Pt and Ir following reduction. It therefore appears that small differences leading to and/or during reduction determine whether bimetallic clusters are obtained. Presumably most of the investigators have not been conducting their experiments under conditions that reproduce the Exxon catalyst preparation and activation conditions.

### B. Platinum and Partially Reduced Metal

The Pt−Sn catalyst system has been the most widely studied member of this class of catalysts. However, considerable data exist for two other combinations of Pt-Group IVA metals, Pt−Ge and Pt−Pb. These two systems will be covered briefly in this section.

Romero et al. (331,332) studied the hydrogenolysis of $n$-butane and the hydrogenation of benzene over fluorided platinum-germanium catalysts and characterized them using infrared spectroscopy, electron microscopy, and H$_2$−O$_2$ adsorption. They concluded that germanium enhanced the selectivity for the isomerization of $n$-butane while hydrogenation of benzene went through a maximum at about 0.4% Ge. Electronic effects were used to account for these results. A marked decrease in the dispersion of the platinum was found, as well as an increase in the CO stretching frequency, when germanium was added to the catalyst. Electron microscopy showed a bimodal distribution of particle size for the bimetallic catalyst. The authors suggested that the decrease in the dispersion of the metal as the amount of germanium increased was related to competitive exchange during preparation.

Goldwasser et al (333) prepared two series of Pt−Ge-alumina catalysts. In one series the Ge-containing alumina was calcined at 873 K prior to the addition of Pt (series B); both series of catalysts were oxidized at 673 K after both metals had been added. For the series A catalysts (Ge added, then dried at 383, and then Pt added) there was a slight decline in platinum dispersion, calculated from the amount of hydrogen adsorbed, as the germanium content increased. For series B, the dispersion did not depend on Ge content. For higher contents of Ge in series A, it appeared that Ge as well as Pt was reduced, in contrast to the lower loading. Thus, the catalysts containing 0.3% Ge and no Ge gave similar TPR curves, whereas the one containing 1.0% Ge showed a second reduction peak when the temperature was increased above about 1000 K. Since a reforming catalyst would normally not be reduced or operated at 1000 K or higher, the data suggest that Ge would not be reduced in a reforming catalyst.

Bouwman and Biloen (334) showed by XPS that a Pt/Ge catalyst contained $Ge^{2+}$ and $Ge^{4+}$ after reduction at 823 K, whereas after reduction at 923 K it contained $Ge^{2+}$ and Ge(0) species; the latter was found to be alloyed with platinum. de Miguel et al. (335) reported Ge reduction to Ge(0) at 450°C ranged from 15 to 26%; however, these Pt−Ge catalysts were prepared using $GeO_2$. It is therefore quite likely that discrete $GeO_2$ particles remained intact during preparation and oxidation and may be responsible for the extent of reduction. However, a later report (336) utilized $GeCl_4$ in the preparation procedure, and again Ge was reduced. From the hydrogen consumption, the authors calculated that 42% of the germanium was $Ge^{2+}$ and 58% was Ge(0) and for samples reduced to 1123 K, 80–100% of the Ge was present as Ge(0).

The major difference between the series B catalysts utilized by Goldwasser et al. (333) and de Miguel et al. (336,337) was that the former utilized an oxidation temperature of 673 K whereas the latter utilized a temperature of 773 K. It is likely, just as was the case with Pt−Ir catalysts, that the formation of $GeO_2$ particles during oxidation leads to a different extent of reduction of Ge.

Lead-containing Pt−$Al_2O_3$ catalysts have been characterized by a variety of techniques. Völter et al. (337) found that with increasing amounts of either Sn or Pb in a Pt−$Al_2O_3$ catalyst, the weak adsorption of hydrogen decreased and the adsorption of oxygen increased. There was a correlation between the amount of weakly adsorbed hydrogen and the activity of the catalyst for dehydrocyclization. It was found that the addition of Pb to Pt−$Al_2O_3$ catalysts decreased of the adsorbed CO stretching frequency (338), consistent with the transfer of *s* electrons from Pb to Pt. Their IR and catalytic results indicated that both ligand and ensemble effects were important for this catalyst system. Lieske and Völter (339) concluded, on the basis of TPR results, that the major part of Pb is not reduced but exists as carrier-stabilized $Pb^{2+}$ species; a minor amount of the lead exists as Pb(0) in an alloy with Pt.

## VIII. CHARACTERIZATION OF ALKALINE L ZEOLITES–Pt

The availability of high-resolution TEM permitted the extension of this technique to the characterization of small metal clusters in zeolites. Gallezot and co-workers (224) were among the early workers to utilize this technique for this purpose. For example, they reported on the states of Pt dispersed in a Y-zeolite: isolated atoms in sodalite cages, 10-Å-diameter agglomerates fitting into the supercages, and 15–20-Å crystallites occluded in the zeolite crystals. In at least some cases it appeared that the metal crystallites had ruptured the zeolite structure. This mode of characterization was of interest for naphtha reforming catalysis, particularly following the observation by Bernard (353) that Pt-L exchanged with alkaline cations had exceptional activity and selectivity for $n$-hexane dehydrocyclization to benzene.

Besoukhanova et al. (340) utilized a combination of XRD, electron micrography, and IR studies of CO adsorption to characterize Pt–L zeolite catalysts containing alkali cations. They concluded that there were four types of Pt particles: large 100–600-Å particles outside the channels, crystals 10–25 Å in diameter and small metallic cylinders inside and outside the channels, and very small particles in the cavities of the zeolite structure. They related the catalytic activity of their catalysts to the 10–25 Å crystals and the small cylinders of Pt. From the infrared band at 2060–2065 cm$^{-1}$ they concluded that the Pt particles have an excess of electrons and/or atypical faces, corners, or edges.

Hughes et al. (341) determined the Brönsted acid site density of alkaline earth–exchanged L-zeolite from the infrared spectrum of adsorbed pyridine. Increasing the oxidation temperature decreased the acidity, and a BaK-L sample oxidized at 866 K had insufficient acidity accessible to pyridine for it to be determined. Shifts in frequency during oxidation permitted them to show that oxidation at 866 K caused the alkaline earth ions to migrate to the locked position in the small cage (inaccessible to pyridine) and the potassium ions to occupy predominantly the open sites in the large channels of zeolite L (342). Newell and Rees (343) reported a similar movement of ions in zeolite L during oxidation. It was contended that any Brönsted acidity present in the BaK-L zeolite was located in the locked sites and could not participate in acidic reactions.

Hughes et al. also reported the apparent dispersion of Pt (0.8 wt %)/BaK-L, assuming that the average metal particle diameter is related to the ratio of surface to total metal atoms by the equation suggested by Anderson (344) (Figure 19). The particles that could be detected by TEM were about 0.8–1 nm in diameter. The dispersion data presented in Figure 19 are not in agreement with the TEM data for higher temperatures of reduction. The authors explain this by assuming that the particles are located within the large intercrystalline zeolite channels and, as the particles grow with increasing reduction time or temperature, the majority of the surface becomes inaccessible to the CO used to mea-

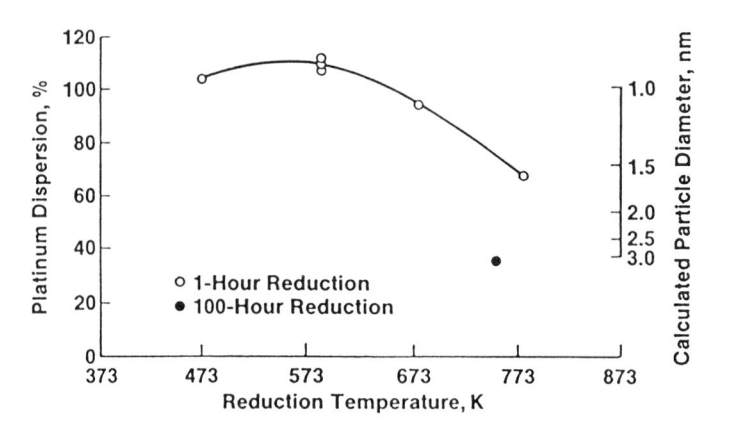

**Figure 19** Effects of reduction temperature and time on Pt dispersion in 0.8 Pt/BaK-L-866. (From Ref. 343.)

sure the dispersion. Thus, the decrease in dispersion shown in Figure 19 is considered to be due to this restricted accessibility caused by the Pt particle packing within the large channels and not to Pt particle growth beyond about 1 nm diameter.

Rice et al. (345) considered the merits of high-resolution bright-field imaging in the TEM and high-angle annular-dark-field imaging in the STEM for the detection and measurement of small noble-metal clusters in zeolites, using Pt−KL−zeolite. They confirmed that high-resolution bright-field imaging is better suited for resolving the zeolite framework. However, even with contrast enhancement, bright-field images are ineffective for detecting clusters containing fewer that about 20 Pt atoms in supports thicker than about 10 nm. This was attributed mainly to phase contrast patterns associated with beam-damaged regions of the zeolite framework. STEM measurements were shown to be capable of detecting single Pt atoms against about 20-nm-thick zeolite support, but the ability to establish the position of the atom relative to the unit cell is limited by beam-damage-induced distortion of the zeolite framework.

Miller et al. (346) utilized TEM as one means of characterizing their Pt-BaK-L zeolite catalyst. EDX spectroscopy was used to analyse several areas of the catalyst to provide evidence that the platinum is present within the interior of the imaged crystallites. Since the platinum particles were not imaged by the electron microscope, the authors inferred that the platinum particles were very small and most likely well dispersed in the zeolite crystallites. Hydrogen chemisorption data indicated that extrapolation to zero pressure showed no adsorption for the KL zeolite and that the hydrogen adsorption on the Pt-BaK-L zeolite sample paralleled the higher-pressure isotherm for the zeolite. The data gave a value of H/Pt equal to 1.3. EXAFS data provide a Pt−Pt coordination number of 3.7 that, for a bulklike fcc packing of the Pt atoms in a cluster, corresponds to a cluster of 5 or 6 atoms. The absence of higher-order Pt−Pt coordination shells is consistent with this nearly unique cluster size. These data showed that extremely small metal particles covered with chemisorbed hydrogen had metal-metal distances equal to those in the bulk metal; this was consistent with earlier data (149,347). These authors compared their data for H/Pt and coordination number determined by EXAFS (Figure 20) with earlier data reported by Kip et al. (150) for a series of Pt supported on $\gamma$-Al$_2$O$_3$. Another feature of the EXAFS data was evidence for a Pt-Ba interaction; this is surprising in light of the observation by Hughes et al. (341) that K preferentially locates in the channel. However, the sample used by Miller et al. (346) was oxidized only at 400 K and then reduced at 773 K, both lower temperatures than used by Hughes et al. (341) (886 K).

Larsen and Haller (348) reported a ratio of CO/H of about 0.7 for the adsorption of these two gases on Pt in KL catalysts and attribute the difference to some degree of reversibility of the hydrogen uptakes at 298 K. These authors preferred the CO/Pt ratio as a measure of the dispersion of platinum for these catalysts. These authors also obtained a linear relationship between the EXAFS coordination number and the H/Pt ratio; however, for a particular value of H/Pt, Larsen and Haller obtain a larger coordination number than Miller et al. (346). Larsen and Haller obtained a dispersion, defined as H/Pt,

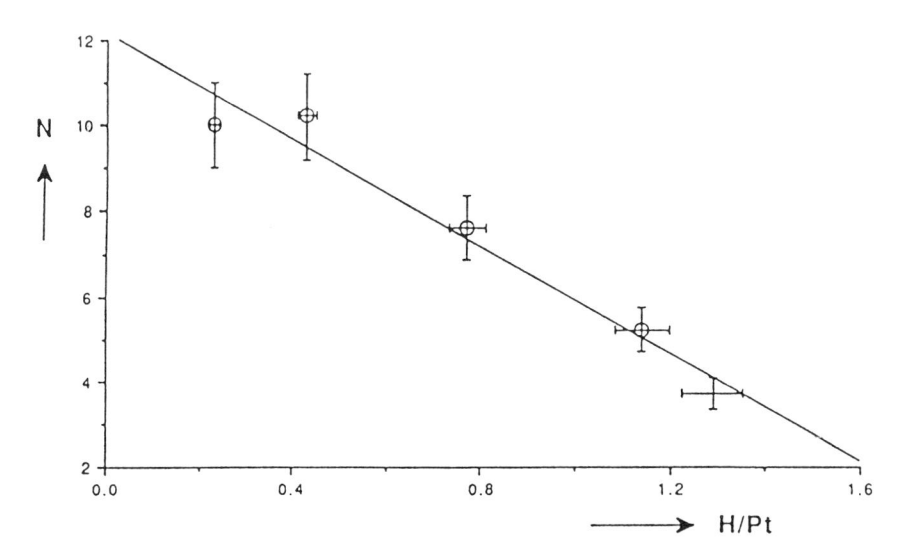

**Figure 20** Correlation of hydrogen chemisorption and first-shell Pt−Pt EXAFS coordination number data. Data represented by circles from Kip et al. (From Ref. 150.)

versus oxidation temperature that was very similar to the one obtained by Hughes et al. (341) suggesting that restricted access must also apply for hydrogen.

McVicker et al. (349) reported on the characterization of both fresh and sulfur-poisoned 0.6% Pt-K-L zeolite catalysts. A catalyst oxidized at 623 K and then reduced at 773 K yielded H(irreversible)/Pt ratios in the range 0.9 to 1.1. Bright-field TEM images showed few Pt clusters > 1.0 nm in diameter within the channels of the fresh catalyst, and this was confirmed by in-depth Z-contrast STEM studies. Infrared spectra of adsorbed CO were consistent with deactivation by sulfur occuring by the systematic loss of accessible Pt sites by a plugging of channels with Pt agglomerates sufficiently large to "double plug" a channel, since only the intensity differed for the two catalyst samples. Furthermore, the benzene selectivity from the conversion of $n$-hexane was the same for the sulfur-free and sulfur-poisoned catalysts. These authors therefore consider pore mouth blockage as the reason for the extreme sensitivity of this catalyst to sulfur poisoning and depict this as illustrated in Figure 21. They offered a histogram, obtained using STEM, of Pt particle size for catalyst samples aged in a sulfur-containing feed that showed particle growth. They obtained bright-field TEM micrographs that showed Pt agglomerates clustered predominantly at the channel entrance (Figure 22). Iglesia and Baumgartner (350) report that Pt crystallites located outside the zeolite channels are large ($\approx$ 100 Å) and are coated with graphitic carbon during $n$-hexane conversion at 783 K; those particles located in the channel retain their activity for $n$-hexane conversion. Vaarkamp et al. (351) found that H/Pt was 1.4 for a fresh catalyst with a Pt–Pt coordination number of 3.7 as determined from EXAFS data, suggesting an average cluster size of 5 to 6 atoms. Following sulfur

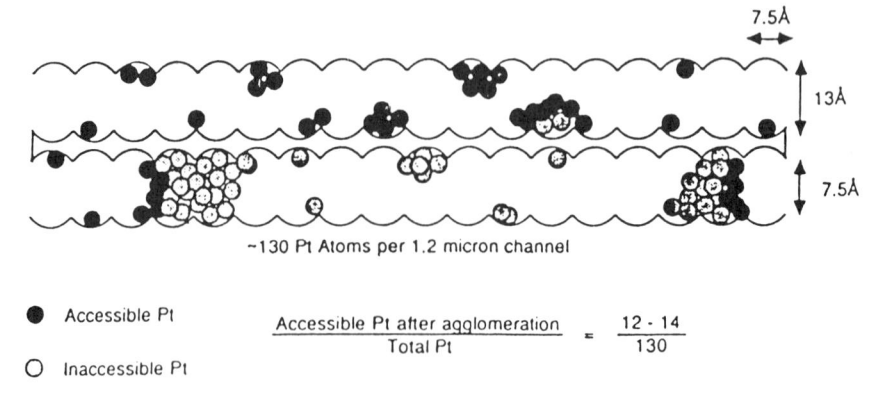

**Figure 21** Proposed model for pore mouth blockage of Pt/Kl catalysts. (From Ref. 349.)

*Characterization of Naphtha Reforming Catalysts* 165

**Figure 22** Bright-field TEM micrograph of an unbound 0.6% Pt-K-L catalyst showing the presence of Pt clusters elongated along the KL channel. The catalyst was recovered from an aromatization experiment after 100 h on a feed containing 50 wppm sulfur. KL-zeolite channel spacings of 15.9 Å are visible and provide an internal standard for Pt cluster size measurement. (From Ref. 349.)

posioning, the first-shell Pt−Pt coordination number increased to 5.5, indicating growth of the average platinum cluster size to 13 atoms and a decrease in H/Pt to 1.0. They conclude that growth of the platinum particle was sufficient to block the pore.

## IX. SUMMARY

The physical properties of a naphtha reforming catalyst are determined primarily by the alumina support. Thus, the surface area and pore structure result from the support. Likewise, the crystal structure and the physical shape of the alumina particles determine the physical features of the catalyst. Plateletlike alumina may therefore provide steps which can provide a sink for trapping Pt or other metal atoms formed during catalyst reduction and use. General features of the morphology of the catalyst can be obtained from the shape of the nitrogen adsorption/desorption isotherm and the hysteresis between the two. The surface area is best obtained from the nitrogen adsorption isotherm using the BET equation. The pore size distribution is a very demanding assignment that produces at best engineering, and not absolute, data. Factors that introduce the major uncertainties in the pore size distribution are the technique used to correct for the thickness of the adsorbed layer, whether the adsorption or desorption isotherm is used, and the assumed shape of the pore. From the engineering point of view, one is usually safe to use any model provided data are obtained for comparative purposes. For most samples, the method of calculation by Broehoff and de Boer (30–31) is most likely to produce a pore size distribution that produces a surface area which closely agrees with the one obtained using the BET equation.

An absolute measure of the acidity of a reforming catalyst is probably not possible. Adsorption data indicate that alumina adsorbs nitrogen bases from either the gas or liquid phase at room temperature to such an extent that it cannot be reliably used to estimate catalytic activity. One potential approach to measuring the acidity responsible for the bifunctional character of the reforming catalyst is to utilize the chemisorption of a sterically hindered base, e.g., 2,6-dimethylpyridine, at high temperatures (300–400°C). With chloride introducing additional or enhanced acidity, the use of chemisorption to measure acidity becomes much more difficult. For the engineering approach, chemisorption of a base is most likely the technique that will provide data to aid in evaluating the potential of a catalyst formulation or following changes in catalyst performance. It must be emphasized that the technique is not absolute and will therefore provide reliable performance prediction or postmortem examination only after reliable correlations of the plant performance with chemisorption data have been developed. Furthermore, the ability to extrapolate correlations from one catalyst formulation to one that is significantly different, e.g., Pt-alumina and fluorided Pt-alumina, is risky.

Two instrumental techniques that should provide additional understanding of the data for the chemisorption of base are infrared spectroscopy and calorimetry. Calorimetry offers the potential to measure the distribution of the acidity, at least on a relative basis. Calorimetry may be utilized to determine acidity distributions at elevated temperatures, closer to those used in commercial reformers. The infrared measurement of chemisorbed molecules provides a measure of the distribution of the acid sites between those of a Lewis or Brönsted type. It therefore would appear that a preferred method of acidity measurement is to conduct a high-temperature adsorption of a sterically hindered base for quantitation and to augment the primary adsorption data by characterization of the acid site strength distribution with calorimetric measurements as well as acid site type measurements with in situ measurements using infrared techniques.

In a commercial operation the pretreatment and operation procedures appear to provide a catalyst that has a high dispersion of platinum. The combination of high dispersion and low metal loading causes most of the experimental techniques to have little or limited value. For the engineering analysis it appears that chemisorption measurements are the most appropriate. As in the case of the determination of acidity by adsorption, the ability to convert the experimental data to an exact measure of dispersion is limited. This is primarily the result of the lack of knowledge of the "correct" ratio of adsorbate to metal to use in making the calculations as well as the geometry of the "cluster" of platinum atoms. For the engineering approach, a comparison of samples of a given catalyst formulation can be made from a comparison of the value of the ratio of adsorbate to metal, e.g., the H/Pt ratio for the chemisorption of hydrogen. For a catalyst suitable for commercial operation, the ratio should closely approach or exceed one.

High-resolution electron microscopy is applicable, under suitable conditions, for dispersions that approach the atomic level. Even under conditions where the contrast between the crystalline support and dispersed metal is not ideal, it should be possible to provide an upper limit for the size of the metal particles. TEM is most suited for the examination of materials where the supported material is of a rather uniform size distribution. Great care must be exerted by spectroscopists to prevent a focus on structures that are of unusual shape or large sizes as representative the state of the metal, because it is not, in general, possible to define quantitatively the fraction of metal represented by the large particles. The tendency will be to consider the larger particles as representative of the metal function when in fact they may represent only a minor fraction of the whole sample.

The definition of the state of two or more metals in the polyfunctional reforming catalyst becomes a more demanding task. Consider the case of the Pt−Sn-alumina catalyst. The data show that the Pt is reduced to the metal but that only a fraction of the Sn is reduced from the $Sn^{4+}$ state. This means that one can obtain only an "average" valence of the Sn. For catalyst formulations prepared by impregnation of both tin and platinum compounds, it has been demonstrated that some of the tin is reduced to the metallic state. Furthermore, the extent of tin reduction depends on the Sn/Pt ratio; the higher the ratio, the greater the reduction of tin to the zero valence state. For most catalyst compositions, the major Pt-Sn alloy is found to be PtSn. For very small Sn/Pt ratios one can observe some Pt-rich alloys; likewise, for high Sn/Pt ratios, higher tin alloys may be present. Considering the data from the characterization studies using a variety of techniques (Mössbauer, XRD, TEM, microdiffraction, XPS) for a series of these catalysts, a consistent model can be constructed for this catalyst system. A fraction of the tin is present in an oxidized state ($Sn^{2+}$, $Sn^{4+}$); some of the tin is found to be present as a material approximating a tin aluminate. Depending on the Sn/Pt ratio, some or all of the Pt will be present as an alloy with the dominant species being PtSn. As the surface area of the support increases, the fraction of tin present in an oxidized form increases. Thus, for the same Sn/Pt ratio, less alloy will be present when the support has a high surface area than when it is low. The extent of tin reduction also depends on the support; it is much easier to reduce tin to the zero valence state when the support is silica than when it is alumina. A variety of species were proposed by Srinivasan and Davis (316). A fraction of the surface is covered with a "monolayer" structure which resembles tin aluminate; presumably this tin species retards agglomeration of the Pt species present in the catalyst. Pt is distributed between metallic Pt and PtSn alloy forms; the fraction of each form of Pt depends on the Sn/Pt ratio and the surface area of the support.

In some commercial catalyst formulations, platinum may be supported on a tin/aluminum oxide prepared by coprecipitation. Catalysts prepared this way may have a very different structure than ones prepared by coimpregnation. In the case of one type of coprecipitated catalyst, essentially all of the Pt is present in an unalloyed form even for a Sn/Pt ratio of 3. It appears that for coprecipitated supports, one key role of tin is to provide surface traps to retard Pt diffusion and agglomeration. Alloy formation may be less of a factor.

In other cases, Pt-Re or Pt-Ir, the model described above is not appropriate. In these catalysts both Pt and the second metal appear to be mostly or completely reduced to the zero valence state but alloy formation does not occur to a significant extent. It does appear, however, that there is significant interaction of the two metals in the case of the Pt-Re catalyst. Some authors consider the Re to form monolayer rafts on the alumina support and these rafts serve as an interface between the alumina support and the Pt metal cluster, thereby altering the properties of the supported Pt and preventing agglomeration. While many authors indicate that some portion of the additional metals utilized together with platinum is not completely reduced, the role of the ions of the second component is not clear. One of the roles most frequently encountered in the literature is that of inhibiting the agglomeration of platinum.

In summary, it appears that many of the instrumental techniques for catalyst metal function characterization can be applied to the commercial naphtha reforming catalyst only with great difficult and with limited success. The techniques that depend on the presence of metallic crystals, e.g., XRD, are applicable only for postmortems to identify the reason for catalyst failure or aging. For the very active commercial catalyst, the characterization technique must be applicable to situations in which the dispersion is at, or near, the atomic level. For these characterizations, it appears that chemisorption offers the "best" characterization capabilities. Some of the newer instrumental techniques, e.g., EXAFS, offer the potential to provide a characterization at these high dispersions but further advances in the ability to derive structures from the experimental data must be attained.

## REFERENCES

1. R. L. Burwell, Jr., Manual of symbols and terminology of physicochemical quantities and units—Appendix II, Part II: heterogeneous catalysis. *Adv. Catal.* 26, 351 (1977).
2. V. Haensel, in *Chemistry of Petroleum Hydrocarbons* (B. T. Brooks, C. E. Boord, S. S. Kurtz, Jr., and L. Schmerling, eds.), Vol. 2, p. 189. Reinhold, New York, 1955.
3. S. Brunauer, P. H. Emmett, and E. Teller, *J. Am. Chem. Soc.* 60, 309 (1938).
4. P. H. Emmett, "Citation classic. Commentary on *J. Am. Chem. Soc.* 60, 309 (1938)." *Current Contents* 35, 9 (Auugst 29, 1977).
5. A. W. Adamson, *Physical Chemistry of Surfaces*, 5th ed. Wiley-Interscience, New York, 1990.
6. B. H. Davis, *Chemtech 21*, 18 (1991).
7. G. Halsey, *J. Phys. Chem.* 16, 931 (1948).
8. D. Avnir, *J. Am. Chem. Soc.* 109, 2931 (1987).
9. B. H. Davis, *Appl. Catal.* 10, 185 (1984).
10. E. Barrett, L. Joyner, and P. Halenda, *J. Am. Chem. Soc.* 73, 373 (1951).
11. A. Wheeler, *Adv. Catal.* 3, 249 (1951).
12. D. Dollomire and G. Heal, *J. Appl. Chem.* 14, 109 (1964).
13. J. C. P. Broekhoff, Adsorption and capillarity. Ph.D. Thesis, University of Delft, 1969.
14. J. C. P. Broekhoff and B. C. Linsen, in *Physical and Chemical Aspects of Adsorbents and Catalysts* (B. G. Linsen, ed.). Academic Press, New York, 1970.
15. B. A. Adkins and B. H. Davis, *Langmuir 3*, 722 (1987).
16. W. B. Innes, in *Experimental Methods in Catalytic Research* (R. B. Anderson, ed.), Vol. 1, pp. 45–99. Academic Press, New York, 1968.
17. W. B. Innes, *Anal. Chem.* 29, 1069 (1957).
18. R. B. Anderson, *J. Catal.* 3, 50 and 301 (1964).
19. L. G. Wayne, *J. Am. Chem. Soc.* 73, 5498 (1951).
20. B. C. Lippens, B. G. Linsen, and J. H. de Boer, *J. Catal.* 3, 32 (1964).
21. C. Schull, P. Elkin, and L. Roess, *J. Am. Chem. Soc.* 70, 1410 (1948).
22. C. Pierce, *J. Phys. Chem.* 57, 149 (1953).
23. R. W. Cranston and F. A. Inkley, *Adv. Catal.* 9, 143 (1957).
24. *Pure Appl. Chem.* 57, 603–619 (1985).
25. S. Brunauer, L. S. Deming, W. S. Deming, and E. Teller, *J. Am. Chem. Soc.* 62, 1723 (1990).
26. B. D. Adkins and B. H. Davis, *Ads. Sci. Technol.* 5, 168 (1988).
27. B. D. Adkins and B. H. Davis, *Ads. Sci. Technol.* 5, 76 (1988).
28. A. G. Foster, *Trans. Faraday Soc.* L8, 645 (1932).
29. L. H. Cohan, *J. Am. Chem. Soc.* 60, 433 (1938).
30. J. C. P. Broekhoff and J. H. de Boer, *J. Catal.* 9, 15 (1967).
31. J. C. P. Broekhoff and J. H. de Boer, *J. Catal.* 10, 368 (1967).

32. H. L. Ritter and L. C. Drake, *Ind. Eng. Chem. Anal. Ed. 20*, 665 (1948).
33. Effect of coating of catalyst supports in mercury porosimetry. Powerder Technology Note 16, Quantachrome Corp.
34. H. A. Benesi and B. H. C. Winquist, *Adv. Catal. 27*, 97 (1978).
35. C. Walling, *J. Am. Chem. Soc., 72*, 1164, (1950).
36. H. A. Benesi, *J. Catal. 28*, 176, (1973).
37. V. Gold and B. W. Hawes, *J. Chem. Soc.,* 2102 (1951).
38. N. C. Deno, J. J. Jaruzelski, and A. Schriescheim, *J. Am. Chem. Soc. 77*, 3044 (1955).
39. N. C. Deno, H. E. Berkheimer, W. L. Evans, and H. J. Peterson, *J. Am. Chem. Soc. 81*, 2344 (1959).
40. K. Tanabe, *Solid Acids and Bases*. Academic Press, New York, 1970.
41. K. Tanabe, in *Catalysis by Acids and Bases* (B. Imelik, C. Naccache, G. Coudurien, Y. Ben Taaret, and J. C. Vedrin, eds.), Vol. 20, p. 1. Elsevier, Amsterdam, 1985.
42. K. Tanabe, M. Misono, Y. Ono, and H. Hattori, *New Solid Acids and Bases*, Vol. 51. Elsevier, Amsterdam, 1989.
43. E. P. Parry, *J. Catal. 2*, 371 (1963).
44. J. B. Peri, *J. Phys. Chem. 69*, 220 (1965).
45. N. D. Parkyns, *J. Phys. Chem. 75*, 526 (1971).
46. P. Fink, *Z. Chem. 7*, 324 (1967).
47. J. W. Highower and W. K. Hall, *Trans. Faraday Soc. 66*, 477 (1970).
48. J. B. Peri, *Int. Congr. Catal. 2nd*, Paris, 1960, p. 1333 (1961).
49. R. P. Eischens and W. A. Pliskin, *Adv. Catal. 10*, 2-95 (1958).
50. H. Knoözinger, *NATO ASE Ser. B 265*, 167 (1991).
51. A. T. Bell, *Springer Ser. Chem. Phys. 35* (Chem. Phys. Solid Surf. 5), 23–38 (1984).
52. E. Paukstis and E. N. Yurchenko, *Usp. Khim. 52*, 426 (1983).
53. P. C. M. Van Woerkom and R. L. De Grott, *Appl. Opt. 21*, 3114 (1982).
54. A. V. Kiselev and V. I. Lygin, *Infrared Spectra of Surface Compounds*. Wiley, New York, 1975.
55. J. B. Peri, *J. Phys. Chem. 69*, 211 (1965).
56. J. B. Peri and R. B. Hannan, *J. Phys. Chem. 64*, 1526 (1960).
57. J. L. Carter, P. J. Lucchesi, P. Cornell, D. J. C. Yates, and J. H. Sinfelt, *J. Phys. Chem. 69*, 3070 (1965).
58. H. Dunken and P. Fink, *Z. Chem. 6*, 194 (1966).
59. W. K. Hall, H. P. Leftin, P. J. Cheselsee, and D. E. O'Reilly, *J. Catal. 2*, 506 (1963).
60. J. B. Peri, *J. Phys. Chem. 69*, 231 (1965).
61. K. Knözinger, in *Catalysis by Acids and Bases* (B. Imelik et al., eds.). Elsevier, Amsterdam, 1985, pp. 111–125.
62. V. B. Kazansky, V. Yu. Borovkov, and L. M. Kustov, *Proc. 8th Int. Congr. Catal.*, Berlin, (1984), Vol III, pp 3–14.
63. E. Escalona Platero and C. Otero Aren, *J. Catal. 107*, 244–247 (1987).
64. C. Morterra,, G. Magnacca, F. Filippi, and A. Giachello, *J. Catal. 137*, 346–356 (1992).

65. M. A. Vannice, L. C. Hasselbring, and B. Sen, *J. Catal.* 95, 57 (1985).
66. A. Gervasini and A. Auroux, *J. Thermal Anal.* 37, 1737 (1991).
67. N. Cardona-Martinez and J. A. Dumesic, *J. Catal.* 128, 23 (1991).
68. N. Cardona-Martinez and J. A. Dumesic, *J. Catal.* 127, 706 (1991).
69. B. C. Gates, J. R. Katzer, and G. C. A. Schuit, in *Chemistry of the Catalytic Processes*. McGraw-Hill, New York, 1979, p. 289.
70. J. C. Musso and J. M. Parera, *Appl. Catal.* 30, 81 (1987).
71. A. Parmaliana, F. Frusteri, A. Mezzapica, and N. Giordano, *J. Catal.* 111, 235 (1988).
72. H. Lieske, G. Lietz, H. Spindler, and J. Völter, *J. Catal.* 81, 8 (1983).
73. S. Sivasanker, A. V. Ramaswamy, and P. Ratnasamy, in *Preparation of Catalysts II* (B. Delmon, P. Grange, P. A. Jacobs, and G. Poncelet, eds.). Elsevier, Amsterdam, 1979, p. 185.
74. A. A. Castro, O. A. Scelza, E. R. Benvenuto, G. T. Baronetti, and J. M. Parera, *J. Catal.* 69, 222 (1981).
75. A. A. Castro, O. A. Scelza, G. T. Baronetti, M. A. Fritzler, and J. M. Parera, *Appl. Catal.* 6, 347 (1983).
76. A. Bishara, K. M. Murad, A. Stanislaus, M. Ismail, and S. S. Hussain, *Appl. Catal.* 7, 337 (1983).
77. J. M. Grau, E. L. Jablonski, C. L. Pieck, R. J. Verderone, and J. M. Perera, *Appl. Catal.* 36, 109 (1988).
78. P. P. Mardilovich, G. N. Lysenko, G. G. Kupchenko, P. V. Kurman, L. I. Titova, and A. I. Trokhimets, *React. Kinet. Catal. Lett.* 36, 357 (1988).
79. F. Arena, F. Frusteri, N. Mondello, N. Giordano, and A. Parmaliana, *J. Chem. Soc. Faraday Trans.* 88 (22), 3353 (1992).
80. A. Bishara, K. M. Murad, A. Stanislaus, M. Ismail, and S. S. Hussain, *Appl. Catal.* 6, 347 (1983).
81. A. Ayame, G. Sawada, H. Sato, G. Zhang, T. Ohta, and T. Izumizawa, *Appl. Catal.* 48, 25 (1989).
82. M. Tanaka and S. Ogasawara, *J. Catal.* 16, 157 (1970).
83. E. G. Garbowski and M. Primet, *J. Chem. Soc. Faraday Trans. 1* 83, 1469 (1987).
84. A. N. Webb, *Ind. Eng. Chem.* 49, 261 (1957).
85. P. Strand and M. Kraus, *Collect. Czech. Chem. Commun.* 30, 1136 (1965).
86. K. Matsuura, T. Watanabe, A. Suzuki, and M. Itoh, *J. Catal.* 26, 127 (1972).
87. V. V. Yusheckenki and T. V. Antipina, *Kinet. Katal.* 11, 134 (1970).
88. P. O. Scokart, S. A. Selim, J. P. Damon, and P. G. Rouxhet, *J. Colloid, Interface Sci.* 70, 209 (1979).
89. F. P. J. M. Kerkhof, J. C. Oudejans, J. A. Moulijn, and E. R. A. Matulewicz, *J. Colloid. Interface Sci.* 77, 120 (1980).
90. V. C. F. Holm and A. Clark, *J. Catal.* 8, 286 (1967).
91. A. Corma, V. Fornés, and E. Ortega, *J. Catal.* 92, 284 (1985).
92. E. R. A. Matulewicz, F. P. J. M. Kerkhof, J. A. Moulijn, and H. A. Reitsma, *J. Colloid. Interface Sci.* 77, 110 (1980).
93. A. E. Hirschler, *J. Catal.* 2, 428 (1963).
94. A. B. R. Weber, Thesis, University of Delft, Netherlands, 1957.

95. V. C. F. Holm and A. Clark, *Ind. Eng. Chem. Product Res. Dev. 2*, 38 (1963).
96. A. Clark, V. C. F. Holm, and D. M. Blackburn, *J. Catal. 1*, 244 (1962).
97. R. I. Hedge and M. A. Barteau, *J. Catal. 120*, 387 (1989).
98. P. O. Scokart and P. G. Rouxhet, *J. Colloid Interface Sci. 86*, 96 (1982).
99. S. Kowalak, *Acta Chim. Acad. Sci. Hung. 107*, 27 (1981).
100. S. Kowalak, *Acta Chim. Acad. Sci. Hung. 107*, 19 (1981).
101. D. E. O'Reilly, *Adv. Catal. 12*, 66 (1960).
102. E. C. DeCanio, J. C. Edwards, T. R. Scalzo, D. A. Storm, and J. W. Bruno, *J. Catal. 132*, 498 (1991).
103. K. W. McHenry, R. J. Bertolacini, H. M. Brennan, J. L. Wilson, and H. S. Seeling, *Proc. 2nd Int. Congr. Catal.*, Paris, 1960, Vol II, p 2293 (1962).
104. L. C. Drake, personal communication.
105. N. R. Bursian, S. B. Kogan, and Z. A. Davydova, *Kinet. Katal. 8*, 1085 (1967).
106. M. F. L. Johnson and C. D. Keith, *J. Phys. Chem. 67*, 200 (1963).
107. P. Putanov, M. Ivanovic, and O. Selakovic, *React. Kinet. Catal. Lett. 8*, 223 (1978).
108. H. E. Kluksdahl and R. J. Houston, *J. Phys. Chem. 65*, 1469 (1961).
109. S. I. Ermakova and N. M. Zaidman, *Kinet. Katal. 10*, 1158 (1969).
110. V. I. Shekhobalova and Z. V. Kuk'yanova, *Russian J. Phys. Chem. 53*, 1551 (1979).
111. G. Lietz, H. Lieske, N. Spindler, W. Hanke, and J. Völter, *J. Catal. 81*, 17 (1983).
112. M. J. P. Botman, L. Q. She, J. Y. Zhang, W. L. Driessen, and V. Ponec, *J. Catal. 103*, 280 (1987).
113. M. J. P. Botman, Ph.D. Thesis, Univ. of Lieden, 1987.
114. J. L. Lemaitre, F. G. Menon, and F. Delannay, in *Characterization of Heterogeneous Catalysts* (F. Delannay, ed.). Marcel Dekker, New York, 1984, pp. 299–366.
115. C. R. Adams, H. A. Benesi, R. M. Curtis, and R. G. Meisenheimer, *J. Catal. 1*, 336 (1962).
116. H. P. Klug and L. E. Alexander, *X-ray Diffraction Procedures*, Chapter 9. Wiley, New York, 1954.
117. R. Srinivasan, L. Rice, and B. H. Davis, *J. Am. Ceram. Soc. 73*, 3528 (1990).
118. H. E. Rhodes, P.-K. Wang, H. T. Stokes, C. P. Slichter, and J. H. Sinfelt, *Phys. Rev. B 26*, 3599 (1982).
119. R. Van Hardevelt and F. Hartog, *Surf. Sci. 15*, 189 (1969).
120. D. White, T. Baird, J. R. Fryer, L. A. Freeman, D. J. Smith, and M. Day, *J. Catal. 81*, 119 (1983).
121. P. L. Gai, M. J. Goringe, and J. C. Barry, *J. Microsc. 142*, 9 (1986).
122. A. K. Datye and D. J. Smith, *Catal. Rev.-Sci. Eng. 34*, 129 (1992).
123. M. M. J. Treacy and S. B. Rice, *J. Microsc. 156*, 211 (1989).
124. J. Schwank, L. F. Allard, M. Deeba, and B. C. Gates, *J. Catal. 84*, 27 (1983).
125. B. Tesche, E. Zeilter, E. A. Delgado, and H. Knözinger, in *Proceedings of the 40th Annual Meeting of the Electron Microscopy Society of America* (G. W. Bailey, ed.). San Francisco Press, San Francisco, 1982, p. 682.
126. I. Langmuir, *J. Am. Chem. Soc. 38*, 2221 (1916).

127. A. F. Benton and T. White, *J. Am. Chem. Soc.* 54, 1820 (1932).
128. H. H. Podgurski and P. H. Emmett, *J. Am. Chem. Soc.* 57, 159 (1953).
129. S. Brunauer and P. H. Emmett, *J. Am. Chem. Soc.* 57, 1754 (1935).
130. S. Brunauer and P. H. Emmett, *J. Am. Chem. Soc.* 59, 310 (1937).
131. G. H. Via, G. Meitzner, F. W. Lytle, and J. H. Sinfelt, *J. Chem. Phys.* 79, 1527 (1982).
132. P. H. Hollaway, *Surf. Interface Anal.* 7, 204–205 (1985).
133. H. A. Benesi, R. M. Curtis, and H. P. Studer, *J. Catal.* 10, 328 (1968).
134. T. A. Dorling, M. J. Eastlake, and R. L. Moss, *J. Catal.* 14, 23 (1969).
135. T. A. Dorling, B. W. J. Lynch, and R. L. Moss, *J. Catal.* 20, 190 (1971).
136. K. Christmann, G. Ertle, and T. Pignet, *Surf. Sci.* 54, 365 (1976).
137. S. F. Adler and J. J. Keavney, *J. Phys. Chem.* 64, 208 (1960).
138. R. H. Herrmann, S. F. Adler, M. S. Goldstein, and R. M. DeBaun, *J. Phys. Chem.* 65, 2189 (1961).
139. V. S. Boronin, V. S. Nikulina, and O. M. Poltorak, *Russian J. Phys. Chem.* 37, 1173 (1963).
140. V. S. Boronin, O. M. Poltorak, and A. O. Turakulova, *Russian J. Phys. Chem.* 48, 156 (1974).
141. A. Frennet and P. B. Wells, *Appl. Catal.* 18, 243 (1985).
142. J. A. Rabo, V. Schomaker, and P. E. Pickert, in *Proceedings of the 3rd International Congress on Catalysis*, Amsterdam, 1964, p. 11–1264.
143. S. Sato, *J. Catal.* 92, 11 (1985).
144. G. B. McVicker, R. T. K. Baker, R. L. Garten, and E. L. Kugler, *J. Catal.* 65, 207 (1980).
145. S. Kirshnamurthy, G. R. Landolt, and H. J. Schoennagel, *J. Catal.* 78, 319 (1982).
146. J. C. Vis, H. F. J. Van't Blik, T. Huizinga, J. Van Grondelle, and R. Prins, *J. Catal.* 59, 333 (1985).
147. J. C. Vis, H. F. J. Van't Blik, T. Huizinga, J. Van Grondelle, and R. Prins, *J. Mol. Catal.* 25, 367 (1984).
148. D. C. Koningsberger and D. E. Sayers, *Solid State Ionics* 16, 23 (1985).
149. B. J. Kip, J. Van Grondelle, J. H. A. Martens, and R. Prins, *Appl. Catal.* 26, 353 (1986).
150. B. J. Kip, F. B. M. Duivenvoorden, K. C. Koningsberger, and R. Prins, *J. Catal.* 105, 25–38 (1987).
151. J. H. Sinfelt, Y. L. Lam, J. A. Cusumano, and A. E. Barnett, *J. Catal.* 42, 227 (1976).
152. J. H. Sinfelt and G. H. Via, *J. Catal.* 56, 1 (1979).
153. A. Sayari, H. T. Wang, and J. G. Goodwin, Jr., *J. Catal.* 93, 368 (1985).
154. C. Yang and J. R. Goodwin, Jr., *J. Catal.* 78, 182 (1982).
155. R. Kramer and M. Andre, *J. Catal.* 58, 287 (1979).
156. R. R. Cavanagh and J. T. Yates, Jr., *J. Catal.* 68, 22 (1981).
157. D. Bianchi, M. Lacroix, G. Pajonk, and S. J. Teichner, *J. Catal.* 59, 467 (1979).
158. S. E. Wanke and N. A. Dougharty, *J. Catal.* 24, 367 (1972).
159. J. A. Konvalinka and J. J. F. Scholten, *J. Catal.* 48, 374 (1977).
160. W. Eberhardt, F. Greuter, and E. W. Plummer, *Phys. Rev. Lett.* 46, 1085 (1981).

161. J. T. Yates, Jr., C. H. F. Peden, J. E. Houston, and D. W. Goodman, *Surf. Sci. 160*, 37 (1985).
162. J. E. Benson and M. Boudart, *J. Catal. 4*, 704, (1965).
163. P. G. Menon, in *Advances Catal. Sci. & Tech.*, Proc. 7th Natl. Catal. Symp., Baroda, India, 1985. Wiley Eastern, New Delhi, 1986, pp. L1-15.
164. J. Prasad, K. R. Murthy, and P. G. Menon, *J. Catal. 52*, 515 (1978).
165. B. H. Isaacs and E. E. Petersen, *J. Catal. 85*, 1 (1984).
166. M. Kobayashi, Y. Inoue, N. Takahashi, R. L. Burwell, J. B. Butt, and J. B. Cohen, *J. Catal. 64*, 74 (1980).
167. J. Freel, *J. Catal. 25*, 149 (1972).
168. D. J. O'Rear, D. G. Löffler, and M. Boudart, *J. Catal. 121*, 131-140 (1990).
169. J. L. Gland, *Surf. Sci. 93*, 487 (1980).
170. G. R. Wilson and W. K. Hall, *J. Catal. 17*, 190 (1970).
171. G. R. Wilson and W. K. Hall, *J. Catal. 24*, 306 (1972).
172. P. Gallezot, A. Alarcon-Diaz, J. A. Dalmon, A. J. Renouprez, and B. Imelik, *J. Catal. 39*, 334 (1975).
173. F. P. Netzer and H. L. Gruber, *Z. Phys. Chem. NF96*, 25 (1975).
174. P. G. Menon and G. F. Froment, *J. Catal. 59*, 138 (1979).
175. G. Porod, in *Small-Angle X-ray Scattering* (O. Glatter and O. Kratky, eds.). Academic Press, New York, 1982, Part II, Chapter 2.
176. H. Brumberger, *Trans. Am. Crystallogr. Assoc. 19*, 1 (1983).
177. H. Brumberger, F. Delaglio, J. Goodisman, and M. Whitefield, *J. Appl. Crystallogr. 19*, 287 (1986).
178. G. L. Brill, C. G. Weil, and P. W. Schmidt, *J. Colloid Interface Sci. 27*, 479 (1968).
179. F. Delaglio, J. Goodisman, and H. Brumberger, *J. Catal. 99*, 383 (1986).
180. P. G. Hall and R. T. Williams, *J. Colloid Interface Sci. 104*, 151 (1985).
181. T. E. Whyte, Jr., P. W. Kirklin, R. W. Gould, and H. Heinemann, *J. Catal. 25*, 407 (1972).
182. G. Cocco, L. Schiffini, G. Strukul, and G. Garturan, *J. Catal. 65*, 348 (1980).
183. G. A. Somorjai, R. E. Powell, P. W. Montgomery, and G. Jura, in *Small-Angle X-ray Scattering* (H. Brumberger, ed.). Gordon & Breach, New York, 1967, pp. 449-466.
184. H. Brumberger, *Makromol. Chem. Macromol. Symp. 15*, 223-230 (1988).
185. J. Goodisman, H. Brumberger, and R. Cupelo, *J. Appl. Cryst. 14*, 305-308 (1981).
186. H. Brumberger, F. Delaglio, J. Goodisman, M. G. Phillips, J. A. Schwartz, and P. Sen, *J. Catal. 92*, 199-210 (1985).
187. H. Brumberger, Y. C. Chang, M. G. Phillips, F. Delaglio, and J. Goodisman, *J. Catal. 97*, 561 (1986).
188. H. Brumberger and J. Goodisman, *J. Appl. Crystallogr. 16*, 83 (1983).
189. R. de L. Kronig, *J. Phys. 76*, 468 (1932).
190. H. Winick and S. Doniach, *Synchrotron Radiation Research*. Plenum, New York, 1980.
191. E. A. Stern, D. E. Sayres, and F. W. Lytle, *Phys. Rev. B 11*, 4825, 4836 (1975).
192. B. K. Teo, *EXAFS: Basic Principles and Data Analysis* Springer-Verlag, Berlin, 1986.

193. B. K. Teo and D. C. Joy, eds., *EXAFS Spectroscopy: Techniques and Applications*. Plenum, New York, 1981.
194. F. W. Lytle, R. B. Greegor, and E. C. Marques, in *Proceedings of the International Catalysis Congress, 9th* (M. J. Phillips and M. Ternan, eds.), Chemical Institute of Canada, Ottowa, 1988, Vol. 5, p. 54.
195. J. H. Sinfelt, G. H. Via, and F. W. Lytle, *Catal. Rev. Sci. Eng.* 26, 81 (1984).
196. R. Prins and D. C. Koningsberger, in *Catalysis*, Springer-Verlag, Berlin (J. R. Anderson and M. Boudart, eds.), 1988, Vol. 8, p. 321.
197. D. Bazin, H. Dexpert, and P. Lagarde, *Topics Current Chem.* 145, 69 (1988).
198. P. Lagarde and H. Dexpert, *Adv. Phys.* 33, 567 (1984).
199. J. H. Sinfelt and G. D. Meitzner, *Accounts Chem. Res.* 26, 1 (1993).
200. S. M. Heald and J. M. Tranquada, *Phys. Methods Chem.* Wiley, New York, 2nd ed., 1990, Vol. 5, p. 189.
201. N. S. Guyot-Sionnest, D. Bazin, J. Lunch, J. T. Bournonville, and H. Dexpert, *Physica B 158*, 211 (1989).
202. B. Lengeler and P. Eisenberger, *Phys. Rev. B21*, 4057 (1980).
203. J. Zhao and P. A. Montano, *Phys. Rev. B40*, 3401 (1989).
204. J. Zhao, Ph.D. Thesis, West Virginia Univ., 1991; personal communication.
205. F. W. Lytle, R. B. Greegor, E. C. Marques, D. R. Sandstrom, G. H. Via, and J. H. Sinfelt, *J. Catal.* 95, 546 (1985).
206. F. W. Lytle, R. B. Greegor, E. C. Marques, V. A. Biebesheimer, D. R. Sandstrom, J. A. Horsley, G. H. Via, and J. H. Sinfelt, in *Catalyst Characterization Science* (M. L. Deveney and J. L. Gland, eds.), *ACS Symp. Series 288*, 280 (1985).
207. E. C. Marques, D. R. Sandstrom, F. W. Lytle, and R. B. Greegor, *J. Chem. Phys.* 77(2), 1027 (1982).
208. R. A. Dalla Betta, M. Boudart, K. Foger, D. G. Löffler, and J. Sánchez-Arrieta, *Rev. Sci. Instrum.* 55, 1910 (1984).
209. T. L. Neils and J. M. Burlitch, *J. Catal.* 118, 79 (1989).
210. H. Dexpert, *J. Phys. Colloqu. C8, Suppl.* 12 47, C8-219, C8-226 (1986).
211. N. S. Guyot-Sionnest, F. Villain, D. Bazin, H. Dexpert, F. Le Peltier, J. Lynch, and J. P. Bournonville, *Catal. Lett.* 8, 283 (1991).
212. P. S. P. Wei and F. W. Lytle, *Phys. Rev. B 19*, 679 (1979).
213. F. W. Lytle, *J. Catal.* 43, 376–379 (1976).
214. G. Meitzner, G. H. Via, F. W. Lytle, and J. H. Sinfelt, *J. Phys. Chem.* 96, 4960 (1992).
215. F. W. Lytle, P. S. P. Wei, R. B. Greegor, G. H. Via, and J. H. Sinfelt, *J. Chem. Phys.* 70, 4849 (1979).
216. J. H. Sinfelt, G. H. Via, F. W. Lytle, and R. B. Greegor, *J. Chem. Phys.* 75, 5527 (1981).
217. I. Yu, A. A. V. Gibson, E. R. Hunt, and W. P. Halperin, *Phys. Rev. Lett.* 44, 348 (1980).
218. L.-C. de Ménorval and J. P. Fraissard, *Chem. Phys. Lett.* 77, 309 (1981).
219. D. Rouabah, R. Benslama, and J. Fraissard, *Chem. Phys. Lett.* 179, 218 (1991).
220. L.-C. de Ménorval, T. Ito, and J. Fraissard, *J. Chem. Soc. Faraday Trans.* 78, 403 (1982).

221. R. A. Dalla Betta and M. Boudart, in *Proceedings, 5th International Congress on Catalysis*, Vol. 2 (J. W. Hightower, ed.). North-Holland, Amsterdam, 1973, p. 1329.
222. M. Boudart and G. Djéga-Mariadassou, *Kinetics of Heterogeneous Catalysis Reactions*. Princeton Univ. Press, Princeton, NJ, 1984.
223. P. Gallezot, I. Mutin, G. Dalmai-Imelik, and B. Imelik, *J. Microsc. Spectrosc. Electron 1*, 1 (1976).
224. P. Gallezot, A. I. Bienenstock, and M. Boudart, *Nouv. J. Chim. 2*, 263 (1978).
225. J. P. Fraissard and T. Ito, *Zeolites 8*, 350 (1988).
226. M. Boudart, R. Ryoo, G. P. Valenca, and R. Van Grieken, *Catal. Lett. 17*, 273 (1993).
227. R. S. Weber, Ph.D. Dissertation, Stanford Univ., Stanford, CA, 1985.
228. R. Ryoo, C. Pak, and B. Chmelka, *Zeolites 10*, 790 (1990).
229. M. Boudart and G. P. Valenca, *J. Catal. 128*, 447 (1991).
230. M. Boudart, M. G. Samant, and R. Ryoo, *Ultramicroscopy 20*, 125 (1986).
231. F. J. Rivera-Latas, R. A. Dalla Betta, and M. Boudart, *AlChE J. 38*, 771 (1992).
232. H. Dexpert, E. Freund, E. Lesage, and J. P. Lynch, *Stud. Surf. Sci. Catal. 11*, 53 (1982).
233. S. Subramanian, *Platinum Metals Rev. 36*, 98–103 (1992).
234. S. Subramanian and J. A. Schwarz, *Appl. Catal. 68*, 131 (1991).
235. C. Defossé in *Characterization of Heterogeneous Catalysts* (F. Delannay, ed.). Marcel Dekker, New York, 1984, pp. 225–298.
236. R. J. Bird and P. Swift, *J. Electron Spectrosc. Relat. Phenom. 21*, 227 (1980).
237. M. Boudart, L.-C. de Ménorval, J. P. Fraissard, and G. P. Valenca, *J. Phys. Chem. 92*, 4033 (1988).
238. C. J. Powell, N. E. Erickson, and T. E. Madey, *J. Electron. Spectrosc. Relat. Phenom. 17*, 361 (1979).
239. W. P. Dianis and J. E. Lester, *Anal. Chem. 45*, 1416 (1973).
240. J. P. Contour and G. Mouvier, *J. Electron. Spectrosc. Relat. Phenom. 7*, 85 (1975).
241. J. L. Ogilvie and A. Wolberg, *Appl. Spectrosc. 26*, 401 (1972).
242. T. E. Madey, C. D. Wagner, and A. Joshi, *J. Electron Spectrosc. Relat. Phenom. 10*, 359 (1977).
243. G. B. Hoflund, D. A. Asbury, C. F. Corallo, and G. R. Corallo, *J. Vac. Sci. Technol. A6*, 70 (1988).
244. D. A. Asbury and G. B. Hoflund, *Surf. Sci. 199*, 552 (1988).
245. D. J. Dwyer, J. L. Robbins, S. D. Cameron, N. Dudash, and J. Hardenbergh, in *Strong Metal-Support Interactions* (R. T. K. Baker, S. J. Tauster, and J. A. Dumesic, eds.). American Chemical Society, Washington, DC, 1986, pp. 21–33.
246. D. J. Dwyer, S. D. Cameron, and J. Gland, *Surf. Sci. 159*, 430 (1985).
247. F. Ewertowski, J. Klimas, and A. Maj, *Rocz. Chem. 49*, 351 (1975).
248. J. Klimas and F. Ewertowski, *Pol. J. Chem. 54*, 867 (1980).
249. W. A. Pliskin and R. P. Eischens, *Z. Phy. Chem. (Frankfurt) NF 24*, 11 (1960).
250. R. Barth, R. Pitchai, R. L. Anderson, and X. E. Verykios, *J. Catal. 116*, 61 (1989).
251. J. B. Lantz and R. D. Gonzalez, *J. Catal. 41*, 293 (1976).

252. B. Sen and M. A. Vannice, *J. Catal.* **130**, 9 (1991).
253. M. F. L. Johnson and V. M. LeRoy, *J. Catal.* **35**, 434 (1974).
254. A. Webb, *J. Catal.* **39**, 485 (1975).
255. M. F. L. Johnson, *J. Catal.* **39**, 487 (1975).
256. B. D. McNicol, *J. Catal.* **46**, 428 (1977).
257. C. Bolivar, H. Charcosset, R. Frety, M. Primet, L. Tournayan, C. Betizeau, G. Leclercq, and R. Maurel, *J. Catal.* **39**, 249 (1975).
258. H. C. Yao and M. Shelef, *J. Catal.* **44**, 392 (1976).
259. N. Wagstaff and R. Prins, *J. Catal.* **59**, 434 (1979).
260. B. H. Isaacs and E. E. Petersen, *J. Catal.* **77**, 43 (1982).
261. R. L. Mieville, *J. Catal.* **87**, 437 (1984).
262. W. F. Ambs and M. M. Mitchell, *J. Catal.* **82**, 226 (1983).
263. P. G. Menon, J. Sieders, F. J. Streefkerk, and G. J. M. van Keulen, *J. Catal.* **29**, 188 (1973).
264. V. Eskinazi, *Appl. Catal.* **4**, 37 (1982).
265. P. S. Kirlin, B. R. Strohmeier, and B. C. Gates, *J. Catal.* **98**, 308 (1986).
266. S. R. Adkins and B. H. Davis, *ACS Symp. Ser.* **228**, 57 (1985).
267. M. S. Nacheff, L. S. Kraus, M. Ichikawa, B. M. Hoffman, J. B. Butt, and W. M. H. Sachtler, *J. Catal.* **106**, 263 (1987).
268. J. H. Onuferko, D. R. Short, and M. J. Kelley, *Appl. Surf. Sci.* **19**, 227 (1984).
269. D. R. Short, S. M. Khalik, J. R. Katzer, and M. J. Kelley, *J. Catal.* **72**, 288 (1981).
270. M. J. Kelley, R. L. Freed, and D. G. Swartzfager, *J. Catal.* **78**, 445 (1982).
271. G. Meitzner, G. H. Via, F. W. Lytle, and J. H. Sinfelt, *J. Chem. Phys.* **87**, 6354 (1987).
272. D. Bazin, H. Dexpert, P. Lagarde, and J. P. Bournonville, *J. Phys.* **47**, C8–293 (1986).
273. F. Hilbrig, C. Michel, and G. L. Haller, *J. Phys. Chem.* **96**, 9893 (1992).
274. M. A. Vuurman, D. J. Stufkens, A. Oskam, and I. E. Wachs, *J. Mol. Catal.* **76**, 263 (1992).
275. Nan-Yu Tôpsoe, *J. Catal.* **128**, 499 (1991).
276. A. M. Turek, E. Decanio, and I. E. Wachs, *J. Phys. Chem.* **96**, 5000 (1992).
277. M. Sibeijn, R. Spronk, J. A. R. van Veen, and J. C. Mol, *Catal. Lett.* **8**, 201 (1991).
278. F. D. Hardcastle, I. E. Wachs, J. A. Horsley, and G. H. Via, *J. Mol. Catal.* **46**, 15 (1988).
279. K. Jothimurugesan, A. K. Nayak, G. K. Mehta, K. N. Rai, S. Bhatia, and R. D. Srivastava, *AIChE J.* **31**, 1997 (1985).
280. M. J. Kelley, D. R. Short, and D. G. Swartzfanger, *J. Mol. Catal.* **20**, 235 (1983).
281. R. Burch, *Platinum Metals Rev.* **22**, 57 (1978).
282. B. A. Sexton, A. E. Hughes, and K. Folger, *J. Catal.* **88**, 1566 (1984).
283. H. Lieske and J. Völter, *J. Catal.* **90**, 46 (1984).
284. R. Burch, *J. Catal.* **71**, 348 (1981).
285. R. Burch and L. C. Garla, *J. Catal.* **71**, 360 (1981).
286. R. Burch and A. J. Mitchell, *Appl. Catal.* **6**, 121 (1981).

287. A. C. Müller, P. A. Engelhard, and J. E. Weisang, *J. Catal.* 56, 65 (1979).
288. S. R. Adkins and B. H. Davis, *J. Catal.* 89, 371 (1984).
289. J. M. Stencel, J. Goodman, and B. H. Davis, *Proc. 9th Int. Congr. Catal.* 3, 1291 (1988).
290. Y.-X. Li, J. M. Stencel, and B. H. Davis, *Reaction Kinet. Catal. Lett.* 37, 273 (1988).
291. Y.-X. Li, J. M. Stencel, and B. H. Davis, *Appl. Catal.* 64, 71 (1990).
292. R. Bouwman and P. Biloen, *Surf. Sci.* 41 348 (1974).
293. G. B. Hoflund, D. A. Asbury, P. Kirszenzatejn, and H. A. Laitinen, *Surf. Interface Anal.* 9, 169 (1986).
294. G. B. Hoflund, in *Preparation of Catalysts III* (G. Poncelet, P. Grange, and P. A. Jacobs, eds.). Elsevier, Amsterdam, 1983, pp. 91–100.
295. S. D. Gardner, G. B. Hoflund, and D. R. Schryer, *J. Catal.* 119, 179 (1989).
296. D. F. Cox and G. B. Hoflund, *Surf. Sci.* 151, 202 (1985).
297. S. D. Gardner, G. B. Hoflund, D. R. Schryer, and B. T. Upchurch, *J. Phys. Chem.* 95, 835 (1991).
298. H. A. Laitinen, J. R. Waggoner, C. Y. Chan, P. Kirszensztejn, D. A. Asbury, and G. B. Hoflund, *J. Electrochem. Soc.* 133, 1586 (1986).
299. G. B. Hoflund, D. A. Asbury, and R. E. Gilbert, *Thin Solid Films* 129, 139 (1985).
300. R. Bacaud, P. Bussiere, and F. Figueras, *J. Catal.* 69, 399 (1981).
301. V. H. Berndt, H. Mehner, J. Völter, and W. Meise., *Z. Anorg. Allg. Chem.* 429, 47 (1977).
302. R. Bacaud, P. Bussiere, F. Figueras, and J. P. Mathieu, *C. R. Acad. Sci. Paris Ser. C 281*, 159 (1975).
303. R. Bacaud, P. Bussiere, and F. Figueras, *J. Phys. Colloq.* 40, C2-94 (1979).
304. J. S. Charlton, M. Cordey-Hayes, and I. R. Harris, *J. Less-Common Met.* 20, 105 (1970).
305. Y.-X. Li, Y.-. Zhang, and K. J. Klabunde, *Langmuir* 4, 385 (1988).
306. K. J. Klabunde, Y.-X. Li, and K. F. Purcell, *Hyperfine Interact.* 41, 385 (1988).
307. L. Lin, R. Wu, J. Zang, and B. Jiang, *Acta Pertrol. Sci. (China)* 1, 73 (1980).
308. N. A. Pakhomov, R. A. Buyanov, E. N. Yurchenko, A. P. Cherynshev, G. R. Kotel'nikov, E. M. Moroz, N. A. Zaitseva, and V. A. Patanov, *Kinet. Katal.* 22, 488 (1981).
309. P. R. Gray and F. E. Farha, in *Mössbauer Effect Methodology* (I. J. Grunerman and C. W. Seidel, eds.). Plenum, New York, 1976, Vol. 10, p. 47.
310. E. N. Yurchenko, V. I. Kuznetsov, V. P. Melnikova, and A. N. Startsev, *React. Kinet. Catal. Lett.* 23, 137 (1983).
311. P. Zhang, H. Shao, X. Yang, and L. Pang, *Cuihua Xuebao* 5, 101 (1984).
312. Y.-X. Li, Y.-F. Zhang, and Y.-F. Shia, *Cuihua Xuebao* 5, 311 (1985).
313. S. Zhang, B. Xie, P. Wang, and J. Zhang, *Cuihua Xuebao* 1, 311 (1980).
314. Y.-X. Li, K. J. Klabunde, and B. H. Davis, *J. Catal.* 128, 1 (1991).
315. V. I. Kuznetsov, A. S. Belyi, E. N. Yurchenko, M. D. Smolikov, M. T. Protasova, E. V. Zatolokina, and V. K. Duplyakin, *J. Catal.* 99, 159 (1986).
316. R. Srinivasan and B. H. Davis, *Platinum Metal Rev.* 36, 151 (1992).
317. R. Srinivasan, R. J. De Angelis, and B. H. Davis, *J. Catal.* 106, 449 (1987).

318. Y.-X. Li, N.-S. Chiu, W.-H. Lee, S. H. Bauer, and B. H. Davis, Characterization and catalyst development. An interactive approach. *ACS Symp. Ser. 411*, 328 (1989).
319. N.-S. Chiu, W.-H. Lee, Y.-X. Li, S. H. Bauer, and B. H. Davis, in *Advances in Hydrotreating Catalysts* (M. L. Occelli and R. G. Anthony, eds.). Elsevier, Amsterdam, 1989, pp. 147-163.
320. G. Metizner, G. H. Via, F. W. Lytle, S. C. Fung, and J. H. Sinfelt, *J. Phys. Chem. 92*, 2925 (1988).
321. R. Srinivasan and B. H. Davis, *Appl. Catal. (A General) 87*, 45 (1992).
322. J. H. Sinfelt, *Bimetallic Catalysts. Discoveries, Concepts, and Applications.* Wiley, New York, 1983.
323. S. C. Chan, S. C. Fung, and J. H. Sinfelt, *J. Catal. 113*, 164 (1988).
324. S. Subramanian and J. A. Schwarz, *Appl. Catal. 68*, 131 (1991).
325. J.-J. Huang, S. C. Fung, W. E. Gates, and G. B. McVicker, *J. Catal. 118*, 192 (1989).
326. S. Subramanian and J. A. Schwarz, *Appl. Catal. 61*, L15 (1990).
327. Y.-J. Huang and S. C. Fung, *J. Catal. 131*, 378 (1991).
328. G. B. McVicker and J. J. Ziemiak, *Appl. Catal. 14*, 229 (1985).
329. H. von Brandis, F. E. Wagner, J. A. Sawicki, K. Marcinkowska, and J. H. Rolston, *Hyperfine Interact. 57*, 2127 (1990); J. A. Sawicki, F. E. Wagner, and J. H. Rolston, *Mater. Res. Soc. Symp Proc. 111*, 219 (1988).
330. J. H. Sinfelt, G. H. Via, and F. W. Lytle, *J. Chem. Phys. 76*, 2779 (1982).
331. T. Romero, B. Arenas, E. Perozo, C. Bolivar, G. Bravo, P. Marcano, C. Scott, M. J. Pérez Zurita, and J. Goldwasser, *J. Catal. 124*, 281 (1990).
332. T. Romero, J. Tejeda, D. Jaunay, C. Bolivar, and H. Charcoset, in *Proceedings, VIIth Iberomaerican Congress in Catalysis*, La Plata, Argentina, 1980, p. 453.
333. J. Goldwasser, B. Arenas, C. Bolivar, G. Castro, A. Rodriquez, A. Fleitas, and J. Giron, *J. Catal. 100*, 75 (1986).
334. R. Bouwman and P. Biloen, *J. Catal.* 48, 209 (1977).
335. S. R. de Miguel, O. A. Scelza, and A. A. Castro, *Appl. Catal. 44*, 23 (1988).
336. S. R. de Miguel, J. A. Martinez Correa, G. T. Baronetti, A. A. Castro, and O. A. Scelza, *Appl. Catal. 60*, 47 (1990).
337. J. Völter, H. Lieske, and G. Lietz, *React. Kinet. Catal. Lett. 16*, 87 (1981).
338. A. Palizov, Kh. Bonev, G. Kadinov, D. Shopov, G. Lietz, and J. Völter, *J. Catal. 71*, 1 (1981).
339. H. Lieske and Völter, *React. Kinet. Catal. Lett. 23*, 403 (1983).
340. C. Besoukhanova, J. Guidot, and D. Barthomeuf, *J. Chem. Soc. Faraday Trans. 1 77*, 1595 (1981).
341. T. R. Hughes, W. C. Buss, P. W. Tamm, and R. L. Jacobson, in *New Developments in Zeolite Science and Technology* (Y. Murakami, A. Iijima, and J. W. Ward, eds.). Elsevier, Amsterdam, 1986, pp. 725-732 and discussion (pp. 120-126).
342. J. W. Ward, *J. Catal. 10*, 34 (1968); in *Zeolite Chemistry and Catalysis* (J. A. Rabo, ed.). ACS Monograph Series, Vol. 171. American Chemical Society, Washington, DC, 1976, pp. 229-234.
343. P. A. Newell and L. V. C. Rees, *Zeolites 3*, 22, 28 (1983).

344. J. R. Anderson, *Structure of Metallic Catalysts*. Academic Press, New York, 1975.
345. S. B. Rice, J. Y. Koo, M. M. Disko, and M. M. J. Treacy, *Ultramicroscopy 34*, 108 (1990).
346. J. T. Miller, D. J. Sajkowski, F. S. Modica, G. S. Lane, B. C. Gates, M. Vaarkamp, J. V. Grondelle, and D. C. Koningsberger, *Catal. Lett. 6*, 369 (1990).
347. J. B. A. D. van Zon, D. C. Koningsberger, H. F. J. van't Blik, and D. E. Sayers, *J. Chem. Phys. 83*, 5742 (1985).
348. G. Larsen and G. L. Haller, *Catal. Today* 15, 431 (1992).
349. G. B. McVicker, J. L. Kao, J. J. Ziemiak, W. E. Gates, J. L. Robbins, M. M. J. Treacy, S. B. Rice, T. H. Vanderspurt, V. R. Cross, and A. K. Ghosh, *J. Catal. 139*, 48 (1993).
350. E. Iglesia and J. E. Baumgartner, *10th Int. Congr. Catal.*, Budapest, 1992, paper 0-65.
351. M. Vaarkamp, J. T. Miller, F. S. Modica, G. S. Lane, and D. C. Koningsberger, *J. Catal. 138*, 675 (1992).
352. O. Glatter and O. Kratky, *Small-Angle X-ray Scattering*. Academic Press, New York, 1982.
353. J. R. Bernard, *Proc. Fifth Intern. Conf. Zeolites,* Heyden, London, 1980, p. 686.

# 6
# Evaluation of Catalysts for Catalytic Reforming

## S. Tiong Sie
*Technical University Delft, Delft, The Netherlands*

## I. INTRODUCTION

Catalytic reforming of naphtha, i.e., a petroleum fraction boiling between approximately 80 and 180°C, is a key process in the manufacture of gasoline of high octane quality, as evidenced by the figures presented in Table 1. The large total capacity of catalytic reformers implies significant economic returns with even modest improvements in catalyst performance. Therefore, accurate assessment of catalyst performance which provides information pertinent to industrial operations is of paramount importance, e.g., in development of improved catalysts, in establishing optimal operating conditions, and in monitoring industrial applications.

As discussed in previous chapters of this book, the catalytic reforming process uses a bifunctional catalyst in which an acidic function (generally provided by a chlorided alumina carrier) is combined with a hydrogenation-dehydrogenation function (provided by platinum or platinum with a second metal). The improvement of octane quality is mainly obtained by formation and concentration of aromatic hydrocarbons as a result of a variety of reactions. These reactions are discussed in the previous parts of this book as well as in the literature, e.g., in the review article by Ciapetta et al. (2).

**Table 1** Major Gasoline Production Processes

| Process | Worldwide capacity in 1990 ($10^6$ bbl/d) |
| --- | --- |
| Catalytic cracking | 10.187 |
| Catalytic reforming | 8.730 |
| Hydrocracking | 2.556 |
| Alkylation + polymerization | 1.613 |

*Source*: From Ref. 1.

An important feature of the catalytic reforming process is that the catalyst deactivates in the course of the process, mainly as a result of deposition of carbonaceous material on the catalyst. The catalyst may be regenerated by burning off this material with diluted oxygen, optionally with redispersal or redistribution of the metal(s) by chlorination, followed by reduction. The cycle length between regenerations depends on the type of process: in the semiregenerative type of process the operation may last several thousand to more than 10,000 hours before it is interrupted for regeneration, whereas in the fully regenerative (swing-reactor type) or continuously regenerative (moving-bed reactor type) process the cycle may last only a few hundred hours or even less. Factors that determine the cycle length are the nature of the catalyst and feedstock, octane target, space velocity, hydrogen-to-oil ratio, and, above all, the operating pressure. The operating pressure of semiregenerative reformers is typically 25–30 bar; however, the advent of the more stable bimetallic catalysts allowed semiregenerative reformers to be designed for operation at about 15 bar or even lower (3). Typical pressures in fully regenerative and continuously regenerative reformers are between 7 and 14 bar, and for the latter type of reformers even lower pressures are possible. A breakdown of catalytic reforming capacity according to process type is shown in Figure 1, based on data from Gerritsen (4).

Another feature of the catalytic reforming process is that the overall conversion is highly endothermic. The process is therefore carried out in a number of adiabatic reactors in series (typically three or four) with intermediate heating. To compensate for the catalyst deactivation during production of a reformate of constant octane quality, temperatures are gradually increased with time in the semiregenerative mode of operation. In the fully regenerative type of processes, where the most deactivated reactor is taken out of service to be regenerated and replaced by another one in a merry-go-round fashion, temperature programming may or may not be practiced, depending on the switching frequency. In the continuously regenerative processes, e.g., the UOP CCR

Evaluation of Catalysts for Reforming

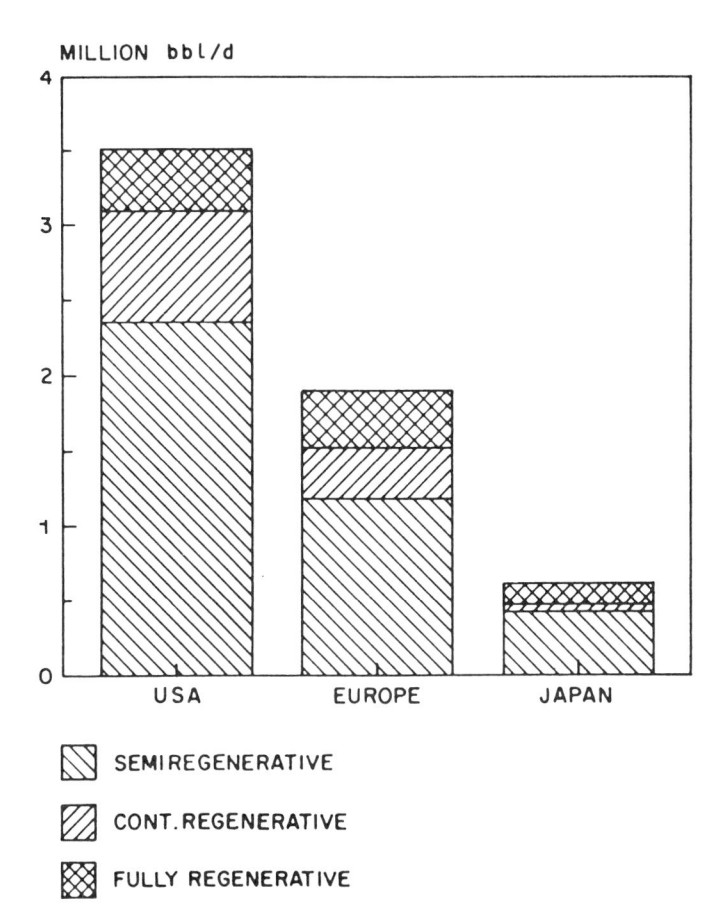

**Figure 1** Breakdown of catalytic reforming capacity according to process type in 1985 (1).

Platforming process (5) and the IFP Aromizing process (6), the temperatures in the reactors remain in principle constant.

These characteristics of catalytic reforming processes have to be taken into account when devising laboratory test procedures to yield performance data that can be considered meaningful for industrial practice. These tests should provide information not only on catalyst activity but also on equally important aspects such as selectivity and deactivation behavior under conditions that are sufficiently representative of commercial operation.

Several laboratory tests are described that meet this requirement, comprising tests in small isothermal reactors with a light naphtha as feed in either a constant or a programmed temperature mode. These tests are relatively simple and capable of yielding useful information on catalyst behavior.

Closer simulation of a commercial catalytic reformer is obtained with a combination of small adiabatic reactors with interreactor heaters, which constitutes a miniature pilot plant giving results closely matching those obtained in a larger pilot plant.

## II. OPTIONS FOR TESTING CATALYSTS IN THE LABORATORY

Performance tests on catalysts in the laboratory are needed in the development of new catalysts, for monitoring the production of commercial catalysts, for selecting the best candidate catalyst for a given application, for defining optimal operating conditions, or for determining the effectivity of a given regeneration procedure. The various possibilities for catalyst testing are discussed below.

### A. Functional Activity Tests with Pure Model Compounds

Tests with pure model hydrocarbons may provide data on the activities for the various catalytic functions needed for the various reactions. For instance, the dehydrogenation activity may be assessed using cyclohexane or methylcyclohexane as feed, whereas the dehydrocyclization activity may be determined using $n$-hexane or $n$-heptane. Although such tests are quite often used in fundamental catalysis studies, they hardly give relevant quantitative data on selectivity and stability under practical conditions.

A principal difficulty in the use of pure model compounds is that the behavior of a given hydrocarbon is affected by the presence of other components in the feed as a result of competitive adsorption. To obtain meaningful kinetic data on pure compounds that are not affected by the product molecules formed, the conversion of the feed hydrocarbon must be kept very low; i.e., differential conversions are called for. Although it may in principle be possible to translate data from such differential functional activity tests to catalyst performance under more practical conditions with the aid of an extensive kinetic computer model, the experimental and computational efforts required would be quite elaborate, whereas results may still have rather doubtful practical value, especially in the case of catalysts of unknown type.

Tests with $n$-hexane, $n$-heptane, cyclohexane, methylcyclohexane, or methylcyclopentane have found rather widespread use in the catalysis literature. An example of the use of model feedstocks to study the role of sulfur in reforming reactions on Pt/alumina and Pt-Re/alumina catalysts can be found in the work reported by Menon and Prasad (7).

### B. Tests in Integrated Pilot Plants with Practical Feedstocks

Tests carried out in an integrated pilot plant that features adiabatic reactors with interheaters, product stabilizer, and gas recycle using a full-range naphtha as feedstock provide the closest simulation of an actual plant. As in practical operation, temperatures are set by monitoring product quality against the octane target. If product octane is monitored by means of a CFR motor (8), the amount of product required calls for relatively large reactors, e.g., of a few liters capacity. A relatively large minimum plant size is also dictated by the requirement of adiabaticity of the reactor, as well as by the relatively large holdup of the on-line product stabilizer and the buffers of the gas recycle compressor.

Although such an integrated pilot plant is capable of giving very meaningful data for commercial practice, it is too expensive to be run for the routine testing of large numbers of catalyst samples.

### C. Small-Scale Tests with Integral Conversion of (Semi)-Practical Feedstocks and Once-Through Hydrogen

A compromise between the two types of tests just discussed that avoids their main drawbacks and yet is capable of providing meaningful information is to apply practical feedstocks under integral conversion conditions but to omit gas recycle, product stabilization, and actual determination of octane number. If gas-liquid chromatography (GLC) techniques are used to monitor product quality, large reactors are no longer needed. The analytical demands may be further reduced by narrowing the boiling range of the feed without detracting too much from its representativeness in terms of the presence of the main types of components. Small reactors are most easily constructed as isothermal reactors, i.e., reactors in which the catalyst bed has a uniform temperature. Such an isothermal reactor makes it possible to attain significant integral conversion in a single catalyst bed because the endothermic heat of reaction is supplied from outside. Thus, there is no need for more than one reactor, in contrast to adiabatic reactors, several of which have to be arranged in series to achieve significant overall conversion without an unduly large temperature drop per reactor. The tests to be discussed in the following belong to the category of isothermal reactors.

## III. TESTS IN ISOTHERMAL MICROREACTORS AT CONSTANT TEMPERATURE

This relatively simple test is carried out with a very small quantity (about 2 mL) of catalyst and uses as feed a 85–102°C fraction of a hydrotreated Middle East naphtha consisting predominantly of $C_7$ hydrocarbons, with minor amounts of $C_6$ and $C_8$. Activity for aromatization is monitored by determination of toluene in the product, while cracking activity is assessed by determination of $C_{4-}$ products.

## A. Equipment and Procedure

A flow diagram of the apparatus is shown in Figure 2. The feed is kept dry and free of dissolved air by storage over 4 Å molecular sieves under nitrogen and is fed to the reactor by a displacement pump via a bed with deoxo catalyst (BASF BTS catalyst) to remove any residual chemically bound and dissolved oxygen. Oxygen-free, dry hydrogen is added to the hydrocarbon stream upstream of a preheater. The reactor has an internal diameter of 10 mm and is provided with a central thermowell of 5 mm outer diameter which houses a movable thermocouple to determine bed temperature and to check its isothermicity. The reactor is heated by a furnace with three separate heating sections, which allow isothermal operation within 2°C. A diagram of the reactor is shown in Figure 3.

To minimize radial temperature gradients in the catalyst bed contained in the annular space between thermowell and reactor wall, the catalyst is diluted

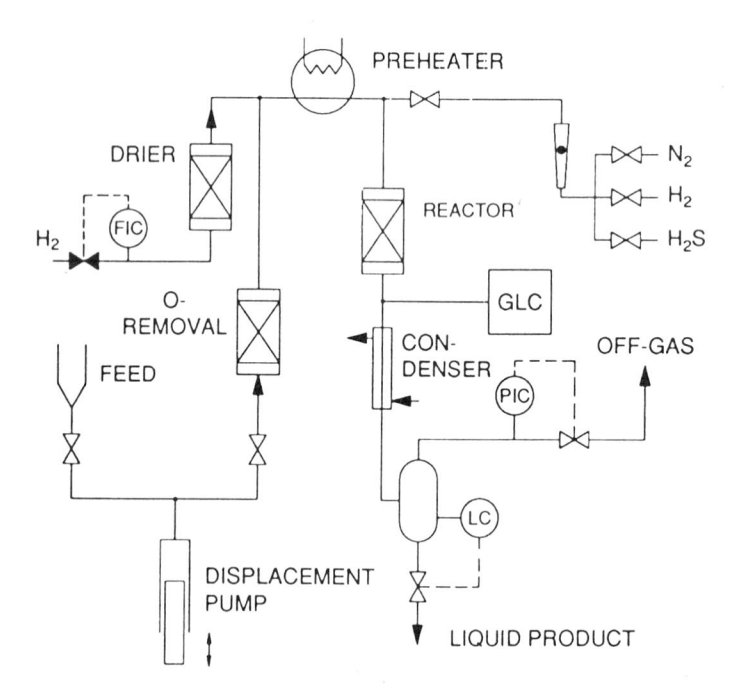

**Figure 2** Flow diagram of the isothermal, constant-temperature test unit.

# Evaluation of Catalysts for Reforming

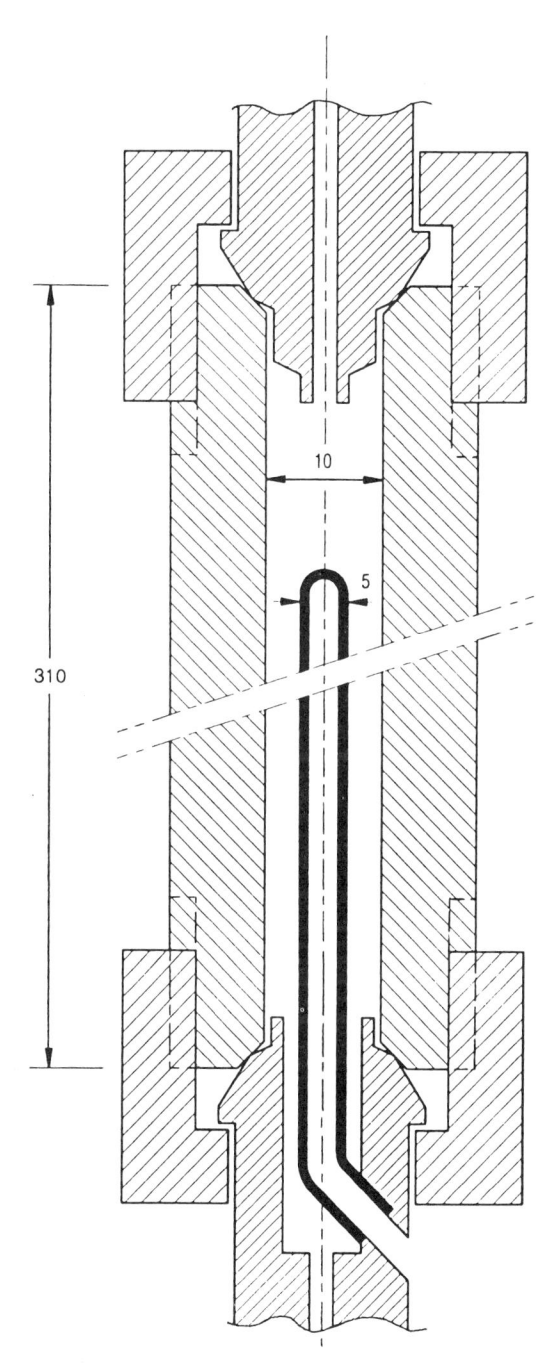

**Figure 3** Diagram of a microreactor. Dimensions are in mm.

**Table 2** Typical Conditions of Catalyst Testing in Isothermal Laboratory Reactors

|  | Constant temperature | Programmed temperature |
|---|---|---|
| Feed |  |  |
| Type | Hydrotreated $C_6$–$C_8$ fraction from Middle East crude | Hydrotreated naphtha from Middle East crude |
| Boiling range, °C | 85–102 | 65–157 |
| P/N/A | 70/23/7 | 69/20/11 |
| Sulfur, ppm | <1 | <1 |
| Operating conditions |  |  |
| T, °C | 510 | 490–535 |
| P, bar | 8.5 | 10 |
| LHSV, L/(L·h) | 2.0 | 1.8 |
| $H_2$/oil, molar ratio | 4.0 | 1.5 |
| Target RON | — | 98 |
| Run length, h | 100 | 100–200 |

with two to three times its volume of small particles of silicon carbide, which are catalytically inert and have a high heat conductivity.

The reactor effluent is analyzed by an on-line GLC (a Becker 407 modified PNA analyzer as described by Boer et al. [9]) and, after cooling, the condensed liquid is drawn off via a sluice vessel. The pressure is regulated by a back-pressure valve in the off-gas line. The apparatus has provisions for in situ presulfiding and reduction of catalyst samples.

Main conditions of the test are listed in Table 2. In view of the measures taken to ensure dry operation (water levels below 10 ppm) and the short duration of the test (100 h), chlorine addition is not practiced. It has been found that under proper conditions chloride losses from the catalyst during the test are low (see Table 3).

**Table 3** Chloride Retention on Catalyst During Tests in Isothermal Reactors

|  |  | $Cl^-$ on catalyst (wt %) | |
|---|---|---|---|
| Catalyst | Test mode | Before test | After test |
| A | Constant temperature | 0.90 | 0.75 |
| B | Constant temperature | 0.90 | 0.76 |
| D | Constant temperature | 1.0 | 1.0 |
| C | Programmed temperature | 0.93 | 0.90 |

## B. Examples of Application

Figure 4 shows the results of comparative testing of four different catalysts, consisting of platinum on chlorided alumina without a second metal (catalyst A) and with different metals (Re, Ge, or Sn) besides platinum (catalysts, B, C, and D, respectively). In this figure, the activity for aromatic formation has

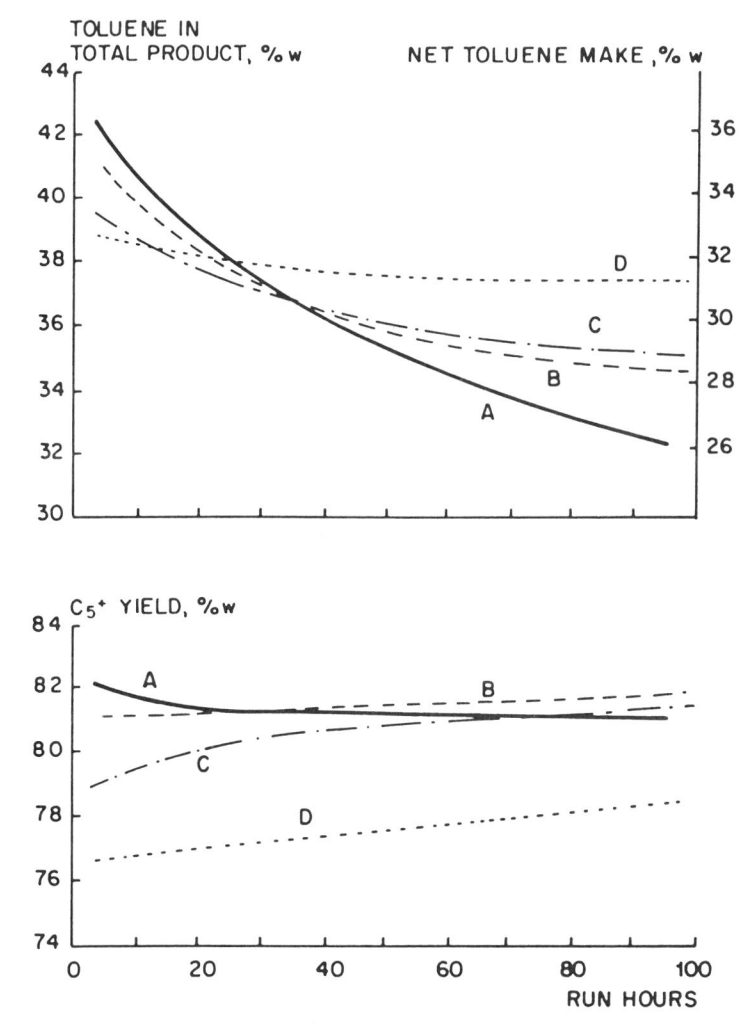

**Figure 4** Results of isothermal, constant-temperature tests on four different catalysts. A is a monometallic platinum catalyst; B, C, and D are bimetallic catalysts with Re, Ge, or Sn combined with platinum.

been expressed as toluene in the total product or as net toluene make, which is in principle the same if allowance is made for the toluene content in the feed. The cracking activity is expressed in a reversed way, as the yield of $C_{5+}$.

It can be seen that the stability of the monometallic catalyst A is least as far as aromatization is concerned, and bimetallic catalyst D is the most stable. The lower part of Figure 4 shows that the latter catalyst is the least selective.

In actual practice, the activity of a catalyst for enhancing the octane value of the liquid product is the result of both aromatic formation and concentration of these aromatics in the liquid by cracking away nonaromatic liquid hydrocarbons to gaseous components. These two contributing factors are combined when the test results are expressed in terms of toluene concentration in the $C_{5+}$ fraction of the product, as shown in Figure 5. The superiority of catalyst D in product quality as well as in stability for product quality maintenance over the other catalysts is very pronounced.

Figure 6 shows the effect of different chloride levels on a (sulfided) Pt/Re-on-alumina catalyst. Although the differences in performance are relatively small for chloride levels between 0.9 and 1.1 wt %, it can be seen that the catalyst with the higher chloride content is initially somewhat more active, but this activity gain is lost in the course of the run, as it is less stable. It also follows that the catalyst with the higher chloride content is somewhat less selective.

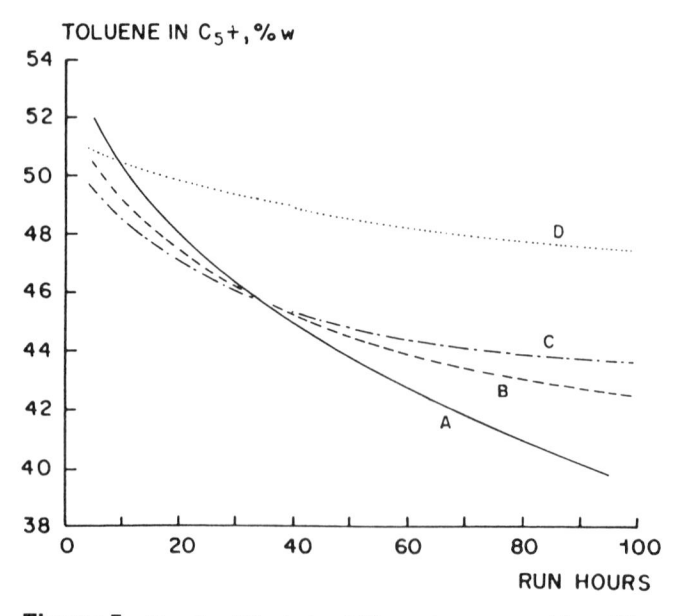

**Figure 5** Results of the tests of Figure 4 represented in a different way.

*Evaluation of Catalysts for Reforming*

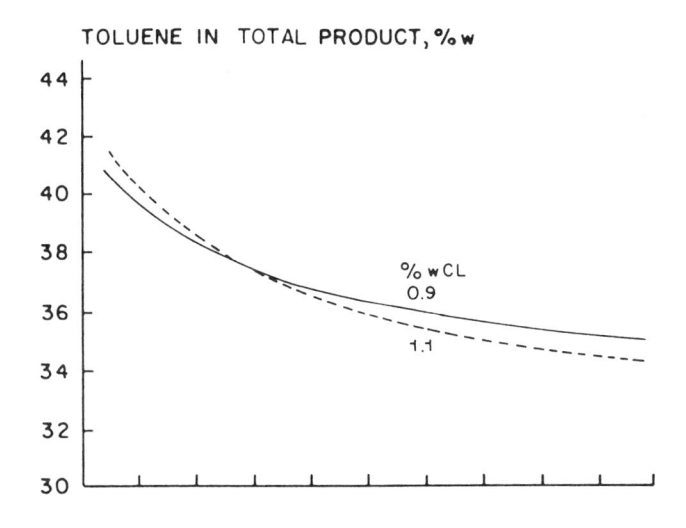

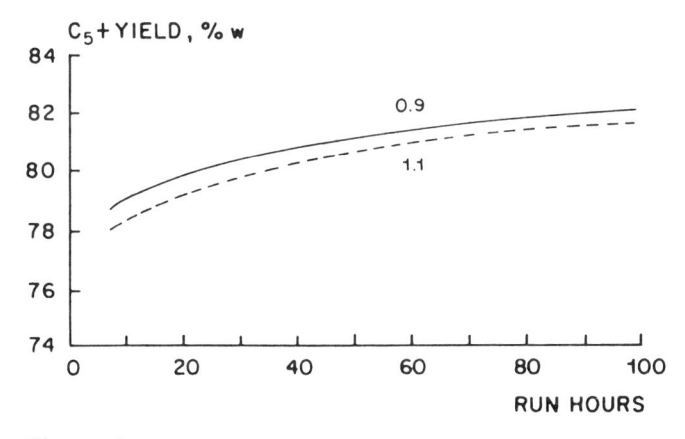

**Figure 6** Comparison of the performances of sulfided Pt/Re catalysts with different levels of chloride in the isothermal, constant-temperature test.

An example of the use of the test in studying the effect of catalyst formulation is shown in Figure 7. This figure compares the performances of three (sulfided) Pt/Re-on-alumina catalysts, all with the same platinum loading but with different amounts of rhenium. The tests show that of this particular series of catalysts, the sample with the highest rhenium content is initially the least active but is the most stable. In consequence, it has the highest activity at the end of the test.

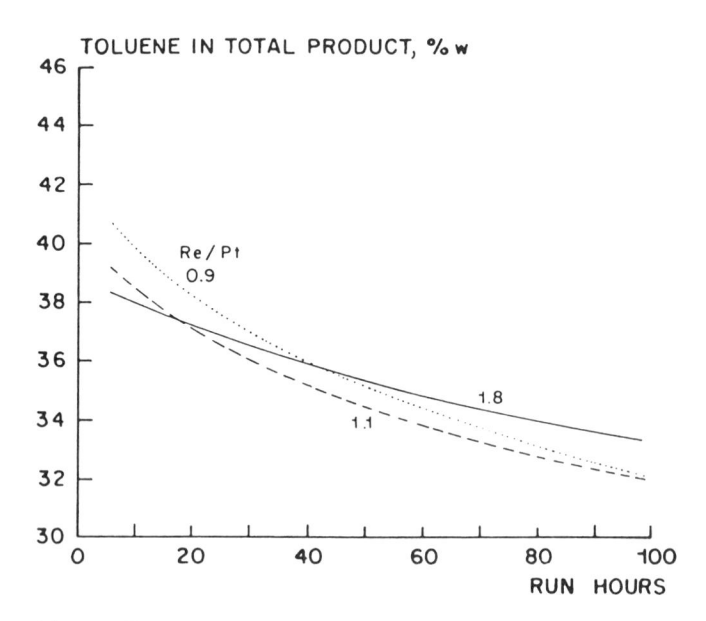

**Figure 7** Comparison of sulfided Pt/Re catalysts of the same platinum load but with different atomic ratios of rhenium to platinum in the isothermal, constant-temperature test.

Figure 8 shows an example of the use of the test for monitoring the quality of the catalyst during operation in a CCR Platformer, where it is subjected to repeated regenerations. It can be seen that this particular catalyst maintained its performance during almost 2 years of operation.

### C. Accuracy of the Tests

When sufficient care is given to details of the test procedure, such as loading of the catalyst in the reactors, pretreatment of the catalyst, maintenance of a sufficiently low water level in the process streams, and accuracy of the product analyzer, the results are quite accurate. The repeatability of the test is illustrated by Figure 9, which shows the results of three tests carried out on the same catalyst sample at different dates with appreciable periods of elapsed time in between. It can be seen that the repeatability is quite satisfactory, which implies that the test allows the assessment of small differences in catalyst performance, as already demonstrated in the previous examples.

## IV. TESTS IN ISOTHERMAL BENCH-SCALE REACTORS WITH TEMPERATURE PROGRAMMING

The previous test at constant temperature does not provide a good indication of the decline of liquid yield with time during operation of a semiregenerative

*Evaluation of Catalysts for Reforming*

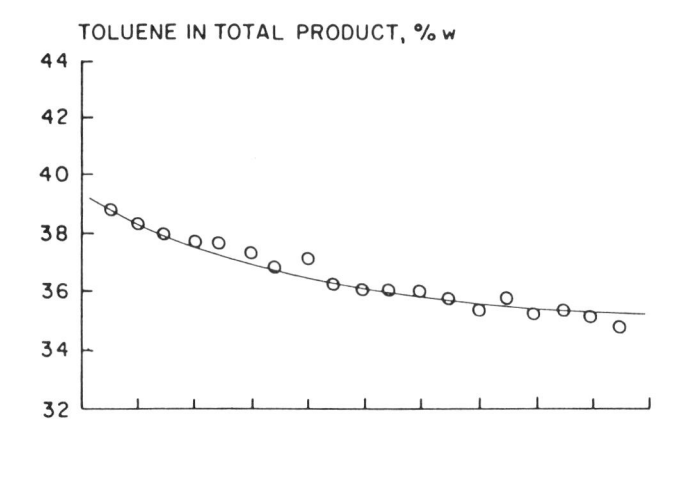

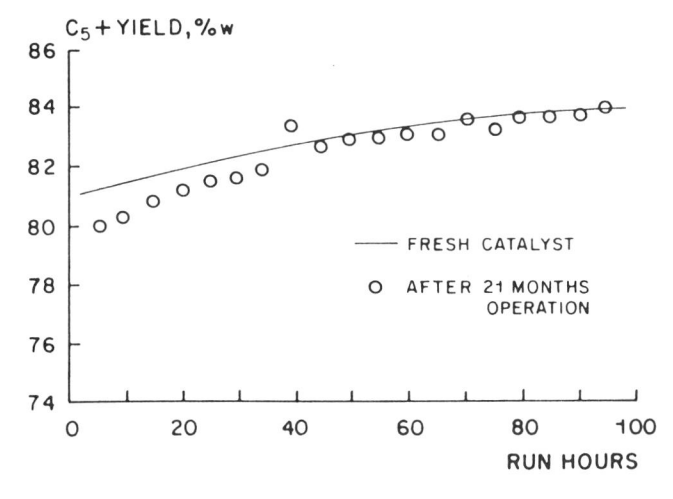

**Figure 8** Comparison of fresh and used bimetallic catalysts from a CCR Platformer in the isothermal, constant-temperature test.

reformer in practice. At a constant temperature, the deactivation of the cracking function may cause an increase of $C_{5+}$ yield with time (see the lower part of Figure 4). In practice, the inlet temperatures to the reactors and therefore the average reaction temperature are increased with time on stream in the semiregenerative mode of operation in order to compensate for the loss of aromatization activity of the catalyst while still producing a product of the desired quality level. This operation at constant product quality gives rise to more pronounced cracking at the higher temperatures with increasing catalyst age, so that liquid yield drops in the course of the run. The end of an operating

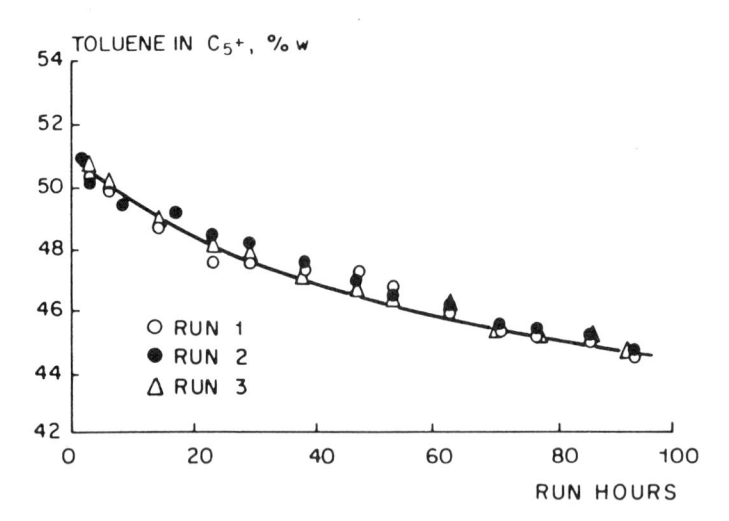

**Figure 9** Repeatability of the isothermal, constant-temperature test.

cycle may be reached when temperature requirements approach hardware design limits but may also be determined by liquid yields becoming uneconomically low. Thus, both activity-stability and selectivity-stability are of importance, as either of these may determine the useful life of a catalyst in practice.

To obtain information on the decline of catalyst performance, it is customary in pilot plant tests to increase the reactor temperature in discrete steps during a run. The timing and height of each step are determined after information on product quality from the previous period has become available. Obviously, this way of manual temperature programming is applicable only if the rate of deactivation is relatively low, and consequently this technique is generally applied in pilot plant runs of relatively long duration. Van Keulen (10) has described an accelerated (60 h) test in an isothermal reactor with 20 g of catalyst and once-through hydrogen gas in which the run was divided into six periods of 10 h each. The reactor temperature was varied during the test, but because this was done rather arbitrarily, the quality of the product did not remain constant and was found afterward to have varied between RON-0 > 102 at the beginning to RON-0 = 93 at the end of the test.

The variable-temperature test developed at the Amsterdam laboratory of Shell, which is to be described in the following, is also a test in an isothermal bench-scale reactor with once-through hydrogen gas, but it differs from the previous one in that the octane level of the liquid product is kept constant during the run by means of an on-line quality measuring instrument that controls the reactor temperature. This test, which allows assessment of activity-stability as well as selectivity-stability of the catalyst, lasts between 100 and 200 h (the actual duration depends on the stability of the catalyst and the severity of operation because a range of temperatures is covered).

## A. Equipment and Procedure

The main conditions of the test are listed in Table 2. The setup of the test equipment is very similar to the one used in the previous test. The main differences are a larger reactor (a bench-scale reactor with 40 mL of catalyst instead of the microreactor), a plunger-type feed pump instead of a displacement pump, and the incorporation of a continuous refractometer cell (Anacon type 47) in the liquid product draw-off line. The liquid holdup of the latter system is kept small (a few mL) and care is taken to avoid gas bubbles in the refractometer cell. The reactor temperature is adjusted automatically to maintain a constant refractive index of the liquid product. The small holdup of the separator-cell system in combination with the larger reactor volume results in a rapid response time of the automatic control system.

A diagram of the equipment is shown in Figure 10. Figure 11 shows the main dimensions of the bench-scale reactor. As in the case of the microreactor, temperatures within the catalyst bed can be measured with a sliding thermocouple. To allow a straight axial temperature profile, the reactor is heated by means of an electric furnace with five independent heating sections. Just as in the isothermal microreactor test, radial temperature differences are minimized by diluting the catalyst with fine particles of catalytically inert silicon carbide.

The refractive index of the liquid product correlates with the aromatic concentration and hence with the octane quality of the product, but the correlation is not exact. Therefore, the product composition is also monitored by on-line GLC, which allows calculation of the octane quality on the basis of the molecular composition of the $C_{5+}$ product, and the raw data obtained during the test are corrected afterward to a constant target octane level. Chlorine addition is not practiced because chloride retention on the catalyst is in general satisfactory (see Table 3).

## B. Examples of Application

Figure 12 shows some results of tests carried out on two different bimetallic catalysts. These tests can be considered as accelerated performance tests for semiregenerative application. It can be seen that catalyst E is far superior to catalyst B in both activity-stability (upper graph) and selectivity-stability (lower graph). The superiority of catalyst E over B follows not only from a comparison at equal space velocity but also from the finding that the performance of catalyst E at a 1.8 times higher space velocity becomes quite comparable to that of B at the lower space velocity.

An example of the use of this type of testing to assess the effects of changes in operating conditions is shown in Figure 13. This figure compares the perfor-

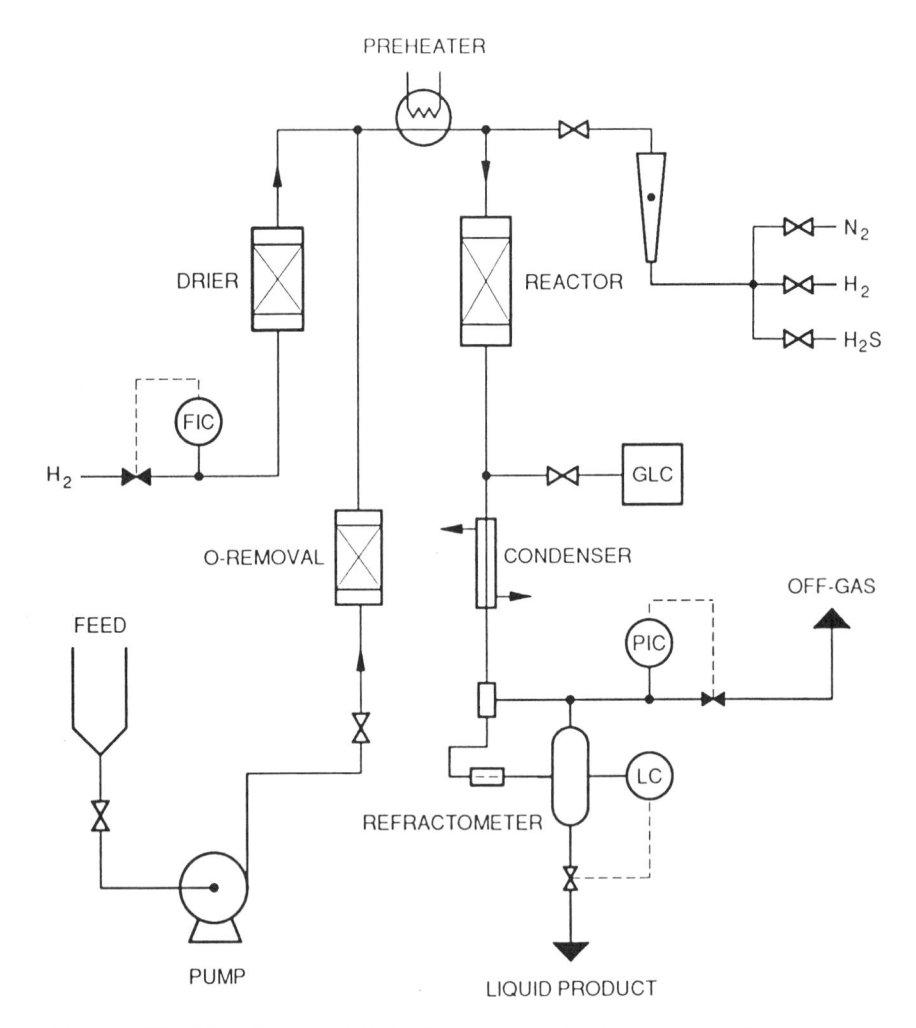

**Figure 10** Flow diagram of the isothermal test unit with programmed temperature.

mance of a bimetallic catalyst at two different pressure levels, 9 and 11 bar. At the lower pressure the liquid yield is significantly higher, but catalyst stability is lower. This observed effect of pressure is in line with practical experience.

### C. Accuracy and Meaningfulness of the Tests

Figure 14 illustrates the repeatability of the test, which is sufficiently high to allow discrimination between catalysts of moderately different performance and to determine the effect of changes in operating variables. The data from these

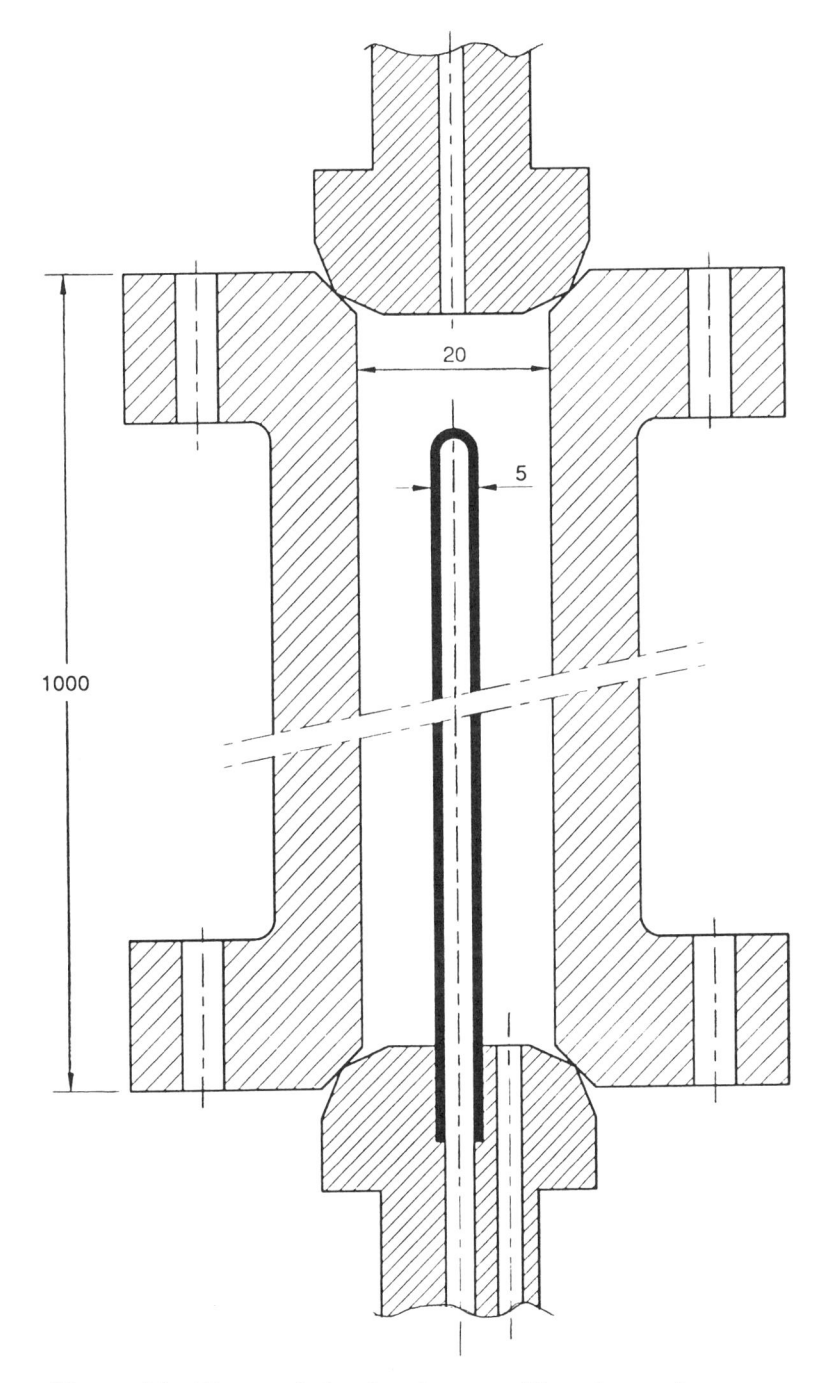

**Figure 11** Diagram of a bench-scale reactor. Dimensions are in mm.

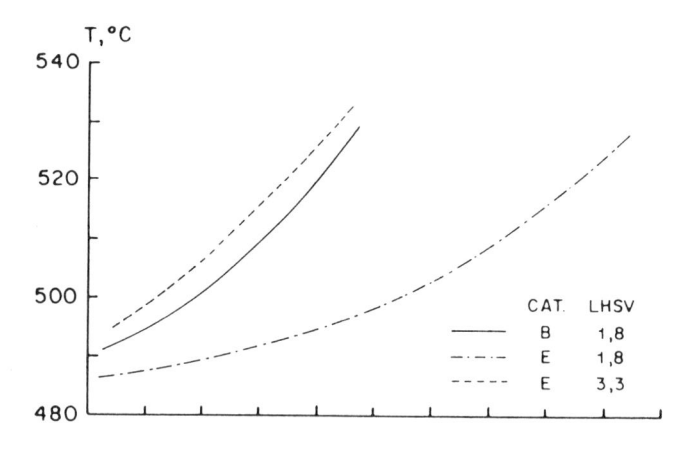

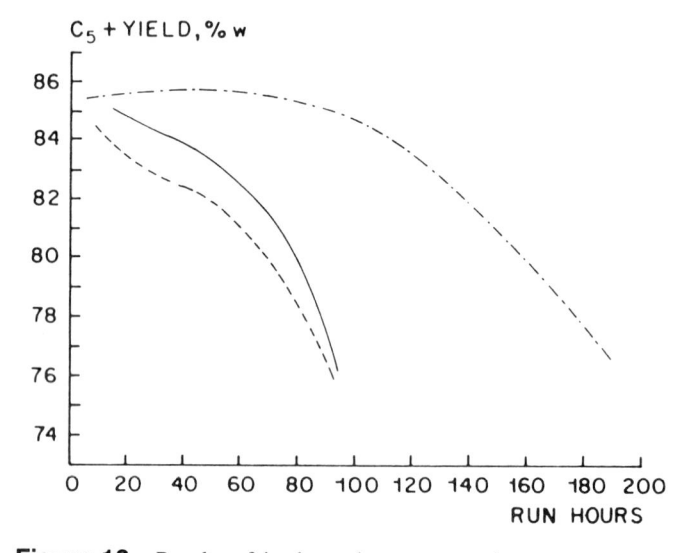

**Figure 12** Results of isothermal, programmed-temperature tests on two bimetallic catalysts. Target RON-0 = 98.

tests can be translated to practical situations by means of a process model. Thus, they allow prediction of, e.g., cycle lengths both when temperature ceilings are limiting (see upper parts of Figures 12 and 13) and when yield considerations determine the end of a cycle (see lower parts of Figures 12 and 13). Such estimates have been found to correlate well with practice so that the need for large pilot plant tests has greatly diminished.

*Evaluation of Catalysts for Reforming*

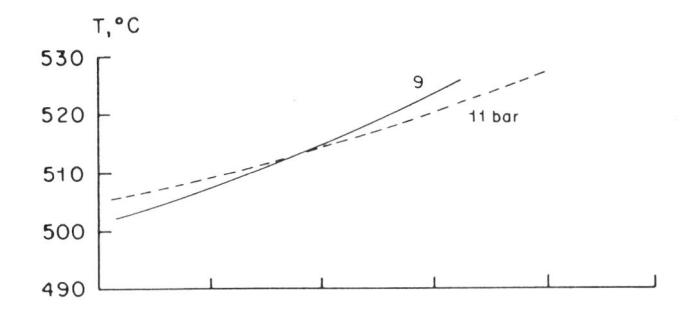

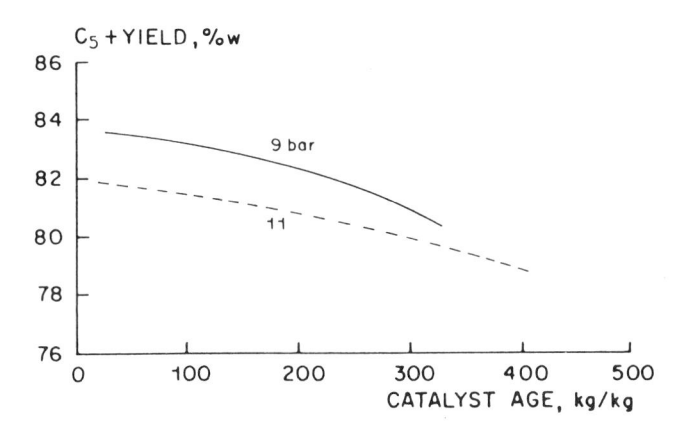

**Figure 13** Effect of operating pressure on the performance of a bimetallic catalyst in the isothermal, programmed-temperature test. Target RON-0 = 100.

## V. TESTING IN A MINIADIABATIC PILOT UNIT

Although the previously described tests in isothermal reactors are capable of providing the bulk of the information on catalyst performance required in practice, some instances remain where a need is felt for testing in an adiabatic system. Cases in point are e.g., when information is required on the temperature drops over catalyst beds (insofar as they cannot be determined by calculation), when the composition of interreactor streams has to be determined, or when the effect of differences in catalyst distribution over the reactors has to be assessed experimentally. For such cases a large integrated pilot plant that closely simulates a practical plant is an appropriate tool for obtaining the desired information. However, the high costs involved provide an incentive for developing a smaller setup that can yield almost the same information. Such a

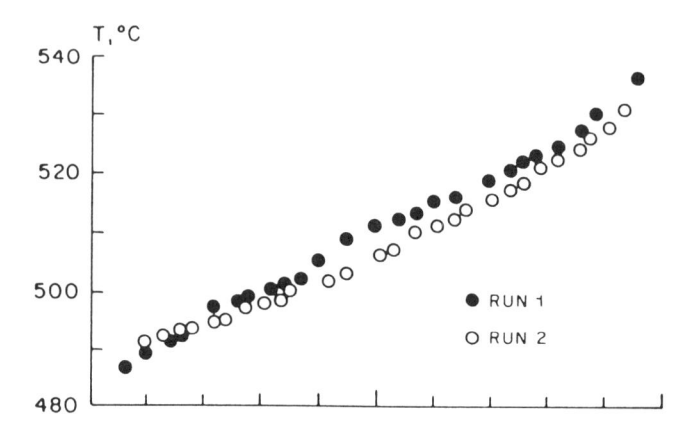

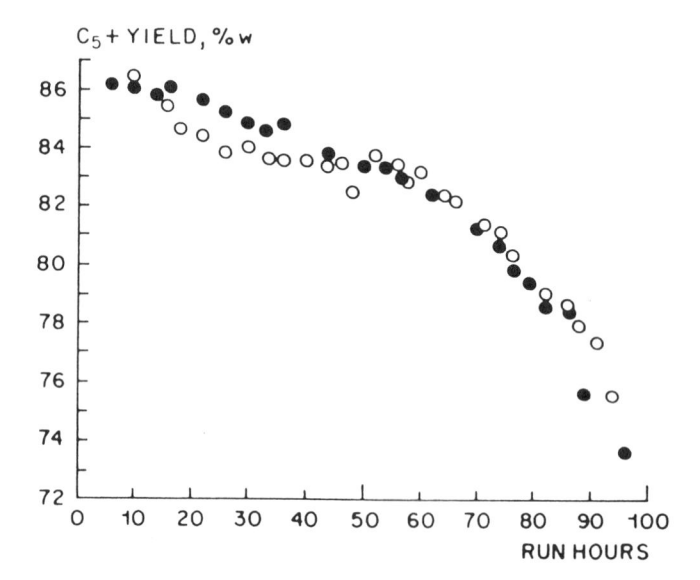

**Figure 14** Repeatability of the isothermal, programmed-temperature test.

system, a miniadiabatic pilot unit, is described here. This system further reduces the gap between the large integrated pilot plants and the micro- and bench-scale isothermal test units described in the preceding sections. It features four adiabatic reactors in series with a total catalyst volume of 100 mL, a factor of 10–25 reduction in scale compared with a conventional integrated pilot plant.

# Evaluation of Catalysts for Reforming

## A. Construction of an Adiabatic Microreactor

In this pilot unit, the highest demands on adiabaticity are placed in the first reactor, since about half of the total drop in temperature occurs in the first 10% of the total catalyst volume. Design of the reactor and temperature control is difficult because the heat effects are only of the order of 5 W in a small reactor of 10-mL catalyst content, the more so since the reactor has to operate adiabatically at a temperature level of about 500°C.

A diagram of this reactor is shown in Figure 15. It functions according to a well-known principle: at some distance from the reactor tube a number of resistance heaters are mounted on a metal screen, and these heaters are controlled with the temperature difference between the catalyst bed and the metal screen as input signal. In this way, the axial temperature profile of the catalyst bed is duplicated on the metal screen so that at any axial position a radial heat

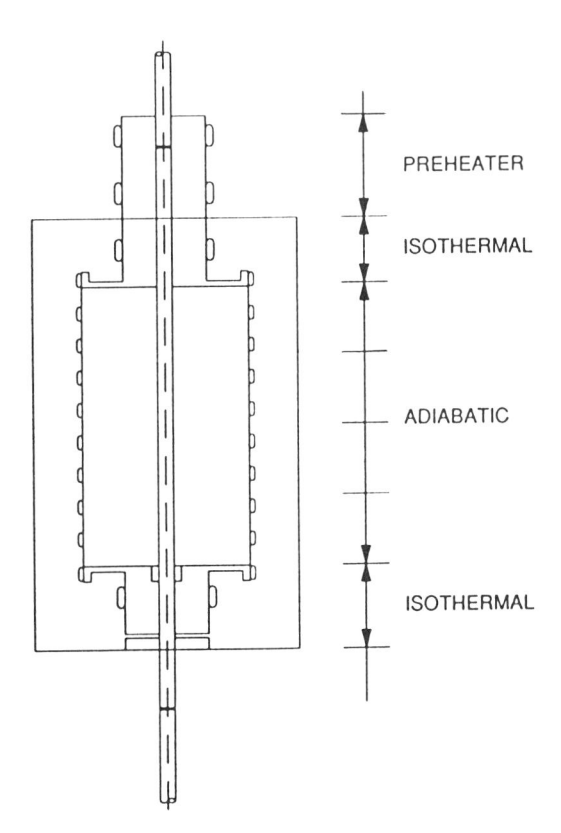

**Figure 15** Diagram of adiabatic microreactor (10 mL of catalyst).

flux is prevented. At both ends of the reactor tube, axial heat flow occurs, causing disturbances, and therefore the catalyst-containing, adiabatic part of the tube is preceded and followed by an isothermal part filled with inert material.

With proper geometric and thermal design of the various parts and suitable location of the thermocouples, a reactor can be obtained that combines the required adiabaticity under steady conditions with a sufficiently fast response to changes in feed inlet temperature. The performance of such an adiabatic microreactor has been tested with a naphtha feed and a known reforming catalyst, assuming the reactor to be the first one containing 13% of the total catalyst inventory. The temperature response proved to be fast: after feed cut-in, it took only 1–1.5 h before the temperature profile was established and lined out. Figure 16 shows the observed temperature profile together with the profile calculated with a catalytic reforming process model. Both profiles are in good agreement and are also very similar to that observed in a large pilot plant containing 2.3 L of catalyst.

## B. Miniadiabatic Pilot Unit

A fully automated miniadiabatic pilot unit has been constructed by combining, in series, four adiabatic reactors of the type just described, with interheaters to

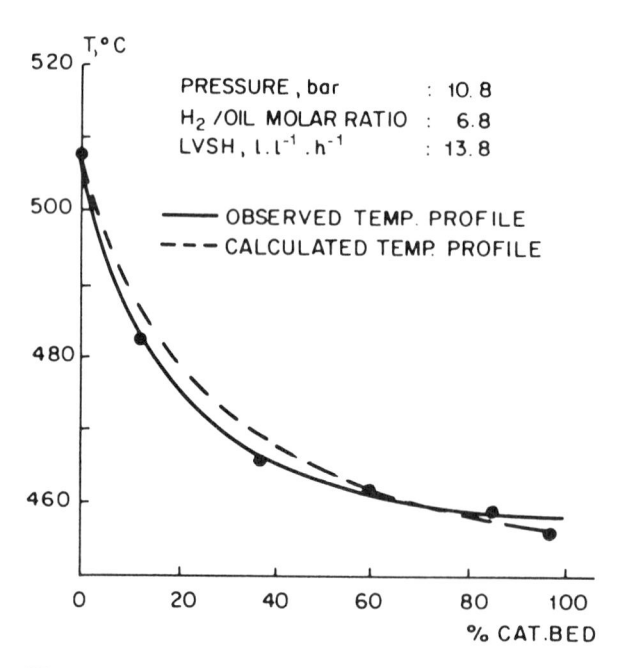

**Figure 16** Temperature profile in adiabatic microreactor operating as the first reactor of a catalytic reformer.

# Evaluation of Catalysts for Reforming

allow adjustment of reactor inlet temperatures. The first, second, third, and fourth reactors (counted in the direction of the reactant stream) have volumes of 10, 15, 25, and 50 mL, respectively, resulting in a total volume of 100 mL. A diagram of this unit is shown in Figure 17. Programming of reactor inlet temperatures can be done on the basis of a setpoint control system acting on the signal of a continuous on-line liquid refractometer, so that the refractive index (and the octane number of the reformate) is kept constant during the run. The unit also features the possibility of on-line GLC analysis, not only of the effluent of the last reactor but also of the streams between the reactors.

Facilities for chlorine injection and control of sulfur levels have been incorporated. However, in view of the small size of the reactors, no gas recycle is applied.

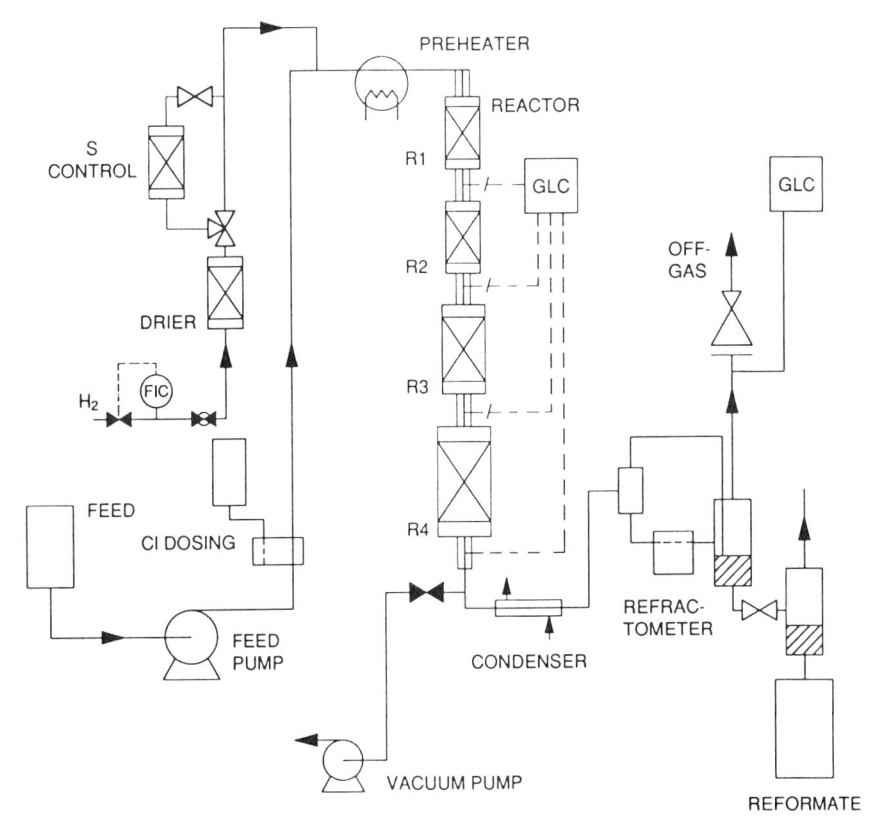

**Figure 17** Flow diagram of the miniadiabatic pilot unit.

## C. Test Results

Figure 18 shows a comparison of test results obtained in the miniadiabatic pilot unit and in a large integrated pilot plant of approximately 2.5 L total catalyst volume. The total catalyst volume in the miniadiabatic unit was 87 mL, to ensure comparable distributions of catalyst over the reactors. To compensate for differences in hydrogen purity between the tests in the large pilot plant with gas recycle and the mini pilot plant operated with once-through hydrogen gas, the total pressures were set at different levels so that the partial pressure of hydrogen was the same at the inlet to the first reactor.

From Figure 18 it can be inferred that the results of the two tests match rather well. Considering the small unavoidable differences in the tests, e.g., gas once-through versus recycle and nonidentical target octane definition (calculated RON-0 from composition of $C_{5+}$ versus RON-0 of stabilized reformate determined with an engine), the agreement may even be considered very satisfactory. This implies that the large integrated pilot plant can be replaced by the mini pilot unit in most investigations of catalyst performance, which entails considerable savings in R & D costs.

## VI. CONCLUSIONS

The three test procedures described in this chapter have in common that they are accurate and reliable tools for evaluating reforming catalysts in terms of activity, selectivity, and stability. They differ, however, in their complexity, feedstock requirements, and the degree to which they simulate commercial practice.

The isothermal, constant-temperature test is the simplest and requires the least feedstock but is farthest away from practical situations. It can, however, provide a qualitative ranking of catalysts and, as a fingerprinting tool, it is very suitable for establishing the identity of catalyst samples. Thus, it can be used to establish whether a regeneration procedure has succeeded in restoring the catalyst to its original state and it may also serve to monitor the maintenance of catalyst quality during commercial operation or in the commercial manufacture of a given catalyst.

The isothermal, programmed-temperature test can be considered an accelerated test for estimating catalyst performance in semiregenerative operation. It yields information on activity as well as selectivity maintenance. It allows semiquantitative ranking of catalysts and determination of the effects of operating variables such as space velocities, pressures, and target octane levels.

The miniadiabatic pilot unit is the most complex and most demanding unit to run. It has, however, the greatest flexibility in the range of feedstocks and operating conditions and simulates practical situations most closely. It can provide almost the same information as that obtained from large integrated pilot

*Evaluation of Catalysts for Reforming* 205

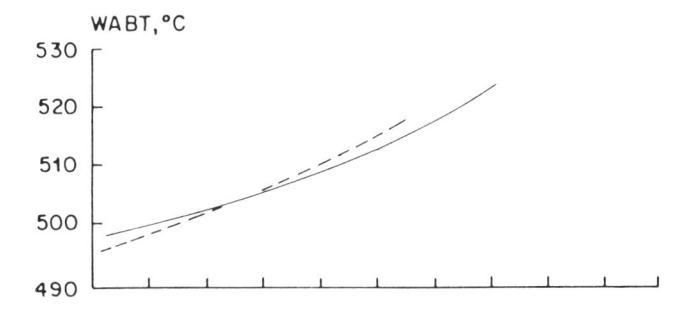

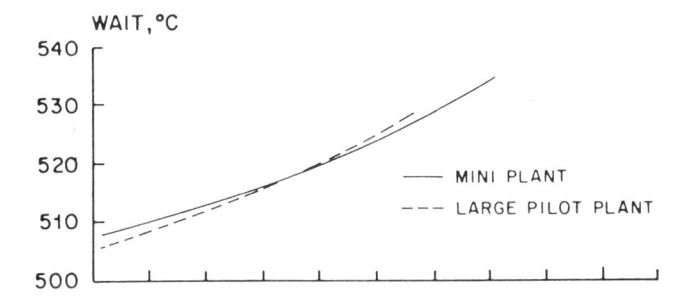

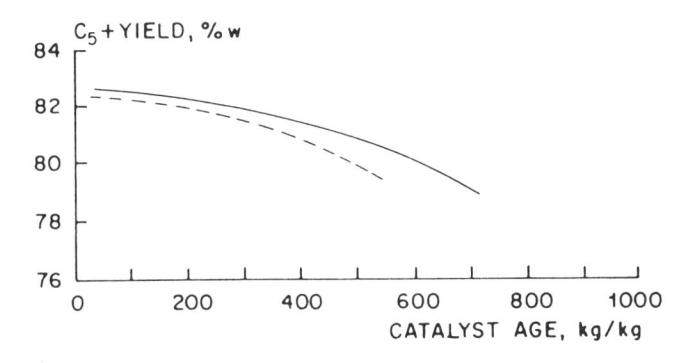

**Figure 18** Comparison of tests in the miniadiabatic pilot unit (87 mL of catalyst) and in a large integrated pilot plant (2.3 L of catalyst). Weighted average bed temperature, weighted average inlet temperature, and liquid yield as a function of catalyst age. Feed = full range naphtha, P/N/A = 66/21/13. Bimetallic catalyst. $H_2$ partial pressure = 6.2 bar, $H_2$/oil = 5.7 molar, LHSV = 1.5, target RON-0 = 100.

plants and has, therefore, replaced the latter, much more expensive plants in R & D on catalytic reforming at Shell.

## ACKNOWLEDGMENTS

The author gratefully acknowledges the contributions of those at the Shell Laboratory in Amsterdam who in the course of many years had a share in the development and application of the tests described here. Thanks are in particular due to Dr. B. D. McNicol, Dr. N. Wagstaff, and Dr. M. M. P. Janssen for their contributions in the development of the isothermal tests, to Ir. H. Gierman and Ing. D. Pultrum for their work in the development of the miniature adiabatic reactor and the miniadiabatic pilot plant, and to Ir. P. P. J. Moorthamer and Dr. G. J. den Otter for studies carried out in the latter plant. The author is also indebted to Dr. Ir. P. M. M. Blauwhoff for helpful discussions.

## REFERENCES

1. *International Petroleum Encyclopedia,* Vol. 23. Penwell Publishing Co., Tulsa, OK, 1990.
2. F. G. Ciapetta, R. M. Dobres, and R. W. Baker, Catalytic reforming of pure hydrocarbons and petroleum naphthas. In *Catalysis* (P. H. Emmett, ed.). Reinhold, New York, 1955, Vol. VI, Chap. 6.
3. R. C. Robinson, L. A. Fredrickson, and R. L. Jacobson, Cat-reforming pressure drops again. *Oil Gas J. 124* (May 15, 1972).
4. L. A. Gerritsen, Recent developments in catalytic reforming, Ketjen Catalyst Symposium '86, Scheveningen, The Netherlands, Paper H-1. Akzo Chemie Ketjen Catalysts, The Netherlands, 1986.
5. E. A. Sutton, A. R. Greenwood, and F. H. Adams, New processing concept. Continuous platforming. *Oil Gas J. 52* (May 22, 1972).
6. P. Bonnifay, B. Cha, J.-C. Barbier, A. Vidal, and R. Huin, IFP regenerative technology for producing high aromatics petrochemical feedstocks, Paper AM-75-18 presented at the 1975 NPRA Annual Meeting, March 23–25, 1975, San Antonio, Texas.
7. P. G. Menon and J. Prasad, The role of sulphur in reforming reactions on Pt-Al$_2$O$_3$ and Pt-Re-Al$_2$O$_3$ catalysts. *Proc. 6th Int. Congr. on Catalysis,* London, July 12–16, 1976. The Chemical Society, London, 1977, p. 1061.
8. ASTM D 2699-88a. *Annual Book of ASTM Standards,* Vol. 05.04. American Society for Testing and Materials, Philadelphia, 1990.
9. H. Boer, P. van Arkel, and W. J. Boersma, An automatic PNA analyser for the under 200°C fraction contained in a higher boiling product. *Chromatographia 13,* 500 (1980).
10. G. J. M. van Keulen, Kinetic model for selectivity decline on reforming catalysts, *Proc. 6th Int. Congr. on Catalysis,* London, July 12–16, 1976. The Chemical Society, London, 1977, p. 1051.

# 7
## Structure and Performance of Reforming Catalysts

**K. R. Murthy, N. Sharma, and N. George**
*Indian Petrochemicals Corporation Ltd., Vadodara, India*

## I. INTRODUCTION

Reforming of hydrocarbons is a process in which molecules in the gasoline range, including those formed from larger hydrocarbons by cracking, are reconstructed, or reformed, ideally without changing their carbon numbers. Industrially, the reforming process is practiced to produce high-octane gasoline and aromatics such as benzene, toluene, and xylenes. In this process a number of reactions—dehydrogenation-hydrogenation, dehydrocyclization, isomerization, and hydrocracking—which are widely different in thermodynamics, kinetics, and equilibria, proceed simultaneously.

The reforming process is generally carried out over bifunctional catalysts which consist of one or more metals, most commonly platinum or platinum combined with rhenium or tin or iridium, dispersed on an acidic support. The catalyst support, such as acidic alumina, and the active metal species, such as promoted or unpromoted platinum, have their own different functions during catalytic reactions. The dehydrogenation-hydrogenation activity is associated with the metal component that is dispersed on the support, whereas the reactions such as isomerization, dehydrocyclization, and hydrocracking are governed predominantly by the acid function. It is important that the same catalyst supports the needs of these widely different reactions.

The performance of any catalyst, in general, comprises three important aspects: activity, selectivity, and stability or life. The activity and selectivity of reforming catalysts are inherent properties controlled by their surface structure. The stability of the activity and selectivity of the catalyst over a period of time is dictated by the nature of deactivation, which may be sintering of the catalyst or poisoning/coking caused by deposition of feed stream impurities or the formation of some surface residues (coke) by other side reactions. In accordance with their surface structure, catalysts have different types of active sites on their surface which are responsible for catalyzing various reactions. For some reactions the active site may be a single atom; for others several surface atoms arranged in some special fashion may be responsible for the activity or selectivity. Recently, research efforts in heterogeneous catalysis have shifted from studies of reaction kinetics and mechanism to investigations leading to understanding of the structure of the catalyst and the chemical composition of the working catalyst on the atomic scale. Several investigations have emerged which deal with the specific surface reactions, identification of surface intermediates, and interpretation of catalytic processes in terms of crystal structure and imperfections, surface coordination, and electronic interaction between substrate and overlayer. The concept of structure sensitivity of reactions is one of these investigations which demands consideration in some detail.

Yet another important factor that determines the catalyst activity is the influence of various operating parameters. The major operating conditions which control the activity and stability of the catalyst are temperature, pressure, space velocity, and hydrogen-to-hydrocarbon mole ratio. Moreover, during the reforming process the catalyst is exposed to a feed containing $C_6$ to $C_9^+$ normal paraffins, naphthenes, and aromatics. The conversion of individual components to the desired products and, thus, the overall performance depend strongly on the feed composition. Surface modifiers such as sulfur, chloride, and other feed impurities also modify the characteristics of the catalyst. This is reflected in the overall performance of the catalyst during the reforming process.

Several groups of investigators have proposed various models and hypotheses which explain to some extent the mechanisms of reforming reactions over the catalyst surface. In this chapter an attempt has been made to put forth systematically the knowledge gained so far through several fundamental investigations in order to explain the basics of reforming catalyst activity, selectivity, and stability in relation to the structure of the catalyst and the operating conditions of the process. The purpose of the present chapter is to correlate the factors which govern the performance of reforming catalysts: catalyst characteristics and operating parameters, with the performance of these catalysts.

## II. CATALYST STRUCTURE, MORPHOLOGY, AND CHARACTERISTICS

The basic catalyst system used in reforming is platinum supported on alumina. Many versions of the catalysts with improved performance in terms of $C_5^+$ reformate yield, activity for particular applications, and life are available worldwide. The ultimate performance of the supported catalyst depends not only on the support or on the active species individually but also on their mutual interaction, which modifies the properties of both. The surface characteristics (acidity/basicity), texture, and porosity of the support are the major factors which determine the catalyst performance.

### A. Support Properties

In reforming catalysis the preferred support is alumina because of its versatility. Alumina supports can be prepared in a wide variety of crystalline forms, surface areas, porosities, and purities. Moreover, the surface acidity can be manipulated easily by incorporation of halogens or alkali metals. The catalytic activity is closely related to both the amounts and strength of acid sites distributed over the catalyst surface. The relation between catalytic activity and the amount of acidity is illustrated by many authors (1–5). Very weak acid sites may not be active catalytically, and very strong acid sites may lead to excessive cracking and carbon deposition on the catalyst. In addition, the catalytic properties of Brönsted acid sites (proton donors) may well be different from those of Lewis acid sites (electron acceptors). It has been suggested that Brönsted sites are much more active than Lewis acid sites for skeletal transformations of hydrocarbons (6). Lewis acid sites are observed frequently on many single and mixed metal oxides (7), whereas Brönsted acid sites are found only on a limited number of mixed oxides, for example, silica-alumina (8–10). Moreover, Brönsted acid sites are rarely observed on pure oxides free from additives such as P, Cl, or S. As alumina does not possess protonic acidity it should have no activity for hydrocarbon conversions in which the formation of a carbonium ion is involved. Nevertheless, if HF is added to alumina it presents protonic acidity and is able to catalyze hydrocarbon cracking reactions. In a reforming catalyst, where the noble metal is supported on alumina, the acidity is promoted with chlorine compounds. In this way the promoted acidity of alumina is responsible for isomerization reactions without inducing excessive cracking. A balanced amount of chloride maintained on the catalyst provides optimum paraffin cracking activity.

The most important aspects of the support which impact the activity are its surface area and porosity. In order to achieve high activity it is necessary to produce a catalyst with very fine particle size, which provides a large surface area for reactant molecules to come into contact with the catalyst surface, react, and then desorb as products. Generally, aggregate particle size of 50 μm to 10 mm are suitable for a catalyst of good activity. These particles must be porous enough to allow diffusional ingress of the reactants and egress of product molecules.

The pore size distribution and pore volume in catalysts depend on the method of preparation. The pores are classified into three categories dependent on their size. The micropores, <20 Å in diameter, are formed in the crystal lattice by removal of water from between the crystal planes, resulting in some type of phase change and a reduction in crystallite size (11). The mesopores 20–500 Å, are the spaces generated by the stacking of gamma alumina crystallites and are formed through loss of water and shrinkage of hydrated particles during calcination (12, 13). Macropores, in the range 500–2000 Å, are typically generated during the final forming operations (11).

In well-dispersed Pt/alumina catalysts the metal particles are usually in the range 5–15 Å. Micropores in the range 2–8 Å are too small in relation to these metal particles to have any significant contribution in maintaining catalyst activity (11). These pores may be blocked by coke or feed impurities during reaction (14). It is the mesopores which contribute to a large degree in providing high surface area and maintaining catalytic activity in reforming reactions (13).

The three ranges of the pore sizes form the boundary between the bulk gaseous diffusion and Knudsen diffusion. Catalysts containing intermediate porosity and operating in gases at pressures below 5 bar fall in the Knudsen regime, in which the gaseous diffusion mean free paths are restricted by the closeness of the catalyst particles. When operating above 50 bar little restriction of mean free path occurs. The extent to which diffusion restricts the egress of the reaction product is particularly important if good selectivity of an intermediate product in a series of consecutive reactions is required.

Another major function of the support is the ability to retain its surface area. Most heterogeneous catalysts undergo sintering as a result of high temperatures during either the start-up of the process or regenerations. The resistance of the support to sintering and hence loss of surface area plays a vital role in determining the performance of these high-temperature heterogeneous catalysts.

Moreover, the physical strength parameters of the catalysts, such as crushing strength and resistance to abrasion, are of significant importance. In industrial reactors, a catalyst with greater abrasion and crumbling tendency can cause severe pressure drop in reactors and pipelines. Changes in the environment inside the reformer caused by physical degradation of the catalyst adversely affect the selectivity, activity, and ultimately the life of the catalyst.

However, it has to be emphasized that no support material is completely inert and the effect of the support on the course of reaction and its mechanism must be considered. Impurities like Na and Fe in the support have a large effect on the performance of the catalyst even when they are present in minute quantities.

## B. Catalyst Structure and Properties

Great improvement in reforming technology was accomplished by the introduction of the bifunctional alumina-supported platinum catalyst by Universal Oil Products Co. (now UOP) in 1949. The metallic function of the catalyst was later improved by the addition of a second metal such as Re, Sn, Ge, or Ir. Due to the commercial success of these catalysts, a number of studies were devoted to the understanding of their surface morphology and activity correlations.

### 1. Active Sites

In 1925 Taylor (15) hypothesized that a heterogeneously catalyzed reaction occurs predominantly at a few distinct places on the catalytic surface called the *active sites*. The active species exist in the form of finely dispersed crystallites on the catalyst surface and the active sites on the faces of these metal-promoter crystallites are responsible for the breaking and forming of bonds, i.e., reaction. These active sites may consist of a single atom or an ensemble of atoms. Identification of the active sites where chemical bond scission or rearrangement occurs necessitates investigation of the structure and the chemical composition of the working catalyst on the atomic scale.

There could be a variety of active sites having different activity and selectivity for different reactions on the crystallites depending on the position and coordinative environment of the individual atoms, i.e., whether they are on crystal planes or atomic steps or kinks. Depending on this, the number of nearest neighbors and hence the coordination number of the active centers changes, which alters the nature of the adsorbate-adsorbent interaction. It has been concluded from studies of single crystal surfaces that the sites for $C-H$, $C-C$, and $H-H$ bond breaking are quite different in their characteristics (16). The surface atoms bind with the adsorbate molecules with sufficient strength that the residence time of the adsorbate is enough for the necessary chemical rearrangement to occur and then permit the product molecules to be desorbed from the surface, leaving the sites available for new reactant molecules. It is obvious that excessively strong chemical bonds between surface atoms and reaction intermediates lead to permanent blocking (poisoning) of active sites. On the other hand, insufficient interaction will lead to ineffectiveness of the catalyst in crucial bond-breaking processes. This adsorption-desorption process is greatly influenced by the electronic environment of the atoms, which in turn is determined by the metal-support interaction and presence of promoters and modifiers. This is described as the *ligand effect*.

Another important aspect which is relevant in determining the performance of reforming catalysts is the *ensemble effect*. Boudart et al. (17) classified the catalytic reactions on Group VIII metals into two categories:

1. Structure-insensitive reactions or facile reactions, for which the reaction rate is proportional to the whole metallic surface area. Under this category, any accessible superficial atom forms an active center whose properties are independent of its crystallographic site and environment.
2. Structure-sensitive reactions or the demanding reactions, whose specific activity (turnover number) varies with the structure of active sites. Under this category of reactions the rate is not proportional to the metallic area but depends on dispersion of the metal and the method of preparation of the catalyst.

It is seen that demanding reactions take place preferentially on specific parts of the surface. Some reforming reactions are structure sensitive, whereas some others structure insensitive (18).

Cyclohexane dehydrogenation and benzene hydrogenation are considered to be structure insensitive (19–25), whereas hydrogenolysis reactions in general are found to be structure sensitive (26–28). The latter type of reactions may require, on the surface, either a particular type of active site or a cluster of atoms with specific coordination and size. The activity and selectivity for such reactions presumably depend on the availability of specific ensembles, and the existence of such special arrangements of atoms that provide specific sites for adsorption has been demonstrated by Boudart (29). Coq and Figueras (30) showed that division of surface platinum into small ensembles, in this case diluted with promoter atoms, decreases the number of contiguous platinum atoms and hence increases the number of active centers. This enhances the activity and reduces coke deposition or deactivation by preventing multipoint adsorption of hydrocarbon molecules on the surface.

Since the structure and chemical composition of the working catalyst surface are extremely complex, it is difficult to distinguish elementary chemical processes associated with specific active sites from macroscopic studies of dispersed catalyst systems. Hence, the correlations between specific active sites and their reactivities obtained through studies of single crystal surfaces have been used to understand the overall mechanisms of reforming reactions on the catalyst surface on an atomic scale. Various crystal surfaces on Pt fall broadly into two categories;

1. Low-Miller-index surfaces having a large number of high-coordination-number Pt atoms
2. High-Miller-index surfaces having a large number of low-coordination-number Pt atoms

These surfaces can be generated by cutting crystals along different planes and can be regarded as building blocks of a dispersed catalyst. Taylor (15) suggested that atomic sites of low coordination number should be especially catalytically active. This idea has been expanded by others. Kesmodel and Falicov (31) argued that high-coordination sites should show unusual activity. Somorjai (16) put forth the idea that there are atoms in various positions on the surface that are distinguishable by their number of nearest neighbors, namely atoms in steps, in kinks, near adatoms, point defects, and vacancies as shown in Figure 1. Of these the terrace, kink and step sites are perhaps most important with regard to heterogeneous catalysis. The relative concentrations of the atoms in terraces, steps, and kinks can be varied by cutting single crystals in various crystallographic planes. When the crystals are cut to the low-Miller-index orientations, most of the surface atoms will be in terrace positions and will have the highest possible coordination number. On cutting high-Miller-index surfaces at some angle with respect to low-Miller-index surfaces, the atomic surface structure changes completely. The terraces have (111) orientations (Figure 2a). The steps have (100) orientations as shown in Figure 2b. A (557) surface exhibits periodic steps of monatomic height, separated by terraces of six-atom width as illustrated in Figure 2b. On cutting a single crystal in the middle of the stereographic triangle, a large density of kinks in steps will be produced (Figure 2c).

*a. Atomic Structure of Low-Miller-Index Surfaces.* The crystal orientations with low-Miller-index surfaces such as (111), (100), and (110) have the lowest surface free energy and highest atomic density (16). The Pt(111) crystal face, as shown in Figure 3, exhibits a sixfold rotational symmetry. Diffraction studies of the clean Pt(100) surface by Morgan and Somorjai (32) revealed that the surface is constituted of a so-called (5 × 1) surface structure. Schematic representations of the clean (100) surface and (100) surface with a hexagonal overlayer are shown in Figure 4a and b, respectively. The Pt(100) surface structure appears to be stable at all temperatures from 25°C to the melting point, although at elevated temperatures adsorbed carbon impurities can diffuse into the surface and cause transformation of the structure to an impurity-stabilized (1 × 1) surface structure.

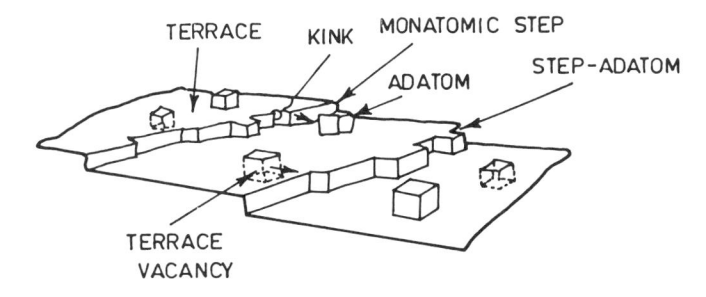

**Figure 1** Schematic representation of the heterogeneous surface on the atomic scale. From Ref. 16.

a) Pt-($\bar{1}$11)

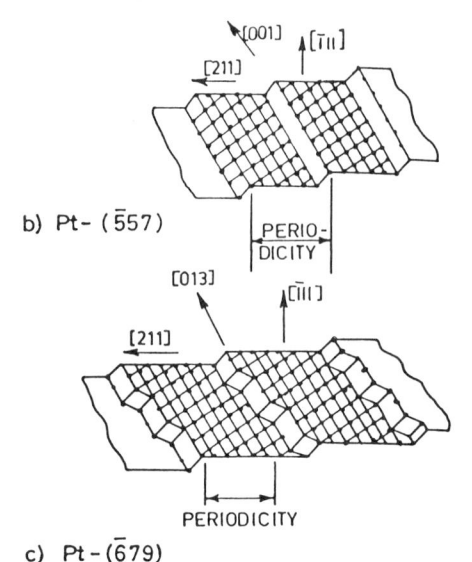

b) Pt-($\bar{5}$57)

c) Pt-($\bar{6}$79)

**Figure 2** Schematic representation of three types of face-centered cubic (fcc) crystal surfaces: (a) ($\bar{1}$11) orientation containing less than $10^{12}$ defects/cm$^2$; (b) ($\bar{5}$57) with $2.5 \times 10^{14}$ step atoms/cm$^2$ and six-atom-wide terraces between steps; (c) ($\bar{6}$79) surface with $2.3 \times 10^{14}$ step atoms/cm$^2$ and $7 \times 10^{13}$ kink atoms/cm$^2$. The average spacing is seven atoms between steps and three atoms between kinks. From Ref. 16.

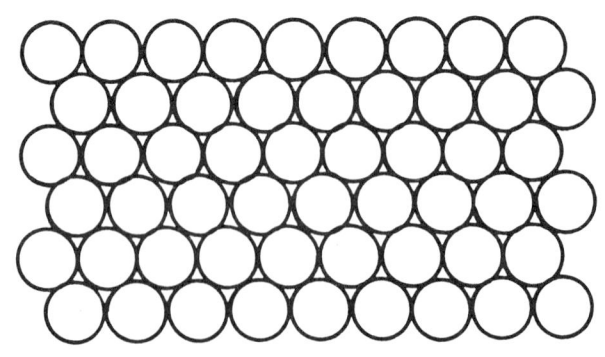

**Figure 3** Schematic representation of the Pt(111) crystal face. From Ref. 16.

*214*

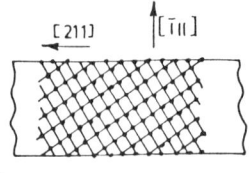

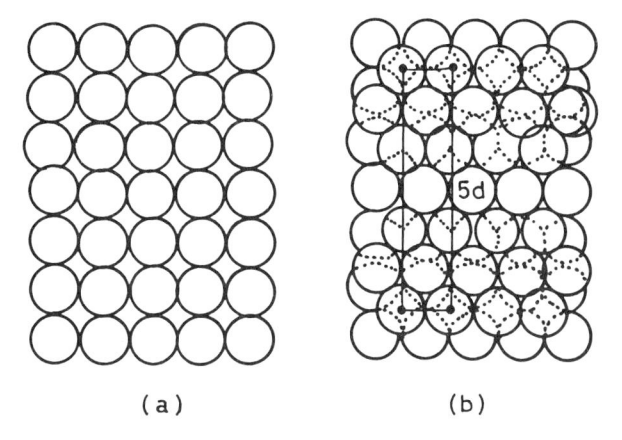

**Figure 4** Schematic representations of (a) Pt(100) surface; and (b) Pt(100) with hexagonal overlayer. From Ref. 16.

The Pt(110) unit cell apparently is a (1 × 2) unit mesh. The chemisorption characteristics of several adsorbates have been studied on this face of a Pt crystal. Adsorbates such as CO and $O_2$ are found to be more strongly bound on the Pt(110) crystal surface than on the (111) and (100) faces (33, 34).

*b. Atomic Structure of High-Miller-Index Surfaces.* Low-energy electron diffraction (LEED) investigations of Pt surfaces indicated that the surfaces of crystals characterized by high Miller indices consist of terraces of low-index planes separated by steps that are often one atom in height as seen in Figure 5. These surfaces, also known as stepped Pt crystal surfaces, are designated as Pt(S)-[$M$(111) × $N$(100)], where $M$(111) designates a terrace of (111) orientation $M$ atomic rows in width, and $N$(100) indicates a stepped (100) orientation $N$ atomic layers high. Lang et al. (35) described the nomenclature of these stepped structures in detail.

Many high-Miller-index surfaces of Pt show extraordinary thermal stability. These surfaces may be heated above 1200°C, where they may disorder, but on cooling to 800°C the ordered step structure is reestablished. The varying degrees of thermal stability of these ordered stepped surfaces play an important role in catalytic surface reactions that take place at temperatures below the temperature at which the surface structure orders by annealing.

The importance of stepped Pt surfaces lies in their great reactivity compared to the low-Miller-index crystal surfaces. These ordered stepped surfaces behave differently from low-Miller-index Pt surfaces for the chemisorption of probe molecules such as hydrogen, oxygen, and carbon monoxide, and even the chemisorption behaviors of various stepped surfaces are found to be different (36). The chemical reactions which involve breaking H−H, C−H,

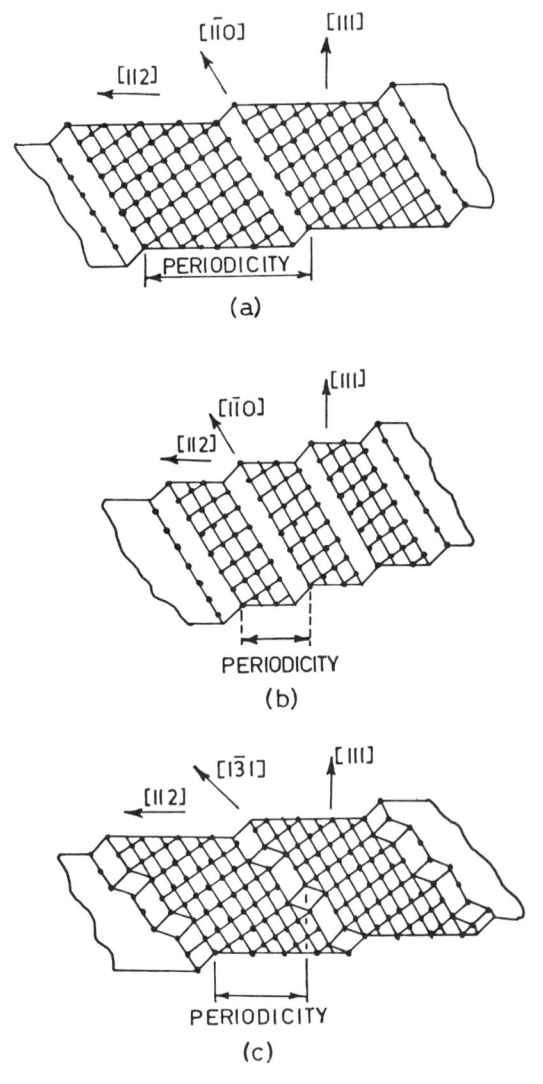

**Figure 5** Schematic representations of Pt stepped surface: (a) Pt (S)-[9(111) × (100)]; (b) Pt(S)-[4(111) × (100)]; (c) Pt(S)-[7(111) × (310)]. From Ref. 16.

and C−C bonds take place on the stepped Pt surfaces and are found to be less facile on low-Miller-index surfaces (37). Structure-sensitive C−H and C−C bond-breaking processes have been reported for the catalytic dehydrogenation and hydogenolysis of cyclohexane and cyclohexene (38). Atomic steps and kinks therefore can be responsible for selective bond breaking of hydrocarbon molecules. Conversely, by blocking active sites such as kinks, the selectivity of the catalysts may be markedly altered.

## 2. Chemisorption of Hydrocarbons on Catalyst Surfaces

During hydrocarbon reactions, the catalyst active surface is covered with a monolayer or near monolayer of carbonaceous deposits. A carbene-like metal carbon structure on the platinum surface was identified by LEED studies when ethylene and acetylene were the adsorbate molecules (39, 40). These molecules form ordered surface structures on the Pt(111) surface. Above 75°C, these molecules assume an ethylidene type of stable structure with the C−C internuclear axis perpendicular to the platinum surface as shown in Figure 6. The carbon atom closer to the metal is bonded to three Pt atoms with a Pt−C bond distance of 2 Å. This short metal-carbon bond is characteristic of carbene-like molecules, which exhibit unique chemical activity in many different displacement reactions. The C−C bond is a single bond with distance of 1.5 Å and is perpendicular to the surface plane. This resembles the structure of several trinuclear metal-acetylene complexes.

In order to understand the adsorption and ordering characteristics of probe molecules, the shape and bonding characteristics of many organic molecules have been studied by Gland and Somorjai (41). Correlations have been made between the charge distribution that occurs on adsorption of various organic molecules and their interaction with metal surfaces. These studies indicate that, at low adsorbate surface density, the adsorption of molecules with molecular dimensions smaller than substrate interatomic distances usually gives rise to the formation of ordered adsorbed structure (42). In this case, the local substrate-adsorbate interactions play an important role in determining the adsorption and

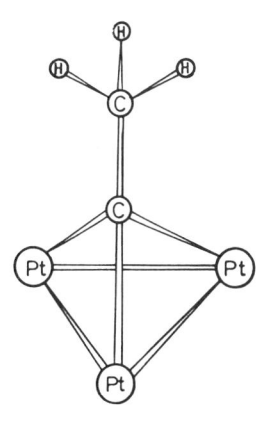

**Figure 6** Ethylidyne molecule that forms upon adsorption of ethylene or acetylene on the Pt(111) crystal face. From Ref. 50.

ordering of the adsorbed overlayer. With an increase in the surface density of the small molecules the adsorbate-adsorbate interactions become increasingly important and may distort the structure of the adsorbed overlayer (43).

When the molecular size of the adsorbate is larger than the interatomic distances in the substrate, the adsorbate-adsorbate interaction is predominant and the structure of the adsorbed overlayer do not necessarily depend on the local substrate-adsorbate bonds. Since the interaction of large molecules upon adsorption is with several surface atoms, the adsorption of the larger molecules is less controlled by a local substrate-adsorbate bond.

$C-H$ and $C-C$ bond breaking was found to be absent during the chemisorption of hydrocarbons on low-Miller-index Pt surfaces below 500 K and at low adsorbate pressures. The $C-H$ bond and $C-C$ bond breaking on the terrace sites requires considerable activation energy and the barrier can be overcome by either raising the surface temperature or increasing the reactant partial pressure. However, raising the temperature beyond 900 K causes the formation of graphitic coke (16).

In contrast to the chemisorption behavior on low-Miller-index surfaces, the $C-H$ and $C-C$ bond breaking takes place most readily over high-Miller-index surfaces of Pt even at 300 K and at low adsorbate pressures (16). The chemisorption characteristics of a number of hydrocarbons over four different stepped crystal faces of Pt, namely Pt(S)-[9(111) × (100)], Pt(S)-[6(111) × (100)], Pt(S)-[7(111) × (310)], Pt(S)-[4(111) × (100)], have been studied (44). The nucleation and growth of ordered carbonaceous surface structures and partial dehydrogenation of carbonaceous deposits through $C-H$ bond breaking take place predominantly on [9(100) × (100)] and [6(111) × (100)] crystal faces, whereas the disordered carbonaceous deposits form on the [7(111) × (310)] surface with a high concentration of kinks in the steps. The rearrangement of substrate by faceting is predominant on the [4(111) × (100)] face. Thus, the lack of reactivity of low-Miller-index surfaces in hydrocarbon reactions indicates the importance of steps in breaking $C-H$ and $C-C$ bonds. Such a marked change in chemical activity from sites of high coordination number on low-Miller-index surfaces to sites of low coordination number associated with high-Miller-index surfaces is one of the major attributes of Pt that is responsible for its unique catalytic activity. However, on low-Miller-index surfaces of Ir, partial decomposition of hydrocarbons may occur at low temperatures and pressures due to stronger adsorbate-substrate (hydrocarbon-metal) bonds.

Nieuwenhuys et al. (45) studied the chemisorption characteristics of various surfaces of Ir and Pt. The chemisorption characteristics of Ir(111) and Pt(111) surfaces were found to be different from each other. Also, the behaviors of the low-Miller-index Ir(111) surface and high-Miller-index stepped Ir[6(111) × (100)] surface were found to be markedly different. These studies show that acetylene and ethylene form surface structures that are better ordered on the stepped iridium than on the low-Miller-index Ir(111) surface. The lack of ordering on iridium surfaces similar to that on the Pt(111) surface indicates either the lack of mobility of hydrocarbons necessary for ordering at that temperature or the occurrence of a chemical reaction, i.e., decomposition. The observation that $C_2H_2$, $C_2H_4$, and $C_6H_{12}$ all yield the same diffraction pattern on the stepped iridium surface regardless of molecular size suggests that decomposition occurs on stepped iridium surface even at 300 K, the degree of decomposition at room temperature being higher on stepped iridium surfaces than on the Ir(111) surface. The poorly ordered (2 × 2) structure that has been identified on Ir[6(111) × (100)] surfaces after adsorption of $C_2H_2$ and $C_2H_4$ at room temperature was found absent on the Ir(111) surface below 500 K, indicating a higher degree of dehydrogenation on the stepped surface than on the Ir(111) surface.

Upon heating the surface to high temperature (>800°C) in the presence of hydrocarbons, the ordering characteristics of Pt(111) and Ir(111) surfaces were as follows: Ir(111) yields an ordered (9 × 9) coincidence carbon structure, which can be attributed to hexagonal overlayers of carbon similar to that of basal plane of graphite or benzene, deposited on the (111) surface. A similar structure was found on the Pt(S)-[6(111) × (100)] surface when heated to a high temperature with various hydrocarbon adsorbates. However, on the Pt(111) surface a diffraction pattern characteristic of a graphite overlayer with a rotationally disordered domain was observed.

Thus, it appears that strong metal-carbon interaction on iridium surfaces imposes the periodicity on the carbon atoms in the overlayer, while the surface of the graphite overlayer on the Pt(111) face is independent of the substrate periodicity and rotational symmetry. The order of ability of Pt and Ir surfaces for $C-H$ and $C-C$ bond breaking as suggested by Somorjai (16) is as follows:

$$Ir(S)-[6(111) \times (100)] > Ir(111) = Pt(S)-[6(111) \times (100)] > Pt(111)$$

Of these the Ir(111) and Pt(S)-[6(111) × (100)] surfaces are suggested to be the most versatile catalysts for reactions where $C-H$ and $C-C$ bond dissociations are necessary steps.

## 3. Reconstruction of Catalyst Surfaces

The atomic surface structure of the fresh reforming catalyst undergoes drastic changes or restructuring under reaction conditions and in the presence of different adsorbates such as hydrocarbons, carbon, hydrogen, oxygen, chlorine, or sulfur. It is suggested that this surface reconstruction arises from the high polarizability of the metal atoms, which provides the driving force to reconstruct the surface under the surface electric field (46, 47). Contraction or expansion in the direction perpendicular to the surface is evident in several unreconstructed low-Miller-index surfaces of clean monatomic and diatomic solids (48). LEED studies of single crystal surfaces of Pt under different environments indicate that most crystal surfaces restructure as the surface composition is changed (49). These studies reveal that some of the surface structures that are stable in ultrahigh vacuum and in the presence of oxygen reconstruct in the presence of carbon, whereas others that are stable when both clean and covered with carbon reconstruct when covered with oxygen (50). The changes of surface structure that mostly occur as a result of changes in the catalyst environment are presented schematically in Figure 7. A surface that exhibits the one-atom-height, step-terrace configuration reconstructs into a multiple-height step structure with wider terraces as the surface composition is changed. Other surfaces develop a hill and valley configuration consisting of large facet planes. Most of these reconstructions are reversible in nature.

There are a few stable surfaces that do not restructure at all under any conditions of the experiment (49). Besides the low-Miller-index surfaces (111), (001), and (011), there are the (112), (133), (122), and (012) crystal faces. These crystal planes are characterized by a very high density of periodic steps one atom in height or by a complete lack of steps. Because of their structural stability, it is expected that they play important roles in the catalytic chemistry of the transition metal surfaces.

One of the useful effects of restructuring in the presence of adsorbates can be visualized in the redispersion of platinum in reforming catalysts. Chlorine is a known adsorbate that in the presence of oxygen aids in the redispersion of Pt. Sulfur is another adsorbate that induces gross reconstruction of platinum sur-

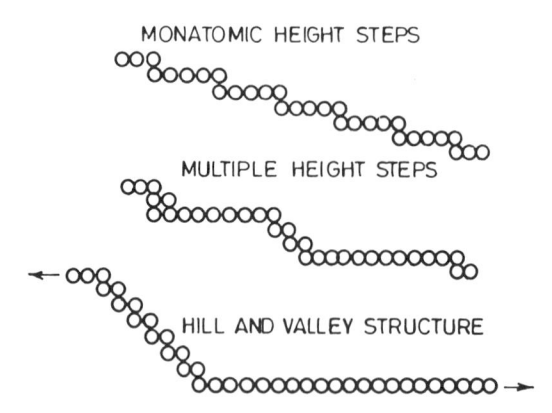

**Figure 7** Schematic representation of surfaces exhibiting one-atom-height step-terrace configuration, multiple-height step structure, and "hill and valley" configuration consisting of large facet planes. Reconstruction from one type to another occurs on adsorption. From Ref. 50.

faces (51). Although sulfur is known to be a poison for Pt catalysts, it acts as a useful modifier of the surface in small concentrations. It has been reported that poisoning of the catalytic activity is often due to the reconstruction of the active catalyst structure induced by adsorbates (51). Sulfiding in reforming catalysts is one example in which the advantageous effect of selective poisoning through reconstruction is exploited in industrial practice. The effects of sulfur and chloride on catalyst structure and performance are discussed in detail in Section II.D.

### 4. Morphology and Activity of Single Crystal Surfaces

Somorjai (52) showed that atoms in steps, kinks, and corners exhibit a higher degree of coordinative unsaturation and are more active than those on the low-Miller-index surfaces. It has been found that crystallite corner atoms have a nearest-neighbor coordination number of 4, edge atoms of 6 or 7, and atoms on the face of the plane of 8 or 9. These surface atoms have incomplete sets of neighbors and hence unbalanced valences which may be taken up by adsorbates and reactants. Various reforming reactions which involve the breaking and formation of different bonds need specific active sites having different coordinative environments. Table 1 lists the coordination requirements for typical reforming reactions (53). As discussed earlier, depending on the coordinative environment of the individual atoms, a variety of sites exists on the crystallites. Thus, atoms of different surroundings such as corner, edge, and face have different stereochemistry because of energetic effects and display distinct adsorption and reaction properties (54).

However, it has been observed that the activity of single crystal surfaces is influenced by the reaction conditions. For example, Somorjai and Blakely (38) observed significant yield differences at two pressure extremes for dehydrogenation and hydrogenolysis of cyclohexane while comparing the structure sensitivity and activation energies on Pt single crystal surfaces. Smith et al. (55) have studied the specific rate of benzene production on (111), (557), and (10,8,7) Pt surfaces at both low and high pressures and found that the initial rate of benzene production depends on the surface structure at high pressure, with the (10,8,7) surface being three times more active than the (111) surface. The apparent activation energy of dehydrogenation at low pressure is less than 4 kcal/mol (56), whereas at high pressure this value is 17 ±2 kcal/mol, indicating a possible difference in reaction mechanism (18).

There is a marked difference in the structure sensitivity of hydrogenolysis of cyclohexane. At low pressure $n$-hexane production from cyclohexane increases with increasing concentration of surface kink atoms. This is in contrast to high pressure, at which hydrogenolysis is faster on the (111) surface. Herz et al. (18) stated that this difference in catalytic behavior of the Pt single crystals may be due to the large variation in the relative concentrations of reversibly adsorbed hydrogen and adsorbed carbonaceous species at the two pressure

**Table 1** Site Requirement on Platinum Crystallites

| Reaction | Example compound | Generic reaction mechanism code | Crystallite size range (where mechanism dominant) | Effect of increasing dispersion on rate[a] | Minimum ensemble requirement | Coordination requirement |
|---|---|---|---|---|---|---|
| Hydrogenolysis | 1. Methyl-cyclopentane | Nonselective, cyclic | < 20 Å | − | 1 | Corner/edge |
|  |  | Bond shift | Full range | − | 1 | Corner/edge |
|  |  | Selective, cyclic | > 25 Å | + | 2 | Corner/edge and face |
|  | 2. Alkanes |  | Small | − | 1 | Corner/edge |
|  |  |  | Large |  | 3 | Face |
| Isomerization | 1. 3-Methyl-pentane | Nonselective, cyclic | < 20 Å | + | 1 | Corner/edge |
|  |  | Bond shift | Full range | − | 1 | Corner/edge |
|  |  | Selective, cyclic | > 25 Å | + | 2 | Corner/edge and face |
|  | 2. 2-Methyl-pentane | Nonselective, cyclic | 12–200 Å | + | 1 | Corner/edge |
|  |  | Bond shift | 12–200 Å | − | 1 | Corner/edge |
| Dehydrocyclization (with 1,5 closure) | Hexane | Nonselective, cyclic | Small | + | 1 | Corner/edge |
|  |  | Bond shift | Small | + | 1 | Corner/edge |
|  |  | Selective, cyclic | Large | − | 2 | Corner/edge and face |
| Ring enlargement (after 1,5 closure) | Methylcyclo-pentane |  | Full range | Negligible | 1 | Corner/edge and face |
| Dehydrocyclization (with 1,6 closure) | Heptane |  | Full range | Negligible | 1–2 | Corner/edge and face (higher on corners) |
| Dehydrogenation | Alkanes and naphthene |  | Full range | Negligible | 1 | Corner/edge and face |
| Self-poisoning | Alkanes |  | > 10 Å | − | 2 | Face |

[a] +, increase of rate; −, decrease of rate.
*Source:* Ref. 53.

extremes. The substantial difference in apparent activation energies at low and high pressures suggests that rate-limiting reaction steps might be different in the two cases. Possibly the rate-limiting step at low pressure is structure insensitive, while the rate-limiting step at high pressure is structure sensitive.

Nieuwenhuys and Somorjai (57) reported that on stepped Pt surfaces the turnover number for dehydrocyclization is 4 to 11 times larger than on the Pt(111) surface. Similar studies on Ir(111), Ir[6(111) × (100)], and Ir(111)-(9 × 9)-C showed no difference in their dehydrocyclization activity. However, the activities for dehydrocyclization on Ir surfaces were found to be much lower than those on similar surfaces of Pt. This is consistent with studies by Carter et al. (58), who reported that at atmospheric pressure Ir powder has a much lower activity for the dehydrocyclization of $n$-heptane than Pt powder.

## 5. Comparison of Studies on Single Crystals and Dispersed Catalysts

Single-crystal platinum catalysts offer the distinct advantage of allowing direct determination of surface structure and composition. Extensive studies of various crystal faces of single crystals under different reaction conditions and environments, such as hydrocarbons, oxygen, and hydrogen, have been carried out by Somorjai and co-workers (37–39, 41, 42, 44) to identify the active sites responsible for individual hydrocarbon conversions. The difference in reactivities of these sites arises from their coordinative environment. The special coordinative environment is intentionally generated on single-crystal surfaces, whereas on a well-dispersed catalyst surface all these sites coexist and their relative concentrations depend on the particle size and the dispersion of the metal, which are controlled by the method of preparation and pretreatments. Obviously, the studies on the structure sensitivity of reactions on dispersed catalysts have generally been reported in relation to the particle size and the dispersion. In polydispersed catalysts, varying the dispersion effectively changes the relative concentrations of these different types of active sites on the surface. Hence, identification of the active sites and understanding of the importance of their relative concentrations will lead to the design of better catalysts. However, due to the complexity of the heterogeneous catalyst surface, obvious correlations can rarely be drawn.

Somorjai (16) identified two types of structural features of platinum surfaces which influence the catalytic surface reactions: (1) atomic steps and kinks, i.e., sites of low metal coordination number, and (2) carbonaceous overlayer, ordered or disordered. The surface reaction may be sensitive to both or just one of these structural features or it may be totally insensitive to the structure. For example, dehydrogenation of cyclohexane to cyclohexene is found to be a structure-insensitive reaction. It takes place even on the Pt(111) crystal face, which has a very low density of steps, and proceeds even in the presence of a disordered carbonaceous overlayer. The dehydrogenation of cyclohexene to benzene is very structure sensitive. It requires the presence of atomic steps and an ordered overlayer and does not occur on the Pt(111) crystal face. However, it has been found that on dispersed metal catalysts the dehydrogenation of cyclohexane to benzene is structure insensitive (19, 21, 22). Dispersed particles of any size always contain many steplike atoms of low coordination and therefore the reaction appears to be structure insensitive. Somorjai and Carrazza (59) have found excellent agreement on structure sensitivity between supported and single-crystal catalysts in case of major reforming reactions as shown in Table 2.

**Table 2** Dependence of Several Hydrocarbon Reactions on the Surface Structure of Supported and Single-Crystal Pt Catalysts

| Reaction | Structure sensitivity | |
|---|---|---|
| | Single-crystal catalyst | Supported catalyst |
| | C–H bond formation and/or breaking | |
| Hydrogenation of olefins | Very little effect | Structure insensitive |
| Dehydrogenation of cyclohexane | Very little effect | Structure insensitive |
| | C–C bond formation and/or breaking | |
| Isomerization of light alkanes ($\leqslant C_4$) | Strongly structure sensitive; rate enhanced by presence of (100) sites; favored by the presence of kinks and steps | Structure sensitive; activity increases with decrease in particle size |
| Isomerization of large alkanes ($\geqslant C_6$) | Structure insensitive | Structure insensitive |
| Hydrogenolysis | Strongly enhanced by the presence of kinks; smaller effect in the presence of steps | Extremely structure sensitive; fivefold activity decrease with increase in particle size (15–60 Å) |
| Dehydrocyclization of $n$-heptane to toluene | Favored by presence of steps in (100) orientation; in the presence of kinks, formation of benzene (not toluene) favored | Structure sensitive; activity increases with decrease in particle size |

*Source*: Ref. 59.

The rates of dehydrogenation of cyclohexane to benzene obtained on Pt single-crystal surfaces have been compared by Herz et al. (18) with the rates obtained by Cusumano et al. (21) on dispersed Pt catalysts. Herz et al. found on a (557) surface a turnover number of 27, which was much higher than the 3.2 molecules/Pt atom/s calculated from the results of Cusumano et al. for the dispersed 2 wt % Pt/alumina catalyst. The higher rate for dehydrogenation of cyclohexane to benzene on the single crystal was attributed to the two different methods of determining the active surface area on the two types of catalysts. Herz et al. (18) found that the specific initial rates with dispersed Pt and single-crystal catalysts vary by at least a factor of 4.

On stepped surfaces the slow step in the dehydrogenation of cyclohexane to benzene is the production of the cyclohexene intermediate at low pressures. Cyclohexene dehydrogenates very rapidly at a step to form benzene. However, on the Pt(111) surface, which is practically free of steps, the rate of dehydrogenation of cyclohexene is slow enough to be rate limiting (60). Sinfelt et al. (61) concluded that the dehydrogenation of methyl cyclohexane to toluene was rate limited by the desorption of toluene. Maatman et al. (19) postulated that the slow step was the formation of an intermediate species. Haensel et al. (62) observed the intermediate cyclohexene species at very high liquid hourly space velocities. This indicates that the intermediate found at atmospheric pressure reaction conditions is very reactive at the step and edge atoms, which must exist on the dispersed metal particles.

It has also been found that on clean Pt single crystals the specific rate for cyclopropane hydrogenolysis is significantly higher than that reported for dispersed Pt catalysts (63, 64). The hydrogenolysis of cyclohexane has not been studied as a function of particle size on supported platinum catalysts. However, Lam and Sinfelt (65) investigated the reaction on ruthenium catalysts of widely varying dispersion and found that the specific rate of methane production increased by over an order of magnitude as the dispersion decreased from close to one to close to zero.

Studies similar to those reported for platinum must be carried out using crystal surfaces of other transition metals to ascertain that these arguments are more broadly applicable in describing the catalytic chemistry of transition elements. There is evidence that the heat of adsorption of hydrogen on palladium and iridium crystal surfaces varies markedly with step density (57, 66).

## 6. Particle Size and Dispersion

Solid catalyst surfaces are obviously heterogeneous on an atomic scale as shown by various LEED and field ion microscopic studies. In a real dispersed catalyst, where the metal is dispersed into small particles of the order 5–500 Å, the crystallites are mostly located in the pores of the high surface area support. The conclusions regarding surface sensitivity of reactions have been derived from studies of the particle size dependence of reaction rate. It is obvious that the rate of dehydrogenation-hydrogenation reaction is directly proportional to the degree of dispersion (67) as shown in Figure 8. Because dehydrogenation-hydrogenation is exclusively controlled by the metal function, the reaction rate depends on the number of exposed Pt atoms, which increases with decreasing particle size. Few data are available on the effect of dispersion on dehydrocyclization (68). Since this reaction is governed mainly by the acid function, it does not necessarily follow a distinct activity-dispersion relationship, as illustrated in Figure 9 (69). However, the maintenance of activity and selectivity of the catalysts depends on various other factors such as surface overlayer of carbon and presence of promoters, which are discussed later.

The Pt atoms on the surface of the support exist as electron-deficient species, which makes them active toward chemisorption (70). The hydrocarbon molecules (adsorbates) usually act as electron donors (71). This electron deficiency is caused by the metal-support interaction (70). The desorption of the olefinic products from the surface is enhanced by this electron deficiency. Highly dispersed small crystallites have higher activity and stability than larger crystallites. On larger crystallites, back donation from metal to olefin could be more effective, leading to increased residence time and thereby deactivation. Moreover, on large crystals, because of the close proximity of Pt atoms, multisite adsorption and formation of polyolefinic coke precursors may also occur.

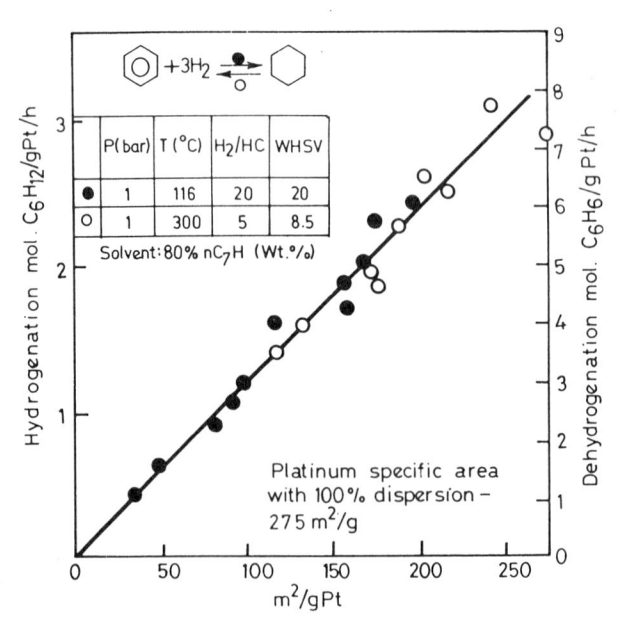

**Figure 8** Influence of platinum dispersion on dehydrogenation of cyclohexane and hydrogenation of benzene. From Ref. 67.

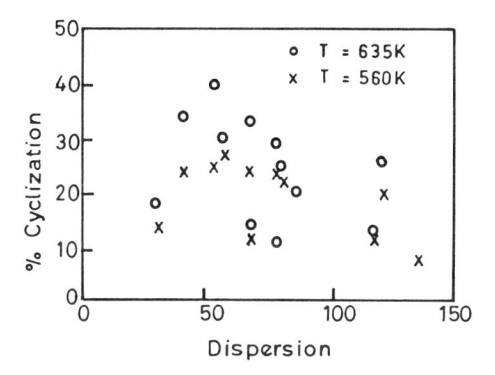

## Structure and Performance of Reforming Catalysts 227

**Figure 9** Effect of platinum dispersion on dehydrocyclization of $n$-hexane. From Ref. 69.

Large crystallites favor coke formation on the alumina due to stabilization of the cyclopentadiene molecule for the Diels-Alder polymerization (72). Small crystallites, because of their electron deficiency, cannot stabilize and catalyze the polymerization of the $C_5$ ring coke precursors (73, 74).

In promoted coimpregnated Pt catalysts, although the size of the aggregates may be larger, the Pt atoms may be diluted by the atoms of the promoter element. This decreases the possibility of contiguous Pt atoms on the catalyst surface and thus reduces multisite adsorption of olefinic hydrocarbons on Pt and excessive coking. The electronic interaction of Pt and some promoter elements, for example, tin (75) and germanium (76), results in more electron-deficient Pt atoms. This also enhances the activity and reduces coke formation. In the case of promoters such as Re and Ir, the electronic interaction may lead to more electron-rich Pt atoms. However, sulfided Re and Ir, the form in which they are present on the working catalyst surface, are reported to behave similarly to Sn, which is claimed to be catalytically inert (77). Evidence with bimetallic alloy catalysts that large Pt ensembles are required for hydrogenolysis has been reported by Dautzenberg and co-workers (78). Ensembles of contiguous Pt atoms are required for hydrogenolysis and subsequent transformation of cracked products into carbonaceous deposits. Alloying reduces the concentration of large ensembles and, hence, the formation of coke deposits (79). The hydrogenolysis activity of tin-promoted catalysts is found to be lower than that of the corresponding monometallic catalysts because of the dilution effect (75). Promoters such as Ir and Re have high inherent activity for hydrogenolysis (75, 80). Hence, in Re- or Ir-promoted Pt catalysts, although the hydrogenolysis activity of Pt is greatly reduced by the ensemble effect, the resultant hydrogenolysis activity of the bimetallic catalysts is found to be higher than that of the monometallic catalysts. Hence, the excessive hydrogenolysis activity of Re- and Ir-promoted catalysts is controlled by sulfiding (77).

## C. Catalyst Promoters and Alloy Formation

A number of promoter elements such as Re, Sn, Ir, and Ge have been used along with Pt in reforming catalysts and are known to prolong the active life of catalysts. The effect of individual promoter elements and the role of each in modifying the surface morphology resulting in the improvement of activity and life of the catalysts are discussed below.

### 1. Rhenium

According to Betizeau and Bolivar (81, 82), Re exists in the metallic state and possibly forms an alloy with Pt. The effect of Pt and Re on each other is not completely additive. This suggests that a portion of each is modified by interaction while the other portions of each act independently of the other. The authors observed that up to an Re atomic fraction of 0.25 the product of cyclopentane hydrogenolysis remains $n$-pentane, beyond which the secondary reactions forming methane, ethane, $n$-propane, and $n$-butane increase.

However, other authors have suggested that Re exists in the oxide form and perhaps interacts with the acidic support (83, 84). Issacs and Peterson (85) reported that in the presence of moisture $Re_2O_7$ migrates to the proximity of Pt centers, which catalyze the reduction of Re during hydrogen treatment. Although there is no difference of opinion about the promotional effect and coke-resisting behavior of Re, there is a controversy regarding the state of Re and its mode of action.

Bertolacini and Pellet (86) used Pt/alumina, Pt-Re/alumina, and mixtures of Pt/alumina and Re/alumina as catalysts and observed that an intimate mixture of Pt/alumina and Re/alumina is just as active and stable as the coimpregnated catalyst. Elemental analysis of the physically separated particles after the reaction has shown that elemental migration does not occur during reaction. This rules out the possibility of alloy formation and confirms the theory of modification and removal of coke precursors on the surface (87).

On the contrary, Shum et al. (88) reported that the physical mixture of Pt/alumina and Re/alumina is markedly different from cosupported catalyst. The latter showed high selectivity for hydrogenolysis, which is attributed to the Pt-Re alloy phase. Wagstaff and Prins (89) demonstrated that at low loading of metals under reaction conditions only Pt(0) and Re(0) exist. These studies were carried out for a relatively short period and the effect of coimpregnation and electronic interaction on stability of these catalysts was not ascertained.

It is however, clear that Re supported on separate particles exerts a promotional effect by removing volatile coke precursors. Hence, in the case of Re-promoted catalysts, besides the ligand and ensemble effects, which are operative only in coimpregnated catalysts, there appears to be some other inherent property of Re which retards the deposition of coke.

The major influence of Re in Pt-Re bimetallic catalysts is in enhancing the life of the catalyst. It has been observed that Pt-Re exhibits higher activity than monometallic catalyst even when the two catalysts contain the same amount of coke. Thus, the role of Re becomes obvious when the catalyst has been exposed to hydrocarbons for a relatively longer period and a substantial carbonaceous deposit has been accumulated on the catalyst (90, 91), since the Pt surface in Pt-Re catalyst is maintained clean and free of adsorbed coke by the effect of Re. Hence, in comparison with monometallic catalysts, the extent to which the severity of commercial operation has to be raised to maintain a certain level of activity remains low in the case of Pt-Re bimetallic catalysts (87).

## 2. Tin

Pt-Sn/alumina catalysts are reported to exhibit a very high degree of dispersion of Pt (20–25 Å size) and high selectivity toward aromatics formation during paraffin dehydrocyclization (92). Bond and Sermon (71) postulated an electron donation from the small metal-supported crystallites to the support and modifications of this by Sn. Davis and co-workers (70) reported that the addition of tin to Pt-alumina greatly enhanced the activity and decreased the rate of deactivation for the dehydrocyclization of alkanes.

Numerous reports are available about the state and role of Sn in Pt-Sn catalysts. It has been observed that, depending on the experimental techniques used and the parameters studied, the inferences differ. Surface studies of Pt-Sn systems by $H_2$-$O_2$ adsorption and electron microdiffraction have revealed that Sn exists in the +2 state and no Pt-Sn alloy exists (93). X-ray photoelectron spectroscopy (XPS) data indicated that Sn atoms are not present in the metallic state irrespective of the state of Pt (94, 95). In situ XRD data showed the presence of Pt-Sn alloy with a atomic ratio composition of unity (96). Based on temperature-programmed reduction (TPR) studies, Lieske and Volter (97) reported that the amount of alloyed tin increases with an increase in Sn content. Moessbauer studies (98–102) provided evidence for the formation of Pt-Sn alloy and for the presence of highly dispersed particles containing Sn(IV), Sn(II), and Sn(0) states. Srinivasan et al. (103) concluded that Sn alters the activity and aromatic selectivity of Pt, which implies intimate contact between some form of Sn and Pt(0). Davis (104) observed a maximum in the activity near an Sn/Pt atomic ratio 3–4 and postulated that it is the dispersed, X-ray amorphous Pt phase rather than the Pt-Sn alloy which is catalytically active for hydrocarbon conversions. Balakrishnan and Schwank (105) studied Pt-Sn/alumina using XPS and reviewed the reports available so far. However, their studies did not give direct evidence for the presence of zero-valent Sn and indicated that there may be small amounts of Sn(0) below the limit of XPS detection. They attributed the changes in reactivity trends to the presence of Sn(0) in close proximity to Pt(0). Li et al. (106) have used the Moessbauer technique to study alloy formation in Pt-Sn/alumina. They found that Pt-Sn alloy exists on the surface and its amount increases with increasing Sn content. Although the latest reports support alloy formation, a controversy still exists and no definite conclusion regarding the state and mechanism of the promoting action of Sn can be drawn.

### 3. Iridium

The Pt-Ir combination, though not widely exploited commercially, has attracted the attention of many investigators. Iridium has many similarities to Pt, but striking differences exist with respect to their melting points, heats of evaporation, and surface tensions. These facts indicate that the metal-metal bond in iridium would be stronger than that in platinum. Limited mutual solubility of the two metals has been reported (107) and the two phases crystallize out extremely slowly. Moreover, the alloys are reported to solidify initially without phase separation. Rasser et al. (80), after reviewing the studies on Pt-Ir alloy systems, concluded that there is no evidence from thermodynamics of the Pt-Ir system for the existence of an ordered phase. Whereas monometallic Pt/ alumina catalysts have outstanding selectivity for naphthene-to-aromatic conversions, Pt-Ir catalysts have higher selectivity for paraffin isomerization and dehydrocyclization and exhibit a higher yield stability (80). However, Anderson and Avery (108) reported that pure Ir metal has no isomerization activity and is more active than Pt for hydrocracking. Ramaswamy et al. (109) attributed superior performance of Pt-Ir catalysts to the dilution of Pt by Ir, which should result in a lower concentration of coke precursors on the metal surface. Rasser et al. (80) ascribed the reduced coke coverage of Pt-Ir systems to the higher hydrogenolysis activity of Ir. The authors observed that once a carbonaceous deposit is formed, iridium and iridium-rich surfaces are more easily freed from carbon by hydrogen treatment. Brunelle et al. (110) stated that iridium has a higher hydrogenolytic activity than platinum. This is also true for iridium-rich platinum alloys. Iridium shows a weaker tendency than Pt to form graphitic deposits and is therefore more easily decarbided by hydrogen. This might be due to stronger Ir−C bonds, which in turn preclude the formation of the C−C bonds from adsorbed carbon atoms. The absence of C−C bonds makes the carbon atoms on the Ir surface more active for hydrogenation to methane. Rasser et al. (80) concluded that carbon atoms deposited on Pt probably migrate to Ir centers, where they are hydrogenated. Nevertheless, they did not observe any improvement in selectivity toward isomerization and aromatization at atmospheric pressure. In the absence of any evidence for the formation of ordered alloy phases and with no improvement in selectivities as a function of Ir-Pt ratio, the authors, however, ruled out the possibility of any strong ligand effect.

## 4. Germanium

The Pt-Ge system has been investigated for surface characteristics, activity, and selectivity trends. XPS studies conducted by Bouwman and Biloen (111) showed the presence of $Ge^{2+}$ and $Ge^{4+}$ species after reducing the Pt-Ge catalyst at 823 K; however, after reduction at 923 K the catalyst contained $Ge^{2+}$ and Ge(0) phases. The Ge(0) phase was found in an alloyed state with platinum. Tomero et al. (112) and Bolivar et al. (113), in studies of the hydrogenation of benzene and hydrogenolysis of n-butane, concluded that Ge enhances the activity for hydrogenation and the selectivity toward isomerization to isobutane for the hydrogenolysis transformation. They observed a considerable reduction in dispersion and an increase in the CO stretching frequency on addition of Ge to Pt/alumina. Gold Wasser et al. (76) observed 30–40% reduction of Ge species to Ge(0) for catalysts containing >0.3 wt % Ge with Pt approximately 1.0 wt %. It appears that there is a minimum amount of Ge at which reduction can be initiated and beyond which activation energy for benzene hydrogenation starts to increase. This reaction is structure insensitive and hence should not depend on such parameters as dispersion and geometry. The activity for benzene hydrogenation may therefore be ascribed to the electronic effect. Bouwman and Biloen (111) have observed that in Pt-Ge catalyst Pt is more electron deficient than in monometallic catalysts. They excluded a geometric effect and concluded that there could be an electron withdrawal from Pt by Ge species, which is the reason for improved performance.

Beltramini and Trimm (114) studied the reaction selectivities on reduced catalysts and reported that the addition of Ge to Pt/alumina has reduced hydrogenolysis and isomerization, increased hydrocracking, and thereby changed the relative importance of aromatization through the dehydrocyclization pathway. The initial high hydrocracking activity dropped rapidly to the level of Pt/alumina catalyst with a corresponding increase in dehydrocyclization, which is suggestive of modification of the surface by coke. The enhancement in aromatic selectivity and reduction in the rate of coke accumulation have been ascribed to electronic modification. The Ge oxide species present on the surface increase the acidity, which is reflected in the high hydrocracking activity of the fresh catalyst.

## 5. General Comments

The activity, stability, and selectivity of promoted catalysts are usually different from those predicted from individual metal properties. The ensemble effects are usually dominant because the adsorption behavior of reactant molecules depends to a large extent on the critical size of the assembly of metal atoms which form the active "landing" site for the adsorbate molecules (115–118). Many factors, such as surface segregation, site elimination by blocking, and metal-support interaction, influence the state of dispersion and stability of bimetallic catalysts under reaction conditions (115). Dispersion and alloying are of prime importance. As discussed earlier, ensemble size of single metal crystallites decreases with an increase in dispersion; consequently, the proportion of multiple bond intermediates decreases. Thus, the catalyst experiences lower coking rates. Simultaneously, the increase in the number of low-coordination-number sites causes an increase in turnover number (TON) of hydrogenolysis.

The active metal ensemble size diminishes when the metal is alloyed with inactive metal with or without a change in geometry or crystallite size, and as a result the rates of hydrogenolysis and metal site coking decrease drastically (119, 120). Thus, the creation of low-coordination-number sites and the reduction in active metal ensemble size have opposite influences on hydrogenonlysis. Adsorbed coke (121-123) and sulfur (124-127) drastically decrease the hydrogenolysis activity, and this has been successfully exploited with Pt-Re/alumina catalysts. Sulfur bound to Re blocks its cracking activity and, interestingly, reduces the size of Pt ensembles (128, 129). This in turn enhances the rate of dehydrogenation, single-site dehydrocyclization, and isomerization. Thus, Pt-Re-S alloys behave similarly to Pt-Sn alloys because sulfided Re acts as an inert diluent. In the case of Pt-Ir catalysts, where both components are active, the improved aromatic yield is due to the high dehydrocyclization activity of the iridium component (130). The higher hydrogenolytic activity due to iridium is controlled by adsorbed sulfur, which reduces Pt-Ir ensemble size and improves selectivity for the single-site 1,6-dehydrocyclization reaction (131).

## D. Surface Modifiers

### 1. Sulfur

Hydrotreated feeds are generally used in catalytic reforming units because sulfur in the feed is a well-known poison for Pt reforming catalysts. Reforming catalysts contain some extremely active sites which impart excessive hydrogenolysis activity and result in faster deactivation (90). Sulfur adsorbs, reversibly and irreversibly, on Pt and promoter atoms and precludes the active sites taking part in the reaction. Presulfidation of the catalyst selectively blocks the hyperactive sites responsible for hydrogenolysis activity. This is why start-up procedures of reformer units frequently include presulfidation.

Even though different sulfur compounds are used for sulfidation, all catalysts are finally subjected to a high-temperature hydrogen treatment. This eliminates part of the sulfur as hydrogen sulfide and the rest remains irreversibly adsorbed on the surface. This corresponds approximately to 0.5 S atoms/Pt (*S*) (132).

In the lower-temperature range (< 700 K) where the catalyst is exposed to a low level of self-poisoning by coking, sulfur acts exclusively as a poison

(124, 127, 132–134). Unsulfided platinum catalysts usually exhibit a very high initial cracking activity that decreases rapidly; at the same time, isomerization and cyclization activity increase steadily. The selectivity for dehydrocyclization remains low and stable. These selectivities are substantially modified by sulfidation. It has been reported by Shum et al. (129) that sulfided Pt/alumina and Pt-Re/alumina catalysts, chlorided or not, show a very significant reduction in total conversion for $n$-hexane reforming. Although the activity of sulfided bimetallic Pt-Re catalyst is lower than that of sulfided Pt/alumina catalyst, the maintenance of activity of the former is excellent, whereas the latter tends to lose its activity rapidly. For monometallic Pt, the selectivity for cracking decreases considerably on sulfiding with a simultaneous increase in the selectivity for isomerization and cyclization (formation of a five-member ring cycloalkane from alkane) reactions. The selectivity for paraffin dehydrocyclization to aromatics decreases.

These selectivity modifications are reported to be dependent on the extent of poisoning of the various reaction paths; hydrogenolysis is far more suppressed than dehydrogenation or aromatization due to adsorbed sulfur (124, 127, 134). Higher levels of sulfur are known to kill active sites responsible for isomerization, cyclization, and dehydrocyclization. For example, it was found that $n$-hexane can be dehydrogenated only over a highly sulfided Pt catalyst (133). The low-coordination-number corner and kink atoms, which are hyperactive and could activate $C-C$ bonds, are preferentially masked by the sulfur atoms. This makes the catalyst unable to accomplish $C-C$ bond fission and thus retards the hydrogenolysis and cracking activity (79). Rhenium is known to have high hydrogenolysis activity. However, in Pt-Re catalyst the ensemble effect prevails and it exhibits low hydrogenolysis activity (77).

High hydrogenation activity is required to hydrogenate olefinic fragments and produce lighter molecules. Because of the formation of inert Re-S species, the Pt-Re ensemble, which is originally active for hydrogenolysis, presumably becomes inactive (135). Shum et al. (88) explained the decreased deactivation or improved stability of Pt-Re-S catalysts by a model assuming sulfur atoms to be preferentially adsorbed on Re atoms that cannot easily be removed by hydrogen under process conditions. Somorjai and Blakely (38, 136) had shown that the reorganization of carbonaceous overlayers into more harmful species distinctly depends on the topography of the adsorbing metal surface. The strongly chemisorbed sulfur species impede the reorganization of the omnipresent hydrocarbonaceous overlayer (soft coke) to pseudographitic species (hard coke) covering the metal particles (135). It has been observed that on presulfided Pt and Pt-Re catalysts, the hydrogen-to-carbon (H/C) ratio of coke deposits remains high. When compared to unsulfided catalysts, the ratio increases from 0.24 to 1.84 and 0.36 to 1.52 for Pt-Re and Pt catalysts, respectively. This indicates the influence of sulfur on the preservation of hydrogenation activity (137).

The effect of sulfur on the activity and selectivity of reforming catalysts is discussed in the literature. Some reports indicate decline in activity due to a poisoning effect (88, 127, 131, 132, 138), whereas others show an improvement in activity on sulfidation (79, 131, 139). In the case of metal-catalyzed nondestructive reactions, the selectivity improves on sulfidation. The effect of sulfur on catalytic activity seems to depend on reaction conditions. Apparently an improvement in activity on sulfidation is manifested under conditions which favor dehydrocyclization due to a strong self-poisoning effect by coking, that is, at high temperature, low pressure, and low $H_2$/HC ratio (140).

Experiments with neo-hexane on Pt-Re-S/alumina showed that the acidic alumina plays a significant role in the isomerization reaction. It is argued that sulfur may have some influence on the acidity of alumina, which modifies the performance of the catalyst. Sulfur is reported to interact with Lewis acid sites and thus deactivate sites for coking (141–143).

Sulfur adsorption leads to reduction of the number of large ensembles by the geometric effect. Besides this, there is an electronic transfer from Pt to S which influences the bonding between Pt and adsorbate. Both effects perhaps may be acting together, culminating in a substantial reduction of hydrogenolysis activity and coke deposition (75).

Thus, for maintenance of the activity of reforming catalysts, presulfidation is an essential step. Sulfur incorporation makes the catalysts superior through various modes. It can be concluded that sulfur preferentially deactivates the hyperactive sites; interacts with Re to make it less active for hydrogenolysis; adsorbs on Pt and promoter metals, reducing the ensemble sizes; and effectively increases dispersion. Moreover, it prevents reorganization of coke into graphitic form; blocks the Lewis acid sites on the support (reducing cracking); preserves the hydrogenation activity of Pt, thereby increasing the H/C ratio of coke, etc. Despite the decline in initial activity due to sulfidation, sulfur induces excellent stability.

## 2. Chlorine

Reforming catalysts are bifunctional in nature. The metal function is provided by platinum and the acid function is provided by the high specific surface area alumina, whose intrinsic acidity is enhanced by chlorine. The optimum chloride content is in the range 0.8 to 1.3 wt %. Figure 10 illustrates the effect of chloride on the activity of the catalyst in terms of reformate RON (144).

For optimum performance of the catalyst, it is crucial that a proper balance between metal and acid function be maintained. If the metal function is too strong, excessive hydrogenolysis to $C_1$–$C_4$ gases and dehydrogenation to polyolefinic coke precursors can occur; however, if it is too weak, then excessive coking also takes place and the catalyst deactivates rapidly. If the acid function is too strong, it leads to excessive hydrocracking and coke laydown on the catalyst; however, if it is too weak, the rate-determining carbonium ion

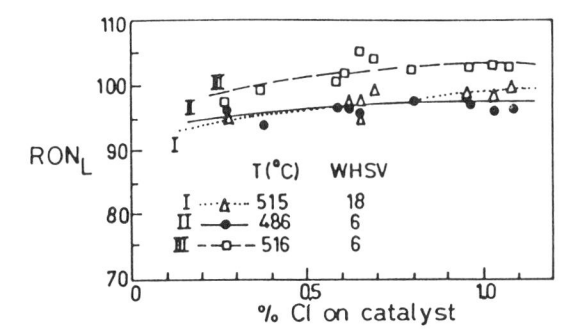

**Figure 10** Octane number as a function of weight percent chloride on the catalyst ($p = 30$ kg/cm$^2$, H$_2$/$n$-heptane = 8). From Ref. 144.

reactions involved in dehydroisomerization and dehydrocyclization do not proceed fast enough, which in turn leads to an increase in C$_1$–C$_4$ gas production and a decrease in the yield of liquid product. The studies of the dehydrocyclization of $n$-heptane in the presence and absence of ammonia compounds, as illustrated in Figure 11, demonstrate that catalyst acidity is compulsory for dehydrocyclization (67).

The two main factors that control the acidity of the catalyst under actual reforming conditions are moisture in the feed and chloride on the catalyst. Excessive moisture at higher temperatures strips off the chloride from the catalysts, thereby lowering the acidity. Loss of chloride can be partially compensated by increased dosing of chloride in the feed, in order to restore the water-chloride balance according to the following reaction:

$$O\begin{matrix}\text{Al}-\text{OH}\\\text{Al}-\text{OH}\end{matrix} + HCl \rightleftharpoons O\begin{matrix}\text{Al}-\text{Cl}\\\text{Al}-\text{OH}\end{matrix} + H_2O$$

It must be emphasized that water is necessary to ensure a homogeneous distribution of chloride on the catalyst. It is observed that the practice of operating under the driest possible conditions, without injecting chloride, is harmful to the acid function. This is because there is no regeneration of acid sites to compensate for those deactivated by coke deposits. In an industrial unit, depending on whether operation at the time is too wet or too dry, the following phenomena can be observed (67).

In the case of too wet operation:

1. All the chloride migrates toward the last reactor.

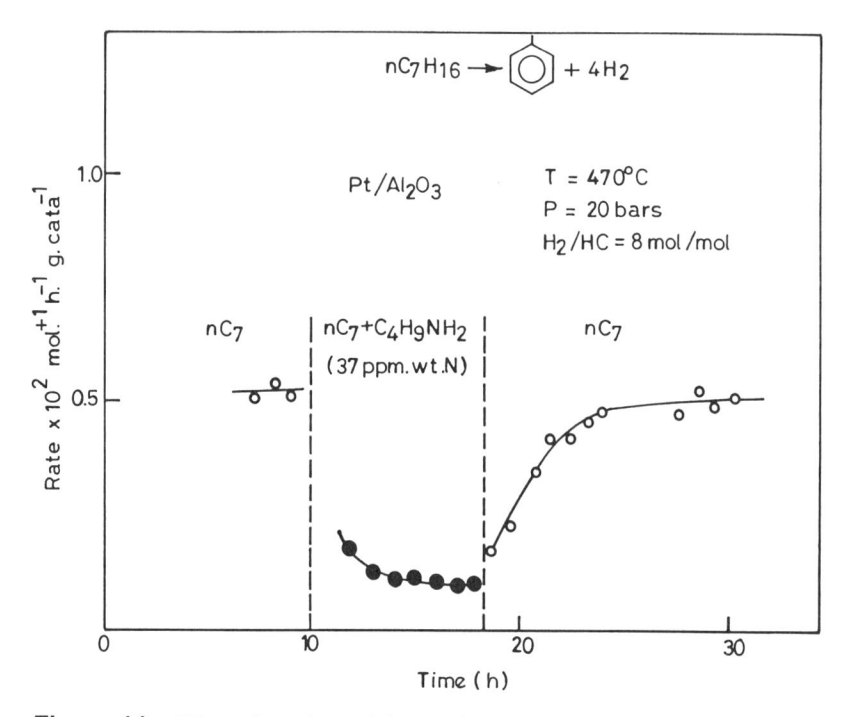

**Figure 11** Effect of catalyst acidity on the dehydrocyclization of $n$-heptane. From Ref. 67.

2. Average temperature of the last reactor increases.
3. Production of light hydrocarbons, especially $C_3$ and $C_4$, increases as long as the last reactor is overchlorinated.
4. Hydrogen content of the recycle gas decreases.

In the case of too dry operation:
1. More chloride adsorbs on the catalyst.
2. Average temperature of the first reactor increases.
3. $C_3$, $C_4$ production increases.
4. Hydrogen content of the recycle gas decreases.

In the presence of hydrogen, halogens and organic halogenated compounds are transformed with the production of HX acid (where X denotes halogen) on the catalyst. In a very dry atmosphere, HX adsorbs mainly at the front of the catalytic bed of the first reactor. In a wetter atmosphere, the whole catalyst inventory is halogenated. In all cases, there is an increase in the acid function of the catalyst, which means an increase in production of propane and butane through hydrocracking.

In reforming catalysts, chloriding is usually performed by injection of $CCl_4$, HCl, chlorocarbon compounds, or chlorocompounds of sulfur. The possibility of increasing the acid function of the catalyst by injecting water and HCl is used on the industrial level in some reforming units.

The chloriding agent is reported to have an influence on the selectivities of the catalyst for different reactions. Menon et al. (145) have studied the effect of chloriding using $CCl_4$ and HCl on the selectivity of Pt/alumina catalyst for hydrogenolysis and isomerization. They observed a marked decline in hydrogenolysis activity on chloriding with $CCl_4$ in $H_2$ and attributed this to the small amount of coke deposited on the catalyst, which attenuates the sites responsible for hydrogenolysis. When the catalyst was chlorided with HCl gas, there was only a minor decrease of the selectivity for hydrogenolysis. Thus, chloriding with $CCl_4$ more effectively influences the Pt sites responsible for hydrogenolysis, which become covered by coke. The isomerization activity in both the cases improves due to an increase in acidity of the alumina.

Somewhat contradictory reports are available in the literature regarding the interaction of chlorine with Pt sites. Massoth and Kiviat (146) observed an identical weight increase in the case of $Al_2O_3$ and $Pt/Al_2O_3$ on chlorination with HCl and concluded that the chlorination reaction per se was independent of the presence of Pt. On the other hand, results of infrared (IR) (147) and electron spin resonance (ESR) studies (148) suggest the existence of Pt-Cl species in the $Pt/Al_2O_3$ catalyst as it is normally used, even before any additional chloriding. It has been reported that chlorine interacts with Pt atoms on the surface and changes the energies with which they can chemisorb or desorb hydrogen as seen from TPD studies of adsorbed hydrogen (149). The authors attributed the reduction in hydrogenolysis activity of Pt/alumina catalysts when chlorided with HCl gas (no coke) to this modification of Pt atoms by chemisorbed chlorine. This has been further confirmed by the fact that driving out most of the chlorine from the catalyst by air oxidation, followed by reduction at 400°C, restores the original hydrogenolysis activity of Pt. Hence, it is concluded that chlorine itself is partly responsible for the attenuation of hydrogenolysis activity of Pt.

XPS studies of Pt/alumina catalysts showed a change in the $Pt_{4d}/Al_{2s}$ signal intensity ratio on chlorination with HCl at 400°C, which indicates uneven coverage of alumina and Pt by chlorine (149). Moreover, addition of chlorine to the catalyst enhances the stability of Pt crystallites on the alumina surface and prevents their sintering (150) and even causes a redispersion of Pt crystallites during rejuvenation (67). The decline of Pt surface area has been studied in a reducing atmosphere with and without chlorine injection at various hydrogen pressures. The results shown in Figure 12 demonstrate that at 550°C the presence of chlorine inhibits sintering (curves Ia and IIa), with the decline of metallic surface area ($S$) expressed as

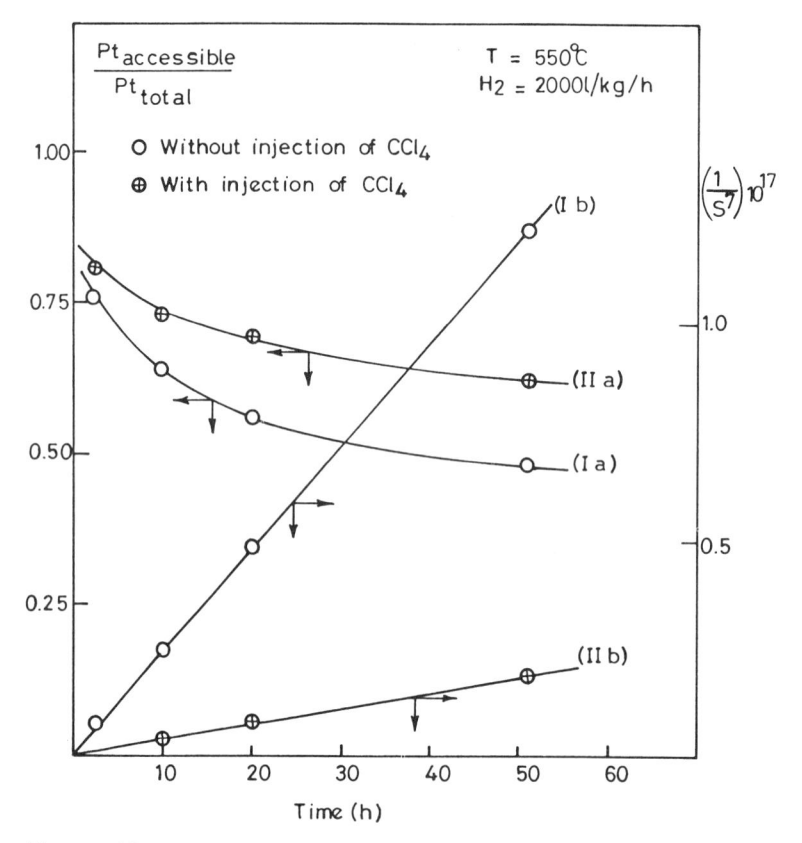

**Figure 12** Effect of chlorine on platinum sintering: S = metallic surface area (m²/g). From Ref. 67.

$$\frac{dS}{dT} = KS^n$$

The order $n$ with respect to the metallic surface area is found to be equal to 8 whether the chlorine content is kept constant or not (curves Ib and IIb) (67). Thus, the sintering mechanism is not affected by chlorine. But the activation energy increases from 33 to 41 kcal/g atom of Pt when the concentration of Cl is kept constant. These results show an increase in metal crystallite adhesion to the support surface. This is in agreement with the results of infrared spectroscopic studies of carbon monoxide chemisorption on Pt/alumina catalyst carried out by Primet et al. (151) showing that chlorine accentuates the metal-support interaction.

Cooper and Trimm (152) claimed that coking can be a function of the catalyst acidity related to the presence of chloride. Parera et al. (153) reported that chloride helps spillover of hydrogen to alumina, which can more easily hydrogenate the coke precursors. From studies with a series of catalysts containing chloride in the range of 0.3 to 1.1 wt % the authors found that the catalyst having 0.89 wt % Cl exhibited minimum coking, and they concluded that this is the optimum chloride level which allows reduction of coke deposits through a mechanism involving activated spillover of hydrogen. Presence of chloride is found to be a prerequisite for this mechanism to work.

Chlorine plays a very significant role during the regeneration of coked catalysts. A coke burned catalyst usually has remaining $\leqslant$ 0.6–0.8 wt % Cl, which has to be increased by chloride precursor injection in the presence of oxygen. This step is followed by oxidation at a higher temperature in the presence of oxygen, which assists in redispersion of the Pt crystallites on the catalyst surface. In the course of the oxychlorination stage, sintering of Pt may be observed if oxygen, $CCl_4$, and water partial pressures are not properly adjusted. In the presence of excess moisture, Cl may not be sufficiently retained by the catalyst, while an excess of $CCl_4$ can cause Pt leaching. The chlorination under hydrogen or inert gas such as nitrogen, however, does not result in redispersion. Taking into account the need to have simultaneously oxygen, chlorine, and water to perform efficient Pt dispersion, it has been suggested that the mechanism of dispersion involves volatile complexes of Pt on alumina, such as $PtCl_2(AlCl_3)_2$, which has been synthesized (150, 154). Other authors demonstrated the formation of $[PtO_2Cl_2]^{2-}$ which enables the dispersion of Pt (155).

## III. OPERATING PARAMETERS

Catalytic reformers are designed for flexibility in operation, whether for motor fuel production or to produce aromatics such as benzene, toluene, and xylenes (BTX). To meet these drastically varying demands, the reforming catalyst must respond to changes in unit operating conditions. The variables which affect the performance of the catalyst and change the yield and quality of reformate are feedstock properties, reaction temperature, space velocity, reaction pressure, and hydrogen-to-hydrocarbon mole ratio ($H_2$/HC). Little (156) has compiled the data available in the literature on the effect of various operating parameters on the performance of reforming catalysts.

### A. Feedstock Properties

In reforming operations, the feed properties are parameters which can significantly influence the performance of the catalyst. One feedstock property which a refiner can control is boiling range. Besides the initial boiling point, the final boiling point and the ASTM distillation end point (EP) of the reactor charge are of interest to refiners. Refiners usually specify a maximum ASTM 205°C EP on reformer reactor charge. This is because hydrocarbons boiling above 205°C are known to form polycyclic aromatics and cause coke laydown under reforming conditions. As a rule, a 13°C increase in feed EP costs about 35% of catalyst life between levels of 190 and 218°C feed EP (156).

Straight-run naphthas generally constitute the major feedstocks for reformers. Feedstocks containing an appreciable concentration of unsaturated hydrocarbons such as thermal cracked, catcracked, coker, and pyrolysis naphtha must usually be hydrotreated before reforming in order to prevent undue hydrogen consumption in the reformer and excessive catalyst deactivation. In addition to being high in sulfur and nitrogen, these stocks contain substantial amounts of monoolefins and diolefins. These olefinic species are undesirable in reformer feed for several reasons. First, in a reformer reactor, olefins hydrogenate, consuming hydrogen that might be needed for hydrotreating other stocks. Second, hydrogenation of olefins markedly lowers their octane numbers, resulting in lower-octane reformate. The third reason for not including olefins in the feed is their tendency to polymerize and form coke on the reforming catalyst, which is particularly true of the diolefins. The concentrations of impurities in the feed that can act as catalyst poisons must also be rigorously controlled. Sulfur in the feed poisons the metal function, and organic nitrogen compounds, which can form ammonia, poison the acid function of the catalyst; hence, their concentrations must be maintained at less than about 1 ppm by hydrotreating before reforming. Water and chlorine compounds are undesirable feed components because they alter the acidity of the support, and their concentrations must be carefully controlled to maintain the proper catalyst acidity. Metals such as As, Pb, and Cu must be kept at extremely low concentrations, because they deposit on the Pt component or otherwise deactivate it. The maximum allowable limits for As, Pb, and Cu in the feed as reported in the literature are 2, 10, and 10 ppb, respectively (157).

## B. Reaction Temperature

Reformers are generally designed with three or more reactors in series and each reactor may contain a different quantity of catalyst. Since the bed temperatures change continually as feed flows through the catalyst, the temperatures commonly monitored are the weighted-average reactor inlet temperature (WAIT) and the weighted-average bed temperature (WABT). The WAIT is the sum of the inlet temperature to each reactor multiplied by the weight percent of total catalyst in each reactor. Similarly, WABT is the sum of the average of the inlet and outlet temperatures of each reactor multiplied by the weight percent of the catalyst in the reactor. In day-to-day operations it is always easier to monitor WAIT than WABT.

Reaction temperatures are chosen to balance the advantage of increased catalytic activity and the disadvantage of increased deactivation rate as temperature is increased. The values range from 460 to 525°C and are usually between 482 and 500°C. Low-pressure processes are operated at slightly higher temperatures than the others to optimize conversions to high-octane-number products. As catalysts lose activity in operation, reactor temperature is gradually increased to maintain a constant octane number in the product reformate.

The catalytic reformer has the flexibility to raise the reformate research octane number (clear) (RONC) from 85 to more than 100 by increasing the reactor temperature. As a rule (156), in the range of 90–95 RONC, the WAIT increase should be from 2 to 3°C per octane increase. In the 95–100 RONC range, the WAIT increase may be from 3 to 4°C per octane. Each feedstock has its own temperature-octane relationship. There is no known general correlation that can confidently show the temperature required to produce a given octane number. However, in raising RONC from 85 to 100, the reformate yield drops by about 13 liquid vol % of charge.

Another important parameter is the delta temperature ($\Delta T$) across each reactor. Reforming reactions are predominantly endothermic and the temperature at the exit of a reactor is normally less than that at the inlet. The $\Delta T$ across the first reactor indicates the extent of dehydrogenation of naphthenes. In the last reactor, since dehydrocyclization and hydrocracking reactions predominate, a positive $\Delta T$ (1 to 3°C or even higher during start-up with fresh catalyst) is usually observed. The $\Delta T$ across the reactor depends on the reactions occurring as well as the $H_2/HC$ ratio used.

## C. Space Velocity

Space velocity is an important reforming process variable. It is a measure of the contact time between the reactants and the catalyst. This is expressed as either liquid hourly space velocity (LHSV) or weight hourly space velocity (WHSV). LHSV is the volume per hour of reactor charge per volume of catalyst. WHSV is the weight per hour of reactor charge per unit weight of catalyst.

Modern commercial reformers usually operate between 1.0 and 2.0 $h^{-1}$ LHSV. At LHSV below 1.0 $h^{-1}$, undesired side reactions and increased hydrocracking occur, reducing reformate yield. However, the choice represents a compromise between allowable hydrocracking and desired dehydrocyclization. Aromatization and isomerization in general are not affected by changes in space velocity, because these reactions approach equilibrium even at high space velocities (158). However, from thermodynamic considerations these reactions for low-molecular-weight fractions such as $C_6$ hydrocarbons are not favored to the same extent as for higher hydrocarbons.

Generally, for reformate octane numbers in the range of 90-100 RONC, doubling LHSV requires an increase of 15-20°C in reactor inlet temperature. Higher paraffinic naphthas require a 20-30°C increase in reactor inlet temperature as space velocity doubles. Naphthas which are low in paraffins require only a 8-12°C increase in reactor inlet temperature as space velocity doubles.

## D. Reaction Pressure

The reformer units designed in the 1950s typically operated in the pressure range 260 to 500 psig, while those designed in the 1980s operate in the range 50-150 psig. Reformers have three or four reactors usually in series, and the pressure is different in each reactor. It is preferable to use the average reactor pressure, if it is available, for estimating yields and cycle life. Otherwise, the average of the first reactor inlet pressure and the last reactor outlet pressure is employed.

The pressure under which reforming reactions take place significantly affects yields and cycle length. Decreasing pressures increase the production of aromatics and hydrogen by favoring dehydrogenation of naphthenes and dehydrocyclization of paraffins and reducing hydrocracking. The adverse effects of reduced pressure are increased catalyst coking and shorter cycle life.

The real incentive for reducing pressure in reformer reactors is more reformate yield and increased hydrogen production. It has been shown that at 100 RONC there is an increase in the percent yield of reformate from 79.7 to 83.5 by a reduction in pressure from 300 to 100 psig. In other words, at the 100 RONC level, liquid yield increases or decreases by about 2 vol % of charge per 100 psig change in pressure. At the 90 RONC level, the yield difference is about 1 liquid vol % of charge per 100 psig change in pressure. Although there is an increase in hydrogen yield on pressure reduction, it is not as clearly defined as in the case of reformate.

## E. Hydrogen-to-Hydrocarbon Ratio

In catalytic reformers hydrogen is recycled in order to maintain a certain $H_2$/HC ratio and a high partial pressure of $H_2$ in the reactors. Reformers designed in the 1950s generally operated in the range of 8-10 $H_2$/HC ratio and 260-500 psig; however, reformers designed during the 1980s operate in the range of 2-5 $H_2$/HC ratio and 50-150 psig.

The hydrogen reacts with coke precursors, removing them from the catalyst before they form significant amounts of polycyclic aromatics and deactivate the catalyst. A high hydrogen-to-hydrocarbon ratio adversely affects aromatization, increases hydrocracking, and reduces catalyst deactivation rates. A lower hydrogen partial pressure favors dehydrogenation of naphthenes and dehydrocyclization of paraffins. However, lowering the hydrogen-to-hydrocarbon ratio adversely affects the life of the catalyst due to increased coking. By reducing $H_2/HC$ from 8 to 4 and 4 to 2, the carbon on catalyst is increased by a factor of 1.75 and 3.6, respectively (140). However, Edgar (159) suggests that a reduction of the $H_2/HC$ ratio from 5 to 4 shortens cycle length by about 20%. Because high ratios require high hydrogen recycle rates, reduction in the $H_2/HC$ ratio leads to savings in terms of energy costs. Therefore, the value chosen is usually governed on the lower bound by the desired amount of hydrocracking and the maximum acceptable deactivation.

## IV. MONITORING THE PERFORMANCE OF REFORMING CATALYSTS

Catalytic reforming units operate best when the process variables are adjusted to produce the maximum yield of the desired products within the limits of acceptable catalyst cycle life. The primary concerns of a refiner are the yield and the quality of the reformate and the catalyst performance in general. The starting point is a unit material balance, which is seldom found to be 100%. If the material balance is in the range of 98 to 102 wt %, the yield data can be used after adjusting the charge or reformate volumes to bring the balance to 100%. To allow optimization of the process conditions for best performance, any loss or gain in material should be viewed carefully.

The performance of a catalytic reformer depends on a combination of controllable process variables. Reactor temperatures and pressures, space velocity, chloride-water balance, and the level of feed contaminants all influence catalyst activity, selectivity, and stability. When the catalyst is performing properly, the reformer may be operated to attain the maximum yield of the desired product. A close monitoring of the process parameters during reformer operation is required to understand the variations in the catalyst performance. The major parameters used to monitor reformer-catalyst performance are as follows:

1. WAIT/WABT (weighted average inlet/bed temperature)
2. Delta temperatures across the reactors
3. Octane number of feed and reformate
4. Precursor conversion/recovery
5. Aromatics production
6. $C_5^+$ reformate yield
7. $C_3/C_1$ ratio
8. Recycle rate and $H_2/HC$ ratio
9. Hydrogen production and purity

The reactor inlet temperatures, WAIT and WABT, are invariably the most effective parameters which can provide information about the relative activity of the catalyst. Generally, higher inlet temperatures indicate lower catalyst activity. This means that the catalyst has been deactivated and the increase in temperature is required to maintain the desired level of activity. Hence, these temperatures are used to monitor the performance of the catalyst.

However, reactor temperature differentials convey the most crucial information regarding the sequence of reactions taking place in individual reactors/beds. Reactor temperature differentials are strongly dependent on feedstock composition and circulating gas rate, which can be modified as required. These temperature variables, along with recycle gas composition and specific gravity, provide instantaneous indication of any change in reformate yield. Reactor pressure differentials are used to monitor catalyst bed fouling by particulate material. Iron is the most common particulate material and not only causes bed plugging but also interferes with catalyst chloride retention and acidity.

An increase in feed rate requires that reactor temperatures be increased to maintain a specified level of catalyst activity. For example, a 10% increase in feed rate necessitates approximately 3 to 4°C increase in temperature in order to maintain octane. Conversely, reduction in feed rate warrants reduction in WABT to maintain octane. Although the feedstock composition is not an easily controllable variable, any change in the feed quality will be reflected in the performance of the catalyst. To ascertain the feed quality, Universal Oil Products Inc. (UOP) developed an expression N+2A (naphthenes plus two times aromatics; liquid vol %) using correlations of ASTM distillation and API gravity with feed composition in terms of paraffins, naphthenes, and aromatics. This formula essentially gives a measure of the aromatic precursors available in the feed which should contribute to the production of aromatics. This expression also enables the calculation of precursor conversion or recovery to judge the overall performance of the catalyst. Table 3 shows briefly the effect of changes in feed quality and the process variables on the catalyst performance.

The hydrogen-to-hydrocarbon ratio has a significant effect on catalyst deactivation. The cycle life varies directly with hydrogen-to-hydrocarbon ratio to approximately the 1.6 power. The purity of recycle hydrogen varies with separator pressure and temperature, octane, and feedstock characteristics. Lower purity of recycle hydrogen indicates a lower hydrogen-to-hydrocarbon ratio and greater deactivation of the catalyst. Yield of aromatics in the product increases with increase in temperature and as a result the reformate octane also increases. But reformate yield decreases with increasing temperature. Thus, increasing octane results in lower reformate yields. Hydrogen yield also increases as octane rises, as do the other light ends. Good catalyst performance results in high hydrogen and reformate yields.

Composition of recycle gas and stabilizer overhead liquids and gases are best used in conjunction with accurate flow rates to determine overall methane ($C_1$), ethane ($C_2$), propane ($C_3$), and butane ($C_4$) yields. The molar ratio of

**Table 3** Effect of Feed Properties and Process Variables on Catalyst Performance

| Increase | $H_2$ purity[a] | $H_2$ yield | $C_5^+$ yield | Delta temperature | Reactor temperature |
|---|---|---|---|---|---|
| Feed property | | | | | |
| API | − | − | − | − | + |
| Paraffins | − | − | − | − | + |
| Naphthenes | + | + | + | + | − |
| Aromatics | − | − | + | − | 0 |
| Initial boiling point | − | − | − | − | − |
| Final boiling point | 0(−) | + | + | + | + |
| Process variable | | | | | |
| Separator pressure | + | 0 | + | 0 | 0 |
| Separator temperature | − | 0 | − | 0 | 0 |
| Reactor pressure | − | − | − | − | − |
| Reformate RONC | − | + | − | + | + |

[a] +, increase; −, decrease; 0, no effect.

propane to methane ($C_3/C_1$) obtained from these overall yields is an indication of cracking activity and is frequently monitored. Ideally, the $C_3/C_1$ mole ratio during normal operation should be in the range 0.7–1.2. Performance outside this range indicates that the catalyst is out of acid-metal balance. During a normal operating cycle, metal activity is not an easily moderated variable; therefore, the acid activity is the primary alterable variable. Consequently, if the performance of the catalyst indicates acid-metal imbalance, optimization of water-chloride injection rates will be essential. However, if the modification of the water-chloride balance is inadequate to restore the activity, the temperature is the only parameter which can be adjusted to achieve the desired performance.

## V. CONCLUSIONS

The preceding discussions make it clear that the performance of reforming catalysts depends significantly on the factors of surface structure, surface reconstruction under reaction conditions, and the operating parameters. Surface characteristics of a fresh reforming catalyst are determined by the texture and properties of the support, dispersion of the active metal, and the presence of promoters and surface modifiers. Various effects such as metal-support interactions, metal-metal interactions, and electronic and ensemble effects operate on the surface. The interplay of these various factors makes it extremely difficult to draw definite and generalized conclusions applicable in a universal sense.

For example, in the case of promoters, although the ensemble effect is favorable in all the known cases, each promoter imparts its own characteristic property, which may not necessarily be desirable. The importance of surface modifiers such as chloride, sulfur, and adsorbed coke, in conveniently modifying the surface through reconstruction to yield desirable results has been demonstrated. The studies on single-crystal surfaces and their application in understanding the active sites responsible for individual reforming reactions have provided a new insight into catalyst performance. Thus, understanding of the structure-performance relationship offers new insights into the catalyst and process, which ultimately will allow the refiner to derive the best from the catalytic reformer.

## ACKNOWLEDGMENTS

The authors wish to express their sincere thanks to the management of Indian Petrochemicals Corporation Limited, Baroda for permission to publish this chapter. Also the authors express their gratitude to Dr. I. S. Bhardwaj, Director (R&D), IPCL for his keen interest and support during the execution of this work. Thanks are also due to Dr. A. B. Halgeri, Senior Manager (R&D), IPCL for his constant encouragement. Assistance from all our colleagues in the Research Centre in completing the manuscript is gratefully acknowledged.

## REFERENCES

1. K. Tanabe and M. Katayama, The acidity of surfaces. *J. Res. Inst. Catal. Hokkaido Univ.* 7, 106 (1959).
2. Y. Watanabe and K. Tanabe, Propylene polymerization activity and acidic properties of metal sulfates. *J. Res. Inst. Catal. Hokkaidao Univ* 12, 56 (1964).
3. B. V. Romanovskii and Yu N. Kartashev, Correlation of acid and catalytic properties of zeolites. *Kinet. Katal. 24(3)*, 758 (1983).
4. Y. Yoneda, Linear free energy relationships in heterogeneous catalysis. IV. Regional analysis for solid acid catalysts. *J. Catal.* 9, 51 (1967).
5. K. Tanabe, The catalytic activity of some solid acids for depolymerization of paraldehyde. *J. Res. Inst. Catal. Hokkaido Univ.* 7, 114 (1959).
6. H. A. Benesi and B. H. C. Winquist, Surface acidity of solid acid catalysts. *Adv. Catal.* 27, 98 (1978).
7. E. P. Parry, An infrared of pyridine adsorbed on acidic solids. Characterization of surface acidity. *J. Catal.* 2, 371 (1963).
8. A. E. Hirschler and J. E. Hudson, The reaction of triphenylmethane and perylene with silica-alumina. The nature of acid sites. *J. Catal.* 3, 239 (1964).
9. M. R. Basila, An infrared study of silica-alumina surface. *J. Phys. Chem.* 66, 2223 (1962).
10. C. L. Thomas, Chemistry of cracking catalysts. *Ind. Eng. Chem.* 41, 2564 (1949).

11. S. P. S. Andrew, Theory and practice of the formulation of heterogeneous catalysts. *Chem. Eng. Sci. 36(9)*, 1431 (1981).
12. T. Kotanigawa, M. Yamamoto, M. Utiyama, H. Hattori, and K. Tanabe, The influence of preparation methods on the pore structure of alumina. *Appl. Catal. 1*, 185 (1981).
13. V. J. Lostaglio and J. D. Carruthers, New approaches in catalyst manufacture. *CEP*, March, p. 46 (1986).
14. J. Biswas and D. D. Do, A unified theory of coking deactivation in a catalyst pellet. *Chem. Eng. J. 36*, 175 (1987).
15. H. S. Taylor, A theory of the catalytic surface. *Proc. R. Soc. London Ser. A 108*, 105 (1925).
16. G. A. Somorjai, Active sites in catalysis. *Adv. Catal. 26*, 1 (1977).
17. M. Boudart, A. Aldag, J. E. Benson, N. A. Dougharty, and C. Girvin Harkins, On the specific activity of platinum catalysts. *J. Catal. 6*, 92 (1966).
18. R. K. Herz, W. D. Gillespie, C. E. Peterson, and G. A. Somorjai, The structure sensitivity of cyclohexane dehydrogenation and hydrogenolysis catalyzed by platinum single crystals at atmospheric pressure. *J. Catal. 67*, 371 (1981).
19. R. W. Maatman, P. Mahaffy, P. Hoekstra, and C. Addink, The preparation of Pt-alumina catalyst and its role in cyclohexane dehydrogeneration. *J. Catal. 23*, 105 (1971).
20. M. Kraft and H. Spindler, Relations between structural and catalytic properties of platinum, applied on aluminum oxide. *Osn. Predvideniya Katal. Deistviya, Tr. Mezhdunar. Kongr. Katal.*, 4th, 2, 268 (1968) (Russian) (Ya. T. Eidus, ed.). Nauka, Moscow, 1970.
21. J. A. Cusumano, G. W. Dembinski, and J. H. Sinfelt, Chemisorption and catalytic properties of supported platinum. *J. Catal. 5*, 471 (1966).
22. A. N. Mitrofanova, V. S. Boronin, and O. M. Poltorak, Dehydrogenation of cyclohexane on platinized silica gels of various types. *Vestn. Mosk. Univ. Ser. II 21(5)*, 40 (1966). (Russian)
23. T. A. Dorling and R. L. Moss, The structure and activity of supported metal catalysts. I. Crystallite size and specific activity for benzene hydrogenation of platinum/silica catalysts. *J. Catal. 5*, 111 (1966).
24. J. N. Basset, G. Dalnai-Imelik, M. Primet, and R. Muten, Gas phase benzene hydrogenation on a nickel-silica catalyst. *J. Catal. 37*, 22 (1975).
25. P. C. Aben, J. C. Platteeuw, and B. Stouthamer, Hydrogenation of benzene on coated platinum, palladium and nickel catalysts. *Osn. Predvideniya Katal. Deistviya, Tr. Mezhdunar. Kongr. Katal.*, 4th, 1, 322 (1968) (Russia) (Ya. T. Eidus, ed.). Nauka, Moscow, 1970.
26. G. Maire, C. Corolleur, D. Juttard, and F. G. Gault, Comments on a dispersion effect in hydrogenolysis of methylcyclopentane and isomerization of hexanes over supported platinum catalysts. *J. Catal. 21*, 250 (1971).
27. D. J. C. Yates and J. H. Sinfelt, The catalytic activity of rhodium in relation to its state of dispersion. *J. Catal. 8*, 348 (1967).
28. J. H. Sinfelt, J. L. Carter, and D. J. C. Yates, Catalytic hydrogenolysis and dehydrogenation over copper-nickel alloys. *J. Catal. 24*, 283 (1972).
29. M. Boudart, Catalysis by supported metals. *Adv. Catal. 20*, 153 (1969).

30. B. Coq and F. Figueras, Conversion of methylcyclopentane on platinum-tin reforming catalysts. *J. Catal. 85*, 197 (1984).
31. L. L. Kesmodel and L. M. Falikov, The electronic potential in a metal close to a surface edge. *Solid State Commun. 16*, 1201 (1975).
32. A. E. Morgan and G. A. Somorjai, Low energy electron diffraction studies of gas adsorption on the platinum(100) single crystal surface. *Surf. Sci. 12(3)*, 405 (1968).
33. C. W. Tucker, Low energy electron diffraction studies of gas adsorption on Pt(100), (110) and (111) surfaces. *J. Appl. Phys. 35(6)*, 1897 (1964).
34. H. P. Bonzel and R. Ku, Mechanisms of the catalytic carbon monoxide oxidation on Pt(110). *Surf. Sci. 33(1)*, 91 (1972).
35. B. Lang, R. W. Joyner, and G. A. Somorjai, Low energy electron diffraction studies of chemisorbed gases on stepped surfaces of platinum. *Surf. Sci. 30*, 440 (1972).
36. B. Lang, R. W. Joyner, and G. A. Somorjai, Low energy electron diffraction studies high index surfaces of platinum. *Surf. Sci. 30*, 454 (1972).
37. G. A. Somorjai, The mechanism of hydrocarbon catalysis on platinum crystal surfaces, Battelle Inst. Mater. Sci. Colloq. 9, 395 (1974) (published 1975).
38. G. A. Somorjai and D. W. Blakely, The dehydrogenation and hydrogenolysis of cyclohexene and cyclohexane on stepped (high Miller index) platinum surfaces. *J. Catal. 42*, 181 (1976).
39. L. L. Kesmodel, L. H. Dubois, and G. A. Somorjai, Dynamical LEED study of acetylene and ethylene chemisorption on platinum(111): evidence for the ethylidene group. *Chem. Phys. Lett. 56*, 267 (1978).
40. H. Ibach, H. Hopster, and B. Sexton, Analysis of surface reactions by spectroscopy of surface vibrations. *Appl. Surf. Sci., 1*: 1 (1977).
41. J. L. Gland and G. A. Somorjai, Low energy electron diffraction and work function studies of adsorbed organic monolayers on the (100) and (111) crystal faces of platinum. *Adv. Coll. Interface Sci. 5(3)*, 205 (1976).
42. F. J. Szalkowski and G. A. Somorjai, Simple rules to predict the structure of adsorbed gases on crystal surfaces. *J. Chem. Phys. 54(1)*, 389 (1971).
43. J. C. Tracy and P. W. Palmberg, Structural influences on adsobate binding energy. I. Carbon monoxide on (100) palladium. *J. Chem. Phys. 51*, 4852 (1969).
44. K. Baron, D. W. Blakely, and G. A. Somorjai, Low energy electron diffraction studies of the surface structures of adsorbed hydrocarbons (*n*-heptane, toluene, benzene, ethylene and cyclohexane) on stepped (high Miller index) platinum surfaces. *Surf. Sci. 41*, 45 (1974).
45. B. E. Nieuwenhuys, D. I. Hagen, G. F. Rovida, and G. A. Somorjai, Low energy electron diffraction, Auger electron spectroscopy and thermal desorption studies of chemisorbed carbon monoxide and oxygen on the (111) and stepped [6(111) × (100)] iridium surfaces. *Surf. Sci. 57(2)*, 632 (1976).
46. D. G. Fedak and N. A. Gjostein, Anomalous surface structures of gold. *Surf. Sci. 8*, 77 (1967).
47. A. Ignatiev, A. V. Jones, and T. N. Rhodin, LEED (low energy electron diffraction) investigations of xenon single crystal films and their use in studying the iridium (100) surface. *Surf. Sci. 30*, 573 (1972).

48. M. R. Martin and G. A. Somorjai, Determination of the surface geometry for the aluminium (100) and (111) surfaces by comparison of low energy electron diffraction calculations with experiment. *Phys. Rev. B7*, 3607 (1973).
49. G. A. Somorjai and D. W. Blakely, The stability and structure of high Miller index platinum crystal surfaces in vacuum and in the presence of adsorbed carbon and oxygen. *Surf. Sci. 65(2)*, 419 (1977).
50. G. A. Somorjai, Catalysis on the atomic scale (Emmet award lecture of 1977). *Catal. Rev. Sci. Engg. 18(2)*, 173 (1978).
51. G. A. Somorjai, On the mechanism of sulfur poisoning of platinum catalysts. *J. Catal. 27*, 453 (1972).
52. G. A. Somorjai, Reactions on single crystal surfaces. *Acc. Chem. Res. 9(7)*, 248 (1976); *Chem. Abs.* 85, 99716c.
53. J. Biswas, G. M. Bickle, P. G. Gray, and D. D. Do, The role of crystallite structure on mechanisms of coke and sulfur poisoning in catalytic reforming in *Catalyst Deactivation*, Proc. 4th Int. Symp., Antwerp, Sept–Oct., 1987 (B. Delmon and G. F. Froment, eds.). Elsevier, Amsterdam, 1987, p. 553.
54. J. R. Anderson, K. Foger, and R. J. Breaksphere, Adsorption and temperature programmed desorption of hydrogen with dispersed platinum and platinum-gold catalysts. *J. Catal. 57*, 458 (1979).
55. C. E. Smith, J. P. Biberian, and G. A. Somorjai, The effect of strongly bound oxygen on the dehydrogenation and hydrogenation activity and selectivity of platinum single crystal surfaces. *J. Catal. 57*, 426 (1979).
56. C. E. Smith, Catalysis by platinum single crystal surfaces: low pressure hydrocarbon reactions and the effects of introducing strongly bound oxygen at the surface. Ph.D. dissertation, University of California, Berkeley, 1978.
57. B. E. Nieuwenhuys and G. A. Somorjai, Dehydrogenation of cyclohexane and dehydrocyclization of *n*-heptane over single crystal surfaces of iridium. *J. Catal. 46*, 259 (1977).
58. J. L. Carter, J. A. Cusumano, and J. H. Sinfelt, Hydrogenolysis of *n*-heptane on unsupported metals. *J. Catal. 20*, 223 (1971).
59. G. A. Somorjai and J. Carrazza, Structure sensitivity of catalytic reactions. *Ind. Eng. Chem. Fundam. 25*, 63 (1986).
60. J. L. Gland, K. Baron, and G. A. Somorjai, Low-energy X-ray diffraction, work function change and mass spectrometric studies of chemisorption and dehydrogenation of cyclohexane, cyclohexene and 1,3-cyclohexadiene on the Pt(111) surface. *J. Catal. 36*, 305 (1975).
61. J. H. Sinfelt, H. Hurwitz, and R. A. Shulman, Kinetics of methyl cyclohexane dehydrogenation over Pt-Al$_2$O$_3$. *J. Phys. Chem. 64*, 1559 (1960).
62. V. Haensel, G. R. Donaldson, and F. J. Riedl, Flouride-free platinum group-reforming catalysts. *Proc. Int. Congr. Catal.*, 3rd, 1964, Vol. 1, 1965, p. 294.
63. D. R. Kahn, E. E. Petersen, and G. A. Somorjai, The hydrogenolysis of cyclopropane on a platinum stepped single crystal at atmospheric pressure. *J. Catal. 34*, 294 (1974).
64. R. W. Rice and Kang Lu, Comparison of platinum and platinum-iridium catalysts for heptane reforming at different pressures. *J. Catal. 77*, 104 (1982).
65. Y. L. Lam and J. H. Sinfelt, Cyclohexane conversion on ruthenium catalysts of widely varying dispersion. *J. Catal. 42*, 319, (1976).

66. H. Conrad, G. Ertl, and E. E. Latta, Adsorption of carbon monoxide on palladium single crystal surfaces. *Surf. Sci. 40*, 435 (1974).
67. J. P. Franck and G. Martino, Deactivation and regeneration of catalytic reforming catalysts. In *Progress in Catalyst Deactivation* (J. Figueiredo, ed.). Martinus Nijhoff, The Hague, 1982, p. 355.
68. H. J. Matt and L. Moscou, A study of the influence of platinum crystallite size on the selectivity of platinum reforming catalysts. *Proc. 3rd Int. Congr. Catal.* (W. M. H. Sachtler, G. C. A. Schuit, and P. Zwietering, eds.). Amsterdam, 1964, Vol II, p. 1277.
69. M. J. P. Botman and V. Ponec, Dehydrocyclization of *n*-hexane and the possible role of Pt-ions. *Proc. Symp. Adv. Naphtha Reforming*, ACS, New Orleans Meet., Sept. 1987, p. 722.
70. B. H. Davis, G. A. Westfall, J. Watkins, and J. Pezzanite, Jr., Paraffin dehydrocyclization. VI. The influence of metal and gaseous promoters on the aromatic selectivity. *J. Catal. 42*, 247 (1976).
71. G. C. Bond and P. A. Sermon, Gold catalyst for olefin hydrogenation. Transmutation of catalytic properties. *Gold Bull. 6(4)*, 102 (1973).
72. J. Biswas, G. M. Bickle, P. G. Gray, D. D. Do, and J. Barbier, The role of deposited poisons and crystallite surface structure in the activity and selectivity of reforming catalysts. *Catal. Rev. Sci. Eng. 30(2)*, 161 (1988).
73. J. Barbier, G. Corro, Y. Zhang, J. P. Bournonville, and J. P. Franck, Coke formation on platinum-alumina catalyst of wide varying dispersion. *Appl. Catal. 13*, 245 (1985).
74. J. Barbier, G. Corro, Y. Zhang, J. P. Bournonville, and J. P. Franck, Coke formation on bimetallic platinum/rhenium and platinum/iridium catalysts. *Appl. Catal. 16*, 169 (1985).
75. J. M. Parera, J. N. Beltramini, C. A. Querini, E. E. Martinelli, E. J. Churin, P. E. Aloe, and N. S. Figoli, The role of Re and S in the Pt-Re-S/Al$_2$O$_3$ catalyst. *J. Catal. 99*, 39 (1986).
76. J. Gold Wasser, B. Arenas, C. Bolivar, G. Castro, A. Rodriquez, A. Fleitas, and J. Giron, A. study of platinum-germanium catalytic system. *J. Catal. 100*, 75 (1986).
77. P. Biloen, J. N. Helle, H. Verbeek, F. M. Dautzenberg, and W. M. H. Sachtler, The role of rhenium and sulfur in platinum based hydrocarbon conversion catalysts. *J. Catal. 63*, 112 (1980).
78. F. M. Dautzenberg, J. N. Helle, P. Biloen, and W. M. H. Sachtler, Conversion of *n*-hexane over monofunctional supported and unsupported Pt-Sn catalysts. *J. Catal. 63*, 119 (1980).
79. R. W. Coughlin, A. Hasan, and K. Kawakami, Activity yield patterns and coking behaviour of Pt and Pt-Re catalysts during dehydrogenation of methylcyclohexane. II. Influence of sulphur. *J. Catal. 88*, 163 (1984).
80. J. C. Rasser, W. H. Beindroff, and J. J. F. Scholten, Characterization and performance of platinum-iridium reforming catalysts. *J. Catal. 59*, 211 (1979).
81. C. Bolivar, H. Charcossett, R. Frety, M. Primet, and L. Tournayan, Platinum-rhenium-alumina catalysts. II. Study of the metallic phase after reduction. *J. Catal. 45*, 163 (1976).

82. C. Betizeau, G. Leclereq, R. Maurel, C. Bolivar, R. Frety, and L. Tournayan, Platinum-rhenium-alumina catalysts. III. Catalytic properties. *J. Catal.* **45**, 179 (1976).
83. M. F. L. Johnson and V. M. Le Roy, The state of rhenium in Pt/Re/alumina catalysts. *J. Catal.* **35**, 434 (1974).
84. M. F. L. Johnson, The state of rhenium in Pt/Re/alumina catalyst. *J. Catal.* **39**, 487 (1975).
85. B. H. Issacs and E. E. Peterson, The effect of drying temperature on the temperature-programmed reduction profile of a platinum/rhenium/alumina catalyst. *J. Catal.* **77**, 43 (1982).
86. R. J. Bertolacini and R. J. Pellet, The function of rhenium in bimetallic reforming catalysis. In *Catalyst Deactivation*, Proc. 4th Int. Symp., Antwerp, Oct. 1980 (B. Delmon and G. F. Froment, eds.). Elsevier, Amsterdam, 1980, p. 73.
87. J. Barbier, H. Charcosset, G. De Periera, and J. Riviere, Determination of metallic surface areas of platinum-rhenium catalysts, *Appl. Catal.* **1**, 71 (1981).
88. V. K. Shum, J. B. Butt, and W. M. H. Sachtler, The effects of rhenium and sulfur on the maintenance of activity and selectivity of platinum/alumina hydrocarbon conversion catalysts—Part I. *J. Catal.* **96**, 371 (1985).
89. N. Wagstaff and R. Prins, Alloy formation and metal oxide segregation in Pt-Re/$\tau$-Al$_2$O$_3$ catalysts as investigated by temperature programmed reduction. *J. Catal.* **59**, 434 (1979).
90. H. E. Kluksdahl, Reforming of a sulfur free naphtha with a platinum-rhenium catalyst. *U. S. Patent* 3,415,737, 1968.
91. R. W. Coughlin, K. Kawakami, and A. Hasan, Activity, yield patterns and coking behaviour of Pt and Pt-Re catalysts during dehydrogenation of methylcyclohexane. I. In the absence of sulphur. *J. Catal.* **88**, 150 (1984).
92. V. N. Seleznev, Y. V. Fronichev, and M. R. Levinter, Effect of tungsten oxide additions on the activity and stability of an aluminum oxide–platinum catalyst in the aromatization of hydrocarbons. *Neftekhimiya 14(2)*, 201 (1974). (Russian)
93. A. C. Muller, P. A. Engelhard, and J. E. Weisang, Surface study of platinum-tin bimetallic reforming catalysts. *J. Catal.* **56**, 65 (1979).
94. S. R. Adkins and B. H. Davis, The chemical state of tin in platinum-tin-alumina catalysts. *J. Catal.* **89**, 371 (1984).
95. B. A. Sexton, A. E. Hughes, and K. Foger, An X-ray photoelectron spectroscopy and reaction study of Pt-Sn catalysts. *J. Catal.* **88**, 466 (1984).
96. R. Srinivasan, R. J. De Angeles, and B. H. Davis, Alloy formation in Pt-Sn-alumina catalysts: in situ X-ray diffraction study. *J. Catal.* **106**, 449 (1987).
97. H. Lieske and J. Volter, State of tin Pt-Sn/Al$_2$O$_3$ Reforming catalysts investigated by TPR and chemisorption. *J. Catal.* **90**, 96 (1984).
98. R. Bacaud, P. Bussiere, and F. Figueras, Mossbauer spectra investigation of the role of tin in platinum-tin reforming catalysts. *J. Catal.* **69**, 399 (1981).
99. V. I. Kuznetsov, E. N. Yurchenko, A. S. Belyi, E. V. Zatolokina, M. A. Somolikov, and V. K. Duplyakin, Mossbauer sepctroscopic studies of the state of Pt-Sn/Al$_2$O$_3$ catalyst for hydrocarbon conversion. *React. Kinet. Catal. Lett.* **21**, 419 (1982).
100. Zhang Su, Bing-Yan, Xie, Jing-Ling Zang, Guang-Hou Wang, and Zhang Jian,

Tin-119m Mossbauer spectroscopic study of the platinum-tin catalyst system. *Ts'Ui Hua Hsueh Pao 1(4)*, 253 (1980).
101. Yongxi, Li, Zhang Yongfu, and Yuanfu Shia, Valence states and behaviour of tin in platinum-tin/$\tau$-alumina catalysts prepared by co-precipitation. *Cuihua Xuebao 5(4)*, 311 (1984) (Chinese)
102. V. I. Kuznetsov, A. S. Belyi, E. N. Yurchenko, M. D. Sonolikov, M. T. Protasova, E. V. Zatolokina, and V. K. Duplyakin, Mossbauer spectroscopic and chemical analysis of the composition of Sn containing components of Pt-Sn/Al$_2$O$_3$ (Cl) reforming catalyst. *J. Catal. 99*, 159 (1986).
103. R. Srinivasan, R. J. De Angelis, and B. H. Davis, Structural studies of Pt-Sn catalysts on high and low surface area alumina supports. *Catal. Lett. 4*, 303 (1990).
104. B. H. Davis, Bimetallic catalyst preparation. *U. S. Patent* 3,840,475 (1974).
105. K. Balakrishnan and J. Schwank, Chemisorption and XPS study of bimetallic Pt-Sn/Al$_2$O$_3$ catalysts. *J. Catal. 127*, 287 (1991).
106. Yong-Xi Li, K. J. Klabunde, and B. H. Davis, Alloy formation in supported Pt-Sn catalysts: Mossbauer studies. *J. Catal. 128*, 1 (1991).
107. E. Raub and W. Plate, Tempering and decomposition of platinum-iridium alloys, *Z. Metallka. 47*, 688 (1956).
108. J. R. Anderson and N. R. Avery, The isomerization of aliphatic hydrocarbon over evaporated films of platinum and palladium. *J. Catal. 5*, 446 (1966).
109. A. V. Ramaswamy, P. Ratnasamy, S. Sivasanker, and A. J. Leonard, Structure and catalytic properties of bimetallic reforming catalysts. *Proc. 6th ICC*, London, 1976. Chemical Society, London, 1977, Vol. 1, p. 855.
110. J. P. Brunelle, R. E. Montarnal, and A. A. Sugier, Association effects in hydrogenolysis and hydrogenation over bimetallic supported catalysts. *Proc. 6th ICC*, London, 1976 (G. C. Bond, P. B. Wells, and F. C. Tompkins eds.). Chemical Society, London, 1977, Vol. 2, p. 844.
111. R. Bouwman and P. Biloen, Valence state and interaction of platinum and germanium on $\tau$-Al$_2$O$_3$ investigated by X-ray photoelectron spectroscopy. *J. Catal. 48*, 209 (1977).
112. T. Tomero, J. Tejeda, D. Jaunay, C. Bolivar, and H. Charcosset, Activity of platinum-germanium/alumina catalyst, *Simp. Iberoam. Catal. [Trab] 7th*, La Plata, Argentina, 1980, p. 453.
113. C. Bolivar, H. Charcosset, M. Primet, B. Arenas, and R. Torrellas, *Proc. VIIIth Iberoamerican Congr. Catalysis*, Huelva, Spain, 1982, p. 162.
114. J. N. Beltramini and D. L. Trimm, Effect of germanium on the activity maintenance and selectivity of Pt/Al$_2$O$_3$ reforming catalyst. *Proc. 9th ICC*, Calgary, Canada, 1990, p. 1268.
115. L. Guczi, Mechanism of reactions on multimetallic catalysts. *J. Mol. Catal. 25*, 13 (1984).
116. G. C. Bond and Xu Yide, Silica-supported ruthenium-copper catalysts: structural studies and activities for *n*-butane hydrogenolysis. *J. Mol. Catal. 25*, 141 (1984).
117. A. Frennet, G. Lienard, A. Crucq, and L. Degols, Effect of multiple sites and competition in adsorption on the kinetics of reactions catalyzed by metals. *J. Catal. 53*, 150 (1978).

118. J. A. Dalmon and C. Mirodatos, Exchange with deuterium of methane on nickel catalysts. *J. Mol. Catal.* 25, 161 (1984).
119. J. H. Sinfelt, Supported bimetallic cluster catalysts. *J. Catal.* 29, 308 (1973).
120. P. G. Menon, J. Sieders, F. J. Streefkerk, and G. J. M. Van Keulen, Separate determination of Pt and Re in Pt-Re-Al$_2$O$_3$ reforming catalyst. *J. Catal.* 29, 188 (1973).
121. J. Barbier, P. Marecot, N. Martin, L. Elassal, and R. Maurel, Selective poisoning by coke formation on platinum/alumina. *Stud. Surf. Sci. Catal.* 6 *(Catal. Deact.)*, 53 (1980).
122. D. L. Trimm, Poisoning of metallic catalysts. In *Deactivation and Poisoning of Catalysts* (J. Oudar and H. Wise, eds.). Marcel Dekker, New York, 1985, p. 151.
123. J. M. Parera, R. J. Verderone, C. C. Pieck, and E. M. Traffano, Influence of sulfurization on coke formation over catalysts for naphtha reforming. *Appl. Catal.* 23, 15 (1986).
124. C. R. Apesteguia and J. Barbier, The role of catalysts presulfurization in some reactions of catalytic reforming and hydrogenolysis. *J. Catal.* 78, 352 (1982).
125. J. C. Hayes, R. T. Mitsche, E. L. Pollitzer, and E. H. Homeier, Sulfur as a tool in catalysis. *Prepr. Div. Pet. Chem. Am. Chem. Soc. 19(2)*, 334, (1974).
126. C. R. Apesteguia, T. F. Garetto, C. E. Brema, and J. M. Parera, Sulfurization of Pt/Al$_2$O$_3$-Cl catalyst. V. Pt-particle-size effect on sulfur adsorption. *Appl. Catal.* 10, 291 (1984).
127. P. A. Van Trimpont, G. A. Martin, and G. F. Froment, Activities and selectivities for reforming reactions on unsulfided and sulfided commercial platinum and platinum-rhenium catalysts. *Appl. Catal.* 17, 161 (1985).
128. M. A. Pacheco and E. E. Peterson, Effect of pretreatment on platinum/alumina and platinum+rhenium/alumina catalyst deactivation. *J. Catal. 96(2)*, 499 (1985).
129. V. K. Shum, J. B. Butt, and W. M. H. Sachtler, The effects of rhenium and sulfur on the maintenance of activity and selectivity of platinum/alumina hydrocarbon conversion catalysts. II. Experiments at elevated pressure. *J. Catal.* 99, 126 (1986).
130. J. H. Sinfelt, Specificity in catalytic hydrogenolysis by metals. *Adv. Catal.* 23, 91 (1973).
131. G. B. McVicker, P. J. Collins, and J. J. Ziemiak, Model compound reforming studies: a comparison of alumina supported platinum and iridium catalysts. *J. Catal.* 74, 156 (1982).
132. P. G. Menon and J. Prasad, The role of sulfur in reforming reactions on Pt-Al$_2$O$_3$ and Pt-Re-Al$_2$O$_3$ catalysts. *Proc. 6th ICC* (G. C. Bond, P. B. Wells, and F. C. Tompkins, eds.). London, 1976, paper B45, p. 1061.
133. E. Baumgarten and H. Hoeffkes, Reaction kinetics measurements on selectively poisoned reforming catalysts. *Z. Phys. Chem. (Wiesbaden) 121(1)*, 107 (1980). (German)
134. P. G. Menon, G. B. Marin, and G. F. Froment, Effect of sulfur poisoning on the hydrogenolysis activity of Pt in Pt-Al$_2$O$_3$ catalysts. *Ind. Eng. Chem. Prod. Res. Dev.* 21, 52 (1982).

135. W. M. H. Sachtler, Selectivity and rate of activity decline of bimetallic catalysts. *J. Mol. Catal. 25*, 1 (1984).
136. G. A. Somorjai and D. W. Blakely, Mechanism of catalysis of hydrocarbon reactions by platinum surfaces. *Nature 258*, 580 (1975).
137. S. M. Augustine, G. N. Alameddin, and W. M. H. Sachtler, The effect of Re, S, and Cl on the deactivation of Pt/$\tau$-Al$_2$O$_3$ reforming catalysts. *J. Catal. 115*, 217 (1989).
138. Z. Schay, K. Matusek, and L. Guczi, The effect of pretreatment on fresh and spent Pt/Al$_2$O$_3$ and Pt-Re/Al$_2$O$_3$ reforming catalysts. *Appl. Catal. 10, 173 (1984)*.
139. J. M. Parera, J. N. Beltramini, E. J. Churin, P. E. Aloe, and N. S. Figoli, Efecto del azufre en el reformado de compuestos de C$_6$ sobre catalizadores mono y bimetalicos. *Proc. 9th Ibero-Amer. Symp. Catal.*, Lisbon, 1984, p. 259.
140. N. S. Figoli, J. N. Beltramini, E. E. Martinelli, M. R. Sad, and J. M. Parera, Operational conditions and coke formation on Pt-Al$_2$O$_3$ reforming catalyst. *Appl. Catal. 5*, 19 (1983).
141. M. D. Rosynek and F. L. Strey, The nature of active sites on catalytic alumina: information from site poisoning by sulfur containing molecules. *J. Catal. 41*, 312 (1976).
142. A. V. Deo, I. G. Dalla Lana, and H. W. Habgood, Infrared studies of the adsorption and surface reactions of hydrogen sulfide and sulfur dioxide on some aluminas and zeolites. *J. Catal. 21*, 270 (1971).
143. J. H. Lunsford, L. W. Zingery, and M. D. Rosynek, Exposed aluminium ions as active sites on $\tau$-alumina. *J. Catal. 38*, 179 (1975).
144. N. S. Figoli, M. R. Sad, J. N. Beltramini, E. S. Jablonski, and J. M. Parera, Influence of the chlorine content on the behavior of catalysts for *n*-heptane reforming. *Ind. Eng. Chem. Prod. Res. Dev. 19*, 545 (1980).
145. P. G. Menon, R. P. De Pauw, and G. F. Froment, The influence of chloriding on isomerization and hydrogenolysis of *n*-pentane on Pt-Al$_2$O$_3$ reforming catalysts. *Ind. Eng. Chem. Prod. Res. Dev. 18(2)*, 110 (1979).
146. F. E. Massoth and F. E. Kiviat, Reaction of HCl with an alumina catalyst at elevated temperatures. *Proc. 5th Int. Congr. Catal.* (J. W. Hightower, ed.), Florida, Paper 55. North-Holland, Amsterdam, 1972, p. 807.
147. M. Primet, M. Dufaux, and M. V. Mathiew, Platinum-chlorine interaction during the chlorination of platinum/alumina. *Acad. Sci. Ser. C 280(7)*, 419 (1975) (Fr.).
148. S. Sivasanker, A. V. Ramaswamy, and P. Ratnasamy, Electron spin resonance observation of platinum and iridium catalysts species on reforming catalysts. *J. Catal. 46*, 420 (1977).
149. F. Delannay, C. Defosse, B. Delmon, P. G. Menon, and G. F. Froment, Chloriding of Pt-Al$_2$O$_3$ catalysts, studies by transmission electron microscopy and X-ray photoelectron spectroscopy. *Ind. Eng. Chem. Prod. Res. Dev. 19(4)*, 537 (1980).
150. J. P. Bournonville and G. Martino, Sintering of alumina supported platinum. In *Catalyst Deactivation*, Proc. Int. Symp., Antwerp, Oct. 1980 (B. Delmon and G. F. Froment, eds.). Elsevier, Amsterdam, 1980, p. 159.
151. M. Primet, J. M. Basset, M. V. Mathieu, and M. Prettre, Infrared study of CO

adsorbed on Pt/Al$_2$O$_3$. A method for determining metal-adsorbate interactions. *J. Catal. 29*, 213 (1973).
152. B. J. Cooper and D. L. Trimm, The coking of platinum/alumina reforming catalysts. In *Catalyst Deactivation*, Proc. Int. Symp. (B. Delmon and G. F. Froment, eds.), Antwerp, Oct. 1980. Elsevier, Amsterdam, 1980, p. 63.
153. J. M. Parera, N. S. Figoli, E. L. Jablonski, M. R. Sad, and J. N. Beltramini, Optimum chlorine on naphtha reforming catalyst regarding deactivation by coke formation. In *Catalysts Deactivation*, Proc. Int. Symp., Antwerp, Oct. 1980 (B. Delmon and G. F. Froment, eds.). Elsevier, Amsterdam, 1980, p. 571.
154. H. Scheiffer and M. Z. Frenkel, *Anorg. Allg. Chem. 414*, 137 (1975).
155. G. Lietz, H. Lieske, H. Spindler, W. Wanke, and J. Volter, Reactions of platinum in oxygen and hydrogen treated Pt/$\tau$-Al$_2$O$_3$ catalysts. II. Ultraviolet-visible studies, sintering of platinum and soluble platinum. *J. Catal. 81*, 17 (1983).
156. D. M. Little, Process variables and unit operation. In *Catalytic Reforming*. Penwell, Tulsa, 1985.
157. J. T. Pistorius, Analysis improves catalytic reformer trouble shooting. *Oil Gas J. 146*, June 10 (1985).
158. B. C. Gates, J. R. Katzer, and G. C. A. Schuit, *Chemistry of Catalytic Processes*. McGraw-Hill, New York. 1979, p. 190.
159. M. D. Edgar, Catalytic reforming of naphtha in petroleum refineries. In *Applied Industrial Catalysis* (B. E. Leach ed.). Academic Press, New York, 1983, Vol. 1, p. 123.

# 8
## Naphtha Hydrotreatment

### H. J. Lovink
*AKZO Catalysts, Amersfoort, The Netherlands*

### I. INTRODUCTION

Present-day reforming operations at severities up to 102–105 RON require very clean feedstocks for obtaining continuously high-quality reformate with a long catalyst life. Naphtha hydrotreatment is practiced in order to lower the level of poisons for the platinum reforming catalysts. These poisons are sulfur and nitrogen components (temporary poisons) and As, Pb, and other metals (permanent poisons).

Proper conditions of $H_2$ pressure, temperature, and space velocities are able to reduce these poisons to acceptably low levels with modern catalysts. The definition of acceptable has changed with time, e.g.,

| 1970–1980 specs. | | 1990s specs. | |
|---|---|---|---|
| S | < 1 ppm | S | < 0.5 ppm |
| N | < 1 ppm | N | < 0.5 ppm |
| Cl | < 1 ppm | Cl | < 1 ppm |
| | | Metals: | |
| | | As, Pb, Cu | < 1 ppb |
| | | ($H_2O$) | < 10 ppm |

The reasons for the more stringent specifications of the 1990s are the lower pressures of operation in reformers, particularly in the CCRs, the bimetallic catalysts, and the higher octane numbers required for the lead-free motor gasolines. For these reasons, naphtha hydrotreaters are now an integral part of reformers and also of hydroisomerization units for $C_5$–$C_6$ light naphtha, where Pt/alumina or Pt/zeolite catalysts are employed. The process itself, using Co-Mo-alumina catalysts, stems from the late 1930s, with the original inventor H. Hufman of Union Oil of California (now UNOCAL). Many general features of reforming and naphtha desulfurization were described earlier in Don Little's excellent book on reforming (1).

Hydrotreating catalysts have a long history dating back to the years 1930–1940, when brown coal hydroliquefaction and coal liquids were considered essential strategic materials. Background on hydrotreating catalysts and their function may be found in the literature (2, 3). It should be noted that all major and even minor international catalysis conferences have a cluster of lectures on hydroprocessing catalysts. Catalysts for naphtha hydrotreating are *not* specialties or different from those utilized for other atmospheric petroleum distillates; the same catalysts are used for kerosene, jet fuels, and gas oils.

## II. FEEDSTOCKS FOR NAPHTHA HYDROTREATERS

### A. Straight-Run Naphthas

The first and original feedstocks for reformers were straight-run naphthas from practically all of the world's crude oils. Their qualities vary significantly depending on the crude oil origin and boiling range. Table 1 gives a description of the naphthas from a large number of crudes derived from the well-published crude assays. The usual data mention naphthas with a range of cut points; however, all data have been corrected to 85–165°C cut points, which facilitates a quick comparison of qualities and is also more in line with present and future requirements. The 85°C IBP has been chosen because $C_5$–$C_6$ light naphtha is now excluded, since the $C_6$ portion can enhance the benzene make by some 1–2 vol %. This $C_5$–$C_6$ fraction can better be used in hydroisomerization or in olefin plants. A usual end point is 165°C; higher values give more deactivation and do not increase the $C_7$–$C_8$ aromatics.

The major Middle Eastern crudes contain about 10–20% reformer feedstock of S content averaging 300–500 ppm. The N content of straight-run naphthas, not often reported, is usually around 1–2 ppm. This is so low that the hydrotreater designs do not take it into account. Sometimes, however, this may lead to $NH_4Cl$ in the reformer.

**Table 1** Typical Naphtha Properties, 85–165°C Cut

|  | API° crude | Naphtha specific gravity (g/mL) | S (ppm) | P + O/N/A (wt %) | Vol % of crude |
|---|---|---|---|---|---|
| A. North Sea | | | | | |
| Brent | 37 | 0.769 | 30 | 42/34/24 | 18 |
| Statford | 38 | 0.760 | 37 | 49/37/14 | 18 |
| Ekofisk | 42 | | 20 | 53/32/15 | 20 |
| B. Major 27°–34° API Middle East crudes | | | | | |
| Arabian Light | 33 | 0.742 | 350 | 65/20/15 | 15 |
| Arabian Heavy | 27 | 0.737 | 450 | 70/20/10 | 11 |
| Iranian Light | 34 | 0.746 | 600 | 55/30/15 | 14 |
| Iranian Heavy | 31 | 0.748 | 1000 | 50/35/15 | 14 |
| Abu Dhabi | 31 | 0.747 | 600 | 65/20/15 | 12.5 |
| C. "Extra light" Middle East crudes (37°–50° API) | | | | | |
| Berri | 37 | 0.746 | 500 | 65/20/15 | 16 |
| Murban | 40.5 | 0.750 | 40 | 54/20/15 | 23 |
| Dukhan | 41.7 | 0.753 | 250 | 66/17/17 | 20 |
| Margham | 50.3 | 0.767 | 100 | 45/25/30 | 28 |
| D. African crudes | | | | | |
| Bonny Light, Nigeria | 37 | 0.746 | 500 | 65/20/15 | 16 |
| Brega, Lybia | 40 | 0.750 | 40 | 65/20/15 | 23 |
| Es Sider, Lybia | 37 | 0.753 | 250 | 66/17/17 | 20 |
| Zarzaitina, Algeria | 43 | 0.767 | 100 | 45/25/30 | 28 |
| E. South American crudes | | | | | |
| Tijuana | 32 | 0.745 | 30 | 60/28/12 | 16 |
| Maya | 22 | — | 400 | 60/27/13 | 8 |
| F. USA/Canada | | | | | |
| West Texas Interm. | 37 | 0.757 | 200 | 48/45/12 | 22 |
| West Texas Sweet | 39 | 0.754 | 500 | 47/13/10 | 20 |
| Alaska NSl | 26 | — | 130 | 40/45/15 | 16 |
| Lloyd Minister Canada | 21 | — | 8000 | 55/32/13 | 10 |
| G. Others | | | | | |
| Shengli, China | 24 | — | 120 | 46/40/14 | 5 |
| Urals, Russia | 32.3 | 0.761 | 290 | 38/52/10 | 12 |
| Tapis, Malaysia | 44 | | 13 | 40/40/20 | 22 |

**Table 2** Properties of Cracked Naphthas

| Origins | Br no. (g/100 mL) | Diene no. | S (ppm) | Basic N (ppm) | P + O/N/A (wt %) | wt % on feed |
|---|---|---|---|---|---|---|
| Visbreaker | | | | | | |
|   High S | 90 | 2 | 15,000 | 500 | 33/45/11/11 | 4–5 |
|   Low S | 80 | 2 | 1,000 | 200 | Similar | 4–5 |
| Coker | | | | | | |
|   Low S | 80 | 5 | 2,000 | 450 | 30/38/17/15 | 20 |
| FCC | 50 | <0.5 | 1,200 | 100 | 28/25/22/25 | 10–15 |
| Hydroprocessing | | | | | | |
|   VGO hydrocracked | 0 | 0 | 0–10 | 0–2 | 30/0/55/15 | 10–40 |
|   Resid hydrocracking | 2 | 0 | 20–100 | 50–200 | 35/0/50/15 | 5–10 |
|   Gasoil HDS and mild hydrocracking | 0–2 | 0 | 2–50 | 1–10 | 40/5/30/25 | 1–5 |

## B. Cracked Naphthas

Cracked naphthas are another source of reformer feedstock. Increasing quantities, on average 5–15% of the various feedstocks but locally frequently more, are produced in the conversion units of modern and projected refineries. Table 2 is a summary of the important characteristics of naphthas from several processes. The most important ones are visbreaker, coker, and heavy hydrocracked naphthas. Today the use of the others (FCC, mild hydrocracked, etc.) is more incidental. When comparing these feeds with the straight-run naphthas, much higher sulfur content, N content, olefins, and even diolefins are observed in the thermally cracked naphthas. These are the remnants of the breakdown of N- and S-containing resins and asphaltenes and are mostly ring compounds which are difficult to treat.

The hydrocracked naphthas, produced in small quantities as the naphtha range by-products of gas oil and heavier feedstock hydrotreating, need further processing but probably can be blended into straight-run naphtha when the N content is sufficiently low. FCC naphtha is a special case. Typically, only the middle part of FCC naphtha is low in octane number, as demonstrated in Table 2, but occasionally the octanes are low all the way to the FBP so that an FCC heavy naphtha is created for reforming. This is exemplified in Table 3 (4).

## III. FLOW SHEETS OF NAPHTHA HYDROTREATERS

Figure 1 presents a general scheme of a hydrotreater for cracked feedstocks with two stages of hydrotreating. The first stage is for a low-temperature treat-

**Table 3** FCC Naphtha Cuts

| Property | IBP-75°C | 75–125°C | 125–150°C | 150°C to FBP | Full range |
|---|---|---|---|---|---|
| wt % | 22 | 30 | 16 | 32 | (100) |
| S, ppm | 15 | 20 | 40 | 120 | 50 |
| N, ppm | 5 | 6 | 20 | 75 | 35 |
| Br no., g/100 mL | 100 | 75 | 45 | 25 | 70 |
| PONA, wt % | | | | | |
| P+N | 33 | 38 | 30 | 20 | 30 |
| O | 65 | 50 | 30 | 20 | 45 |
| A | 2 | 12 | 40 | 60 | 25 |
| RON | 95 | 87 | 93[a] | 93[a] | 92 |
| MON | 81 | 77 | 81[a] | 81.5[a] | 80 |

[a] At lower conversion (middle distillate mode) these octane numbers can be even lower than those of the 75–125°C fraction, making heavy FCC naphtha the best choice for reforming; otherwise only the "heart cut 75–125°C."

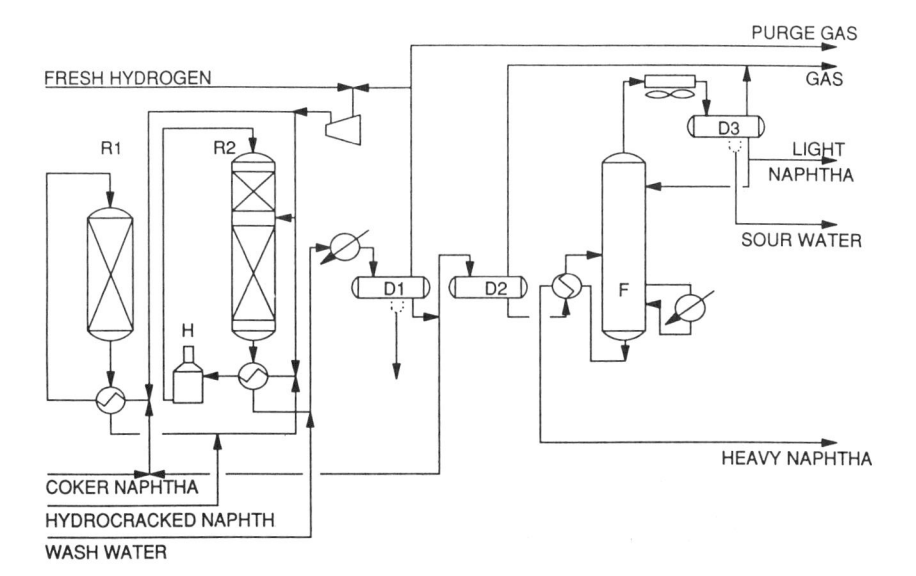

**Figure 1** Hydrotreater for cracked naphthas. R1, diolefins reactor; R2, HDS/HDN/Olef. reactor; D1, D2, D3, separators; F, stripper/fractionator; H, furnace.

ment ($\approx$ 100–150°). It is used to hydrogenate the diolefins and acetylenes that otherwise would polymerize and block the second reactor with "gums" that are not sufficiently soluble in the naphtha. The second stage operates at much higher temperature in order to saturate the olefins and hydrocrack the S and N compounds to $H_2S$ and $NH_3$.

Very deep HDS and HDN are required under the present circumstances: HDS > 99.95% and HDN > 99.9%. This poses extra demands on the reactors, such as

Very even packing with no channeling
Low and stable pressure drop
High, approximately 30 bar, $H_2$ pressure

Both stages mostly use the same catalyst, CoMo or NiMo on alumina. The flow scheme for other feedstocks, such as straight-run naphthas and those with 5% or less visbreaker naphthas, can be much simpler, with only one reactor needed. The space velocity can also be higher.

## IV. CATALYST PROPERTIES

The catalysts used for hydrotreating of naphtha are part of a much larger group which are used in hydrotreating kerosine, gas oils, vacuum gas oils, other distillates, and even residue (5, 6). They consist of freshly prepared oxide mixtures, typically $CoO-MoO_3$ or $NiO-MoO_3$, on activated alumina in the form of extrudates or beads of roughly 1.5 mm, and they are produced in separate dedicated catalyst plants for the petroleum refining industry. The production capacity was more than 60,000 tons in 1993 worldwide.

Table 4 presents typical properties of a naphtha hydrotreating catalyst. Other properties such as the length distribution and pore size distribution are discussed in Ref. 6. The American Society for Testing and Materials (ASTM) has an active group that works on defining methods for the properties of extrudates. Bradley (7) described the status of the so-called bulk crushing strength test, a possible future specified item for fixed-bed reactors, particularly for those working under a high mass velocity and pressure drop; ASTM agreement had still not been reached as of mid-1993.

The active forms of hydrotreating catalysts are not the metal oxides but the sulfides that are formed under the influence of sulfur compounds and $H_2S$ at low temperatures, 200–250°C. At present, presulfided catalysts are available that can be started up as if the unit had already been operating for a few weeks. This eliminates the time-consuming presulfiding with special sulfiding agents like dimethyl disulfide + $H_2$ under low-temperature conditions for 10–24 h. These presulfided catalysts are now used extensively (8). With presulfided catalyst the optimum activity can be obtained immediately after start-up; this

## Naphtha Hydrotreatment

**Table 4** Naphtha Hydrotreating Catalyst—Typical Properties

| Properties | CoMo alumina | NiMo alumina |
|---|---|---|
| Chemical composition (wt %) | | |
| CoO | 3–5 | — |
| NiO | — | 2.5–4 |
| $MoO_3$ | 12–15 | 10–20 |
| $Al_2O_3$ | Balance, which is approx. 80% | |
| $SiO_2$ | 0–10 | |
| $Na_2O$ | Less than 0.5 | |
| $Fe_2O_3$ | Less than 0.2 | |
| Other alkalis | Less than 0.5 | |
| Earth alkalis | Less than 0.5 | |
| $SO_4^{2-}$ | Less than 1.5 | |
| Other anions e.g., $PO_4^{2-}$, $BO_3$ | Several %, sometimes | |
| Physical properties | | |
| Surface area, BET, $m^2/g$ | 150–250 | |
| Pore volume, mL/g | 0.3–0.5 | |
| Size of extrudates,[a] mm | 1–2 | |
| Length, mm | 3–6 | |
| Form of extrudates[a] | Cylindrical, or 3–4 lobes | |
| Reactor density, $kg/m^3$ | 600–750 | |
| Strength of extrudates, kg/mm | 2–4 | |

[a] Also beads of 1.5–3 mm; "top layers" are larger sized.
*Source*: Refs. 9 and 10.

saves more than 1 day in start-up time. Regenerated catalyst can also be supplied in this presulfided form.

## V. REACTION RATES AND KINETICS

Naphtha desulfurization is considered a straightforward, catalytically easy reaction. Assuming first-order kinetics, a pressure of 20 bar, and 330°C, the comparative relative first-order rate constants for the various petroleum feedstocks are:

| | |
|---|---|
| Naphtha | 100 (by definition) |
| Kerosene | 20 |
| Light gas oil | 10 |
| Vacuum gas oil | 1 |

A more extensive review of typical operating conditions and results for commercial installations are presented in Ref. 2. CoMo catalysts are generally used where sulfur removal is critical. NiMo catalysts are used where nitrogen removal and hydrogenation are the most important functions.

A number of bench-scale testing studies have been carried out and reported in the literature (9–11). The results are summarized in Table 5 and in Figures 2–4. However, some remarks have to be considered concerning the testing of naphtha for desulfurization in these rather small reactors and units, as these entail a number of special problems:

1. The complete unit should be extraordinarily clean. A level of 0.5 ppm sulfur is a minute amount; the usual gas oil test units can hardly be cleaned to this low level without taking them completely apart.
2. The pumps should be proper for (nonlubricating) naphtha service.
3. Reactor effluent should be cooled quickly and stripped *immediately*, and continuously, in order to avoid recombination or sulfur formation; see Section VIII.B.

**Table 5** Feedstocks and Operating Conditions of the Kinetic Studies of Removal of Sulfur, Nitrogen, and Olefinic Compounds from Naphthas

| Properties | Straight run | 50/50 blend of FCC and SR naphtha | Coker gasoline |
| --- | --- | --- | --- |
| A. Feedstocks | | | |
| S, ppm | 252 | 1950 | 3700 |
| N, ppm | < 1 | 40.5 | 45 |
| Br no., g/100 mL | — | 45 | 55 |
| FBP, °C | 180 | 220 | 200 |
| B. Operating conditions[a] | | | |
| $pH_2$, bar | 7 | 12 | 30 |
| LHSV, $h^{-1}$ | 5 | 6 | 1.5 |
| Temperature | | Varied, see Figure 2 | |
| Catalyst | KF 752 | ← KF 840 → | |
| | High activity CoMo | High activity NiMo | |

[a] These operating conditions have been derived from proven commercial ones that show catalyst deactivations in the range of 0.5–1.0°C max. per month.

Naphtha Hydrotreatment

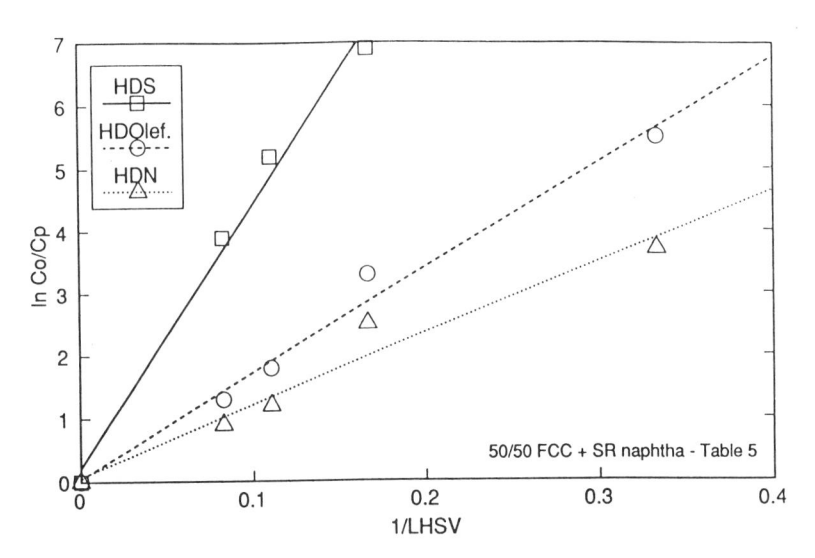

**Figure 2** First-order kinetics apparently apply to naphtha hydrotreating.

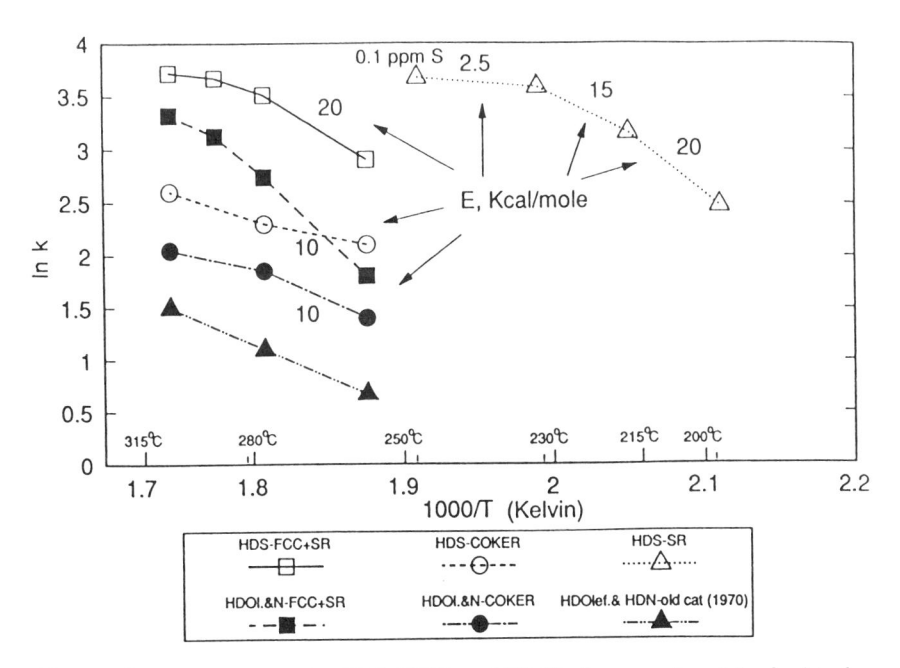

**Figure 3** Rate constants for HDS, HDN, and HDOL for several naphtha feedstocks, over 1990 catalysts.

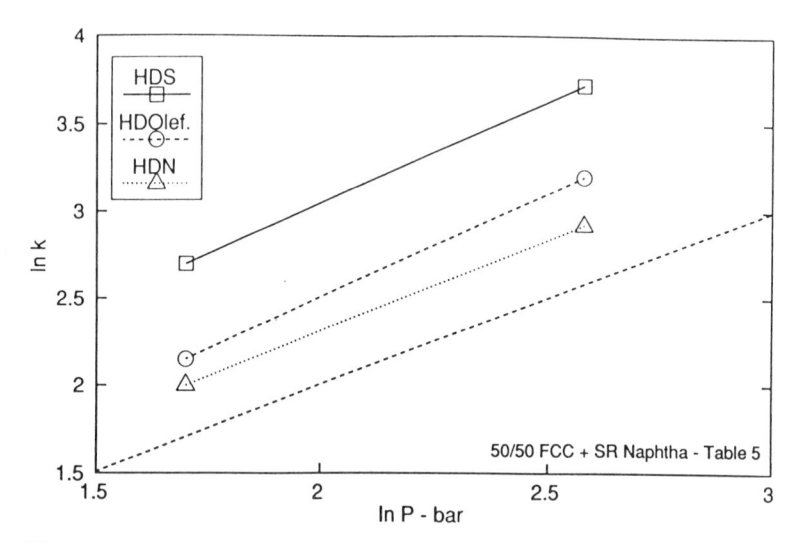

**Figure 4** Influence of hydrogen pressure on HDS, HDN, and HDOL of cracked naphtha mixture.

4. The best analysis methods should be used by experienced technicians:
   | | |
   |---|---|
   | Sulfur | ASTM, D3120 |
   | | UOP Method Number 357/80 |
   | Nitrogen | ASTM, D3431 |
   | Diolefins | ASTM, D1961 |
   | Bromine number | ASTM, D11319 |

Even with all of these controlled carefully, it is not easy to assess exactly the differences between the various straight-run naphthas listed in Table 1 and to check out the best conditions for FCC naphtha or coker/visbreaker naphthas. Moreover, the high desulfurization level of 99.99% demands attention to the $L/D$ ratio of the reactor ($>10$) and the $L/dp$ ratio, $dp$ being particle size, in view of the axial dispersion of reactants in the reactor.

In well-packed commercial reactors with several meters of bed height this is not a problem, but in reactors of 1 inch diameter and with 20–30-cm bed heights this needs due attention. Several publications (12, 13) have dealt with the principles involved, particularly the Mears criterion, which is applied in such cases.

When the following formula is applied:

$$\frac{L}{dp} > \frac{20n}{Pe} \ln(1-x)^{-1}$$

## Naphtha Hydrotreatment

where  $L$ = length of reactor bed
   $dp$ = particle size
   $n$ = order of reaction (usually one for naphtha HDS)
   Pe = Peclet number (about 2 in this case)
   $x$ = conversion

it follows that $L/dp$ for 99.99% conversion of S in naphtha (2000 ppm reduced to <0.5 ppm) should be at least 100.

In catalyst beds of 30-40 cm bed height, catalyst particles of 1.5-2 mm are usually mixed with SiC granules of <0.5 mm in order to fill the space between the catalyst particles. Some of the work reported here, on coker gasoline, has been carried out with larger bed heights.

Other aspects of the study on the three feedstock cases are the following (Table 5):

1. *Straight-run naphtha*: this was one of the most difficult tests because of the high reaction rate. The reactor effluent was washed immediately after continuous depressurization in order to remove the dissolved $H_2S$ as quickly and completely as possible.
2. *A 50/50 mixture of FCC gasoline and straight-run naphtha*: this needs a higher $H_2$ pressure in order to achieve 0.5 ppm S and basic N with a sufficient run length.
3. *Coker gasoline*: this material can have very high S, N, olefin, and even diolefin contents and is one of the worst materials as reformer feedstock. Many older refineries still mix it as such into the pool with sufficient additives for stabilization, but for large cokers this is a less attractive option.

Naphtha HDS/HDN are best represented by simple first order kinetics and Figure 2 has been constructed from the measured data on that basis:

$$\frac{C_0}{C_p} = \frac{k}{\text{LHSV}}$$

and

$$k = Ae^{-E/RT}$$

where  $C_p$ = S, N, or bromine number in product
   $C_0$ = S, N, or bromine number in feed
   $A$ = frequency factor
   $E$ = activation energy

Figure 2 is a graphical presentation of the three functions under the conditions of Table 5. First-order kinetics obviously give a reasonable fit of the data for HDS, HDOlef., and HDN.

Figure 3 is a plot of ln $k$ versus $1000/T$ absolute and demonstrates the great differences between straight-run naphtha and the cracked stocks. The following are remarked:

1. At the highest temperature and for all test runs, products with < 0.5 ppm S have been obtained.
2. HDOlef. and HDN proceed at approximately the same rate, and less than 1 ppm N was obtained at the highest temperatures.
3. The activation energy for HDS is 20 kcal/mol at the lowest temperatures, but at higher temperatures it decreases because of diffusion effects, even with a catalyst of approximately 1.0 mm diameter. A value of 10 has been reported in the literature, but actually at the most severe conditions a value of less than 10 has been found (14).
4. The catalysts used in these cases were high-activity CoMo (Akzo KF 752) for SR naphtha and NiMo (Akzo KF 840) for the cracked stocks. Catalyst type is still important as the data labeled "old cat" indicate; this was a catalyst of 1970 vintage, NiMo alumina type 153, which was still good for HDS but was 50% less active for HDN and olefin removal. For cracked stocks NiMo catalysts are preferred over CoMo.

Figure 4 is a plot of the influence of pressure on HDS, HDN, and olefin removal. HDS and HDOlef. obviously are linear with pressure, but the highest point of HDS is in the diffusion range and therefore somewhat low. $H_2$ pressure is a very important parameter in naphtha hydrotreating and, actually, the pressures and LHSV values of Table 5 have been chosen so that they correspond to commercial operating conditions at which the low deactivation rates, 0.3–1.0°C per month, have been proved.

## VI. CURRENT DESIGN OF NAPHTHA HYDROTREATERS

The basic design data for naphtha hydrotreaters has not changed much during the past few decades as the required higher HDS/HDN activities have been met by improvements in the catalysts. The design basis is as follows:

Good HDS/HDN activities
Low deactivation rate, approximately 0.3–1°C/month
Sufficient temperature "range" between the start of run and the maximum operating temperature where sulfur recombination begins (see Section VIII)
Capacity for some metals (see Section VIII)
Low pressure drop of 1–2 bar across the reactor
Some capacity for scale and deposits

Table 6 compares the operating hydrotreating conditions for straight-run naphtha and cracked naphtha. In the case of cracked naphtha, special care is required when the olefin content is high, as the reactor exotherm can easily surpass the temperature range between SOR and EOR. Delta temperatures of 20–30°C are usual, making recycle of product, quench with $H_2$ gas or product, or blending with straight-run naphtha desirable for cases anticipated to have a $\Delta T$ of more than 35°C (see also Figure 1).

Although good-quality product is obtained regularly, some refiners occasionally experience breakthroughs of S and N and traces of olefins from these cracked naphtha units, which can be very serious for the fixed-bed reformers. This is why a number of those refiners choose to reprocess such streams blended with straight-run feeds in larger naphtha units in order to be sure of adequate pretreatment.

The use of sulfur traps (15) or guardbeds is also advisable in such cases. These are installed downstream of the stripper and are either Ni, Cu, or MnO based; all can retain $H_2S$ and S, but mercaptans are only partially adsorbed.

## VII. $H_2$ CONSUMPTION

The $H_2$ consumption of the straight-run naphthas of Table 1 is very low, 3–6 $Nm^3$/ton, compared to the $H_2$ production of a typical reformer of sometimes 100–150 $Nm^3$/ton. Cracked stocks, with many olefins and higher S contents, consume more $H_2$. A reasonable estimate of the $H_2$ consumption is possible

**Table 6** Operating Conditions of a Hydrotreater for Different Feedstocks

| Parameter | Straight-run naphtha | Cracked naphtha (coker and visbreaker) |
|---|---|---|
| Total pressure, bar | 20–30 | 45–55 |
| LHSV, $h^{-1}$ | 3–10 | 2–5 |
| Temperature, °C | 250–330 | 100–200 [a] |
|  |  | 280–320 [b] |
| $H_2$/oil, $Nm^3/m^3$ | 50–100 | 300–1000 |
| Component distribution (exit reactor): |  |  |
| naphtha/$C_1$–$C_4$/$H_2$ |  |  |
|   Mole ratio | 0.65/0.05/0.30 | 0.30/0.05/0.65 |
|   Partial pressure (bar) | 15–20/1–2/5–8 | approx. 15/2/33 |

[a] First bed.
[b] Second bed.

based on a correlation that the author composed years ago when the cracked stocks began to come into focus:

| $H_2$ required in $Nm^3$/ton feed: | |
|---|---|
| For each 1 wt % of S | 21 |
| Per 10 bromine numbers | 13.8 |
| Per 1000 ppm N | 6.4 |
| For each 1 wt % hydrocracked stock | 4.5 |

Furthermore some $H_2$ is lost on depressurizing and stripping the product; for naphthas at the usual pressure this is approximately 1.5 $Nm^3$/ton per 10 bar of pressure. A similar set of semi-empirical rules is discussed in Ref. 16.

## VIII. PITFALLS IN HYDROTREATING OF NAPHTHA

### A. Metals Removal

For clean Middle Eastern feedstocks metals are not a problem, as their straight-run naphthas and well-distilled cracked stocks generally do not contain any metals. In some U.S. and Russian crudes arsenic may be present.

Contamination with Pb from leaded gasolines, however, still occurs, as does contamination with $SiO_2$ in coker and visbreaker naphthas. At what level will these contaminants break through and poison the Pt catalyst in the reformer, or when will they decrease the HDS/HDN activity? In the NPRA records numerous contamination cases can be found with descriptions of catalyst analyses (11). The conclusion is that it is always recommended to analyze spent or used catalyst for possible contaminants when the reactor is opened. For a straight-run naphtha when the Pb averages 1000 ppm, the As averages 500 ppm, or the Na averages 1000 ppm (increase over the value for fresh catalyst), it is time to change the catalyst and also to check the origin of these contaminants.

The metals and $SiO_2$ have a strong tendency to deposit first in the top layers of the catalysts and later move to the lower sections (Table 7). HDS activity is also affected by Pb, As, and Na. A reduction of 50% in activity with the levels in Table 7 is to be expected. The $SiO_2$ can actually block the pores and also cause pressure drop.

Physical blocking of the catalyst bed by Fe sulfides and scale should be prevented by the installation of proper "baskets," top layers of larger inert balls and larger size catalyst (1, 2). Special Fe-S-trapping catalysts such as Akzo KG-1 are now used very successfully as top layers in hydrotreaters (9).

*Naphtha Hydrotreatment* 271

**Table 7** Distribution of Metal Contaminants in the Catalyst Layers

| Level | Pb (ppm) | Na (ppm) | As (ppm) | Fe (ppm) | SiO$_2$[a] (%) |
|---|---|---|---|---|---|
| Top dust | 3000 | 4000 | 50 | > 10,000 | 10 |
| Top 20% | 300 | 650 | 1500 | 2,000 | 4 |
| Middle | 50 | 600 | 300 | 500 | 0 |
| Bottom 20% | < 5 | 500 | < 50 | 300 | 0 |

[a] With coker and visbreaker naphtha only, stemming from the silicone defoaming agents.

## B. Monitoring of Activity and Apparent Activity Loss

### 1. Monitoring Methods

Monitoring of catalyst activity can be carried out in several ways. There are sophisticated programs for sale for hydrotreating units, but simpler plots of S and N versus time or the first order K plots (see Section V) versus time can also be used. Many good commercial hydrotreaters are run only until the reformer needs regeneration, even though the units have needed only 5–10°C temperature increase. However, checking the catalyst quality, at least that of the top layers when the reactor is opened, is always a wise practice.

### 2. Recombination of Sulfur and Olefin

During long periods of operation, the activity (real or apparent) may have fallen and the reactor temperature was raised for varying reasons above 320°C. The sulfur content, however, does not decrease but increases at higher temperatures of ≥ 330°C. This has been described extensively in the literature and duplicated in laboratory and commercial operations (17). The proper course of action is to lower the reactor temperature, which should have the following effect, e.g.:

| Reactor, °C | S in product, ppm |
|---|---|
| 321 | 1.1 |
| 315 | 0.5 |
| 310 | 0.3 |
| 307 | 0.1–0.2 |

Timmermans (18) looked at the dehydrogenation equilibrium of paraffins → olefins as a function of the pressure of H$_2$S and H$_2$ and came up with the following results, proving this concept (temperature 300°C):

| Bromine number product, g/100 mL | 1.0 | 2.0 | 3.0 |
| --- | --- | --- | --- |
| Corresponding equilibrium S level, ppm | 0.6 | 1.2 | 2.5 |

In the case of a coker + straight-run naphtha blend, sulfur recombination was also noticed (Table 8). Recombination is a real effect, also very well known in vacuum gas oil hydrocrackers. That is why some of those naphthas need additional hydrotreatment before reforming.

A final bottom layer of NiMo catalyst may also help in retarding this kind of sulfur breakthrough because NiMo catalysts are more active in olefin hydrogenation. So, undue cracking, insufficient olefin hydrogenation, or local dehydrogenation may cause this "recombination" of $H_2S$ and olefins of the reactor effluent in the heat exchanger or cooler.

### 3. Heat Exchanger Leakage

The heat exchangers of naphtha hydrotreaters are probably the ones in refineries that are the most critical regarding small leaks, as 99.99% of the feed should pass without leakage to product. The same holds for the stripper. Both of these units should be checked out first before more elaborate product analyses are carried out (mercaptans, S types, etc.).

### 4. Chlorides, Corrosion

Chlorides, particularly $NH_4Cl$, can cause problems in the heat exchangers as they deposit at temperatures from 300°C downward. The chloride stems from the (heavy) chloriding of the reforming catalyst and the $NH_4^+$ from N compounds in the feedstock. A water wash can keep this under control. Precautions must be taken due to the corrosive nature of this effluent.

**Table 8** Recombination in Hydrotreating a Blend of Coker and Straight-Run Naphtha, Catalyst Ketjenfine 840 NiMo/alumina

| Reactor temperatures | 1 | 2 |
| --- | --- | --- |
| Inlet, °C | 305 | 300 |
| Outlet, °C | 333 | 327 |
| Product properties | | |
| Sulfur, ppm | 0.8 | 0.4 |
| Mercaptan S, ppm | 0.8 | < 0.4 |
| N, ppm | < 0.3 | < 0.3 |

## 5. Other Causes

Quite a number of other primary causes can be found quickly, such as higher than usual pressure drop, which may cause maldistribution in the reactor; insufficient hydrogen makeup; and low S in the feed. The latter may not happen often, but it is a fact that lack of sulfur in the feed (< 2 ppm) can gradually deactivate the catalyst by causing the sulfur-bearing active sites on the catalyst to decompose. Another cause is that the catalyst may have been coked up by a flow reversal, $H_2$ starvation, etc.

## C. Catalyst Regeneration

Hydrotreating catalysts are usually regenerated in situ by means of the steam-air method. However, because of the hazards and air pollution involved, this practice has largely been replaced by ex situ regeneration. Specialized companies can discharge the reactor and replace the catalyst by a fresh or a freshly regenerated batch. The original charge should be analyzed and sent to an outside company that has specialized in catalyst regeneration and sieving.

Naphtha catalyst is a service item; it protects the delicate reforming catalysts. Refiners cannot take much quality risk regarding reuse as it is the only protector of these reforming catalysts. Therefore, due attention should be given to the following:

Physical properties
    Strength
    Density
    Length

If any of these properties has suffered 20–30% loss, it is better to discard the catalyst and start with fresh material.

Poison levels
    Lead
    Arsenic
    Sodium

Activity for removal of S, N, O, olefins, and metals removal.

A check of catalyst activity should be carried out in case of doubt; not more than 20% loss relative to fresh catalyst should have occurred.

In general, naphtha catalysts are regenerated only once or twice during a total life of 3–6 years. Typical properties of spent and regenerated catalyst after sieving extensively are listed in Table 9.

**Table 9** Typical Properties of Spent and Regenerated Hydrotreating Catalyst

| Property | Spent | Regenerated |
|---|---|---|
| S, wt % | 6.35 | 0.40 |
| C, wt % | 5.0[a] | 0.08 |
| LOI, wt % | 33.1 | < 0.3 |
| Average length, mm | 3.01 | 3.05 |
| BCS, kg/cm$^2$ | — | 12 |
| SCS, lb/mm | — | 3.7 |
| Surface area, m$^2$/g | — | 230 |

[a] 5–10% in general but much higher for cracked stocks and other distillates.
*Source*: Ref. 19.

## D. Dense Loading of the Catalyst in the Reactor

When catalyst handling is performed completely by outside companies, it is also wise to consider having the reactor filled by the so-called dense loading technique, the results of which are described in Ref. 20. During the usual "sockloading" method, the catalyst is guided by hand to its proper place in the reactor by means of a sock through which it gently flows. In "dense loading" a rotating machine causes the catalyst flow to "rain down," and by contact with the reactor atmosphere, the extrudates preferentially fall down in a horizontal position, which has a number of advantages: (1) the packing of the reactor is higher (more kg/m$^3$) and more homogeneous, and (2) the bed does not settle further when in operation.

Some other proprietary techniques are available, such as ARCO's COP loading, Total's Densicat, and UOP's dense loading system. Although the pressure drop initially is somewhat higher, the final $\Delta P$ is the same or even lower. Table 10 presents some operational data for a dense-loaded naphtha hydrotreating unit (21).

## E. Miscellaneous

### 1. Catalyst Safety

Catalyst manufacturers usually provide the user with complete data on the health implications for the handling and use of CoMo and NiMo catalysts. Extensive product safety data are provided based on consultation with health officers and official authorities. It is stated that:

1. Catalysts of CoMo types are expected to be no more than irritants to the eyes and skin. The MAC values are 0.1–1.5 mg/m$^3$ for the various compositions.

**Table 10** Operational Data for a Dense-Loaded Naphtha Hydrotreating Unit

| Feed type/ | Naphtha |
| Catalyst type | KF-742-1.3Q |
| --- | --- |
| Catalyst bed height, m | 6.0 |
| Diameter, m | 2.5 |
| Volume, m$^3$ | 27 |
| Catalyst content, tons | 20 |
| Loaded density, kg/m$^3$ | 741 |
| Increment over sockloading, % | 23 |
| LHSV, h$^{-1}$ | 6.37 |
| $P$, bar | 33 |
| $T$, °C | 325 |
| H$_2$/HC, Nm$^3$/m$^3$ | 92 |
| Gas rate, kg/m$^2$/s | 0.38 |
| Oil rate, kg/m$^2$/s | 6.9 |
| $\Delta P$ over reactor, bar | 2.2 |

2. Suitable respiratory equipment should be worn when handling the material, and containers with fresh catalysts should be stored in warehouses and not under the open sky or rain, mud, or other wet conditions.

Detailed safety and handling instructions are available for all the different types of catalysts.

### 2. Spent Catalyst

This is a materal that can no longer be "landfilled" because many of its components are water leachable. Several companies in the United States, Japan, and Europe can take these wastes and completely recycle all the components or at least the heavy metals for reuse. The major fresh-catalyst manufacturers can provide a list of such companies. In times of shortages, spent CoMo and NiMo catalysts may generate some cash, but in general the recuperation costs are such that some additional disposal costs (approximately U.S. $500/ton 1993) are involved.

### F. The Future

Hydrotreating is already a large section of industrial catalysis and the environmental aspects of petroleum products are leading to more severe hydrotreating processes. This has an impact on the catalyst research, which continues worldwide at high intensity. A fallout for naphtha HDS will certainly be better catalysts. More cracked stocks will become available for reforming and more isomerization capacity will certainly be built.

One future option might be combined hydrotreating of naphtha, kerosene, and diesel in one (large) unit, followed by subsequent fractionation. It should be possible then to produce 0.5 ppm S in the naphtha and 500 ppm S in the diesel. Another option is a catalyst combination with a zeolite that imparts partial isomerization, which will be good for reformer yields. Furthermore, mild posttreatment HDS of heavy FCC naphtha with 300–700 ppm S to less than 50 ppm S may become a new naphtha hydrotreating process for reformulated gasolines.

## REFERENCES

1. D. M. Little, *Catalytic Reforming*. Pennwell, Tulsa, 1985, pp. 175–194.
2. J. F. Le Page, *Applied Heterogeneous Catalysis*. Ed. Technip, Paris, 1987, pp. 357–460.
3. Workshop Meeting on Hydrotreating Catalysts, Louvain-La-Neuve, Belgium. 1st, *Bull. Soc. Chim. Belge 90*, 12 (1981); 2nd, *Bull. Soc. Chim. Belge 93*, 8–9 (1984); 3rd, *Bull. Soc. Chim. Belge 96*, 11–12 (1987); 4th, *Bull. Soc. Chim. Belge 100*, 11–12 (1991).
4. AKZO Catalysts Symposium, 1986, section F-3.
5. Datasheets of CoMo and NiMo catalysts of major manufacturers: Datasheets of Ketjenfine 700 series (AKZO), Technical Bulletin *Shell Chemical* SC 340-80, and others.
6. F. H. Puls, in *Characterisation and Catalyst Development* (S. A. Bradley et al., ed.). ACS Symposium Series 411, American Chemical Society, Washington, DC, pp.
7. S. A. Bradley, in *Characterization and Catalyst Development* (S. A. Bradley et al., eds.). ACS symposium Series 411, American Chemical Society, Washington, DC, pp.
8. NPRA Q&A book, 1992, p. 114.
9. AKZO Catalysts Symposium, 1991, Workshop on Hydrotreating.
10. AKZO Catalysts Symposium, 1986, section H-3, Application on Ketjenfine 840.
11. NPRA books of 1989, p. 116; 1979, p. 85; 1975, p. 126; 1974, p. 61.
12. D. E. Mears, *J. Catal. 20*, 127 (1971).
13. J. Donald Carruthers et al. Pilot testing of hydrotreating catalysts. Symposium on Catalyst Performance Testing, Unilever Research Lab, Vlaardingen, Holland, March 28–29, 1988, p. 45.
14. J. M. Thomas and K. I. Zamaraev, *Perspectives in Catalysis*. Blackwell, Oxford, 1992, p. 252.
15. NPRA Q&A Session, pp. 147–148, 1992 (several speakers).
16. D. Edgar, NPRA Q&A Session, pp. 72–73, 1985.

17. J. Mauleon, NPRA Q&A Session, question 29, p. 107, 1979.
18. C. H. Timmermans, NPRA Q&A Session, question 1, p. 42, 1975.
19. AKZO Catalysts Symposium, 1984, p. 180.
20. AKZO Catalysts Symposium, 1984, p. 135.
21. R. E. Maples, *Petroleum Process Economics*. Pennwell, Tulsa, 1993, pp. 201–209.
22. AKZO Catalysts Technical Service Information 92/134E, Naphtha Hydrotreatment.
23. ASTM Standards on Catalysts, committee D32, 3rd ed., 1988, Philadelphia.

# 9
# Deactivation by Coking

**Patrice Marécot and Jacques Barbier**

*Université de Poitiers, Poitiers, France*

## I. INTRODUCTION

In the oil refining and petrochemical industry, deactivation of the catalyst by carbonaceous deposits is an important technological problem. Deactivation in the reforming process is attributed to coke formation and deposition. In fact, for reforming reactions of naphthas, the thermodynamics are such that it would be desirable to work at high temperature and low pressure. However, such operating conditions favor coke formation, and many reforming units operate under high pressures in order to increase the lifetime of the catalyst. It is obvious that reducing the rate of coke formation would represent a desirable improvement, particularly at a time when production of lead-free gasoline is requiring more severe operating conditions (higher temperatures and lower pressures).

In recent years, the coking reaction on bifunctional catalysts has been studied. The purpose of this chapter is to highlight the modifications of the characteristics (quantity, localization, chemical nature, and toxicity) of coke induced by a change in the nature of the catalyst, in either the support or metallic phase (dispersion, alloying, presulfurization), and by a change of the experimental conditions (e.g., time, temperature, pressure, feed). In spite of the complexity of the coking reaction, which is the resultant balance between the formation of coke precursors, their polymerization, and their destruction by metallic or acidic mechanisms, it appears that many conclusions are well established. Nevertheless, some questions remain open to discussion.

## II. CHARACTERIZATION OF COKE

Coke deposited on bifunctional catalysts is characterized by (1) quantitative analysis and localization of coke (on the metallic or acidic function) and (2) chemical nature.

### A. Quantitative Analysis and Localization of Coke

The simplest technique for the quantitative analysis of hydrocarbon residues is by the combustion under oxygen of coked catalysts (1–7). In the case of reforming catalysts, temperature-programmed oxidation (TPO) shows two peaks which are particularly well resolved when the partial pressure of oxygen is low in the course of the TPO (8). The low-temperature combustion (250–300°C) corresponds to coke deposited on the metallic phase, whereas the second one (400–450°C) is ascribed to coke deposited on the support (5, 7, 9–11).

Thus, the combustion at 300°C leads to the oxidation of a small amount of carbon but can regenerate the accessibility and the activity of metal atoms (10). Furthermore, it was shown by electron energy loss spectroscopy (EELS) associated with electron microscopy that, after partial combustion at 290–300°C, the coke is always present in the immediate vicinity of the metallic particles (12, 13).

Examination of coked catalysts by a large number of techniques (14–16) has shown great heterogeneity in the coke distribution on the catalyst. That heterogeneous character of the deposit was demonstrated on a macroscopic scale by electron and ion microprobes as well as on a microscopic scale by transmission electron microscopy (TEM) and EELS. It was found that the coke is not deposited in the form of a monolayer but, on the contrary, three-dimensional deposits appear from the onset of coking, and many zones which are only slightly coked remain on very highly coked catalysts.

### B. Composition and Structure of Coke

A documented technique for characterizing the composition of coke is by the analysis of coke combustion products to determine the H/C ratio (1–4, 8, 17). Measuring the amounts of oxygen consumed and carbon dioxide produced during temperature-programmed oxidation (Figure 1) indicates that the coke deposited on Pt/Al$_2$O$_3$ corresponds to the formula CH$_x$ with $x$ increasing when the

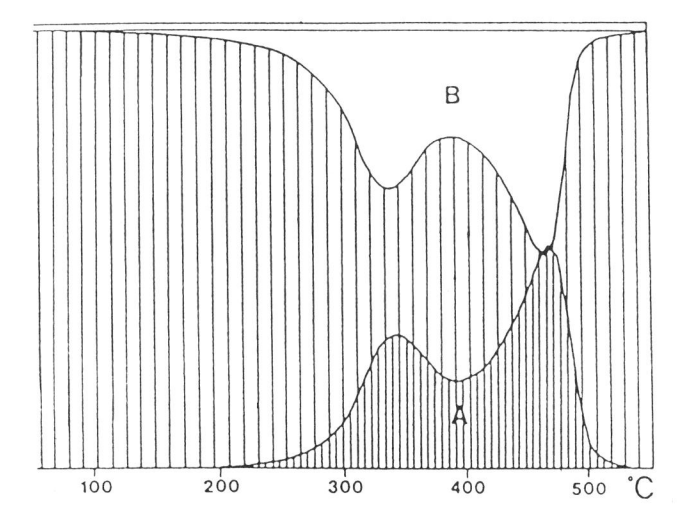

**Figure 1** Temperature-programmed combustion of a coked Pt/Al$_2$O$_3$ catalyst (A) CO$_2$ production; (B) O$_2$ consumption. (From Refs. 8, 47.)

metallic function/acidic function ratio of the catalyst increases (Table 1) (8). In other words, coke deposited on the metal is less dehydrogenated than coke deposited on the acidic support.

A number of studies on hydrocarbon deposits have been carried out by infrared (IR) spectroscopy (17–20). Whatever the catalyst used or the nature of the coking agent, the overall results obtained show the presence of aromatic C–H bonds, of methylene groups, and of aromatic rings in all instances. Extraction of coke with various organic solvents confirms its polyaromatic nature (17, 21, 22).

**Table 1** Determination of the H/C Ratio of Coked Catalysts

| Catalyst | Pt (%) | Dispersion (%) | Time of coking (min) | Coking agent | H/C |
|---|---|---|---|---|---|
| Pt black | 100 | — | 60 | Cyclopentane + 10% cyclopentadiene | 1.05 |
| Al$_2$O$_3$ | — | — | 60 | Cyclopentane + 10% cyclopentadiene | 0.50 |
| Pt/Al$_2$O$_3$ | 0.6 | 54 | 30 | Cyclopentane | 0.80 |
| Pt/Al$_2$O$_3$ | 0.6 | 54 | 180 | Cyclopentane | 0.50 |
| Pt/Al$_2$O$_3$ | 0.185 | 72 | 180 | Naphtha | 0.50 |
| Pt/Al$_2$O$_3$ | 0.37 | 70 | 180 | Naphtha | 0.64 |
| Pt/Al$_2$O$_3$ | 0.98 | 67 | 180 | Naphtha | 0.70 |

*Source*: Ref. 8.

Used catalysts were also characterized by X-ray diffraction, Raman spectroscopy, and electron microscopy (13, 15, 16, 23–27). These techniques proved that coke deposits are heterogeneous, with the presence of three-dimensional aggregates which are well crystallized (pregraphitic or graphitic coke) and of polyaromatic molecules in a disordered arrangement. Only these last polymers can be extracted and analyzed (17, 21, 22).

## III. EVOLUTION OF COKE CONTENT WITH TIME ON STREAM

Generally, after very rapid initial deposition, the amount of coke increases slowly with time on stream (10, 28, 29), following a simple relationship of the type

$$C_c = kt^{1/n}$$

where $C_c$ is the weight percent of coke, $t$ is the reaction time, and $k$ and $n$ are two correlation factors. Such an equation was first introduced by Voorhies (30) for the coking reaction on acidic catalysts. Many other researchers have extended the use of this relationship to other catalytic systems (3, 20, 31–35), either in its integral form or in its differential form, with the latter yeilding the rate of coke deposition. However, even if such an analysis of the phenomenon is of some practical interest, it does not allow characterization of the deactivation of each of the two functions present on a bifunctional catalyst.

Previously, it was pointed out that the amount of coke deposited on the metal can be determined by partial combustion at moderate temperatures (270–300°C). Figure 2 shows the evolution of the number of carbon atoms deposited on the metal per accessible atom of platinum, as a function of time on stream in the course of the reaction of cyclopentane at 400°C on various catalysts (10) using TPO at 270°C for the measurement. The curves obtained show that in the very first moments, when the catalyst is brought into contact with hydrocarbon, the metal attains a coke coverage that will remain constant, whereas the coke continues to increase on the support. This result is in agreement with Figure 3, which shows that the hydrogenation activity of the metal decreases very rapidly with increasing time of coking and then becomes stable, although the amount of carbon deposited on the catalyst composite continues to increase.

The occurrence of coke deposition first on the metallic function and then on the acidic function was demonstrated in laboratory tests (10, 21, 36, 37) as well as during commercial runs (38, 39). However, different conclusions were drawn about the nature of the controlling function in long-term deactivation: some authors concluded that the deactivation of the acid function determines the length of the cycle (39, 40), whereas Shum et al. (41, 42) consider that

*Deactivation by Coking*

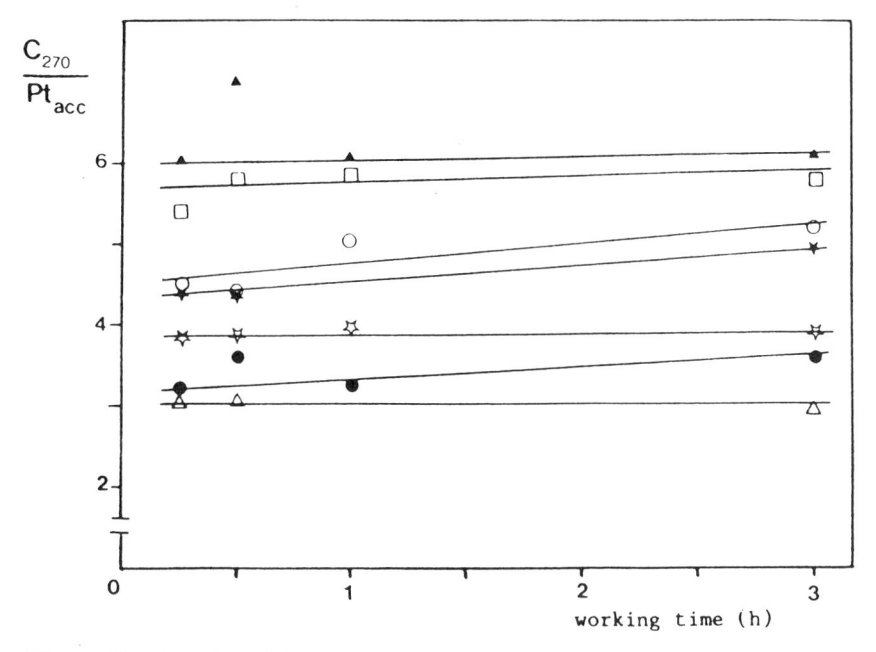

**Figure 2** Evolution of the number of carbon atoms deposited on the metal per atom of accessible platinum with the operating time. (From Ref. 10.) Catalysts with different platinum loadings (wt %): (△) 0.1; (☆) 1.0; (●) 1.9; (○) 3.6; (□) 4.7; (★) 6.6; (▲) 8.7.

long-term deactivation is controlled by the deactivation of the metal function. A possible explanation for this disagreement is that the results were obtained under various conditions. These different conclusions were recently discussed by Parera and Figoli (43), and they concluded that although the acid function deactivation determines the long-term deactivation and the end of the cycle, these values can be different for catalysts dependent on the composition of the metal function.

Recent information obtained by X-ray absorption fine-structure spectroscopy (EXAFS) bears out the rapid formation of a carbon-platinum bond when a hydrocarbon-hydrogen mixture is injected over a Pt/Al$_2$O$_3$ catalyst at 460°C (44). When time on stream increases, the carbon-platinum bond remains unmodified while turnover rates and selectivities indicate evidence for deactivation. The authors concluded, in agreement with previous work (45), that the deactivation with time is due to buildup of a multilayer of carbon.

Coke is made up of polynuclear aromatic compounds (46, 47) and the number of rings in these compounds increases as a function of time in opera-

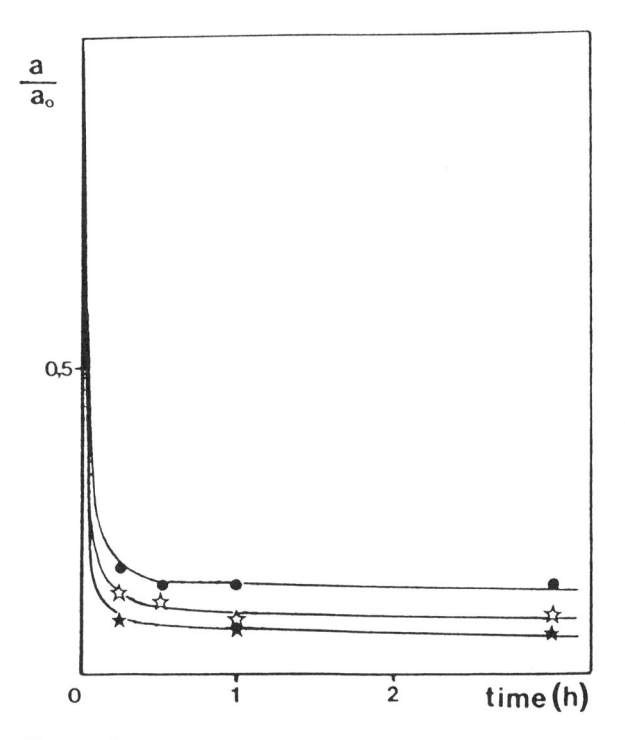

**Figure 3** Evolution of the relative activity in benzene hydrogenation with the time of coking. (From Ref. 10.) Catalysts with different platinum loadings (wt %): (●) 1.9; (☆) 1.0; (★) 6.6%

tion (46). Consequently, the hydrogen content of the hydrocarbon deposit decreases when the coking time increases (8, 21, 47) and the coke gradually becomes more graphitic (48, 49).

The crystalline evolution of coke deposited on the catalyst during a commercial cycle of a naphtha reformer was characterized by X-ray diffraction, TEM, and determination of the H/C ratio (25). Increasing time on stream decreases the hydrogen-to-carbon ratio and the degree of graphitization of the coke at high coverage. This was explained by limitation of the number of parallel aromatic layers in the growing coke arrangements due to the decrease in the pore diameters induced by these arrangements. This result is in accordance with the pore plugging at high degrees of coke deposition (14, 50).

## IV. ROLE OF THE METALLIC PHASE IN COKING OF BIFUNCTIONAL CATALYSTS

In the new generation of reforming catalysts, platinum is always modified by the addition of a second metallic element, such as Re, Ir, Sn, or Ge. The essential contribution of these new catalysts has been greater stability of the catalyst composite. However, the individual characteristics of the platinum function (accessibility, loading) remain important parameters for the coking reaction on bifunctional catalysts.

### A. Platinum Accessibility

Figure 4 shows that the number of carbon atoms deposited on the metal per accessible atom of platinum depends on the dispersion of the metallic phase. Thus, small platinum particles are more resistant to coke deposition than are large particles (10, 51, 52).

In the same way, Lankhorst et al. (53) showed that $Pt/SiO_2$ catalysts with low metallic dispersion are more sensitive to deactivation than are well-dispersed ones. This result is in agreement with that of Somorjai and co-workers (54,55), who proved that coke is stabilized more easily on planes than on the corners and the edges of metallic crystallites.

The metal dispersity also has an influence on the coke deposits on the whole catalyst (metal and support), because the amount of coke is higher on the well-dispersed catalysts (10, 56). Under certain conditions, it was shown that the amount of coke on the catalyst is nearly proportional to the number of accessible platinum atoms (10).

The toxicity of coke for the metal function also depends on platinum accessibility; indeed, at the same carbon content, coke is more toxic for the metal

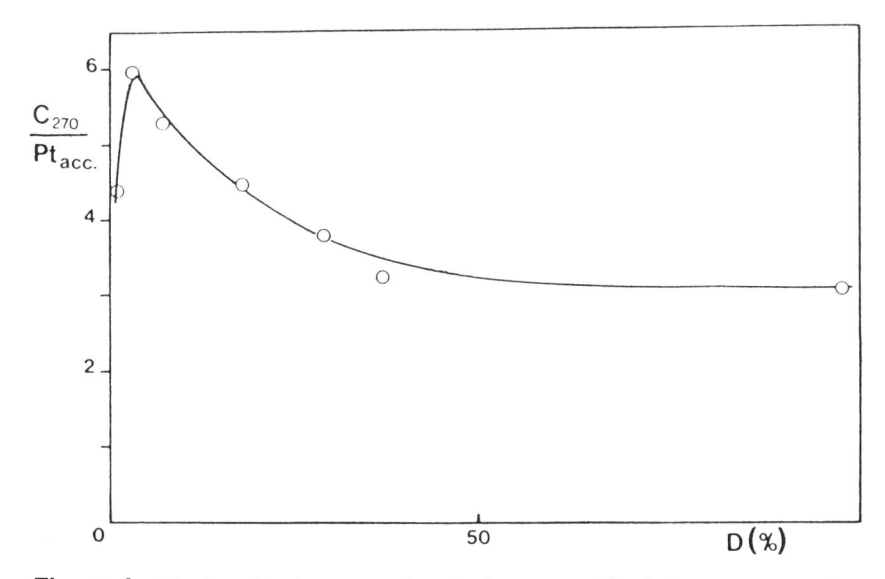

**Figure 4** Number of carbon atoms deposited per accessible platinum atom as a function of the metallic dispersion. (From Ref. 10.)

activity on the less dispersed samples (10, 56). Extractions of these catalysts coked under cyclopentane have shown that the highest toxicity of carbonaceous residues was linked to the presence on the metallic function of greater amounts of light polyaromatic compounds, which are the precursors of more graphitic and more toxic coke.

## B. Platinum Loading

Addition of the platinum function to alumina promotes coke deposition on the catalyst for typical hydrocarbon reactions (51, 57, 58). Thus, the metallic function of a bifunctional catalyst acts on the coking reaction, first by its catalytic dehydrogenation properties, yielding olefins that can polymerize on the support, and then by further elimination of hydrogen from the coke deposits. The hydrogen could migrate from the support to the metal by an inverse spillover effect (51).

Increasing the platinum content, with the same metallic dispersion, increases the amount of coke (22, 37, 47, 58, 59) on the whole catalyst; however, such an effect was found to be less important in the coking reaction by cyclopentane at high temperature (500°C) (22,47). Likewise, the hydrogen-to-carbon ratio is higher when the platinum loading increases (59).

## C. Modification of Platinum by Additional Components

On monometallic catalysts supported on alumina, Pt leads to more hydrocarbon deposits than Ir, Re, Ge, or Sn (60,61). Ponec and co-workers (62) stated that it is more difficult to self-poison Ir with carbon than Pt. Furthermore, Pt and Ir produce poorly organized coke mainly on metal sites, while for Re, Ge, and Sn the carbon is more graphitic and is mainly located on the support (60). The most remarkable progress that has been made in improving the stability of reforming catalysts is due to the use of these different components with platinum.

### 1. Platinum-Rhenium Catalysts

The metallic function of the bifunctional $Pt/Al_2O_3$ catalyst was modified by the addition of rhenium (63), which decreases the coke formation but increases the gas formation (29,51,61,64–74). On the other hand, comparison of a balanced (0.3% Pt, 0.3% Re) and a skewed (0.2% Pt, 0.4% Re) metal function catalyst coked commercially in a naphtha reformer showed that coke deposition is less on the metal of the skewed-metal catalyst (75). This catalyst has a better stability, allowing a 40% increase in the length of the operation cycle as compared with the balanced one under similar operation conditions (76).

The lower coke formation on bimetallic Pt-Re catalysts has been ascribed by some authors to the ability of Re to destroy coke precursors by hydrogenolysis (69,71,74). Zhorov et al. (72) and Burch and Mitchell (73) stated that Pt-Re is more active than Pt in hydrogenating the active precursor cyclopentadiene into cyclopentene, which is less active for the formation of coke. Barbier et al. (66) and Parera et al. (64, 68) suggested that the role of rhenium is to decrease the dehydrogenation capacity of platinum, thus decreasing coke formation. Another explanation for sulfided Pt-Re catalysts was proposed by Sachtler (49) and other authors (77–80): ReS entities divide large platinum ensembles into smaller ones and hence inhibit the coking reaction, because small particles are less sensitive to coke fouling.

When comparing the activities for benzene hydrogenation of coked Pt/$Al_2O_3$ and Pt-Re/$Al_2O_3$ catalysts, it appears that coke deposited on Pt-Re shows weaker toxicity toward the metallic activity of the catalyst than coke deposited on Pt (51, 61). This result can be explained according to various authors (49,54,81,82) by the alteration of the platinum site, which could cause a change in the nature of the deposited coke. Thus, rhenium could induce a rearrangement of the coke deposits that would become less toxic. Determination of the H/C ratio confirms that the nature of carbon deposits is different on Pt-Re and Pt catalysts, as the coke is more dehydrogenated on Pt-Re (29,65).

## 2. Platinum-Iridium Catalysts

Iridium, used as a component for improved reforming catalysts (74, 83–87), decreases coke deposition on bifunctional catalysts (61,65,66,74,88,89). Furthermore, it has been shown that the amount of carbon deposited on the catalyst during dehydrocyclization of $n$-heptane diminishes when the iridium/platinum atomic ratio increases at constant platinum content (28). This result agrees with the observations of several authors (88,90) and can be interpreted in terms of an increased ability of the Pt-Ir couple to destroy coke precursors by hydrogenation (89,90) or hydrogenolysis (74). Indeed, iridium is known to favor hydrogen spillover (91), and as a result, the availability of adsorbed hydrogen on the surface should increase and therefore the chance to hydrogenate coke precursors should increase. It has also been shown that iridium is more active than platinum for the reaction of coronene hydrogenolysis (92), coronene being a polynuclear aromatic coke intermediate (21). Ramaswamy et al. (88) suggested that the "dilution" of Pt by an element with a relatively lower dehydrogenation activity, such as Ir, leads to lower surface concentrations of coke precursors on Pt-Ir/$Al_2O_3$ and hence to reduced fouling rates.

Nevertheless, after coking under normal pressure, the toxicity of coke, defined by its fouling effect on the pure metallic reaction of benzene hydrogenation, is higher on Pt-Ir than on Pt, although the carbon deposits are lower on the bimetallic catalyst (61,65,92). Thus, decreasing the amount of coke is not adequate to diminish the effects of coking. Localization and nature of coke must also be important factors.

Determination of the H/C ratio has shown that coke deposited under normal pressure on the whole catalyst is more dehydrogenated on Pt-Ir than on Pt (65).

However, extractions by benzene of these two coked catalysts showed that more light polyaromatic compounds are present on the bimetallic catalyst. These polyaromatic compounds would be adsorbed on the metallic surface and would be responsible for the catalyst deactivation as coke precursors. Indeed, it was shown by secondary ion mass spectrometry and Auger electron spectroscopy studies of Pt, Ir, and Pt-Ir foils that the carbonaceous deposits on Pt-Ir contain more hydrogen and are less graphitic than on Pt (93).

### 3. Platinum Modified by Other Components

The other components used to improve the stability of platinum have been mainly tin (73,94–100), germanium (28), and lead (66,94). Generally, the beneficial effect of the second component is not related to a decrease of the amount of carbonaceous deposits but rather to modification of the coke deposition. Indeed, the proportion of carbon deposited on the support of the bimetallic catalysts is greater than that on the platinum catalysts. Lieske et al. (97) showed that the addition of tin to platinum gave coke precursors which were more mobile and could migrate more easily to the alumina. However, the reduction of coke formation by addition of tin also depends on the extent of interaction between platinum and tin (96).

## V. EFFECT OF THE SUPPORT ON COKING OF BIFUNCTIONAL CATALYSTS

In the course of reforming, the catalyst support plays a large part in the reaction of coke deposition, since it has been shown by temperature-programmed combustion (TPC) that coke is deposited mainly on the alumina support (5,7,9–11). Table 2 gives the amounts of coke accumulated on different platinum catalysts with comparable metallic accessibilities but supported on various oxides. The amount of carbon deposited during the reaction of cyclopentane parallels the acidity of the support.

The effect of the acidity of the support was observed during the coking of a bifunctional Pt/Al$_2$O$_3$ catalyst (58). Figure 5 compares TPC curves for various catalysts. The combustion of coke deposited under the same experimental conditions on the Pt/Al$_2$O$_3$ catalyst after the basic sites have been neutralized by

**Table 2** Influence of the Nature of the Support on the Coking Reaction of Various Pt Catalysts by Cyclopentane

| Catalyst | Pt/Al$_2$O$_3$ | Pt/TiO$_2$ | Pt/SiO$_2$ | Pt/MgO |
|---|---|---|---|---|
| Amount of C (%) | 2.28 | 1.56 | 0.09 | 0.08 |

*Source*: Ref. 51.

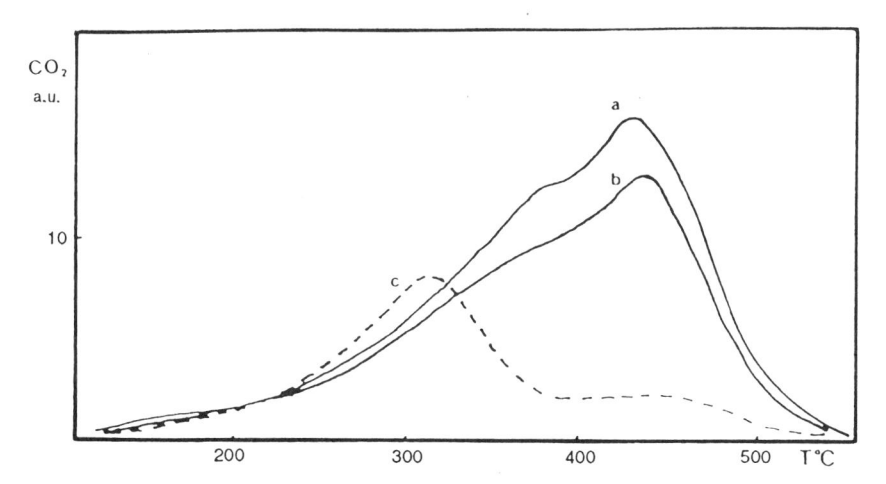

**Figure 5** Envelopes of the $CO_2$ peaks during the oxidation of coked catalysts. (a) 0.59% $Pt/Al_2O_3$; (b) the same catalyst treated with $H_3BO_3$; (c) the same catalyst treated with KOH. (From Ref. 58.)

boric acid is comparable to the combustion of coke deposited on $Pt/Al_2O_3$, the only difference being the result of a slight decrease in the amount of deposited carbon. Finally, for the catalyst neutralized by potassium hydroxide, a clearly defined peak at 310°C (coke on the metal) is observed but the high-temperature peak (coke on the support) has almost disappeared.

The chloride content of the catalyst also affects the acidity of the support and consequently the coke deposition. Industrial catalysts work with a chloride content which is typically maintained in the range of 0.8–1.1 wt %. Chloride is necessary for the initial dispersion of platinum and in the maintenance of its dispersion (28). However, Svajgl (101) demonstrated that an excess of chloride leads to excessive cracking and coking activity. The same effect on coke deposition was displayed during the coking by anthracene, at 400°C, of pure alumina modified by chloride or potassium hydroxide addition. Table 3 shows that the coking level increases with the acidity of the alumina.

**Table 3** Effect of the Acidity of the Alumina on the Amount of Coke Deposited During the Coking Reaction by Anthracene at 400°C

| $Al_2O_3$ | 0.6% K | 0.2% K | Pure | 0.5% Cl | 1.0% Cl |
|---|---|---|---|---|---|
| Acidity (meq $H^+$/g × $10^2$) | 33.2 | 37.1 | 44.1 | 51.2 | 57.8 |
| % C | 0.10 | 0.28 | 0.50 | 0.76 | 1.14 |

According to various authors, the minimum amount of coke on a reforming catalyst is maintained when the catalyst chloride content is about 0.7–0.9 wt % for Pt/Al$_2$O$_3$ and Pt-Re/Al$_2$O$_3$ (6,102–105) and 0.9–1.1% for Pt-Ir/Al$_2$O$_3$ (106). With this chloride content, the deactivation rate is kept at a minimum. This behavior has been attributed to a maximum in hydrogen spillover on the surface of the alumina, thus hydrogenating coke precursors (107).

On the other hand, it is important to note that, in addition to the large increase in the acidity, many studies (108–111) indicate that Cl$^-$ influences the metal function as well; thus, the presence of Cl$^-$ will most likely affect coking on both metal and support. Sachtler and co-workers (112) reported that chloride modifies the localization of coke; the higher the Cl$^-$ loading the higher the ratio of coke deposited on the support with respect to coke deposited on the metal. In the same way, it was pointed out that during the coking reaction by phenanthrene of Pt/Al$_2$O$_3$ at 400°C (113) (Figure 6), addition of 1.0 wt % Cl$^-$ decreases the amount of coke deposited on the metallic function (peak at low temperature).

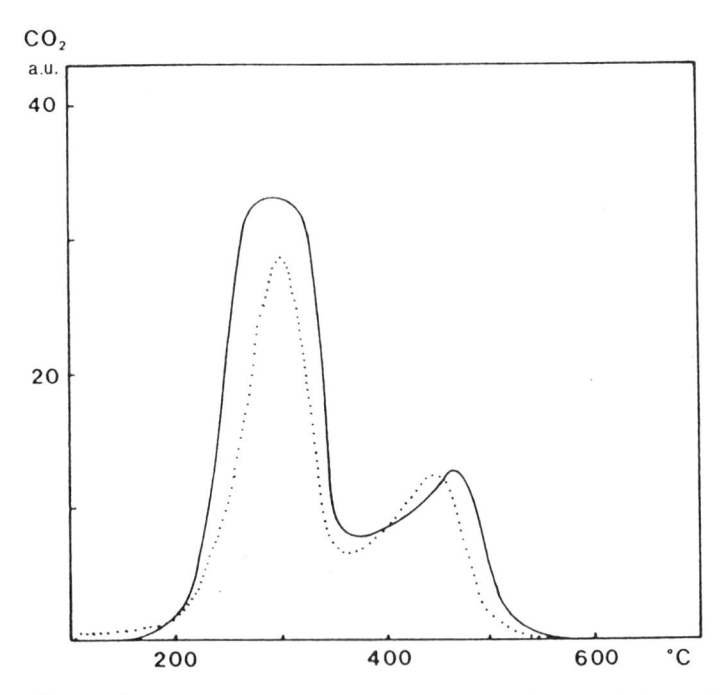

**Figure 6** Effect of chloride on the amount of coke deposited on a Pt/Al$_2$O$_3$ catalyst during the coking reaction by phenanthrene at 400°C. (From Ref. 13.) (—) Without chloride; (· · · ·) with chloride.

## VI. EFFECTS OF THE REACTION CONDITIONS ON COKING DEACTIVATION

The main parameters that may affect the deactivation of reforming catalysts by coke fouling are the hydrogen and hydrocarbon pressures, the space velocity, the temperature of the reaction, and the composition of the feed.

### A. Hydrogen and Hydrocarbon Pressures, Space Velocity

From a strictly thermodynamic point of view, it would be best to operate reforming units at the lowest possible hydrogen pressure; however, catalysts are unstable under such operating conditions. Figure 7 shows that although deactivation is relatively slow at 50 bar, it becomes very fast under 10 bar of hydrogen pressure. This result is in accordance with the negative effect of the hydrogen pressure on the rate of coke fouling, with a negative power of 2 in the range 10–30 bar (114). In the same way, Barbier et al. (22,65) and Figoli et al. (115,116) have determined a negative order for the reaction of coke deposition with respect to the overall pressure, keeping the hydrogen-to-hydrocarbon ($H_2$/HC) ratio constant. Furthermore, the inhibiting effect of high pressure is found to be greater on Pt-Ir/$Al_2O_3$ than on Pt/$Al_2O_3$ and Pt-Re/$Al_2O_3$ catalysts (65).

On the other hand, high pressure alters the localization and the nature of coke. Barbier et al. (22,47,65) have shown that coke deposited at high pressure is more dehydrogenated (Figure 8) and more graphitic (Table 4), suggesting that it is located mainly on the support (8). Indeed, the results reported in Table 5 bear out that less coke is deposited on the metal at high coking pressure and

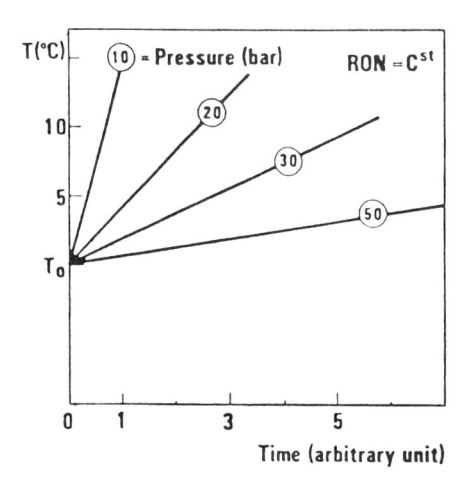

**Figure 7** Typical influence of pressure on stability. (From Ref. 22.)

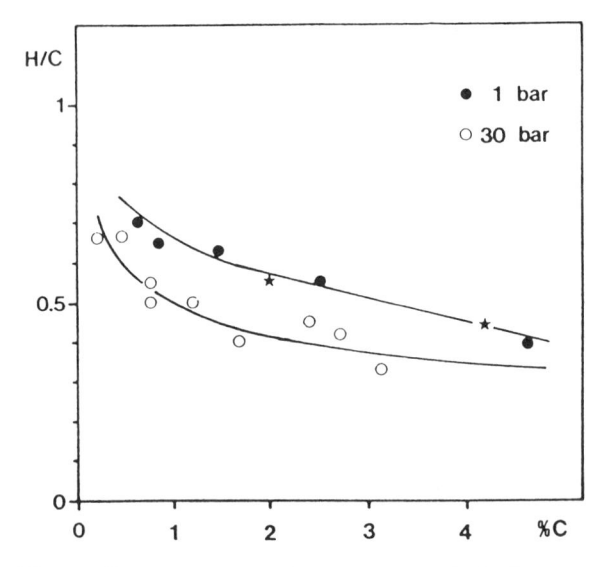

**Figure 8** Effect of working pressure on the hydrogen content of coke deposits during the coking reaction by cyclopentane on Pt/Al$_2$O$_3$. (From Ref. 22.) (●) 1 bar, 440°C; (★) 1 bar, 480°C; (○) 30 bar, 440°C.

**Table 4** Effect of the Coking Pressure on the Graphitization of Coke (Coking Agent, Cyclopentane)

| Coking conditions | C (%) | Extractable coke (%) | Graphitic coke (%) |
|---|---|---|---|
| 30 bar 440°C, 10 h | 0.50 | 8 | 92 |
| 10 bar 440°C, 1/2 h | 0.60 | 25 | 75 |

*Source*: Ref. 22.

**Table 5** Effect of Operating Pressure on the Toxicity of Coke for the Metallic Activity

| Cooking pressure (bar) | Time on stream (min) | C (%) | $a/a_0$ (relative activity in benzene hydrogenation) | Cm (coke on the metal, at/at metal) |
|---|---|---|---|---|
| 30 | 900 | 0.9 | 0.70 | 0.8 |
| 10 | 360 | 1.0 | 0.35 | 1.4 |
| 1 | 60 | 0.9 | 0.18 | 2.8 |

*Source*: Ref. 47.

that its toxicity is lower for the metallic activity, measured for the benzene hydrogenation reaction. The same conclusions were drawn by Pieck and Parera (117) when comparing a commercially coked catalyst (high pressure) to laboratory-coked catalysts (low pressure). The coke on the laboratory-coked catalysts is richer in hydrogen and covers the metallic function in a higher proportion, and as a consequence the catalytic activity for hydrogenation (metallic reaction) is more decreased. However, the decrease in the acidic activity produced by coking is similar on both catalysts (117).

At constant overall pressure, carbon deposition and therefore catalyst aging increase when the $H_2$/HC ratio decreases (115,118). However, a decrease in hydrogen pressure not only promotes aging but also modifies the reaction rate. For example, in the course of the dehydrocyclization of $n$-heptane, there is a maximum rate according to hydrogen pressure and that maximum depends on the hydrocarbon pressure (119,120). The structure of the coke also depends on the $H_2$/HC ratio, since Espinat et al. (26) reported that the increase in the hydrogen pressure on a monometallic Pt/$Al_2O_3$ catalyst gives carbonaceous deposits with a Raman spectrum quite similar to the one obtained for a Pt-Ir bimetallic catalyst. Thus, the increase in the hydrogen pressure would have the same effect as the addition of a second metal to platinum. This conclusion is in accordance with that of Barbier et al. (65), who stated that at atmospheric pressure addition of iridium or rhenium to platinum acts in the same way on coke deposits as increasing the working pressure on pure Pt/$Al_2O_3$ catalysts.

Finally, when decreasing the space velocity but keeping the $H_2$/HC ratio constant, the amount of coke increases mainly on the support (36,116). Thus, the effect of the decrease of the space velocity is similar to the effect of a decrease of the $H_2$/HC ratio at constant pressure.

## B. Temperature of the Reaction

Temperature is the operating variable used for activity maintenance in an industrial unit and therefore it is important to know its influence on coke fouling.

On platinum single-crystal surfaces it was shown that the morphology of the carbonaceous deposit appears to vary continuously from two-dimensional at low reaction temperatures (<277°C) to three-dimensional for temperatures higher than about 327°C (54). Thus, in the range of reforming temperatures, coke deposits will be always three-dimensional.

The amount of coke formed increases with increasing temperature (21,22,36,47,120–122) and this evolution can be explained by the effect of high temperatures on the number of unsaturated products which are coke precursors. The determination of the apparent energy of activation of coke deposition gives values from 85 to 170 kJ/mol. which depend on pressure (121) and the nature of the catalyst (22,120). For example, on Pt/$Al_2O_3$ catalysts with constant metal dispersions, the apparent activation energy decreases as the number of accessible metal atoms increases (Figure 9). As a consequence, an isokinetic temperature appears at around 520°C for the coking of Pt/Al$_2$O$_3$ by cyclopentane under a total pressure of 30 bar and with an H$_2$/HC ratio of 2.33.

Increasing temperature also accelerates the graphitization process of coke (36,53,123). Figure 10 shows that the fraction of reversible coke (partially hydrogenated) decreases as the temperature and the amount of coke increase. In the same way, Barbier et al. (21) pointed out that coke grows mainly on the support with increasing temperature and that it is more dehydrogenated. But on catalysts with comparable coke coverages (Pt/Al$_2$O$_3$ catalysts coked at different

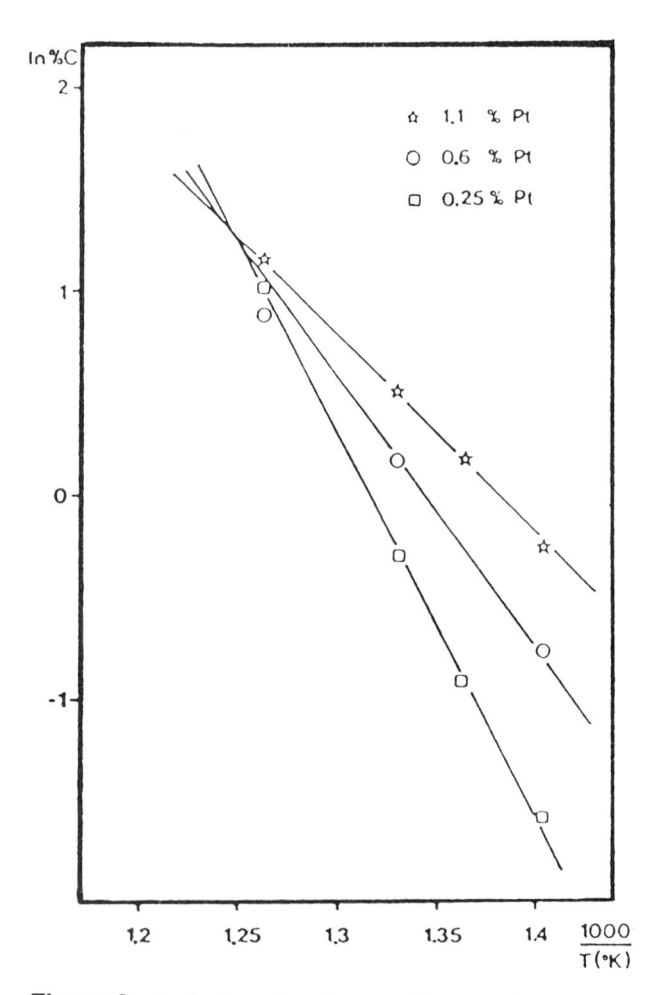

**Figure 9** Arrhenius plot of the coking reaction on different Pt/Al$_2$O$_3$ catalysts. (From Ref. 22.)

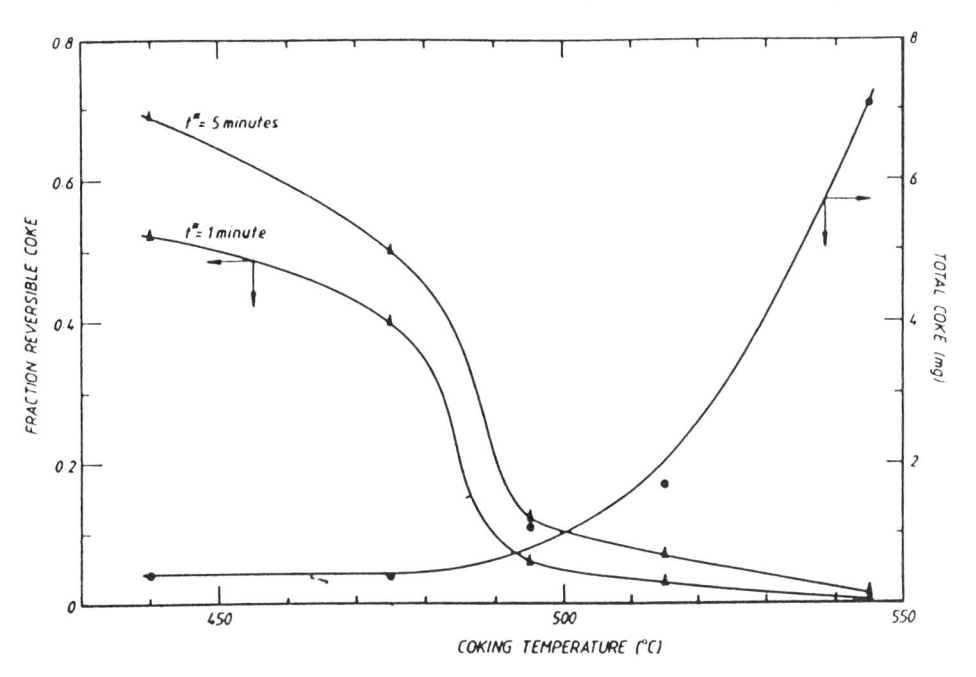

**Figure 10** Fraction of reversible coke on a Pt/Al$_2$O$_3$ catalyst at different temperatures. (From Ref. 23.)

temperatures for various times), the chemical nature of carbonaceous deposits is the same, as indicated by either the H/C ratio or the level of graphitization (Table 6). Moreover, an increase in temperature does not change the location of the coke (on the metal or on the support), because the catalytic activities of the metal, for a constant coke coverage, are comparable on the different samples whatever the coking temperatures may be.

**Table 6** Effect of the Coking Temperature on the Characteristics of Coke

| Coking temperature (°C)[a] | Carbon (%) | Time on stream (min) | H/C | $a/a_0$[b] | Extractable coke (%) | Graphitic coke (%) |
|---|---|---|---|---|---|---|
| 440 | 1.9 | 210 | 0.56 | 0.18 | 12 | 88 |
| 460 | 2.0 | 90 | 0.54 | 0.20 | 10 | 90 |
| 480 | 1.9 | 45 | 0.55 | 0.19 | 12 | 88 |

[a] Coking with cyclopentane at atmospheric pressure; hydrogen/hydrocarbon ratio 2.33.
[b] Relative activity in benzene hydrogenation.
Source: Ref. 22.

## C. Nature of Hydrocarbons in the Feed

Reformer feedstocks vary widely in their boiling range and hydrocarbon type distribution (paraffins, naphthenes, and aromatics), depending on their origin. Generally, heavier cuts produce more coke. It was stated (124) that for cuts of around 204°C end boiling point there is a 1.6 to 2.3% decrease in cycle length per °C increase in feed end point, whereas at an end boiling point around 216°C the decrease in cycle length is 2.1–2.8% per °C. The same conclusions were drawn by Figoli et al. (125), with higher coke depositions and catalyst deactivations for cuts of very high boiling points. But these authors also showed that cuts with a very low boiling point range produced the same effect and they concluded that coke deposition is not determined solely by the boiling point range of the feed.

Indeed, the cuts with the lowest boiling point range have a high content of cyclopentanic compounds, which are great coke producers (118, 126–129). It was pointed out that certain hydrocarbon structures had a particularly high coking effect; for example, Myers et al. (118) showed that hydrocarbons such as cyclopentane, $n$-dodecane, and methylnaphthalene turned out to be very good coke precursors. Likewise, adding low amounts of dicyclic aromatics to $n$-paraffins ($n$-hexane or $n$-heptane) increases catalyst aging significantly (120). Zhorov et al. (129) classified various hydrocarbons according to their likelihood to cause carbon deposition: cyclohexane (1 = reference) < benzene (3) < ethylbenzene (23) < $n$-hexane (35) < $n$-nonane (41) < methylcyclopentane (130) < indane (250) < methylcyclopentene (370) < cyclopentadiene (670).

The same scale was determined by Cooper and Trimm (127), who studied the coke fouling on $Pt/Al_2O_3$ catalysts for different hydrocarbons having six carbon atoms (Figure 11): methylcyclopentane > 3-methylpentane > $n$-hexane > 2-methylpentane > benzene > cyclohexane.

Beltramini et al. (130) and Parera et al. (68) examined the effect of adding different pure hydrocarbons to a naphtha cut on deactivation by coke. When doping with normal paraffins, the authors found that coke deposition and catalyst deactivation were at a minimum for $n$-heptane. Considering aromatics with a paraffinic chain, the tendency to produce coke increases with the possibility that the chain can form a five-carbon-atom ring (indenic structure). When doping with naphthenes, if the ring has five carbon atoms, coke formation and catalyst deactivation are higher than for six-carbon-atom rings. Furthermore, the coke deposition increases in the order cyclopentane < cyclopentene < cyclopentadiene (130). These results are in accordance with those of Barbier et al. (66), who showed that cyclopentadiene is the coking agent responsible for the deposition of coke during the cyclopentane reaction.

It should be noted that coke deposition on bifunctional catalysts occurs mainly on the acidic sites of the support. Therefore, according to Appleby et

*Deactivation by Coking* 297

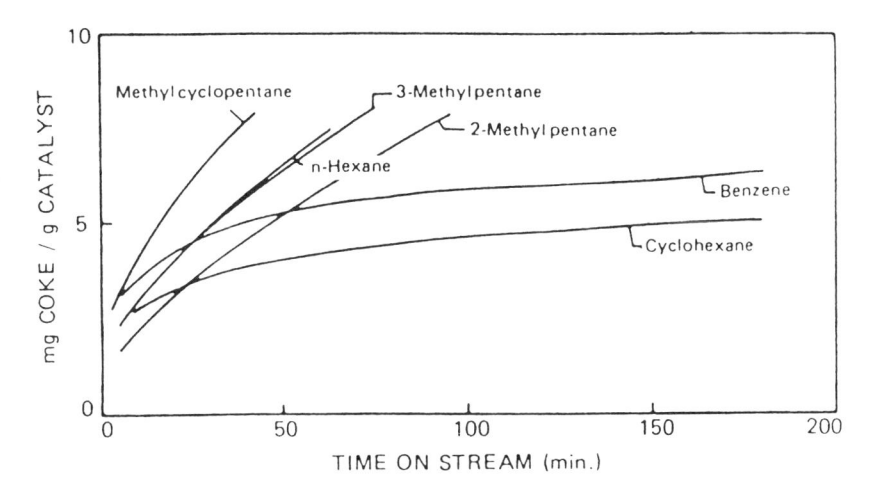

**Figure 11** Coke deposition on Pt/Al$_2$O$_3$ for various hydrocarbons at 500°C. (From Ref. 27.)

al. (2), the acid-base properties of the catalyst-reagent pair would be involved in defining the importance of coke deposits. The amount of coke deposited on a cracking (acid) catalyst increases with an increasing basicity of the processed hydrocarbons (Figure 12). In studying the coking reaction on supports of variable acidities with hydrocarbons of various basicities (anthracene, phenanthrene), it was shown that the higher the molecule basicity (anthracene), the higher the coke deposits (113). The addition of platinum to the various supports enhances coke deposits whatever the hydrocarbon may be. However, the nature and the location of coke depend on the nature of the coking agent. Thus, on Pt/Al$_2$O$_3$, carbonaceous deposits are less dehydrogenated and are located mainly on the metal or in its vicinity as in the case of phenanthrene. However, when coke is more graphitic, it is located on the support as in the case of anthracene (113) (Figure 13).

To conclude, the deactivation by coke is not only related to the molecular weight of hydrocarbons; it appears that cyclopentane structures and polycyclic aromatics with paraffinic chains are very effective coking agents. Moreover, it seems that coke deposition on the whole catalyst increases when the hydrocarbon basicity is high.

## VII. EFFECT OF SULFUR ON THE COKING REACTION

Sulfur compounds are recognized as poisons for reforming catalysts and they can cause a complete loss of activity (131). In order to avoid such deactivation, sulfur concentrations of less than about 1 ppm have been specified for reform-

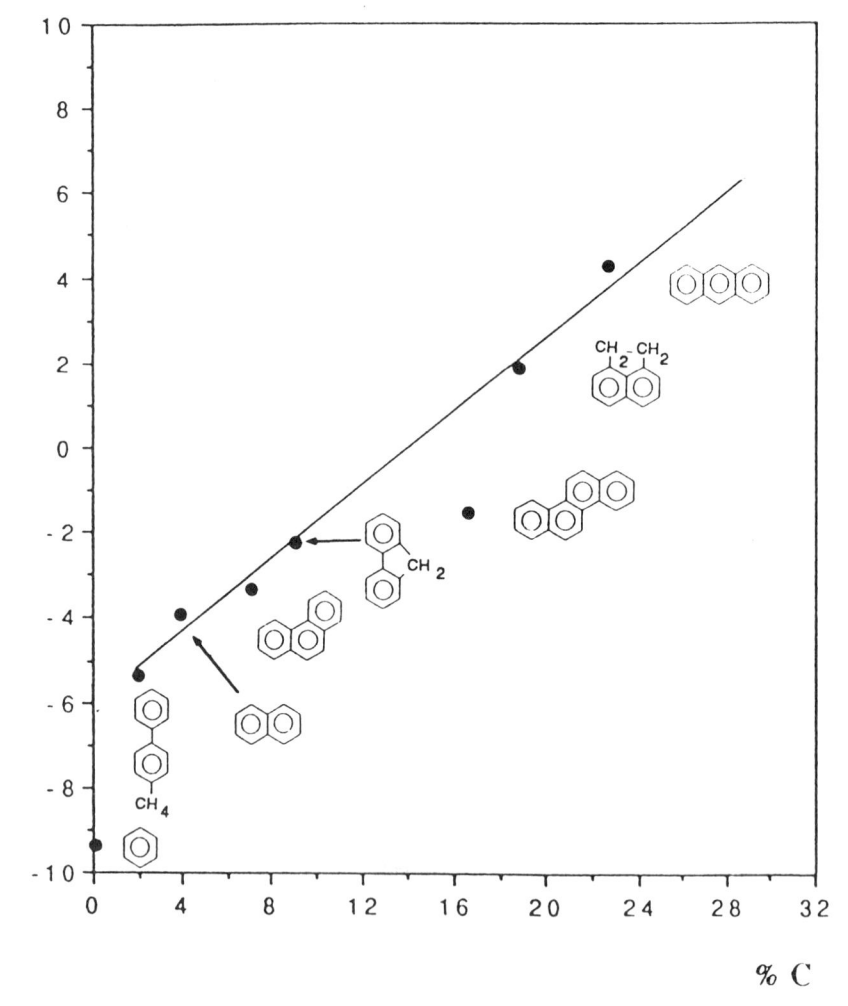

**Figure 12** Influence of the basicity of the hydrocarbon on coke formation in catalytic cracking. (From Ref. 2.)

ing feeds (63, 132, 133). On the other hand, sulfur is not totally unwanted, because a low concentration of sulfur in the feed or during the start-up of units can reduce excessive hydrogenolysis at the beginning of the run and improve the catalyst stability (63, 134–137). However, the presence of sulfur implies a higher reaction temperature for the same performance (Figure 14) (28), and therefore a higher coke-fouling rate.

# Deactivation by Coking

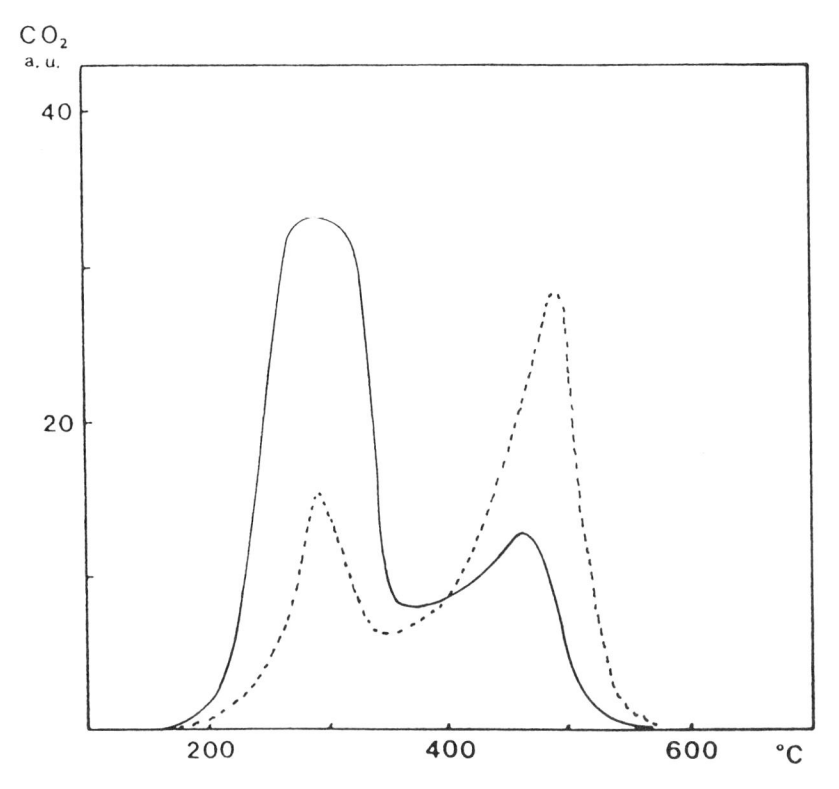

CO₂
a.u.

**Figure 13** Temperature-programmed combustion of coke deposited on a 0.6% Pt/Al$_2$O$_3$ catalyst during the coking reaction at 400°C by (—) phenanthrene and (···) anthracene. (From Ref. 113.)

With respect to the effect of sulfur on coke fouling, one must take into consideration, particularly, the severity of the reaction conditions (temperature and pressure), the nature of the feedstock, and the nature of the catalyst (102). For example, it is well documented that the beneficial effect of sulfur on catalyst stability is greater for Pt-Re/Al$_2$O$_3$ than for Pt/Al$_2$O$_3$ (79, 80, 112, 138–141). The superior activity maintenance of PtRe(S)/Al$_2$O$_3$ was explained by Sachtler and co-workers (49, 79, 112) using a model which assumes that adsorbed sulfur, fixed on Re, impedes the reorganization of hydrocarbonaceous fragments into pseudographitic entities which cause irreversible deactivation of the Pt function by carbonaceous deposits. It was also reported that Pt-Re-S/Al$_2$O$_3$ is more "tolerant" to coke deposition, with commercial operating coke levels of 20–25 wt % being obtained (142, 143).

Reports on the influence of sulfur on the coking rate are conflicting, as sulfur addition has been reported to result in enhanced (64, 144–146), unchanged

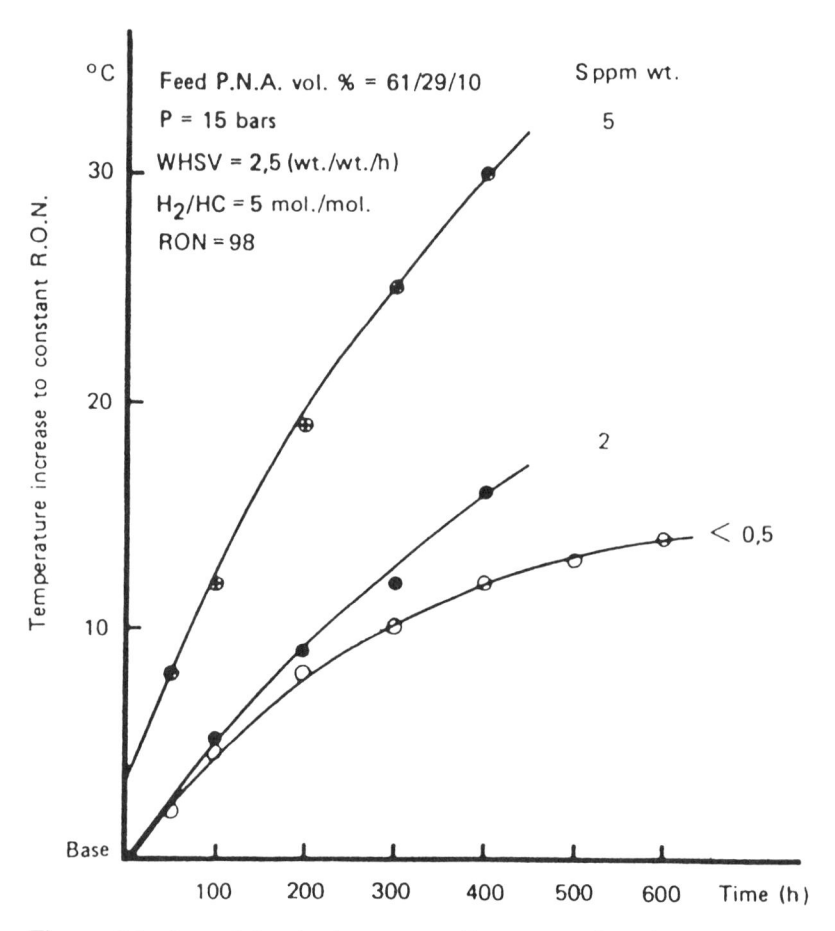

**Figure 14** Rate of deactivation versus sulfur content of the feed; monometallic Pt/Al$_2$O$_3$ at approximately 500°C. (From Ref. 28.)

(64, 146), or decreased (112, 141, 146) coke deposition. This apparent discrepancy can be explained, first, by the time on stream; for example, Figure 15 shows that Pt-Re-S/Al$_2$O$_3$ produces less coke during the first 3 h but more coke at the end of the run compared to the unsulfided catalyst (147). Second, the effect of sulfur on coke deposits is dependent on the nature of the feedstock; for example, with methylcyclopentane, sulfurization increases the amount of coke whatever the nature of the metallic phase of the bifunctional catalyst (64, 146). According to these results, Barbier and Marécot (145) showed that presulfurization of Pt/Al$_2$O$_3$, Ir/Al$_2$O$_3$, and Pt-Ir/Al$_2$O$_3$ catalysts increases coke deposition during the coking reaction with cyclopentane. On the

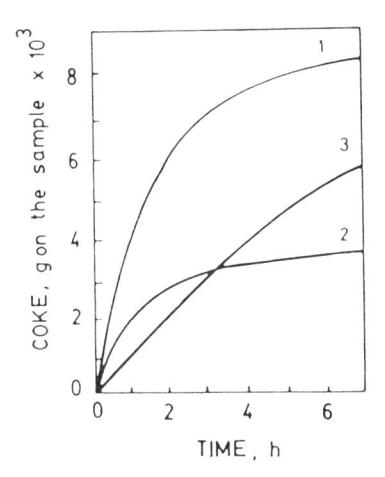

**Figure 15** Methylcyclopentane reforming. Coke formation on 0.2 g of catalyst as a function of time (1, Pt/Al$_2$O$_3$; 2, Pt-Re/Al$_2$O$_3$; 3, Pt-Re-S/Al$_2$O$_3$). (From Ref. 47.)

other hand, n-hexane and n-heptane lead to lower carbonaceous deposits when the catalysts are presulfided (112, 146). The difference between cyclopentane cycles and n-hexane or n-heptane can be explained by the effect of sulfur on the formation of coke precursors. The dehydrocyclization of paraffins is a reaction that is more sensitive to sulfur poisoning than dehydrogenation, and therefore the formation of unsaturated cyclic compounds, which are the most efficient coke precursors, will be more strongly inhibited by sulfur with n-heptane or n-hexane than with cyclopentane. Using a commercial hydrotreated naphtha containing less than 1 ppm sulfur, Parera et al. (144) showed that, on all the catalysts, initial sulfurization increases the total amount of coke, bearing out the main role played by the C$_5$ naphthenic compounds in the coking reaction.

The nature and the location of coke deposits are also affected by the presence of sulfur on the metallic phase. Augustine et al. (112) reported that on Pt-Re/Al$_2$O$_3$, the H/C ratio of the carbonaceous layer is much higher for presulfided samples, in accordance with lower reorganization of deposits into pseudographitic entities. On the other hand, Parera et al. (144, 147) showed that, after sulfurization, the coke is less hydrogenated and is localized mainly on the support. The same conclusion was drawn by Barbier and Marécot (145), who stated that adsorption of the coke precursor (cyclopentadiene) is less stabilized on sulfided platinum, leading to a higher production of olefinic compounds in the gas phase and therefore to more extensive coke deposition on the support as a result of polymerization on the acidic sites of the alumina. In addition, such lower binding energy between cyclopentadiene and the sulfided metallic phase decreases the lifetime of adsorbed unsaturated hydrocarbons and, therefore, their polymerization on the metallic surface area. Thus, sulfur inhibits the deactivation of the metal and hence increases the stability of bifunctional catalysts when chloride addition in the feed allows constant regeneration of acidic sites.

## VIII. EFFECT OF CARBON DEPOSITS ON REFORMING REACTIONS

A comparative study of different metallic reactions such as hydrogenolysis of cyclopentane, hydrogenation of benzene, and exchange of benzene with deuterium on partly deactivated Pt/Al$_2$O$_3$ catalysts has shown that the toxicity of coke is higher for the hydrogenolysis reaction (5). This result agrees with other studies (148–150) and can be explained by a decrease in the number of suitable ensembles for this reaction due to site blocking by coke, especially on the more active high-coordination-number face atoms on crystallites.

The transformation reactions of $n$-heptane, methylcyclohexane, and toluene were studied at 430°C (151). Among the reactions that occur only on metal sites, Table 7 shows that those which induce the cleavage of the carbon-carbon bond (dealkylation of toluene) are the most sensitive to coke. Considering the reactions that occur on acidic sites, their sensitivity to coke deposits is greater if they occur on highly acidic sites. For instance, in the dismutation of toluene, 210 g of catalyst is deactivated by the deposition of 1 g of coke, compared to only 30 g of catalyst being deactivated in the isomerization of 2,3-dimethylbutene-1.

**Table 7** Initial Toxicity of Coke (in g of Deactivated Catalyst per g of Deposited Coke) at 430°C

| Reagent | Reaction | Toxicity | Mechanism[a] |
|---|---|---|---|
| $n$-Heptane | Dehydrogenation | 430 | M |
| | Cracking | 300 | B + M |
| | Isomerization | 125 | B + M |
| | Coke formation | 360 | B + M |
| Methylcyclohexane | Dehydrogenation | 110 | M |
| | Coke formation | 720 | B + M |
| Toluene | Dealkylation | 700 | M |
| | Dismutation | 210 | A |
| | Coke formation | 750 | B + M |
| 3,3-Dimethylbutene-1 | Isomerization | 30 | A |

[a] B, bifunctional catalysis; M, metallic catalysis; A, acidic catalysis.
*Source*: Ref. 151.

*Deactivation by Coking*

To conclude, coke deposits are poisons that can alter the selectivity of bifunctional catalysts, as they show different toxicities according to the reaction taken into consideration. This can be explained by the fact that, on the metal, the coke settles preferentially on the hydrogenolysis sites and then presents major toxicity toward this reaction. On the support, the coke is formed on the sites with the greatest acidity, strongly inhibiting reactions such as dismutation and dehydrocyclization while isomerization of *n*-alkanes is less disturbed.

## IX. MECHANISM OF COKE FORMATION

Coking of bifunctional catalysts results from accumulation of coke on the metal and on the support, and therefore two mechanisms of coke formation have to be proposed.

On metal sites, two typical models for production of coke were set up. The first model involves a series of fragmentation reactions and successive dehydrogenation reactions which lead to the formation of carbon atoms, and these atoms (or partially hydrogenated intermediates) may combine to form more graphitic and more toxic coke deposits (148, 152). The second mechanism suggests that the routes of coke deposition are based on polymerization reactions with the formation of different types of carbonaceous deposits on the metal sur-

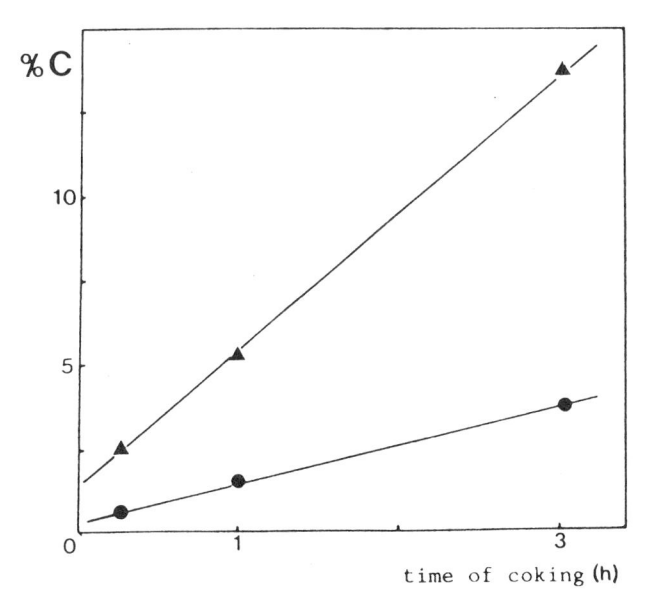

**Figure 16** Coking with a mixture of cyclopentane, cyclopentene, and cyclopentadiene in thermodynamic equilibrium. (From Ref. 51.) (●) $Al_2O_3$; (▲) $Pt-Al_2O_3$.

face (10, 61, 148, 152, 153). For example, large platinum crystallites increase the rate of coke deposition because of their ability to give higher stabilization for adsorbed cyclopentadiene precursor produced during the reaction (10).

On acid sites, it is assumed that coke arises from the polymerization of dehydrogenated intermediates generated by the metallic function (46, 51, 58, 129, 148, 154). In the case of the coking reaction with cyclopentane, Barbier et al. (51, 58) proposed an adaptation of the mechanism of condensation of cyclopentene (155) to explain the formation of polyaromatic compounds. Thus, cyclopentane can lead to naphthalene or to heavier polyaromatic compounds through the following sequence of reactions:

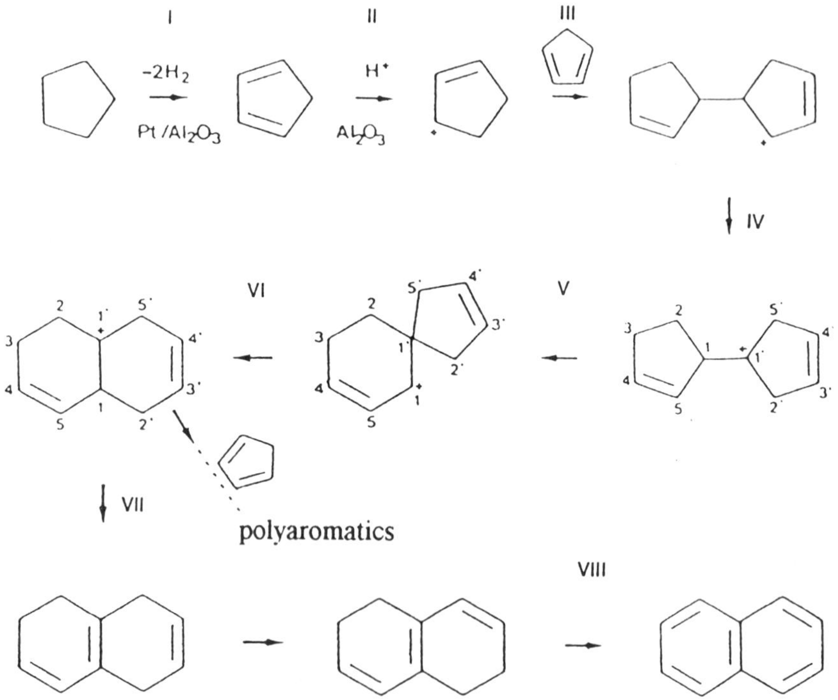

To conclude, the metallic function of a bifunctional catalyst affects the coking reaction, first by its catalytic dehydrogenation properties, yielding olefins that can polymerize on the support. The second role of the metal is to destroy coke precursors at high temperatures by its catalytic hydrogenolysis activity. Finally, Figure 16 shows that the metal can act in a third way on coke deposition. A chlorinated alumina and a Pt/Al$_2$O$_3$ catalyst were coked at 400°C with a mixture of cyclopentadiene, cyclopentene, and cyclopentane at thermodynamic equilibrium concentrations. Under these conditions, the dehydrogenation properties of platinum are unable to explain the increased coke deposition on Pt/Al$_2$O$_3$ in comparison with the coke deposited on pure chlorided Al$_2$O$_3$. Such a result can be explained by assuming that the metal promotes coke accumulation on the support, for example, by elimination of hydrogen (step VIII in the preceding sequence of reactions), yielding stable dehydrogenated carbonaceous deposits. The hydrogen would migrate from the support to the metal by an inverse spillover effect.

## X. CONCLUSIONS

Although numerous studies of the deactivation by coking of industrial or model reforming catalysts have been carried out, interpretations of the results sometimes diverge. The reason is that coking reactions are very complex and their rate can be defined only by the total amounts of coke deposited. However, this total coke loading is due to a combination of the formation of coke precursors, their polymerization (on the support or on the metal), and their destruction by metallic or acidic mechanisms.

In spite of such complexity, the following conclusions are well established: Coke can be deposited on the metal and on the acidic support.

The metallic surface, at the steady state of the reaction, is covered by an amount of coke that remains constant when coke continues to be accumulated on the acidic support.

The higher the metallic dispersion, the lower the coverage by coke of the metallic phase at steady state.

The catalytic properties of the metal are a responsible factor in the coking of bifunctional catalysts:

1. Olefins are produced and can be polymerized on the acidic sites of the support.
2. Polymers produced on the support can be stabilized through dehydrogenation by a reverse spillover of hydrogen.
3. Precursors of coke can be destroyed at high temperature by the metal (and also on the support by spillover of hydrogen).

Coke deposition on the support is defined by the acid-base properties of the catalyst-coking precursor pair.

Location of coke (on the metal or on the support) and nature of coke (light or graphitic) are more important parameters for catalyst stability than coke content.

Modification of Pt by Re or Ir decreases the amount of deposited coke and induces preferential deposition of coke on the support.

Modifications of Pt by Group IV components (Sn, Pb, Ge) or sulfurization of the metallic function also improve the reforming catalyst stability. This improvement is linked to preferential deposition of coke on the support.

In conclusion, in spite of the very low selectivity of catalytic reforming for the coking reaction (only one atom of carbon out of 200,000 activated by a Pt/$Al_2O_3$-Cl catalyst is transformed into nondesorbable coke), catalyst lifetime in the reforming process can still be enhanced by modification of the metallic phase. In fact, such improvements allow the use of reforming catalysts under the more severe reaction conditions needed for the production of lead-free gasoline.

## REFERENCES

1. R. G. Haldemann and M. C. Botty, *J. Phys. Chem. 63*, 489 (1959).
2. W. G. Appleby, J. W. Gibson, and G. M. Good, *Ind. Eng. Chem. Process Des. Dev. 1*, 102 (1962).
3. H. Noda, S. Tone, and T. Otake, *J. Chem. Eng. Jpn. 7*, 110 (1974).
4. N. V. Voikina, L. A. Makhlis, S. L. Kiperman, and O. K. Bogdanova, *Kinet. Katal. 15*, 657 (1974).
5. J. Barbier, P. Marécot, N. Martin, L. Elassal, and R. Maurel, in *Catalyst Deactivation* (B. Delmon and G. F. Froment, eds.). Elsevier, Amsterdam, 1980, p. 53.
6. J. M. Parera, N. S. Figoli, E. L. Jablonski, M. R. Sad, and J. N. Beltramini, in *Catalyst Deactivation* (B. Delmon and G. F. Froment, eds.). Elsevier, Amsterdam, 1980, p. 571.
7. R. Bacaud, H. Charcosset, M. Guénin, R. Torrescas-Hidalgo, and L. Tournayan, *Appl. Catal. 1*, 81 (1981).
8. J. Barbier, E. Churin, J. M. Parera, and J. Riviere, *React. Kinet. Catal. Lett. 29*, 323 (1985).
9. N. S. Figoli, in *Coke Formation on Metal Surfaces* (L. F. Albright and R. T. K. Baker, eds.). ACS Symposium Series no. 202, 1982, p. 239.
10. J. Barbier, G. Corro, Y. Zhang, J. P. Bournonville, and J. P. Franck, *Appl. Catal. 13*, 245 (1985).
11. J. M. Parera, N. S. Figoli, and E. M. Traffano, *J. Catal. 79*, 481 (1983).
12. E. Freund, J. P. Franck, and J. P. Bournonville, IFP, unpublished results.
13. P. Gallezot, C. Leclercq, J. Barbier, and P. Marécot, *J. Catal. 116*, 164 (1989).
14. D. Espinat, Thesis, ENSPM Rueil-Malmaison, France (1982).
15. D. Espinat, E. Freund, H. Dexpert, and G. Martino, *J. Catal. 126*, 496 (1990).
16. T. S. Chang, N. M. Rodriguez, and R. T. K. Baker, *J. Catal. 123*, 486 (1990).
17. E. S. Brodskii, I. M. Lakashenko, and V. G. Lebedevskaya, *Neftepererab. Neftekhim. 2*, 5 (1976).
18. D. A. Best and B. W. Wojiciechowski, *J. Catal. 47*, 11 (1977).
19. D. Eisenbach and E. Gallei, *J. Catal. 56*, 377 (1979).
20. P. E. Eberly, C. N. Kimberlin, W. H. Miller, and H. V. Drushel, *Ind. Eng. Chem. 5*, 193 (1966).

21. J. Barbier, E. Churin, and P. Marécot, *Bull. Soc. Chim. Fr. 6*, 910 (1987).
22. J. Barbier, E. Churin, P. Marécot, and J. C. Ménézo, *Appl. Catal. 36*, 277 (1988).
23. R. A. Bakulin, M. E. Levinter, and F. G. Unger, *Neftekhimiya 14*(5), 707 (1974).
24. M. Massai, S. Shimadzu, T. Sashiwa, S. Sawa, and M. Minura, *Proc. Int. Symp. on Catalyst Deactivation, Antwerp, 1980* (B. Delmon and G. F. Froment, eds.). Elsevier, Amsterdam, 1980, p. 261.
25. F. Caruso, E. L. Jablonski, J. M. Grau, and J. M. Parera, *Appl. Catal. 51*, 195, (1989).
26. D. Espinat, H. Dexpert, E. Freund, G. Martino, M. Gouzi, P. Lespade, and F. Cruege, *Appl. Catal. 16*, 343 (1985).
27. R. A. Cabrol and A. Oberlin, *J. Catal. 89*, 256 (1984).
28. J. P. Franck and G. P. Martino, in *Progress in Catalyst Deactivation*, (J. L. Figueredo, ed.). NATO Advanced Study Institute Series, E. App. Sci., no 54. Martinus Nijhoff, The Hague, 1982, p. 355.
29. J. M. Parera and J. N. Beltramini, *J. Catal. 112*, 357 (1988).
30. A. Voorhies, Jr., *Ind. Eng. Chem. 37*, 318 (1945).
31. C. G. Ruderhausen and C. C. Watson, *Chem. Eng. Sci. 3*, 110 (1954).
32. F. H. Blanding, *Ind. Eng. Chem. 37*, 318 (1945).
33. I. Dybkjaer and A. Bjorkman, *Proc. First Int. Symp. Chem. React. Eng.*, 219 (1970).
34. A. D. Afanaslev and R. A. Buganov, *Kinet. Katal. 15*, 666 (1974).
35. D. Van Zoonen, *Proc. 3rd Int. Congr. Catal. 2*, II, 1319 (1965).
36. J. M. Parera, N. S. Figoli, E. M. Traffano, J. N. Beltramini, and E. E. Martinelli, *Appl. Catal. 5*, 33 (1983).
37. J. N. Beltramini, T. J. Wessel, and R. Datta, in *Catalyst Deactivation* (B. Delmon and G. F. Froment, eds.). Elsevier, Amsterdam, 1991, p. 119.
38. C. A. Querini, N. S. Figoli, and J. M. Parera, *Appl. Catal. 52*, 249 (1989).
39. J. M. Parera, C. A. Querini, and N. S. Figoli, *Appl. Catal. 44*, L1 (1988).
40. J. Margitfalvi and S. Göbölös, *Appl. Catal. 36*, 331 (1988).
41. V. K. Shum, J. B. Butt, and W. M. H. Sachtler, *Appl. Catal. 11*, 151 (1984).
42. V. K. Shum, J. B. Butt, and W. M. H. Sachtler, *Appl. Catal. 36*, 337 (1988).
43. J. M. Parera and N. S. Figoli, *Catalysis, 9*, 65 (1992).
44. N. S. Guyot-Sionnest, F. Villain, D. Bazin, H. Dexpert, F. Le Peltier, J. Lynch, and J. P. Bournonville, *Catal. Lett. 8*, 283 (1991).
45. Hu Zi-Pu, D. F. Ogletree, M. A. Van Hove, and G. A. Somorjai, *Surf. Sci. 180*, 433 (1987).
46. B. C. Gates, J. R. Katzer, and G. C. Schuit, *Chemistry of Catalytic Processes*. McGraw-Hill, New York, 1979.
47. J. Barbier, in *Catalyst Deactivation* (B. Delmon and G. F. Froment, eds.). Elsevier, Amsterdam, 1987, p. 1.
48. J. Biswas, G. M. Bickle, P. G. Gray, and D. D. Do, in *Catalyst Deactivation* (B. Delmon and G. F. Froment, eds.). Elsevier, Amsterdam, 1987, p. 553.
49. W. M. H. Sachtler, *J. Mol. Catal. 25*, 1 (1984).
50. J. Biswas and D. D. Do, *Chem. Eng. J. 36*, 175 (1987).

51. J. Barbier, *Appl. Catal. 23*, 225 (1986).
52. J. Barbier, G. Corro, P. Marécot, J. P. Bournonville, and J. P. Franck, *React. Kinet. Catal. Lett. 28*(2), 245 (1985).
53. P. P. Lankhorst, H. C. de Jongste, and V. Ponec, in *Catalyst Deactivation* (B. Delmon and G. F. Froment, eds.). Elsevier, Amsterdam, 1980, p. 43.
54. S. M. Davis, F. Zaera, and G. A. Somorjai, *J. Catal. 77*, 439 (1982).
55. G. A. Somorjai and D. W. Blakely, *Nature 258*, 580 (1975).
56. P. Marécot, E. Churin, and J. Barbier, *React. Kinet. Catal. Lett. 37*(1), 233 (1988).
57. J. M. Parera, N. S. Figoli, G. E. Costa, and M. R. Sad, *React. Kinet. Catal. Lett. 22*(1–2), 231 (1983).
58. J. Barbier, L. Elassal, N. S. Gnep, M. Guisnet, W. Molina, Y. R. Zhang, J. P. Bournonville, and J. P. Franck, *Bull. Soc. Chim. Fr. 9–10*, 1250 (1984).
59. J. N. Beltramini, E. J. Churin, E. M. Traffano, and J. M. Parera, *Appl. Catal. 19*, 203 (1985).
60. J. N. Beltramini and D. L. Trimm, *React. Kinet. Catal. Lett. 37*(2), 313 (1988).
61. J. Barbier, G. Corro, Y. Zhang, J. P. Bournonville, and J. P. Franck, *Appl. Catal. 16*, 169 (1985).
62. J. G. Senden, E. H. Van Broekhoven, C. T. J. Wreesman, and V. Ponec, *J. Catal. 87*, 468 (1984).
63. H. E. Kluksdahl, U. S. Patent 3,415,737 (1968).
64. J. M. Parera, J. N. Beltramini, C. A. Querini, E. E. Martinelli, E. J. Churin, P. E. Aloe, and N. S. Figoli, *J. Catal. 99*, 39 (1986).
65. J. Barbier, E. Churin, and P. Marécot, *J. Catal. 126*, 228 (1990).
66. J. Barbier, L. Elassal, N. S. Gnep, M. Guisnet, W. Molina, Y. R. Zhang, J. P. Bournonville, and J. P. Franck, *Bull. Soc. Chim. Fr. 9–10*, I 245 (1984).
67. R. W. Coughlin, K. Kawakami, and A. Hasan, *J. Catal. 88*, 150 (1984).
68. J. M. Parera, R. J. Verderone, and C. A. Querini, in *Catalyst Deactivation* (B. Delmon and G. F. Froment, eds.). Elsevier, Amsterdam, 1987, p. 135.
69. R. J. Bertolacini and R. J. Pellet, in *Catalyst Deactivation* (B. Delmon and G. F. Froment, eds.). Elsevier, Amsterdam, 1980, p. 73.
70. W. J. Doolittle, N. D. Skoularikis, and R. W. Coughlin, *J. Catal. 107*, 490 (1987).
71. J. Margitfalvi, S. Göbölös, E. Kwaysser, M. Hegedus, F. Nagy, and L. Koltai, *React. Kinet. Catal. Lett. 24*, 315 (1984).
72. Yu. M. Zhorov, G. M. Panchenkov, and Yu. N. Kartashev, *Kinet. Catal. 22*, 1058 (1981).
73. R. Burch and A. J. Mitchell, *Appl. Catal. 6*, 121 (1983).
74. J. L. Carter, G. C. Mc Viker, M. Weisshan, W. S. Kmak, and J. H. Sinfelt, *Appl. Catal. 3*, 327 (1982).
75. J. M. Grau and J. M. Parera, *Appl. Catal. 70*, 9 (1991).
76. P. A. Larsen, in Question and Answer Session. Hydroprocessing, Ketjen Catalysts Symposium (H. Th. Rijnten and H. J. Lovink, eds.). AKZO Chemie Ketjen Catalysts, The Netherlands, 1986, p. 75.
77. D. A. Dowden, *Chem. Soc. Spec. Publ. Catal. 2*, 1 (1978).
78. S. R. Tennison, *Chem. Br.*, 536 (1981).

79. V. K. Shum, J. B. Butt, and W. M. H. Sachtler, *J. Catal.* **96**, 371 (1986).
80. V. K. Shum, J. B. Butt, and W. M. H. Sachtler, *J. Catal.* **99**, 126 (1986).
81. F. M. Dautzenberg, J. N. Helle, P. Biloen, and W. M. H. Sachtler, *J. Catal.* **63**, 119 (1980).
82. K. H. Ludlum and R. P. Eischens, Symposium on Catalysis on Metals, Division of Petroleum Chemistry, American Chemical Society, New York meeting, April 4-7, 1976, p. 375.
83. N. S. Koslov, J. A. Mostovaja, L. A. Kupcha, and G. A. Zhishenko, *Neftekhimiya* **15**(6), 814 (1975).
84. B. Spurlock and R. L. Jacobsen, U. S. Patent 3,507,781 (1970).
85. J. H. Sinfelt, B. Heights, and A. E. Barnett, U. S. Patent 3,835,034 (1974).
86. N. S. Koslov, J. A. Mostovaja, E. A. Skrigan, and G. A. Zhishenko, *Neftekhimiya* **17**(2), 211 (1977).
87. H. Charcosset, *Rev. Inst. Fr. Pet.* **32**(2), 239 (1979).
88. A. V. Ramaswamy, P. Ratnasamy, S. Sivasanker, and A. J. Leonard, Proceedings, 6th International Congress on Catalysis, London, 1976, p. 855.
89. R. W. Rice and K. Lu, *J. Catal.* **77**, 104 (1982).
90. J. C. Rasser, W. H. Beindorff, and J. F. Scholten, *J. Catal.* **59**, 211 (1979).
91. S. J. Tauster and S. C. Fung, *J. Catal.* **55**, 29 (1978).
92. P. Marécot, S. Peyrovi, D. Bahloul, and J. Barbier, *Appl. Catal.* **66**, 191 (1990).
93. J. W. Niemantsverdriet and A. D. Van Langeveld, *Fuel* **65**, 1396 (1986); A. D. Van Langeveld and J. W. Niemantsverdriet, *Surf. Interface Anal.* **9**, 215 (1986).
94. J. Volter, G. Lietz, M. Uhlemann, and M. Herrmann, *J. Catal.* **68**, 42 (1981).
95. J. Volter and U. Kurschner, *Appl. Catal.* **8**, 167, (1983).
96. Lin Liwu, Zang Jingling, Wu Rongan, Wang Chengyu, and Du Hongzhang, Proceedings 8th International Congress on Catalysis, Berlin, 1984, Vol. 4, p. 563.
97. H. Lieske, A. Sarkany, and J. Volter, *Appl. Catal.* **30**, 69, (1987).
98. J. Beltramini and D. L. Trimm, *Appl. Catal.* **31**, 113, (1987).
99. Lin Liwu, Zhang Tao, Zang Jingling, and Xu Zhusheng, *Appl. Catal.* **67**, 11, (1990).
100. Zhang Tao, Zang Jingling, and Lin Liwu, in *Catalyst Deactivation* (C. H. Bartholomew and J. B. Butt, eds.). Elsevier, Amsterdam, 1991, p. 143.
101. O. Svajgl, *Int. Chem. Eng.* **12**(1), 55 (1972).
102. A. G. Graham and S. E. Wanke, *J. Catal.* **68**, 1 (1981).
103. N. S. Figoli, M. R. Sad, J. N. Beltramini, E. L. Jablonski, and J. M. Parera, *Ind. Eng. Chem. Prod. Res. Dev.* **19**, 545 (1980).
104. R. J. Verderone, C. L. Pieck, M. R. Sad, and J. M. Parera, *Appl. Catal.* **21**, 239 (1986).
105. J. M. Grau, E. L. Jablonski, C. L. Pieck, R. J. Verderone, and J. M. Parera, *Appl. Catal.* **36**, 109 (1988).
106. A. Bishara, K. M. Murad, A. Stanislaus, M. Ismail, and S. S. Hussain, *Appl. Catal.* **7**, 337 (1983).
107. J. M. Parera, E. M. Traffano, J. C. Musso, and C. L. Pieck, in *Spillover of Adsorbed Species* (G. M. Pajonk, S. J. Teichner, and J. E. Germain, eds.). Elsevier, Amsterdam, 1983, p. 101.

108. J. P. Bournonville and G. Martino, in *Catalyst Deactivation* (B. Delmon and G. F. Froment, eds.). Elsevier, Amsterdam, 1980, p. 159.
109. S. Sivasanker, A. V. Ramaswamy, and P. Ramasamy, *J. Catal. 46*, 420 (1977).
110. W. L. Callender and J. J. Miller, Proceedings 8th International Congress on Catalysis, Berlin, 1984, Vol. 2, p. 491.
111. S. M. Augustine, Ph.D. thesis, Northwestern University, Evanston, IL (1988).
112. S. M. Augustine, G. N. Alameddin, and W. M. H. Sachtler, *J. Catal. 115*, 217 (1989).
113. P. Marécot, H. Martinez, and J. Barbier, *Bull. Soc. Chim. Fr. 127*, 57 (1990).
114. *J. Oil Gas*, 88 (Apr 9, 1973).
115. N. S. Figoli, J. N. Beltramini, A. F. Barra, E. E. Martinelli, M. R. Sad, and J. M. Parera, *Div. Petroleum Chem. Am. Chem. Soc. 26*(3), 728 (1981).
116. N. S. Figoli, J. N. Beltramini, E. E. Martinelli, M. R. Sad, and J. M. Parera, *Appl. Catal. 5*, 19 (1983).
117. C. L. Pieck and J. M. Parera, *Ind. Eng. Chem. Res. 28*, 1785 (1989).
118. C. G. Myers, W. H. Lang, and P. B. Weisz, *Ind. Eng. Chem. 53*, 299 (1961).
119. C. Alvarez-Herrera, Thesis, Poitiers (1977).
120. J. P. Franck and G. Martino, Deactivation of reforming catalysts. In *Deactivation and Poisoning of Catalysts* (J. Oudar and H. Wise, eds.). Marcel Dekker, New York, 1985.
121. M. E. Levinter, L. M. Berkovitch, T. V. Kvrchatkina, and N. A. Panfilov, *Kinet. Katal. 16*(3), 797 (1975).
122. G. Martino, in *Catalyse de Contact* (J. F. Lepage, ed.). Editions Technip, Paris, 1978, p. 575.
123. J. Biswas, G. M. Bickle, P. G. Gray, D. D. Do, and J. Barbier, *Catal. Rev. Sci. Eng. 30*(2), 161 (1988).
124. *Hydrocarbon Process*, 159 (March 1979).
125. N. S. Figoli, J. N. Beltramini, E. E. Martinelli, P. E. Aloe, and J. M. Parera, *Appl. Catal. 11*, 201 (1984).
126. J. N. Beltramini, E. E. Martinelli, E. J. Churin, N. S. Figoli, and J. M. Parera, *Appl. Catal. 7*, 43 (1983).
127. B. J. Cooper and D. L. Trimm, in *Catalyst Deactivation* (B. Delmon and G. F. Froment, eds.). Elsevier, Amsterdam, 1980, p. 63.
128. Yu. M. Zhorov, G. M. Panchenkov, and Tu. N. Kovtaskov, *Kinet. Katal. 21*, 580 (1980).
129. Yu. M. Zhorov, G. M. Panchenkov, and Y. N. Kartashev, *Kinet. Katal. 21*, 776 (1980).
130. J. N. Beltramini, R. A. Cabrol, E. J. Churin, N. S. Figoli, E. E. Martinelli, and J. M. Parera, *Appl. Catal. 17*, 65 (1985).
131. F. G. Ciapetta and D. N. Wallace, *Catal. Rev. 5*(1), 88 (1971).
132. F. G. Ciapetta, *Petro/Chem. Eng. 33*(5), C19 (1961).
133. Anonymous, *Oil Gas J.*, 73 (April 19, 1979).
134. V. Haensel, U.S. Patent 3,006,841 (1961).
135. G. Antos, U.S. Patent 4,178,268 (1979).
136. W. K. Meerbott, A. H. Cherry, B. Chernoff, J. Crocoll, J. D. Heldman, and C. J. Kaemmerlen, *Ind. Eng. Chem. 46*, 2026 (1954).

137. W. C. Pfefferle, Division of Petroleum Chemistry, ASC Meeting, Houston, Feb. 22–27, 1970.
138. P. Biloen, J. N. Helle, H. Verbeek, F. M. Dautzenberg, and W. M. H. Sachtler, *J. Catal. 63*, 112 (1980).
139. P. G. Menon and J. Prasad, Proceedings, 6th International Congress on Catalysis, London (G. C. Bond, P. B. Wells, and F. C. Tomkins, eds.), Vol. 2, 1976, p. 1061.
140. L. W. Jossens and E. E. Petersen, *J. Catal. 76*, 273 (1982).
141. R. W. Coughlin, A. Hasan, and K. Kwakami, *J. Catal. 88*, 163 (1988).
142. J. H. Sinfelt, *Bimetallic Catalysts*, Wiley, New York, 1983.
143. M. D. Edgar, in *Applied Industrial Catalysis* (B. E. Leach, eds.). Academic Press, New York, 1983.
144. J. M. Parera, R. J. Verderone, C. L. Pieck, and E. M. Traffano, *Appl. Catal. 23*, 15 (1986).
145. J. Barbier and P. Marécot, *J. Catal. 102*, 21 (1986).
146. M. Wilde, T. Stolz, R. Feldhaus, and K. Anders, *Appl. Catal. 31*, 99 (1987).
147. J. M. Parera and J. N. Beltramini, *J. Catal. 112*, 357 (1988).
148. D. L. Trimm, Catalyst design for reduced coking. *Appl. Catal. 5*, 263 (1983).
149. C. A. Querini, N. S. Figoli, and J. M. Parera, *Appl. Catal. 53*, 53 (1989).
150. Z. Karpinski and T. Koscielski, *J. Catal. 63*, 313 (1980).
151. W. Molina, M. Guisnet, J. Barbier, N. S. Gnep, and L. Elassal, First Franco-Venezuelan Congress, 1983.
152. A. Sarkany, H. Lieske, T. Szilagyi, and L. Toth, Proceedings, 8th International Congress on Catalysis, Berlin, 1984.
153. L. Luck, S. Aeiyack, and G. Maire, Proceedings, 8th International Congress on Catalysis, Berlin, 1984, p. 695.
154. J. M. Parera, N. S. Figoli, J. N. Beltramini, E. J. Churin and R. A. Cabrol, Proceedings, 8th International Congress on Catalysis, Berlin, 1984.
155. H. Pines, *The Chemistry of Hydrocarbon Conversions*. Academic Press, New York, 1981.

# 10
## Deactivation by Poisoning and Sintering

### Jorge Norberto Beltramini
*King Fahd University of Petroleum and Minerals, Dhahran, Saudi Arabia*

## I. INTRODUCTION

During industrial operation, the activity and selectivity of bifunctional naphtha reforming catalysts change as a result of processes like coking, poisoning, and sintering. A previous chapter by Barbier and Marécot (1) covered the effect of coking on the deactivation of reforming catalysts. In this chapter, the poisoning and sintering of platinum on alumina-based naphtha reforming catalysts are examined with special emphasis on modification of the activity, selectivity, and stability.

## II. POISONING OF NAPHTHA REFORMING CATALYSTS

Catalyst poisoning is one of the most severe problems associated with the operation of commercial catalysts. The most common type of catalyst poisoning is caused by an impurity that is either present in the gas stream or formed by some process during the reaction. In both cases, it is known that the poison becomes adsorbed on the active sites of the catalyst, causing either a temporary or permanent decrease in the overall activity. The effect of poisoning on metal catalysis has been the subject of many investigations and many interesting reviews have appears (2,3) since the first pioneering work of Maxted (4).

Poisoning of naphtha reforming catalysts occurs typically as a result of inadequate use of naphtha pretreatment conditions. In commercial practice, the life of the naphtha reforming catalyst may be reduced to a few months or weeks in the presence of only small concentrations of poison contaminants in the feed. In most cases, because of the essentially irreversible adsorption of poison compounds on the metal surface, regeneration is usually impossible or impractical.

The importance of poison molecules in the inhibition of the reforming-type reaction is shown in Table 1 which shows the initial toxicities of different compounds obtained for the hydrogenation of benzene on a $Pt/Al_2O_3$ catalyst by Barbier (5). It seems that the adsorption of a poison deactivates the surface on which it is adsorbed, and the toxicity depends on the number of geometrically blocked metal atoms (5). The authors noted that in the case of a reforming catalyst, the quantity of poison adsorbed on the metal sites cannot be readily measured. It is only possible to know the quantities of impurities injected or deposited on the whole catalyst. The coverage of the metal by the poison molecule can be much lower and not reflective of the real value of the toxicity. As seen in Table 1, as many as 15 atoms of platinum can initially be inhibited by one $Hg^{2+}$ ion. On the other hand, ammonia compounds show abnormally low toxicity, which proves that such compounds are adsorbed on both the metal and the support.

Sulfur is probably the major poison in the case of naphtha reforming catalysts. It is usually present as impurities in the naphtha feed stock in concen-

**Table 1** Initial Toxicities of Various Compounds on $Pt/Al_2O_3$ Catalyst

| Poison | Toxicity, $t^a$ | Relative activity at poison saturation[b] |
|---|---|---|
| $(CH_3)_2SO_4$ | 0 | |
| $SO_2$ | 1.0 | 0 |
| $(CH_3)_2SO_2$ | 0.02 | 0.55 |
| $(CH_3)_2SO$ | 3.0 | 0 |
| Thiophene | 5.0 | 0 |
| Dibenzothiophene | 4.5 | 0 |
| $NH_3$ | 0.10 | 0.67 |
| Pyridine | 0.12 | 0.042 |
| $(CH_3)_2N-NH_2$ | 0.50 | 0.18 |

**Table 1** Continued

| Poison | Toxicity, $t^a$ | Relative activity at poison saturation[b] |
|---|---|---|
| $C_6H_5NO_2$ | 0.24 | 0.10 |
| $CH_3CH_2CH_2CH_2NH_2$ | 0.11 | 0.043 |
| Piperidine | 0.48 | 0.031 |
| Quinoleine | 0.50 | 0.033 |
| Aniline | 0.14 | 0.033 |
| N,N-Dimethylaniline | 0.90 | 0.50 |
| Phenol | 0.075 | 0.55 |
| Resorcinol | 0.27 | 0.035 |
| Diphenyl oxide | 0.50 | 0.55 |
| EtOEt | 0 | |
| (NC)(NC)C=C(CN)(CN) | 0.065 | 0.13 |
| $C_6H_5C_6H_5$ | 0.013 | 0.71 |
| $C_6H_5CH_2C_6H_5$ | 0 | |
| CO | 1.18 | 0 |
| $H_2O$ | 0.00018 | 0.5 |
| $BF_3$ | 0.002 | 0 |
| $Si(CH_3)_4$ | 0.002 | 0.73 |
| $As(C_6H_5)_3$ | 1.9 | 0 |
| $Pb(CH_3)_4$ | 3.4 | 0.05 |
| $Na^+$ | 0 | |
| $K^+$ | 0 | |
| $Ca^{2+}$ | 0 | |
| $Ba^{2+}$ | 0 | |
| $Cr^{3+}$ | 0 | |
| $Mn^{2+}$ | 1.4 | 0 |
| $Fe^{2+}$ | 1.1 | 0 |
| $Co^{2+}$ | 1.0 | 0 |
| $Ni^{2+}$ | 1.0 | 0 |
| $Cu^{2+}$ | 2.0 | 0 |
| $Zn^{2+}$ | 2.5 | 0 |
| $Ag^+$ | 0.7 | 0 |
| $Sn^{2+}$ | 7.6 | 0 |
| $Hg^{2+}$ | 15 | 0 |
| $Tl^+$ | 14 | 0 |
| $Pb^{2+}$ | 5.2 | 0 |

[a] Toxicity defined as number of platinum atoms poisoned by one molecule/atom of poison.
[b] Activity is metal-catalyzed benzene hydrogenation.
*Source*: Ref. 5.

trations of organic sulfide of up to 1500 ppm. It appears that all sulfur compounds are readily converted into $H_2S$ over a platinum catalyst at naphtha reforming conditions. Sulfur apparently bonds so strongly to metal surfaces that marked activity reduction occurs even at extremely low gas phase concentrations of sulfur-containing compounds. Poisoning due to nitrogen compounds is considered to affect the catalyst life to a lesser extent. Nitrogen compounds are known to inhibit principally the acid function of the catalyst. Metal impurities most commonly named as poisons of reforming catalysts are arsenic, sodium, copper, mercury, and lead. It was found that they have a strong affinity for platinum that results in an irreversible poison effect as a result of the formation of stable chemical compounds with platinum. As an example, poisoning due to the presence of sodium can deactivate the catalyst, which in a constant octane mode of operation is extremely deleterious to catalyst life.

In this section, the influence of the sulfur compounds, nitrogen compounds, and metal impurities on the poisoning effect on activity, selectivity, and stability of naphtha reforming catalysts will be investigated.

### A. Poisoning by Sulfur Compound

The poisoning effect of sulfur on platinum catalysts has been studied for a long time. Sulfur is known to adsorb strongly at very low gas phase concentrations and can remain as a very stable adsorbed species under different reaction conditions.

In the case of catalytic reforming, sulfur compounds, in high concentrations in the hydrocarbon feed, have been found to have an adverse effect on the catalytic properties of the catalyst. Catalyst poisoning typically results from strong chemisorption of sulfur compounds on the surface of the metal. However, it is also well known that to increase the octane number or to produce aromatics, the usually dreaded harmful effect of sulfur on $Pt/Al_2O_3$ catalysts, can be beneficial if it is properly controlled because the partially sulfur-poisoned $Pt/Al_2O_3$ catalysts minimize undesired gas production. For this reason, in industrial practice presulfiding of some catalysts is essential before the naphtha is brought in contact with the catalyst (6). However, in commercial operation, it is recommended that sulfur concentrations in the feed must be kept below 20 ppm for monometallic $Pt/Al_2O_3$ catalysts (7). In the case of the bimetallic catalyst ($Pt-Re/Al_2O_3$) the level should be maintained well below 1 ppm. New catalysts containing higher Re/Pt ratios are even more sensitive to sulfur compounds. For these catalysts a maximum sulfur level of 0.5 ppm is allowed (8). Because of these high severity limits regarding sulfur poisoning, an additional sulfur guard unit is often incorporated before the industrial reforming reactors to eliminate almost completely the sulfur compounds in the feed.

To understand the modifications of activity and selectivity of naphtha reforming catalysts in the presence of sulfur compounds, it is necessary to have a clear picture of the nature of the adsorption and the effect of the catalyst properties on the adsorption arrangements on the surface. With this in mind, the poisoning effect of sulfur on reforming catalysts will be discussed within the following area: nature and structure of sulfur adsorption on mono- and bimetallic reforming catalysts, influence of catalyst properties on sulfur poisoning, and the role of sulfur in reforming activity, selectivity, and stability.

### 1. The Nature and Structure of Sulfur Adsorption

In understanding the process of catalytic poisoning by sulfur on reforming catalysts, it is necessary first to elucidate the structure relationship between chemisorbed sulfur and platinum metal surfaces. It is believed that the location of adsorbed sulfur atoms determines the number of sites available for surface reaction (9). Sulfur chemisorbs at high-coordination sites and can deactivate several neighboring adsorption sites. At low coverage, chemisorbed sulfur may block preferentially some of the sites but not others and as a result selective chemisorption and reaction may occur.

The adsorbed sulfur molecule deactivates the surface on which it is adsorbed. In this case, the toxicity, defined as the number of platinum atoms deactivated by one atom of sulfur, depends on the number of geometrically blocked metal atoms. On the other hand, the bond resulting from the chemisorbed sulfur and the metal can modify the electronic properties of the metallic atom responsible for reaction (10).

Bonzel and Ku (11) were the first to elucidate the changes in the adsorption binding energy as a function of sulfur coverage on Pt metals. Using CO adsorption on Pt(110) as a test reaction, they concluded the following:

1. The area of CO adsorption decreases monotonically with increasing sulfur coverage, leading to geometric limitation of the catalytically active area.
2. The binding energy of CO decreases with increasing sulfur coverage, presumably as a consequence of CO-S interaction.
3. The interaction between adsorbed CO and adsorbed sulfur is rather weak and does not lead to a surface reaction.

Similar results were found later by Keleman et al. (12) for Pt(100) and Pt(111). Figure 1 shows that preadsorbed sulfur decreases the initial adsorption binding energy of CO. As the CO concentration increases, its binding energy decreases rapidly due to the repulsive interaction between CO molecules and adsorbed sulfur. These results were confirmed by a thermodynamic study of the enthalpy of formation of chemisorbed sulfur on platinum metal surfaces that shows a decrease in the values of the adsorption energy with increasing sulfur coverage (13).

The previous findings were also confirmed by Oudar et al. (14). From sulfur deactivation studies on Pt(110) using 1,3 butadiene hydrogenation, it was observed that each sulfur atom poisoned one dissociation site for hydrogen

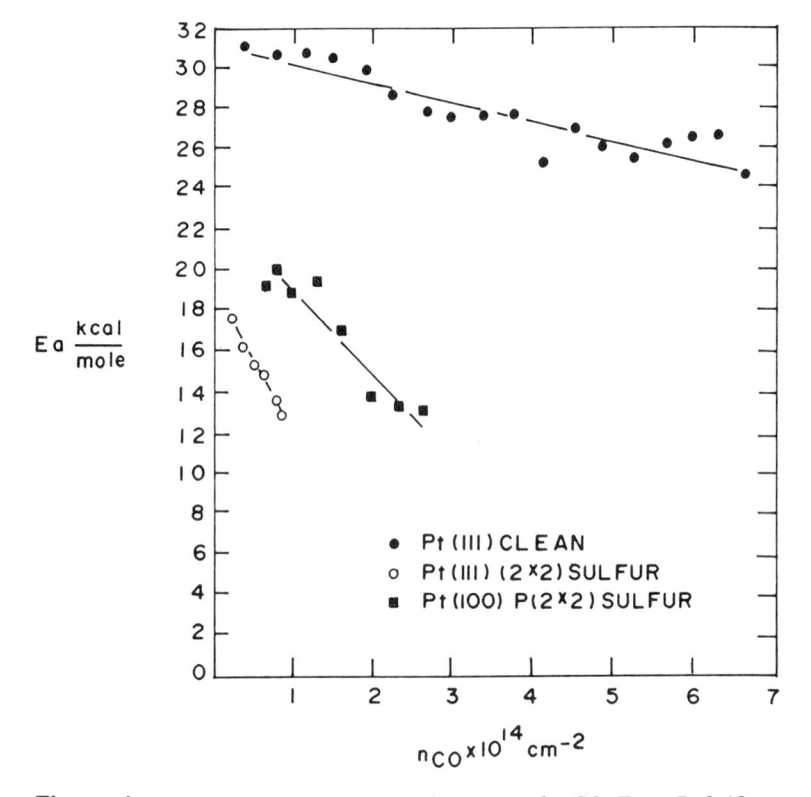

**Figure 1** Coverage-dependent desorption energy for CO. From Ref. 12.

without changing the mechanism of the reations occurring on the sites which are free of sulfur. It was also found that the sulfur-metal binding energy is decreased by 15–20%. As a general conclusion from this study, it appears that the deactivation behavior of sulfur for a catalytic reaction involving unsaturated hydrocarbons depends on the adsorption strength of the hydrocarbon and on its concentration on the surface. Studying the structure of the adsorbed sulfur layer on low-index faces of platinum, the same authors later found that for each two atoms of platinum, about one atom of sulfur remains on the metal (15).

The structure of the adsorbed sulfur layer on platinum metal has been extensively studied (16–18), and as stated by Biswas and co-workers (19), three different structure-related poisoning mechanisms can be identified (where $\theta$ is the sulfur coverage).

1. At low coverage ($\theta < 0.2$), the strong chemical bond of sulfur modifies the chemical properties of the platinum surface and weakens its interaction with adsorbates.

2. When $\theta = 0.25$, molecules can adsorb on the surface but are prevented by sulfur structure from participating in Langmuir-Hinshelwood reactions.
3. When the surface is covered with 1 sulfur atom per 2 platinum surface atoms ($\theta = 0.5$), it is chemically inert.

Sulfur adsorption studies on supported Pt/Al$_2$O$_3$ catalysts led to the introduction of the concept of reversible and irreversible adsorbed sulfur on the metal surface (20–22). Comparison of the sulfur tolerance of Pt/Al$_2$O$_3$ and Pt-Re/Al$_2$O$_3$ on exposure to thiophene at 500°C (20) led to the concept of irreversible sulfur (S$_{irrev}$) on both catalysts in accordance with:

$$Pt + H_2S \longrightarrow PtS_{irrev} + H_2$$

Further interaction of sulfur with the platinum surfaces is defined as a reversible sulfur adsorption (S$_{rev}$), poisoning the catalyst by the following reaction:

$$PtS + H_2S \longrightarrow PtS_{irrev}S_{rev} + H_2$$

The previous results were confirmed by the work of Apesteguia et al. (23), who studied the adsorption of sulfur on alumina support and Pt/Al$_2$O$_3$ catalysts (Figure 2). They found that in the case of alumina support alone, the adsorption of H$_2$S is totally reversible at 500°C. On the other hand, on Pt/Al$_2$O$_3$ at

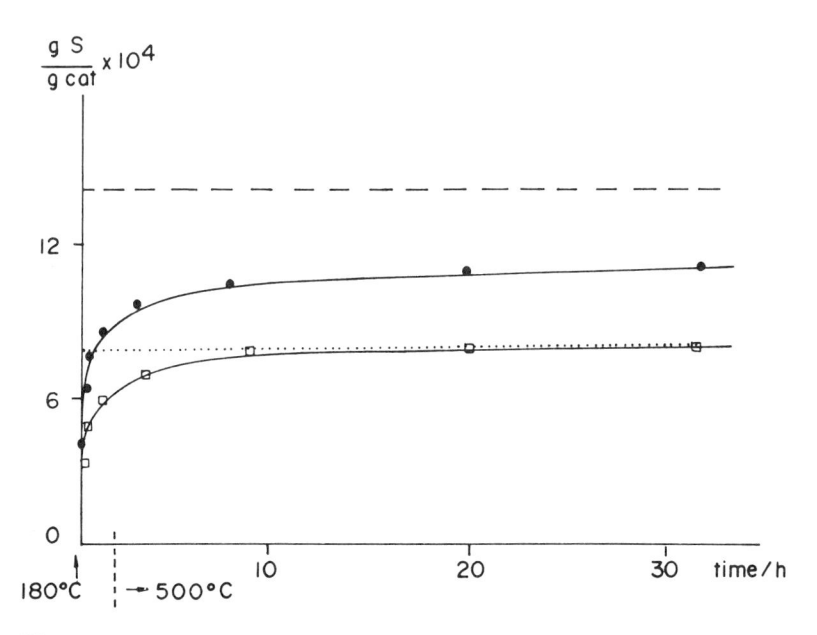

**Figure 2** Desorption of H$_2$S by heat treatment under hydrogen. (●) Pt/Al$_2$O$_3$ 0.37 wt % Pt; ---- initial quantity of sulfur on Pt/Al$_2$O$_3$; (□) Al$_2$O$_3$; ···· initial quantity of sulfur on Al$_2$O$_3$. From Ref. 23.

the same conditions, only a fraction of the adsorbed sulfur is rapidly desorbed. The authors introduced the concept of reversible and irreversible sulfur, in which the irreversible sulfur is the one that interacts totally with the metal. The quantity of irreversible sulfur, determined after 30 h of desorption under hydrogen flow at 500°C, was found to be independent of the sulfiding conditions, as seen in Table 2. It was also stated that the sulfided catalyst has a coverage in the range of 0.40–0.45 atom of sulfur for each atom of accessible platinum (24). On bimetallic Pt-Re/Al$_2$O$_3$ and Pt-Ir/Al$_2$O$_3$ catalysts, Apesteguia and Barbier (24) found that the coverage is always close to 0.4. On the other hand, Van Trimpont et al. (25) showed that in the case of sulfur addition in the liquid feed during the reforming reaction, the amount of strongly adsorbed sulfur is three times lower than that observed by Apesteguia et al. (23). The lower value was attributed to competition of sulfur with coke, hydrogen, and normal heptane feed for the metal sites. In more recent work, Bickle et al. (26) found no difference in the irreversible adsorbed sulfur between the case of sulfur-doped liquid feeds and presulfided catalysts irrespective of the variety of hydrocarbon feed used.

The influence of the alumina support on the nature of adsorption sites of H$_2$S was also studied by Apesteguia et al. (27) for Al$_2$O$_3$, Al$_2$O$_3$-Cl, and Pt/Al$_2$O$_3$-Cl. They found that on Pt/Al$_2$O$_3$-Cl at low coverages, adsorption occurs mainly on the metal and on the strong Lewis acid sites of the alumina, the latter being partially blocked by the predeposited chloride. It was also noted that sulfiding produces only a slight change in the amount of chloride adsorbed on the support, suggesting the existence of a strong Cl-Al interaction.

**Table 2** Total Sulfur and "Irreversible" Sulfur[a] on Chlorinated Al$_2$O$_3$ and Pt/Al$_2$O$_3$

| Percent H$_2$S in H$_2$-H$_2$S mixture | Al$_2$O$_3$ 1% Cl | | Pt/Al$_2$O$_3$ | |
|---|---|---|---|---|
| | Quantity of S deposited | Quantity of irreversible S | Quantity of S deposited | Quantity of irreversible S |
| | g S/g catalyst × 10$^2$ | | | |
| 8 | 0.120 | 0.004 | 0.172 | 0.026 |
| 3.5 | 0.091 | 0.006 | 0.151 | 0.029 |
| 1.5 | 0.085 | 0.004 | 0.121 | 0.028 |
| 0.5 | 0.059 | — | 0.084 | 0.025 |
| 0.02 | 0.029 | 0.003 | 0.055 | 0.028 |

[a] Irreversible sulfur is the amount that is left on the sample after 30 h of desorption under H$_2$ at 500°C.
*Source*: Ref. 23.

Deactivation by Poisoning and Sintering

The possibility of sulfur modifying the interaction of platinum with other adsorbates during reforming was demonstrated in a kinetic study of cyclopentane hydrogenolysis (28). Figures 3 and 4 shows the influence of the partial pressure of hydrocarbons and hydrogen on the rate of hydrogenolysis on unsulfided and sulfided $Pt/Al_2O_3$ catalysts. It can be seen from Figure 3 that the reaction order on unsulfided catalyst is much lower than that on sulfided catalyst. In conjunction with Figures 3 and 4, Table 3 shows that sulfur can modify both the rate constant and the ratio of the adsorption equilibrium constants of hydrogen and cyclopentane, demonstrating that sulfur adsorption affects more strongly the adsorption of hydrocarbon than hydrogen. On the other hand, Apesteguia et al. (29) also showed that sulfur-metal interactions increase the electrophilic character of platinum, thus increasing the adsorption of toluene with respect to benzene, with toluene being a good electron donor.

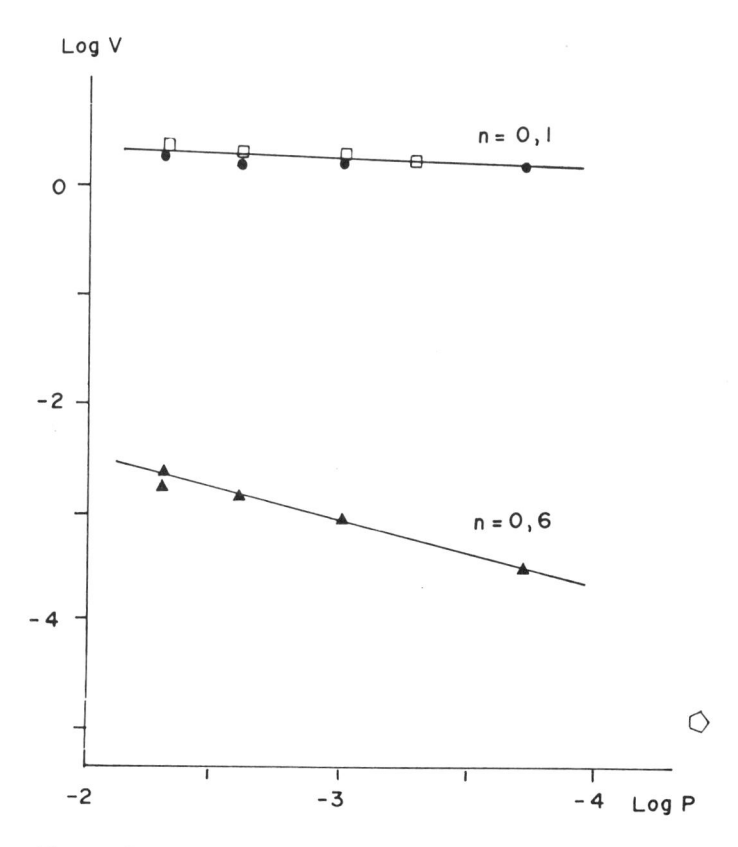

**Figure 3** Effect of the partial pressure in cyclopentane on the rate of hydrogenolysis of this hydrocarbon. (●) and (□) $Pt/Al_2O_3$; (◯) $Pt$-$(S)/Al_2O_3$. From Ref. 28.

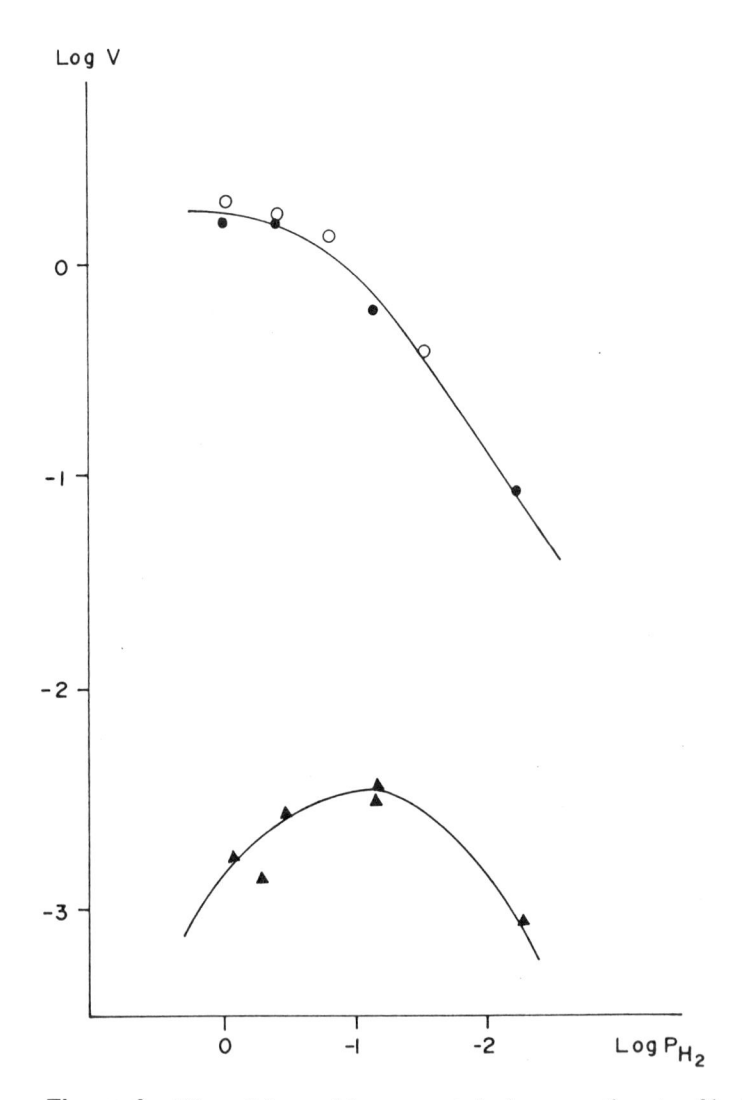

**Figure 4** Effect of the partial pressure in hydrogen on the rate of hydrogenolysis of cyclopentane. (▲) Pt/Al$_2$O$_3$; (○) and (●) Pt(S)/Al$_2$O$_3$. From Ref. 28.

**Table 3** Influence of Sulfided Platinum on the Rate Constant and the Ratio of the Equilibrium Constants of Adsorption of Hydrocarbon and Hydrogen During the Hydrogenolysis of Cyclopentane

| Catalyst | Rate constant | $(b_c/b_H)^{m/2}$ |
|---|---|---|
| Pt/Al$_2$O$_3$ | 1100 | 5.5 ± 0.3 |
| Pt(S)Al$_2$O$_3$ | 130 | 1.5 ± 0.3 |

*Source*: Ref. 28.

## 2. The Influence of Catalyst Properties

The particular properties of the naphtha reforming catalyst such as the type and degree of dispersion of the metal sites, the acidity, and the nature of the support, have a great influence on the interaction with the sulfur molecules and as a consequence will influence the degree of poisoning to different extents. Barbier et al. (30) found that sulfur adsorption on Pt (and Ir) supported on alumina is markedly sensitive to the electronic properties of the metal. When the nature of the metal, support, and metallic dispersion result in a deficiency in electrons, the metal will exhibit reduced ability to adsorb compounds such as sulfur. Figure 5 shows that on $Pt/Al_2O_3$, the evolution of the amounts of the total sulfur ($S_{tot}$) and reversible sulfur ($S_{rev}$) as a function of metal dispersion follows a linear relationship. It appears that whatever the dispersion value, the number of sulfur atoms irreversibly adsorbed by accessible platinum atom is constant and close to 0.45. However, when platinum is supported on an acidic alumina support, with chloride contents ranging between 1.2 and 1.5 wt %, the large metallic particles are more sensitive to sulfur adsorption than small particles, as seen in Table 4. Figure 6 shows that sulfur adsorption on $Ir/Al_2O_3$ is structure sensitive; the S/Ir ratio increases with the metallic dispersion, showing greater sensitivity of small iridium particles. In the case of $Pt-Ir/Al_2O_3$ catalyst,

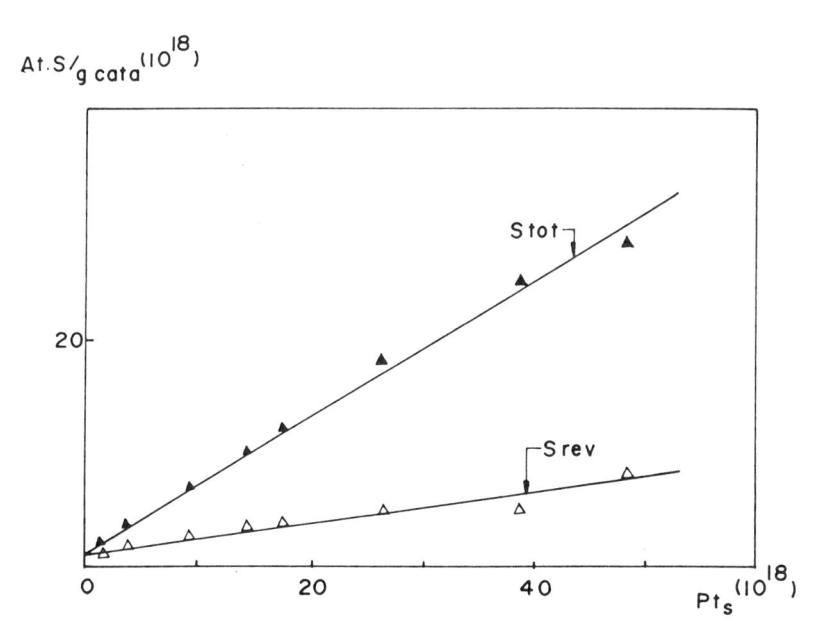

**Figure 5** Total and reversible sulfur adsorption ($S_{tot}$ and $S_{rev}$) as a function of the number of superficial platinum atoms at 500°C. From Ref. 30.

**Table 4** Irreversibly Adsorbed Sulfur on Pt/Al$_2$O$_3$ Catalysts[a]

| Catalyst | D (%) | d (A) | Cl (wt %) | S$_i$   S$_{tot}$   S$_{rev}$ (at S/g × 10$^{18}$) | S$_i$/Pt |
|---|---|---|---|---|---|
| G$_1$ | 47 | 18 | 1.3 | 6.77 − 2.64 = 4.13 | 0.51 |
| G$_2$ | 22 | 38 | 1.2 | 4.98 − 1.90 = 3.08 | 0.81 |
| G$_3$ | 12 | 71 | 1.5 | 3.29 − 1.52 = 1.77 | 0.83 |

[a] S$_i$, irreversible sulfur; S$_{rev}$, reversible sulfur; S$_{tot}$, total sulfur; G$_1$, G$_2$, G$_3$, various Pt supported on γ-alumina of 210 m$^2$/g surface area. D, dispersion; d, average crystallite size.
*Source*: Ref. 30.

the results presented in Figure 7 show that the amount of sulfur adsorbed on one accessible metallic atom increases from pure platinum to pure iridium. Sulfur chemisorption on the metal catalyst depends on the electronic properties of the metal and, as found by Bernard and Busnot (31), the adsorption of an acceptor such as sulfur is promoted on metals of lower electronic affinity. This explains the results of Barbier et al. (30) showing greater reactivity with sulfur or iridium over platinum.

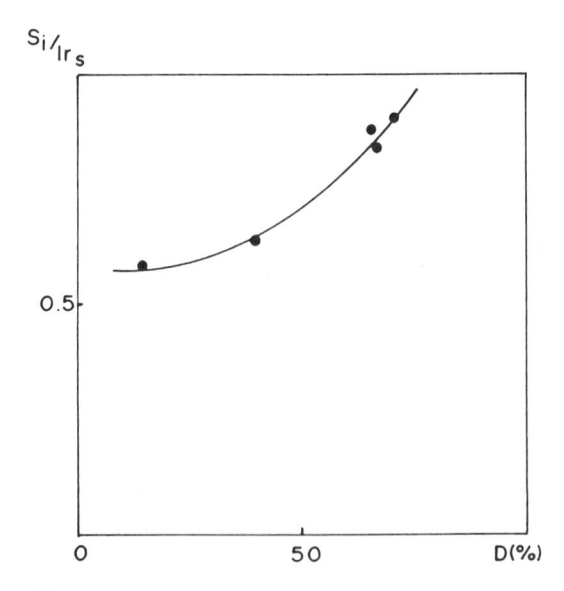

**Figure 6** Irreversible adsorbed sulfur on Ir(S/Ir) as a function of the metal dispersion D (%) at 500°C. From Ref. 30.

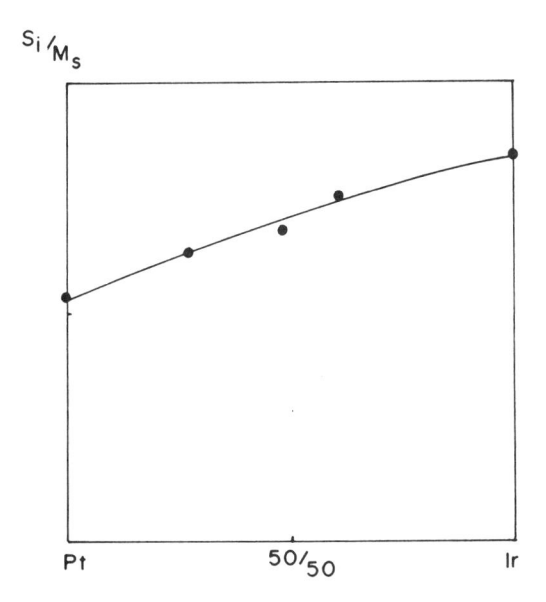

**Figure 7** Irreversible adsorbed sulfur on the metal (S/M) as a function of the iridium content in the bimetallic catalysts Pt-Ir at 500°C. From Ref. 30.

The resistance to sulfur poisoning on platinum catalysts can also be explained by the existence of a metal-support interaction. The strength of this bond increases as the metal is more dispersed and the support is more acidic (32), with the metal becoming less strongly bonded to sulfur, whose adsorption would be reversible in less severe conditions (33). Foger and Anderson (34) and Vedrine et al. (35) agree that the incorporation of small platinum particles on highly acidic supports produces electron transfer from the metal to the support. This results in an electron-deficient character on the metal and hence a decrease in sulfur adsorption, the effect being greater on smaller metallic particles. This result has been confirmed by Gallezot et al. (36), who showed that the toxicity of sulfur on platinum, measured for the hydrogenation of benzene, increases from 6.2 to 10 when the size of the platinum particles increases from 10 to 20 Å. The greater resistance to sulfur adsorption of smaller platinum particles is due to the electronic properties of these small particles, which differ from those of a solid metal (36). Metal-support interactions and metal particle size effects act in an opposite way on electronic properties of a supported metallic catalyst, and its reactivity with sulfur is the result of these two factors.

Some catalytic reactions are structure sensitive and occur on only a fraction of the metallic surface area. Hence poisoning will be seen to be selective if the poison is adsorbed on particular catalytic sites, markedly reducing the activity for reactions occurring on such sites. Sulfur adsorption is structure sensitive on platinum metal surfaces, as confirmed by the results of Oudar and Wise (15). Working on low-index (111), (100), and (110) faces of platinum, the adsorption stoichiometry for sulfur on platinum was found to be different on the three faces. Table 5 shows the effect of adsorbed sulfur on $Pt/Al_2O_3$ catalysts on reaction rate and turnover number for structure-sensitive and -insensitive reactions. The rate of structure-insensitive reactions decreases with coverage of sulfur because of a blocking effect, but the turnover number is not affected (37). On the other hand, the turnover number of hydrogenolysis is attenuated, which is due to the reduction of available ensembles on the catalyst surface, as well as the blocking of high-coordination platinum atoms for hydrogenolysis. From the results, it can be stated that sulfur acts as a path-selective poison (38) disturbing the hydrogenation-dehydrogenation function on the metal less than the hydrogenolysis function. Controlling the presence of sulfur during reaction can result in improvement of the selectivity to isomerization, dehydrogenation, and even dehydrocyclization (39).

From the similar selectivity-enhancing effects of coke and sulfur adsorption on $Pt/Al_2O_3$ catalysts, it can be inferred that sulfur adsorption and coke formation occur preferentially on hydrogenolysis sites (40). Barbier and co-workers (41) pointed out that reactions of coke formation and hydrogenolysis have the same structure sensitivity. In the case of platinum supported on alumina,

**Table 5** Catalytic Activities of Sulfided and Nonsulfided $Pt/Al_2O_3$, $Pt-Re/Al_2O_3$, and $Pt-Ir/Al_2O_3$ Catalysts

| Catalyst | Specific activity (mol/h mol metal) × $10^2$, Reaction[a] | | | | Turnover number ($h^{-1}$), Reaction | | | |
|---|---|---|---|---|---|---|---|---|
| | a | b | c | d | a | b | c | d |
| $Pt/Al_2O_3$ | 13.7 | 150 | 2.54 | 3.9 | 2000 | $25 \times 10^3$ | 400 | 700 |
| $Pt-S/Al_2O_3$—A | 4.3 | 92 | 0.088 | 0.014 | 2150 | $46 \times 10^3$ | 44 | 7.1 |
| $Pt-S/Al_2O_3$—B | 5.1 | 78 | 0.105 | 0.014 | 2500 | $39 \times 10^3$ | 53 | 7.1 |
| $Pt-S/Al_2O_3$—C | 4.7 | 100 | 0.103 | 0.016 | 2300 | $52 \times 10^3$ | 52 | 8.1 |
| $Pt-Re/Al_2O_3$ | 13.3 | 133 | 1.52 | 20.6 | 2400 | $25 \times 10^3$ | 280 | 3,700 |
| $Pt-Re-S/Al_2O_3$ | 13.3 | 133 | 1.52 | 20.6 | 2400 | $25 \times 10^3$ | 280 | 3,700 |
| $Pt-Re-S/Al_2O_3$—D | 1.1 | 65 | 0.032 | 0.057 | — | — | — | — |
| $Pt-Re-S/Al_2O_3$—E | 1.5 | 53 | 0.024 | 0.051 | — | — | — | — |
| $Pt-Ir/Al_2O_3$ | 23.0 | 312 | 7.6 | 99.0 | 2300 | $32 \times 10^3$ | 760 | 10,000 |
| $Pt-Ir-S/Al_2O_3$—K | 7.8 | 204 | 0.6 | 1.4 | 2400 | $63 \times 10^3$ | 190 | 420 |

[a] a, hydrogenation of benzene, 100°C; b, dehydrogenation of cyclohexane, 300°C; c, hydrogenolysis of cyclopentane, 300°C; d, hydrogenolysis of ethane, 420°C. A...K, various sulfiding conditions.
*Source*: Ref. 37.

increasing platinum crystallite size leads to an enhanced catalytic activity for both hydrogenolysis and coking (42). Sulfur, based on its capability to poison hydrogenolysis, should preferentially be adsorbed on large platinum particles, as supported by the already noted strong resistance to sulfur poisoning of small platinum particles (36). From this observation, it follows that the presulfiding of catalysts should decrease the extent of the coking reaction, resulting in an improvement of the catalyst stability, which is observed on commercial, sulfided reforming catalysts. However, these results are in disagreement with those of Barbier and Marecot (42), who found an increase in coke formation with sulfided catalyst. The disagreement in the results can be explained by the fact that, depending on the length of the run, more coke can be deposited on either unsulfided or sulfided $Pt/Al_2O_3$ catalysts, as confirmed by Parera and Beltramini (43).

The effects of platinum particle size and acidity of the support on the resistance to sulfur poisoning of $Pt/Al_2O_3$ catalysts for cyclohexane dehydrogenation, a structure-sensitive reaction, were also studied by Guenin et al. (44). The sulfur resistance level is more affected by the acidity of the support than the variation of Pt particle sizes. As sulfur was removed from the reactant mixture, the authors found the activity recovered to nearly 50% of the initial value. This observation was explained by the following: the main part of the poisoning is due to geometric blockage of the active sites, while the differences in resistance to sulfur are linked primarily to the electronic properties of Pt particles interacting with the support. Even though these latter differences are relatively small, they are important because they are observed in the presence of contaminated feed. The greater resistance to sulfur adsorption is explained as being due to the electronic properties of these small particles. Metal-support interactions cause an electron-deficient character to appear, reducing the capacity for adsorbing an acceptor compound such as sulfur.

## 3. Role of Sulfur in Reforming Activity, Selectivity, and Stability

The way in which sulfur influences the performance of mono- and bimetallic reforming catalysts has been the subject of several investigations. Concerning the influence of sulfur on the stability of reforming catalysts, the situation is still far from clear. In the case of the bimetallic $Pt-Re/Al_2O_3$ catalysts, it was found that sulfur on rhenium restricts the formation of graphitic coke (45), thus improving the stability of the catalyst. However, for $Pt-Ir/Al_2O_3$, the high activity resulting from the addition of iridium allows some degree of sulfur in the feed (46).

Early pioneering work of Meerbott et al. (47) showed the effect of sulfur in the feed on catalyst life. In a series of catalytic experiments for the reforming of an industrial naphtha, the authors found that sulfur has a considerable effect on aromatic yield when its concentration in the feedstock exceeds 0.03 wt % (Figures 8 and 9). It was observed in these experiments that an increase in sul-

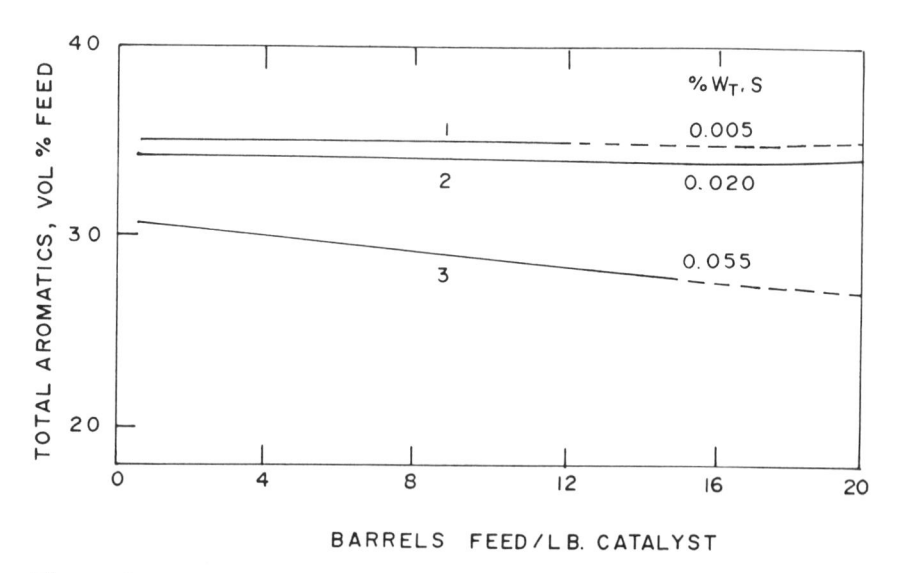

**Figure 8** Effect of sulfur on reforming of naphtha feed at 230 psig hydrogen partial pressure. From Ref. 47.

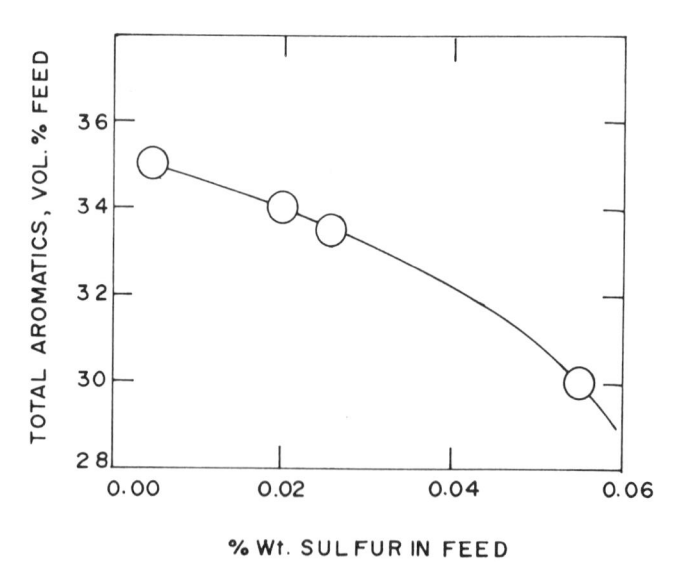

**Figure 9** Effect of sulfur on reforming of naphtha feed at constant catalyst life. From Ref. 47.

fur concentration has two principal effects: initial inhibition of aromatics formation and accelerated catalyst activity decline. The inhibition of the catalyst activity appeared to be a reversible reaction. These conclusive results are illustrated in Table 6. In another study (48), it was found that for *n*-heptane reforming with 0.8 wt % sulfur, dehydrocyclization is reduced and hydrocracking is increased. These results show that sulfur acts to enhance the acidity and inhibit the dehydrogenation activity of the catalyst. The results were confirmed by Oblad et al. (49) by demonstrating that the presence of sulfur reduces the life span of the catalyst. At the same time, a totally different point of view was expressed by Engel (50), who reported that if the catalyst is presulfided the activity can be prolonged. That work was later supported by a patent by Haensel (51), who stated that sulfur introduced with the feed has a stabilizing effect on the overall activity. Pfefferle (52) demonstrated that an optimum sulfur level exists with respect to activity and selectivity for the reforming of hydrocarbons over platinum-alumina catalysts. Minachev and Kondratchev (53) studied the effect of thiophene on dehydrogenation of cyclohexane and dehydroisomerisation of methylcyclopentane on $Pt/Al_2O_3$ catalysts under industrial reforming conditions. The effect of varying concentrations of sulfur on the activity and selectivity of platinum reforming catalysts was reported by Bursian and Maslyanskii (54). They found that sulfiding of $Pt/Al_2O_3$ catalyst increases the thermal stability but decreases the activity for dehydrogenation of cyclohexane. Hayes et al. (55), in studying the reforming of naphtha and pure hydrocarbons, indicated that the beneficial effect on catalyst stability, especially involving high-severity processing conditions, is the consequence of the formation of a platinum-sulfur complex of lower reactivity than the original platinum species to reactions such as coking. They postulated that this is due to an adsorption phenomenon in which only a fraction of the available platinum sites are deactivated by sulfur adsorption and that the identity of those deactivated sites changes constantly because of the dynamic nature of sulfur adsorption.

**Table 6** Effect on Sulfur on Aromatic Yields at Constant Conditions

| Catalyst age (bbl/lb) | Sulfur (wt % feed) | $C_5$ and lighter (wt % feed) | Hexanes + (vol % feed) | Aromatics (vol % feed) Total | Benzene | Toluene |
|---|---|---|---|---|---|---|
| 3.7–4.1 | 0.026 | 10.4 | 85.2 | 37.2 | 7.7 | 23.8 |
| 4.2–5.2 | 0.052 | 8.9 | 87.1 | 33.5 | 6.7 | 21.9 |
| 5.3–6.2 | 0.026 | 9.3 | 86.5 | 35.7 | 7.2 | 23.2 |
| 7.6–8.6 | 0.026 | 9.4 | 86.6 | 34.6 | 7.3 | 22.4 |
| 8.7–9.6 | 0.052 | 9.7 | 86.3 | 32.3 | 6.2 | 21.0 |
| 9.7–10.3 | 0.026 | 9.0 | 86.9 | 34.1 | 7.0 | 22.3 |

*Source*: Ref. 47.

In the case of naphtha reforming, in which more than one reaction is involved, preferential adsorption of sulfur may suppress the activity of one reaction more than another, leading to a change in product selectivity. This leads to the concept of selective poisoning, as noted in the case of suppression of the hydrogenolysis reaction by incorporation of sulfur (41). The concept of selective and nonselective poisoning on reforming reactions was clearly demonstrated by Maurel et al. (56). Working with Pt/Al$_2$O$_3$, they found that certain poisons such as H$_2$S and SO$_2$ are nonselective; they have the same effect in the different reaction studies: hydrogenation of benzene, "nondemanding reaction," and hydrogenolysis of cyclopentane, a "demanding reaction." On the contrary, atomic sulfur is a selective poison that reduces the rate of hydrogenolysis of cyclopentane more than that of hydrogenation of benzene. Their interpretation is based on the concept that nonselective poisons, such as H$_2$S and SO$_2$, just diminish the accessible metallic area, whereas the selective poisons, such as atomic sulfur, act on the metal by an electronic effect which alters the specific activity of the remaining metal surface area for the different reactions.

Sterba and Haensel (57) studied the effect of sulfur on reforming reactions such as hydrogenation, dehydrogenation, and isomerization on Pt/Al$_2$O$_3$. They found that low levels of sulfur on the order of 50 ppm can produce remarkable changes for all reactions, decreasing the overall conversions and changing selectivities.

Menon and Prasad (20) found that suppression of hydrocracking and consequent enhancement of aromatization are achieved for Pt/Al$_2$O$_3$ due to the sulfur adsorbed reversibly over and above that held irreversibly, whereas the Pt-Re/Al$_2$O$_3$ irreversible sulfur alone does the suppression. In the latter case, addition of reversible sulfur to the bimetallic catalyst contributed only to the suppression of the reforming activity. These results are in accordance with those of Weisz (58), who has shown that progressive deactivation of Pt/Al$_2$O$_3$ results in a decrease in hydrocracking activity with an increase in aromatization. However, with time, aromatization also decreases, ultimately giving a totally deactivated catalyst.

Van Trimpont and co-workers (59) found that Pt/Al$_2$O$_3$ catalysts required continuous sulfiding. Sulfur content in the feed should be limited to values corresponding to a molar H$_2$S/H$_2$ ratio of 0.00005 and a molar sulfur-to-surface-platinum ratio close to 1, in order to benefit from the suppression of the hydrogenolysis without a pronounced attenuation of the hydrogenation activity. For Pt-Re/Al$_2$O$_3$ presulfiding alone is sufficient to establish a degree of coverage of the metal function of 0.28, which is sufficient to suppress the hydrogenolysis. Any further addition of sulfur results in a drastic decrease of the dehydrogenation activity and favors an increase in the selectivity for hydrocracking.

Coughlin et al. (60) found that the presence of small amounts of sulfur lowers hydrocracking activity of both Pt and Pt-Re catalysts, and the hydrogenolysis activity appears to be irreversibly poisoned. Excessive amounts of sulfur deactivated the dehydrocyclization reaction on both Pt and Pt-Re catalysts, but the effect was more pronounced with Pt-Re.

These results are in agreement with those of Bickle et al. (26), who showed that for Pt/Al$_2$O$_3$, presulfiding suppressed hydrogenolysis and any further sulfur addition during normal operation caused reduced dehydrocyclization selectivity. The same authors (61) later found that in the case of Pt-Re/Al$_2$O$_3$, irrespective of whether sulfur was added via catalyst presulfiding or in the feed, the steady-state dehydrocyclization and cracking activities are about half the values with Pt/Al$_2$O$_3$. Table 7 compares the effect of presulfiding on the behavior of Pt and Pt-Re catalysts. It was stated that the different effects on both catalysts can be explained in terms of turnover number, the strength of adsorption of sulfur on the rhenium and platinum atoms, the variation in the electronic interactions between platinum and rhenium, and the geometric effect of ensembling.

From the results presented so far, it is clear that presulfiding of the naphtha reforming catalysts can have different deactivation effects on the reactions involved, with the hydrogenolysis reaction strongly influenced by sulfur poisoning. The importance of the presence of sulfur on bimetallic catalysts was also stressed in an interesting study (62) using three different naphtha reforming catalysts, Pt/Al$_2$O$_3$, Pt-Re/Al$_2$O$_3$, and Pt-Ir/Al$_2$O$_3$. It was shown that at comparable sulfur coverage, the bimetallic Pt-Re/Al$_2$O$_3$ catalyst is the most sensitive to the poison. Due to its lower electron affinity (63), sulfur is more

**Table 7** Comparison of $n$-Heptane Reforming on Chlorinated 0.3–0.3% Pt-Re/Al$_2$O$_3$

| Catalyst[a] | Group product yields/100 mol of $n$-heptane feed ||||
|---|---|---|---|---|
| | $n$-Heptane | Isomers | Toluene | Cracked products |
| A | 29.7 | 26.6 | 16.0 | 59.6 |
| B | 39.8 | 33.6 | 7.2 | 35.1 |
| C | 28.2 | 25.7 | 33.0 | 26.5 |
| D | 44.1 | 25.2 | 14.2 | 45.2 |

[a] The catalysts used were sulfided as follows: (A) Pt-Re/Al$_2$O$_3$ without sulfur addition; (B) Pt-Re/Al$_2$O$_3$ after presulfiding in H$_2$S at 480°C during reduction pretreatment followed by reversible sulfur removal using pure hydrogen flow; (C) Pt/Al$_2$O$_3$ without sulfur addition; (D) Pt/Al$_2$O$_3$ presulfided in H$_2$S in the same manner as for catalyst B.
*Source*: Ref. 61.

strongly adsorbed on rhenium than platinum (64), and as a result the majority of rhenium and only a portion of the platinum are in the sulfided state, inhibiting the hydrogenolysis capacity of the rhenium metal. Results in Table 8 show that the introduction of iridium to platinum increases the resistance for hydrogenolysis reactions to sulfur. The authors concluded that the addition of sulfur not only influences the reforming reactions but also modifies the nature of the metallic function. In addition, under industrial conditions, Pt-Re/Al$_2$O$_3$ can tolerate only minimal sulfur content in the feed (45), but Pt-Ir/Al$_2$O$_3$ is more resistant (46).

Improvements in aromatic selectivity for methylcyclopentane conversion on Pt-Ir/Al$_2$O$_3$ due to addition of sulfur was also noted by McVicker and coworkers (65). Adsorption of low amounts of sulfur optimizes aromatization with respect to hydrogenolysis. It was stated in an industrial patent (66) that octane number can be improved by partially poisoning the platinum surface.

On the other hand, the effect of presulfiding on the formation of coke on Pt/Al$_2$O$_3$, Ir/Al$_2$O$_3$, and Pt-Ir/Al$_2$O$_3$ was investigated by Barbier and Marecot (42) for the reforming of cyclopentane. Results presented in Table 9 show that the amount of coke deposited on the catalysts decreases from pure platinum to pure iridium supported catalysts. Furthermore, catalyst sulfiding induces an increase in coke level, which is more important in the case of iridium catalyst. The authors concluded that the increase in coke formation as a result of presulfiding is the consequence of higher production of olefinic compounds in the gas phase, leading to extensive coke deposition on the support as a result of polymerization on the acid sites of the alumina. It was also observed that sulfur prevented the deactivation of the metal site from the coking reaction, explaining the greater stability observed on sulfided reforming catalysts.

**Table 8** Sensitivity to Poisoning by Sulfur on Catalysts Pt/Al$_2$O$_3$, Pt-Re/Al$_2$O$_3$, and Pt-Ir/Al$_2$O$_3$[a]

| | $r/r'$[b] | | |
|---|---|---|---|
| | Pt/Al$_2$O$_3$, S/M = 0.39 | Pt-Re/Al$_3$O$_3$, S/M = 0.44 | Pt-Ir/Al$_2$O$_3$, S/M = 0.45 |
| Hydrogenation of benzene (100°C) | 3.0 | 10 | 3.0 |
| Dehydrogenation of cyclohexane (300°C) | 1.7 | 2.3 | 1.5 |
| Hydrogenolysis of cyclopentane (300°C) | 25 | 50 | 13 |
| Hydrogenolysis of ethane (360°C) | 280 | 400 | 75 |

[a] 0.6% in Pt and 0.06% in Ir.
[b] $r$, nonsulfurized catalysis; $r'$, sulfurized catalysts (units for $r$, $r'$, are mol/h-g metal). $M$ is total number of surface metal atoms.
Source: Ref. 62.

**Table 9** Comparison of Coking Extents on Sulfided and Nonsulfided Pt and Ir Catalysts

| Catalysts | $C_n = \%C$ (nonsulfided catalysts) | $C_s = \%C$ (sulfided catalysts) | $C_s/C_n$ |
|---|---|---|---|
| Pt-/Al$_2$O$_3$ | 1.80 | 2.80 | 1.55 |
| Pt-Ir/Al$_2$O$_3$ (72Pt/28Ir) | 1.30 | 2.18 | 1.68 |
| Pt-Ir/Al$_2$O$_3$ (51Pt/49Ir) | 0.91 | 1.56 | 1.71 |
| Pt-Ir/Al$_2$O$_3$ (39Pt/61Ir) | 0.56 | 1.08 | 1.93 |
| Ir/Al$_2$O$_3$ | 0.34 | 0.93 | 2.73 |

*Source*: Ref. 42.

With the introduction of new catalysts in the market and the increased severity of process operations of commercial reformers, sulfur specifications must be followed in a very strict way. In the case of bimetallic reforming catalysts, sulfur contents of less than 1 ppm in the feed are commonly required. The results presented in Figure 10 show clearly the importance of the effect of sulfur on catalyst deactivation rate expressed as inlet reactor temperature necessary to achieve a constant reformate quality as a function of operating time (67).

It can be concluded that continuously fed sulfur is a poison for the life of the catalyst, but in controlled quantities it can be beneficial, resulting in improved aromatization and catalyst stability. Catalyst presulfiding reduces the coke deposited on the metal sites because of the competition of sulfur adsorption and coke formation reactions taking place on the same sites.

## B. Poisoning by Nitrogen Compounds

Nitrogen compounds are generally present in the form of organic compounds that decompose into ammonia at reforming conditions. They inhibit principally the acid function of the catalyst, but to some extent they can alter the metallic properties of the platinum. In the presence of water, they also result in leaching of chloride through formation of ammonium chloride (67).

Table 10 (68) summarizes the effect of nitrogen as ammonia on the reforming of *n*-heptane, *n*-nonane, and methylcyclohexane. It is clear that nitrogen compounds have little effect on the dehydrogenation reaction. On the other hand, hydrocracking, isomerization, and dehydrocyclization are adversely affected, indicating that acid sites are more involved in these reactions. The authors, however, found that ammonia poisoning is temporary. The effect of nitrogen on dehydrocyclization was also studied by Franck and Martino (67), who found that the dehydrocyclization activity decreases when 37 ppm nitro-

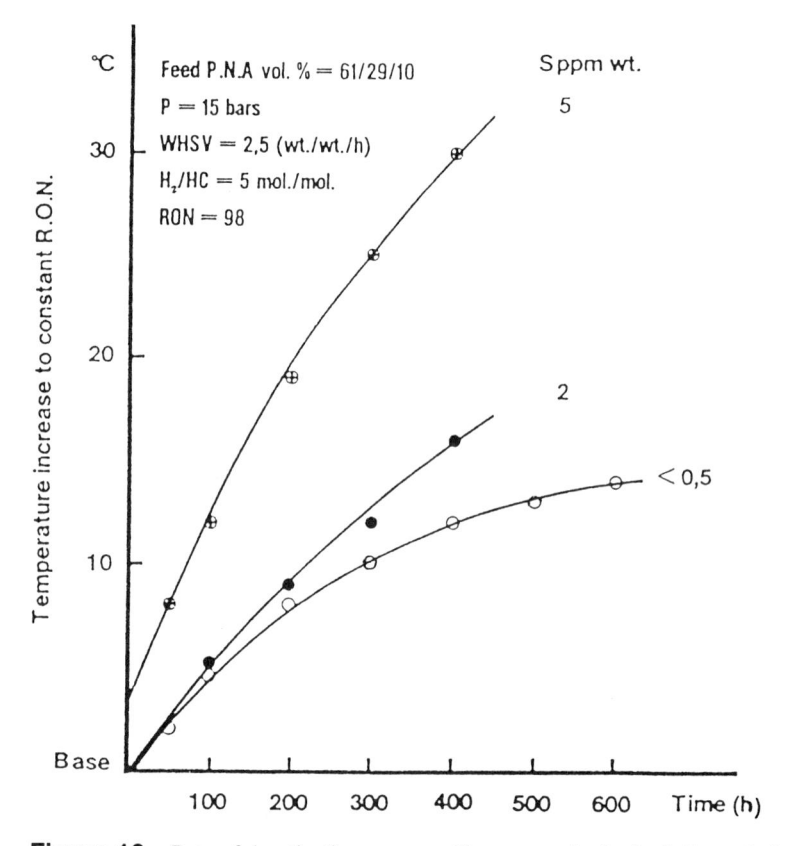

**Figure 10** Rate of deactivation versus sulfur content in the feed. From Ref. 67.

gen is incorporated in the form of butylamine in the feed. The results presented in Figure 11 show that the poison is temporary and the activity of the catalyst can be restored as soon as poison is removed from the feed.

The effect of ammonia poisoning as a function of platinum metal crystallite size was studied by Barbier et al. (69). A series of Pt/Al$_2$O$_3$ catalysts were prepared with metal dispersions ranging from 5 to 80%. Figure 12 shows the results for ammonia inhibition of benzene hydrogenation. The results indicate that benzene hydrogenation on the unpoisoned catalyst is a structure-insensitive reaction; however, it becomes structure sensitive with ammonia poisoning. It was found that small crystallites deactivate more rapidly than the larger ones. The authors concluded that this is the consequence of a heterogeneous surface with nonuniform poison adsorption. Butt et al. (70) gave another explanation for the same results based on the concept that benzene chemisorption requires an ensemble of sites, so that a single ammonia molecule would be a more

**Table 10** Effect of Adding NH$_3$ or Diethylamine to the Feedstock on Hydroforming of $n$-Heptane and $n$-Nonane and on Dehydrogenation of Methylcyclohexane

| | 9751 | | 9450 | | 9450 | |
|---|---|---|---|---|---|---|
| Catalyst: | | | | | | |
| Size, inch: | 5/32 tablets | | 1/16 extrudate | | 1/16 extrudate | |
| Pt concentration, wt %: | 0.6 | | 0.6 | | 0.6 | |
| Temperature, °F | 950 | 950 | 925 | 925 | 925 | 925 |
| Pressure, psig | 200 | 200 | 200 | 200 | 200 | 200 |
| WHSV | 2 | 2 | 2.6 | 2.6 | 100 | 100 |
| H$_2$, HC mole ratio | 5 | 5 | 5 | 5 | 5 | 5 |
| Feed | $n$-C$_7$ | $n$-C$_7$ + 0.046% NH$_3$ | $n$-C$_9$ | $n$-C$_9$ + 0.2% N as DEA | MCH | MCH + 0.2%N as DEA |
| C$_5^+$, vol % of feed | 40.1 | 68.8 | 67.0 | 78.8 | — | — |
| C$_5^+$, RON | 94.9 | 52.3 | 99.0 | 59.6 | — | — |
| Aromatic, vol % of feed | 26.6 | 11.9 | 49.8 | 19.5 | — | — |
| Mole % conversion to aromatics | 37 | 16 | 63 | 24 | 78 | 75 |

*Source*: Ref. 68.

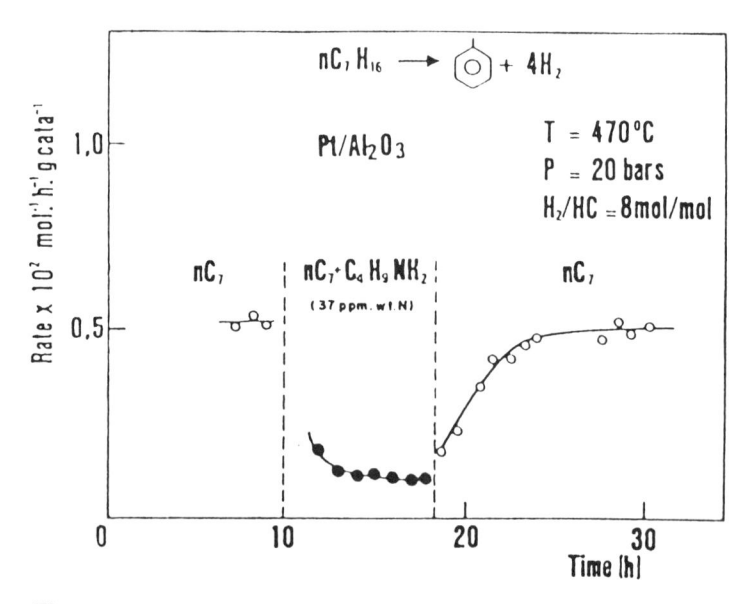

**Figure 11** Effect of nitrogen on dehydrocyclization rate. From Ref. 67.

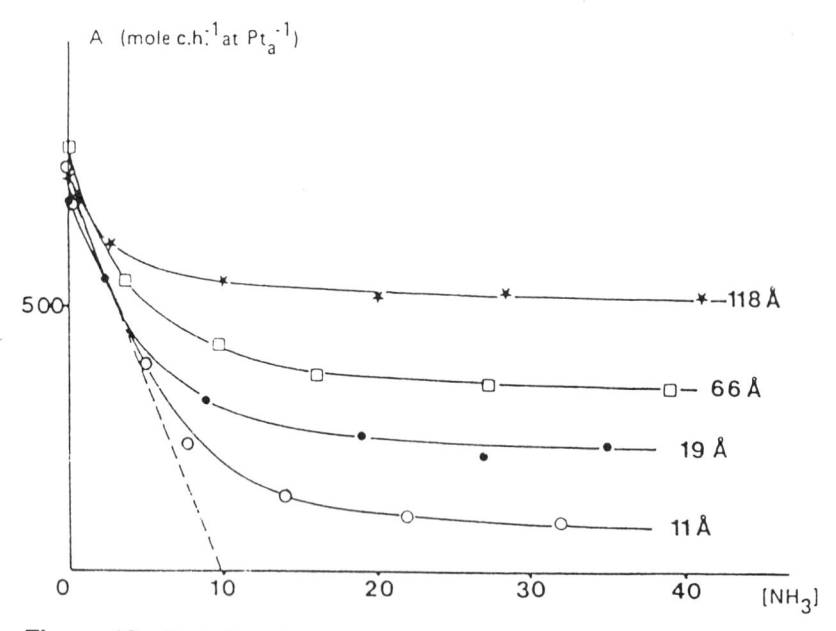

**Figure 12** Evolution of the turnover number in benzene hydrogenation of different platinum catalysts as a function of the amount of ammonia injected in the catalyst. From Ref. 69.

Deactivation by Poisoning and Sintering                                    337

effective poison on smaller crystallites. Overall, the results proved that the presence of ammonia compounds during reforming reactions will affect to a large extent the highly dispersed Pt/Al$_2$O$_3$ catalysts.

Franck and Martino (67) found that an indirect consequence of ammonia poisoning in industrial reformers will be a lowered catalyst life due to an increase in the rate of coke deposition. This is a consequence of constant-octane-number operation and the increase in reactor temperature necessary to compensate for the decrease in hydrocarbon conversion. This effect is exemplified in Figure 13 for the case of 2 ppm nitrogen injection in the hydrocarbon feed.

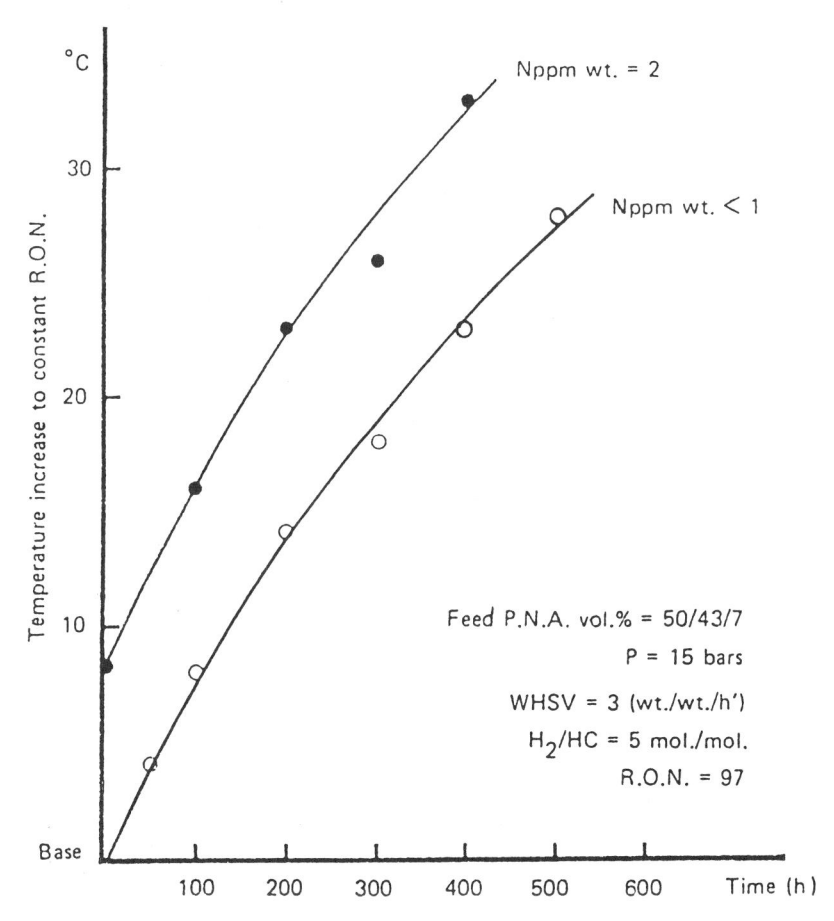

**Figure 13**   Effect of nitrogen on catalyst aging. From Ref. 67.

## C. Poisoning by Metals

Deactivation of naphtha reforming catalysts by metal contaminants has received little attention and only a few fundamental investigations have been reported. It is known that the metallic poisons are rarely eliminated by regeneration, and as a consequence all necessary precautions need to be taken to prevent them from coming into contact with the catalyst. It is known that platinum metal can be deactivated by metallic ions such as mercury, arsenic, lead, bismuth, copper, and iron (71). Table 11 summarizes the effect of poisoning of platinum by metals for reactions that are closely related to the reforming process (69). It can be seen that, except in the cases of iron and arsenic, the effect of poison can be selective to the reaction involved.

Hettinger and co-workers (68) were almost the first to report on the effect of arsenic on $Pt/Al_2O_3$ catalysts. They found 0.01–0.02 wt % of arsenic on the catalyst has very little effect on reforming reactions. An increase in arsenic concentration on the catalyst to 0.5 wt % decreases dehydrogenation activity. It was stated that although arsenic poisoning appears more permanent, the catalyst demonstrated considerable tolerance to the poison.

Barbier and Maurel (72) confirmed that for a toxic effect of metallic ions on platinum catalysts it is necessary that the ions possess d electrons. Table 12 shows the influence of electronic structures on the adsorption and poisoning abilities of various metallic ions on the reforming catalyst. It is also shown that in all cases of metal poisoning, the metals can form a very stable combination with platinum (73). One of the most affected reactions is dehydrogenation. Studies carried out using lead showed this effect (Figure 14). Similarly, Franck and Martino (67) demonstrated that small amounts of iron and copper can accelerate the deactivation in the dehydrocyclization reaction (Figure 15).

**Table 11** Response of $Pt/Al_2O_3$ to Poisoning by Metallic Components[a]

| Poison | $t_d$ | $t_{EP}$ | $t_E$ | $t_{HE}$ | $t_{HC}$ |
|---|---|---|---|---|---|
| $Pb^{2+}$ | 5.0 | 4.8 | 0.58 | 20.0 | 5.1 |
| $Fe^{3+}$ | 1.1 | 1.0 | 1.0 | 1.0 | 1.0 |
| $Zn^{2+}$ | 2.5 | 2.4 | 0.7 | 7.7 | 2.5 |
| $Sn^{2+}$ | 7.6 | 7.5 | 6.3 | 11.1 | 7.6 |
| $Hg_2^{2+}$ | 15.0 |  | 5.2 | 10.0 | 12.5 |
| $As^{3+}$ | 1.9 | 1.92 | 1.9 | 1.9 | 2.0 |

[a] $t_d$, toxicity for benzene deuteration at 85°C; $t_{EP}$, toxicity for epimerization of cis-1,2,-dimethylcyclohexane at 180°C; $t_E$, toxicity for benzene exchange at 85°C; $t_{HE}$, toxicity for ethane hydrogenolysis at 360°C; $t_{HC}$, toxicity for cyclopentane hydrogenolysis at 300°C; toxicity, number of accessible platinum atoms poisoned by one molecule of poison.
Source: Ref. 69.

# Deactivation by Poisoning and Sintering

**Table 12** Influence of Electronic Structure on the Adsorption and Poisoning Abilities on Naphtha Reforming Catalysts of Various Metallic Ions

| Metal ion tested | Precursor salt | Electron occupation of external orbitals | Toxicity |
| --- | --- | --- | --- |
| $Na^+$ | $NaCl$ | $2s^2 2p^6$ | No |
| $K^+$ | $KCl$ | $3d^0 4s^0$ | No |
| $Ca^{2+}$ | $CaCl_2$ | $3d^0 4s^0$ | No |
| $Ba^{2+}$ | $BaCl_2$ | $5d^0 6s^0$ | No |
| $Cr^{3+}$ | $Cr(NO_3)_3$ | $3d^3 4s^0$ | No |
| $Mn^{2+}$ | $MnCl_2$ | $3d^5 4s^0$ | Yes |
| $Fe^{2+}$ | $FeCl_2$ | $3d^6 4s^0$ | Yes |
| $Co^{2+}$ | $CoCl_2$ | $3d^7 4s^0$ | Yes |
| $Ni^{2+}$ | $NiCl_2$ | $3d^8 4s^0$ | Yes |
| $Cu^{2+}$ | $CuCl_2$ | $3d^9 4s^0$ | Yes |
| $Zn^{2+}$ | $ZnCl_2$ | $3d^{10} 4s^0$ | Yes |
| $Ag^+$ | $AgNO_3$ | $4d^{10} 5s^2$ | Yes |
| $Sn^{2+}$ | $SnCl_2$ | $4d^{10} 5s^2$ | Yes |
| $Hg^{2+}$ | $Hg(N_3)_2$ | $5d^{10} 6s^0$ | Yes |
| $Pb^{2+}$ | $Pb(NO_3)_2$ | $5d^{10} 6s^2$ | Yes |

*Source*: Ref. 72.

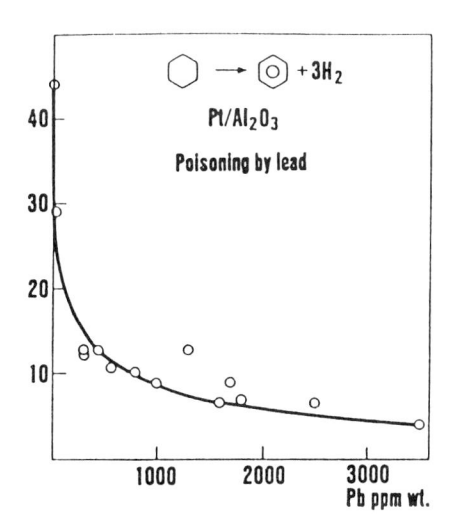

**Figure 14** Poisoning of dehydrogenation by lead. From Ref. 67.

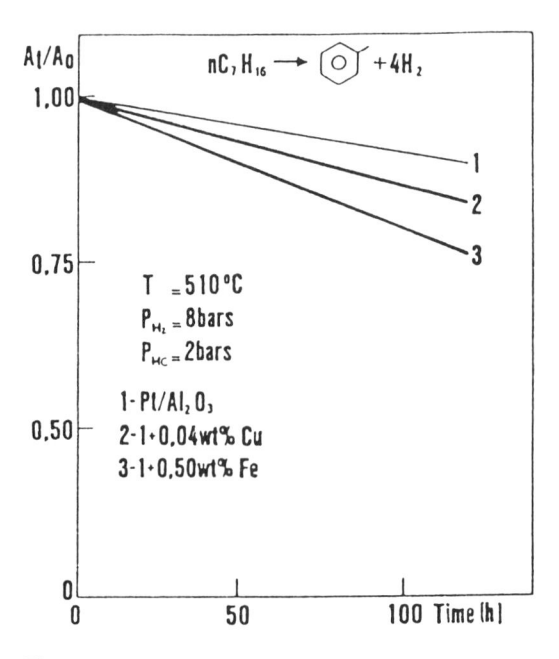

**Figure 15** Poisoning by copper and iron. From Ref. 67.

From the results, it is clear that metal contamination generally causes a decrease in the dehydrogenation and dehydrocyclization functions. As explained by Franck and Martino (67), this is the result of a modification of the adsorption competition between the hydrogen and the hydrocarbons. This factor is explored in certain bimetallic catalysts, where a metal component which is a poison under normal (high-pressure) operating conditions becomes an excellent activity promoter at very low pressures.

## III. SINTERING OF REFORMING CATALYSTS

Sintering is a physical process that is associated with loss of active area of the catalyst due to high temperatures. On bifunctional naphtha reforming catalysts, two types of sintering can be distinguished depending on the type of catalytic active sites affected. On metal sites, the loss of catalytic activity is mainly due to the growth and agglomeration of metal particles. The inverse process, which results in a decrease in metal particle sizes, is called redispersion. On the $Al_2O_3$ support, the operation at high temperatures causes a loss of specific surface area with associated changes in the pore structure, resulting in loss of activity resulting from acid sites. Metallic phase sintering is reversible, whereas alumina sintering is irreversible.

Several empirical mechanistic models for sintering of supported metal catalysts have been postulated in the literature (74–77). Two general mechanisms have been proposed for the sintering of supported metal catalysts: the particle migration mechanism (78) and the atomic migration mechanism (79). The particle migration mechanism describes the sintering process as occurring by random movement of metal particles on the support surface accompanied by collisions and coalescence of migrating particles. The atomic migration mechanism consists of detachment of metal atoms from stationary metal particles, with migration of these atoms on the support surface and their collisions with stationary metal particles or other migrating metal atoms. Unfortunately, sintering results, such as the evolution of metal surface area as a function of time, cannot be used to differentiate between these two models because both of them describe the global behavior well. On naphtha reforming catalysts, metal particle size distribution information as a function of sintering behavior is obtained mostly using transmission electron microscopy (TEM) (80).

This section concentrates on the phenomenon of deactivation of naphtha reforming catalysts by sintering and subsequent regeneration of the metal particle by redispersion. The sintering process can be affected by different variables. The influence of catalyst composition, reaction atmosphere, and the presence of water and chlorine is examined first. Next, the influence of sintering on reforming reactions and the problems related to metal dispersion are discussed. Finally, a mechanistic model for sintering/redispersion of platinum on alumina reforming catalysts is discussed extensively.

### A. Influence of Catalyst Composition

The stability of naphtha reforming catalysts has undergone a marked improvement with the introduction in the market of the bimetallic catalysts. The improvement for bimetallic catalysts appears to result from increased metal activity in some cases (81). Therefore, it is possible that the addition of this second metal can also affect the sintering characteristic of the monometallic Pt/Al$_2$O$_3$ catalyst.

It was found that the presence of a second metal in a Pt/Al$_2$O$_3$ catalyst can decrease the sintering. Ramaswamy et al. (82) and Charcosset et al. (83) agree that addition of iridium can slow down the sintering of platinum. In another TEM study, specimens of Pt/Al$_2$O$_3$ and Pt-Sn/Al$_2$O$_3$ model catalysts were investigated for the sintering of the metal particles in an H$_2$ atmosphere (84). It was found that the metal particles grew with increasing temperature and that the particle size of Pt/Al$_2$O$_3$ was larger than that of Pt-Sn/Al$_2$O$_3$ when the samples were heated at 700°C for 2 h. This suggested that Sn inhibits the sintering of the metal crystallites. In addition, from the electron diffraction patterns of the catalysts, it was observed that only metallic Pt species existed in Pt/Al$_2$O$_3$, but several kinds of Pt-Sn alloys or complexes appear in Pt-Sn/Al$_2$O$_3$. These results suggested that a strong interaction between Pt and Sn might occur during reduction, inhibiting the metallic sintering.

Additives such as germanium and iridium were also studied by Bournonville (85). Figure 16 shows the increase in stability compared with the monometallic platinum alumina catalyst. The explanation for this improvement is not very clear, but the possibility of alloy formation and support modification due to metal-support interactions was proposed (86).

The effect of Re on the sintering of Pt-Re/Al$_2$O$_3$ was explained by considering that in an oxidative atmosphere the dispersion of rhenium in the catalyst is nearly constant due to the high stability of Re$^{7+}$ species (87). Callender and Miller (88) showed that for Pt-Re/Al$_2$O$_3$, the high sintering temperature is due to the instability of the Pt$^{4+}$ species, which forms Pt metal, whereas there is no observation of decomposition of the Re$^{7+}$, which remains dispersed. From these results it can be concluded that the dispersion changes in Pt-Re/Al$_2$O$_3$ catalysts with temperature in an oxidizing atmosphere are mainly changes in the platinum dispersion.

Recent work (89) shows that during the sintering-redispersion processes for Pt-Re/Al$_2$O$_3$ catalyst, the total metal dispersion of the catalyst is influenced by the presence of platinum complexes in either sintering or redispersion steps. This is because, upon reduction of platinum complexes, metallic platinum forms crystallites, where rhenium is reduced to form the platinum-rhenium clusters. Therefore, the higher the platinum dispersion the higher the

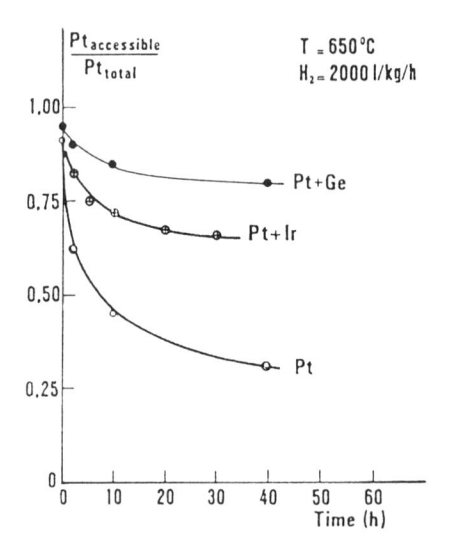

**Figure 16** Platinum sintering under hydrogen: comparison between Pt/Al$_2$O$_3$, Pt-Ir/Al$_2$O$_3$, and Pt-Ge/Al$_2$O$_3$. From Ref. 85.

platinum-rhenium dispersion. Overall, the operating conditions influence the metal dispersion of both mono- and bi-metallic catalysts in a similar way.

## B. Influence of Atmosphere

During commercial operation of naphtha reforming catalysts, most of the sintering of the metal phase and alumina occurs during the regeneration of the catalyst by coke burning. The typical reforming operation is performed in a hydrogen atmosphere and at temperatures ranging from 475 to 525°C. At these conditions, the possibility of metal or support sintering is minimal. On the other hand, during coke burning, the operation is conducted in an oxidizing atmosphere, and the reaction is highly exothermic with additional production of water. In this environment, alumina can sinter because of the high temperature inside the catalyst particles. The metallic phase can also sinter because of the high temperature and the greater tendency to sintering in an oxygen than in a hydrogen atmosphere. Therefore, it is important to examine sintering of the metal particles and support as a function of the existing atmosphere.

The influence of the nature of the atmosphere and of the treatment temperature on the sintering of platinum has been studied by Bournonville and Martino (90). In an air atmosphere, platinum surface area drops as a function of temperature and time, as does the chloride content (Figure 17). Further analysis confirmed that if the chloride content can be maintained at about 1 wt % no sintering occurs (Figure 18). This phenomenon could be the result of an inhibition between sintering and redispersion. When argon was used as an inert atmosphere, sintering was very fast and chloride leaching occurred. Results in Figure 19 show that in a hydrogen atmosphere, the mechanism of sintering is independent of whether the chloride content on the catalyst is maintained constant or not. As a conclusion, the authors explained the differences observed to be a result of the different atmospheres. Different platinum compounds are present at the surface of the alumina and the possible migrating species are not the same. In the case of an oxidizing atmosphere and in the presence of enough chlorine, chloroplatinum aluminate species strongly bonded to the alumina are present. In hydrogen atmosphere, platinum is present as atoms, raft clusters, or small crystallites depending on dispersion. The role of chlorine is also very important: it acts as a redispersion agent in an oxidizing atmosphere and as a stabilizing agent in a reducing atmosphere.

The influences of the treatments in oxidizing atmospheres on the dispersion of $Pt/Al_2O_3$ catalysts reported in the literature were summarized in a comprehensive list by Wanke et al. (91). From Table 13, it can be seen that severe sintering of platinum metal in oxidizing atmospheres appears to be more pronounced when the temperature is greater than 600°C.

Szymura and Wanke (92), using X-ray diffraction (XRD) and hydrogen chemisorption, studied the effect of sintering of $Pt/Al_2O_3$ in $H_2$, $N_2$, and He at

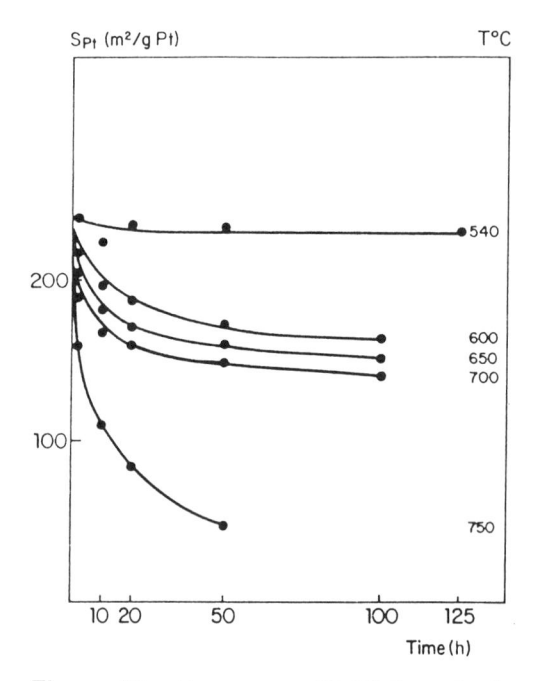

**Figure 17** Air treatment of 0.6% Pt on alumina catalyst. From Ref. 90.

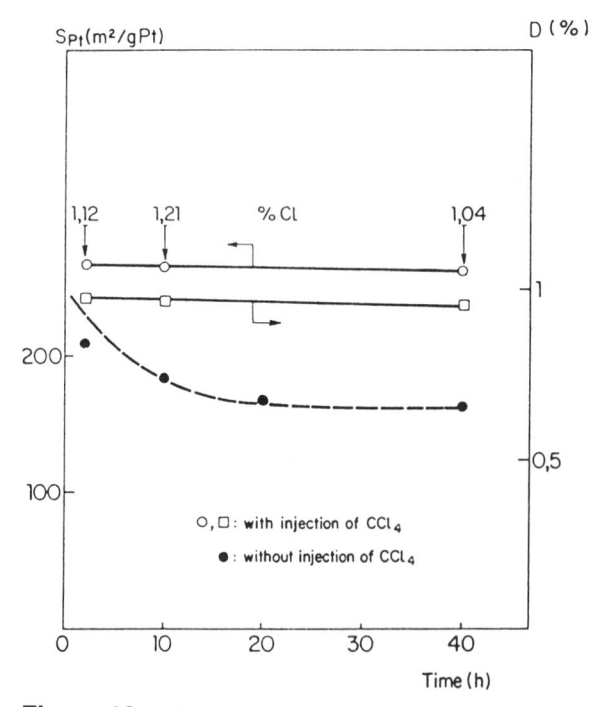

**Figure 18** Air treatment at constant chloride content ($T = 650°C$). From Ref. 90.

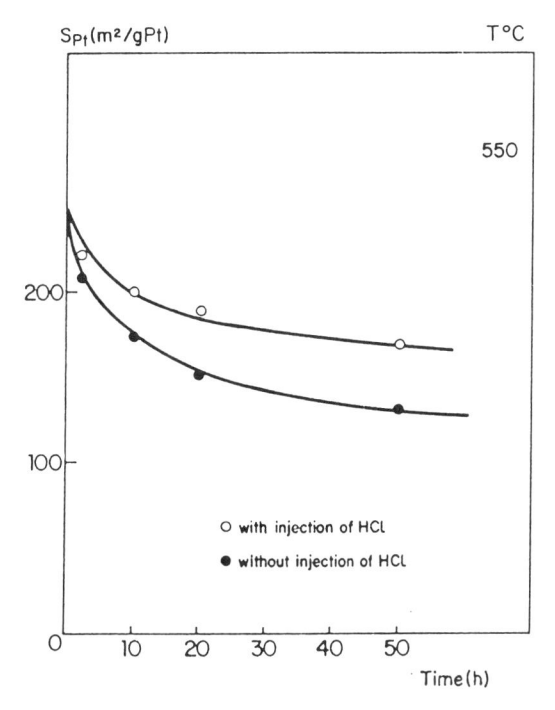

**Figure 19** Treatment in hydrogen with and without chlorine injection. From Ref. 90.

800°C. XRD and chemisorption results, presented in Figure 20 and Table 14, respectively, show that the sintering process occurs under these atmospheres. However, the sintering rate in reducing ($H_2$) and inert ($N_2$, He) atmospheres is much slower than the rate in an oxidizing ($O_2$) atmosphere. The same observation applies to crystallite size distribution. These results are in agreement with those of Lietz et al. (93), who found a bimodal crystallite size distribution for the sintering of platinum in oxygen.

The work of Dautzenberg and Wolters (94) compares the effect of heat treatment in $O_2$ and $H_2$ on the state of dispersion of Pt/$Al_2O_3$ catalyst. Heat treatment in either hydrogen or oxygen led to a decrease in the H/Pt ratio as determined by chemisorption. The authors concluded that the mechanism by which this occurs is not the same in the two cases. During heat treatment in an oxygen-containing atmosphere above 500°C, agglomeration of platinum takes place. The result is a bimodal size distribution of Pt crystallites consisting of uniform large particles of 10–15 nm and a significant fraction of particles below 3 nm, with no particle distribution in between. On the other hand, heat treatments in hydrogen did not lead to the formation of any significant platinum

**Table 13**  Influence of Treatment in Oxidizing Atmospheres on Dispersion of Pt supported on Alumina (All Catalysts Prepared with Chlorine-Containing Pt Precursors)

| Catalyst | Treatment conditions[a] (atm; °C; time, h) | Platinum dispersion | Method |
|---|---|---|---|
| 6.5% Pt/$\eta$-Al$_2$O$_3$ | Fresh, H$_2$, 500°C, 1 h | 0.15 | TEM and H$_2$ ads. |
|  | Air, 600°C, 2 h | 0.10 |  |
|  | Air, 600°C, 8 h | 0.09 |  |
|  | Air, 600°C, 24 h | 0.06 |  |
| 2.75% Pt/$\gamma$-Al$_2$O$_3$ | Fresh, H$_2$, 500°C, 8 h | 0.67 | H$_2$ ads. |
|  | Air, 590°C, 0.5 h | 0.35 |  |
|  | Air, 600°C, 0.7 h | 0.30 |  |
| 0.93% Pt/$\gamma$-Al$_2$O$_3$ | Fresh, H$_2$, 500°C, 8 h | 0.65 | H$_2$ ads. |
|  | Air, 600°C, 5 h | 0.48 |  |
| 0.5% Pt/$\gamma$-Al$_2$O$_3$ | Fresh, H$_2$, 500°C, 1 h | 1.1 | H$_2$ ads. |
|  | O$_2$, 800°C,-: seq[b] | 0.41 |  |
|  | O$_2$, 550°C, 11 h: seq | 0.60 |  |
| 2.0% Pt/$\gamma$-Al$_2$O$_3$ | Fresh, H$_2$, 400°C, 12 h | 0.90 |  |
|  | H$_2$, 700°C, 3 h: seq[b] | 0.39[c] | H$_2$ ads. |
|  | O$_2$, 520°C, 3 h: seq | 0.74 |  |
|  | O$_2$, + HCl, 520°C, 2 h: seq | 1.08 |  |
|  | O$_2$, + HCl, 600°C, 2 h: seq | 0.95 |  |
|  | H$_2$ + HCl, 600°C, 2 h: seq | 0.69 |  |
|  | O$_2$ + HCl, 520°C, 2 h: seq | 0.92 |  |
|  | O$_2$, 800°C, 2 h: seq | 0.07 |  |
|  | O$_2$, 520°C, 2 h: seq | 0.09 |  |

[a] All samples reduced before adsorption measurements.
[b] Seq indicates that runs were done sequentially on the same sample.
[c] H/Pt ratios for these runs probably do not correspond to Pt dispersions.
*Source*: Ref. 91.

agglomeration at temperatures < 675°C. In other work, den Otter and Dautzenberg (95) further investigated the changes in platinum dispersion of a Pt/Al$_2$O$_3$ catalyst treated by hydrogen. In this case, a pronounced decrease in hydrogen chemisorption was observed, while subsequent oxygen and hydrogen titration measurements indicated only a slight decrease in the dispersion values. This behavior was attributed to the formation of a Pt-Al alloy. The results were disputed and are still controversial (96).

Results different from those reported previously on the effect of atmospheres such as nitrogen, oxygen, and hydrogen were found by Hassan et al. (97) for the sintering of Pt/Al$_2$O$_3$ catalysts in the range of 300–800°C. The specific surface area of platinum is illustrated in Figure 21 as a function of sintering temperature and atmosphere. It is clear that the metallic surface area decreases in

Deactivation by Poisoning and Sintering 347

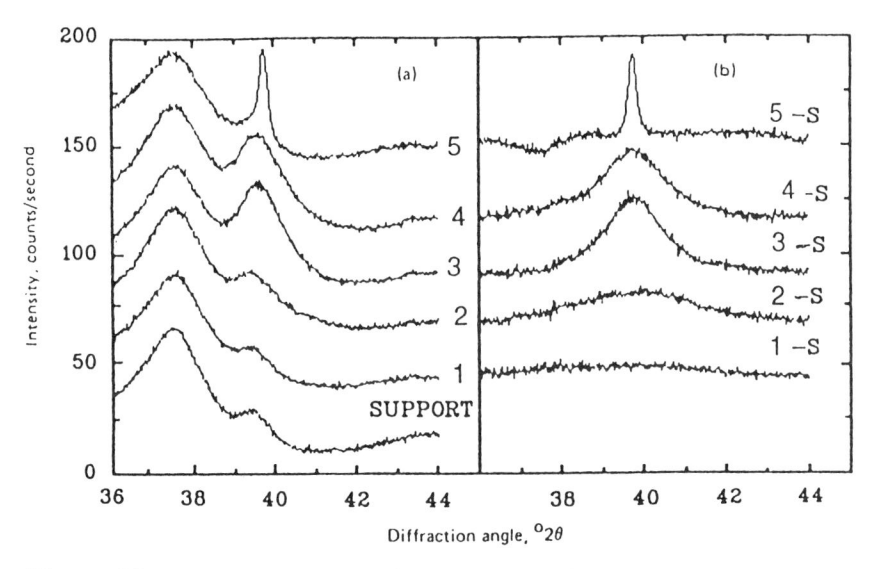

**Figure 20** XRD patterns of 1.16% Pt on alumina after various thermal treatments. (a) Patterns as recorded. (b) Substracted patterns. (Numbers refer to run numbers in Table 14; patterns are offset for clarity.) From Ref. 92.

all cases in the temperature range below 400°C. At higher temperatures, the surface area increases. The surface area of supported platinum increases markedly with an increase of temperature in both hydrogen and nitrogen, but in oxygen the surface decreases continuously with temperature. Moreover, data in Table 15 show that lowering of the H/Pt ratio and consequently an increase in average particle size are observed at temperatures below 400°C, especially

**Table 14** Influence of Treatment in Different Atmospheres on Hydrogen Adsorption by a 1.16 wt % Pt/$\gamma$-Al$_2$O$_3$ Catalyst

| Run | Treatment conditions[a] (atm; °C; time, h) | Chlorine (wt %) | H/Pt | Pt detected by XRD (%) |
|---|---|---|---|---|
| 1 | H$_2$, 500°C, 1 h[a] | 1.2 | 0.93 | 20 |
| 2 | H$_2$, 800°C, 16 h | 0.50 | 0.36 | 60 |
| 3 | He, 800°C, 16 h; H$_2$, 500°C, 1 h | 0.51 | 0.23 | 100 |
| 4 | N$_2$, 800°C, 16 h; H$_2$, 500°C, 1 h | 0.27 | 0.24 | 90 |
| 5 | O$_2$, 700°C, 1 h; H$_2$, 500°C, 1 h | 1.1 | 0.63 | 25 |

[a] Treatment not done sequentially; before each treatment Pt was well dispersed.
*Source*: Ref. 92.

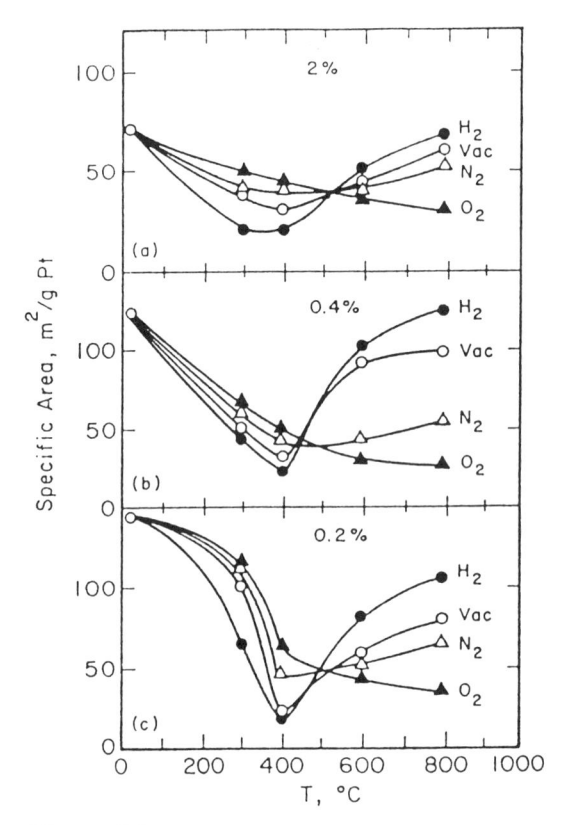

**Figure 21** Specific surface area of supported platinum as a function of temperature of sintering in the following atmosphere: (●) $H_2$ atmosphere; (○) vacuum; (△) $N_2$ atmosphere; and (▲) $O_2$ atmosphere. From Ref. 97.

in a hydrogen atmosphere. Above 400°C, the degree of dispersion increases again and the particle size decreases, which suggests dissociation to separate atoms or ensembles of smaller size. In oxygen, the data indicate that the continuous decrease in surface area is accompanied by a substantial decrease in the degree of dispersion and a continuous increase in the particle size.

Callender and Miller (88) also found that extensive sintering of highly dispersed $Pt/Al_2O_3$ occurred during heating at 620°C for 16 h in nitrogen containing 8 ppm oxygen. On the other hand, they found no sintering when the catalyst was treated under the same conditions in pure nitrogen.

TEM provided evidence for the role of crystal migration and coalescence in the sintering of $Pt/Al_2O_3$ catalysts when heated in hydrogen and alternately in $H_2$ and $O_2$ between 300 and 800°C (98). Results indicate that a large number of phenomena occur, including coalescence of nearby particles, crystal migra-

**Table 15** Degree of Dispersion, H/Pt, and Average Particle Size $a_c$ (Å), as a Function of Sintering Temperature in Different Atmospheres

| Sintering atmospher | Sintering (°C) | Pt/Al$_2$O$_3$ catalyst samples, containing Pt(%) ||||||
| | | 0.2 || 0.4 || 2 ||
| | | H/Pt | $a_c$ | H/Pt | $a_c$ | H/Pt | $a_c$ |
|---|---|---|---|---|---|---|---|
| N$_2$ | 300 | 0.456 | 20 | 0.236 | 42 | 0.175 | 55 |
| | 400 | 0.183 | 48 | 0.178 | 52 | 0.170 | 58 |
| | 600 | 0.212 | 42 | 0.186 | 49 | 0.178 | 57 |
| | 800 | 0.264 | 35 | 0.228 | 42 | 0.225 | 44 |
| H$_2$ | 300 | 0.305 | 31 | 0.186 | 54 | 0.190 | 115 |
| | 400 | 0.081 | 100 | 0.103 | 103 | 0.106 | 121 |
| | 600 | 0.325 | 28 | 0.423 | 24 | 0.430 | 44 |
| | 800 | 0.420 | 22 | 0.505 | 20 | 0.507 | 35 |
| O$_2$ | 300 | 0.476 | 19 | 0.281 | 33 | 0.211 | 50 |
| | 400 | 0.272 | 35 | 0.207 | 47 | 0.186 | 53 |
| | 600 | 0.171 | 55 | 0.145 | 67 | 0.144 | 70 |
| | 800 | 0.134 | 73 | 0.112 | 75 | 0.122 | 87 |

*Source*: Ref. 97.

tion (of both small, 1.5 nm, and large, about 8 nm, particles), migration followed by coalescence, decrease in size and disappearance of small particles near large particles, decrease in size of large particles near unaffected small particles, decrease in size and subsequent migration of particles, collision and inhibited coalescence of particles and collision-coalescence-separation of particles. In view of these phenomena, it appears that two major mechanisms of metal particle sintering for naphtha reforming catalysts exist:

1. Short-distance, direction-selective migration of particles followed by collision and coalescence and direct transfer of atoms between the two approaching particles
2. Localized ripening between a few stationary, adjacent particles

Another interesting study on sintering using TEM was conducted by Harris (99) with Pt/Al$_2$O$_3$ catalysts. Results show that when the specimens were heated in air at 700°C, very rapid sintering of the metal occurs, with the mean particle diameter increasing from 50 to 300 Å after 8 h. From the range of particle size distribution and type of particle structure observed in sintered catalysts, the author suggested that the mechanism involves a combination of migration and coalescence of whole particles. A significant proportion of the particles exhibit abnormally rapid growth, and these particles were often twinned. In some cases, the accelerated growth could be attributed to the presence of reentrant surface features at the twin boundaries; such features would act as preferential nucleation sites in the atomic migration mechanism.

## C. Influence of the Presence of Water

There are few studies (100–102) of the influence of the presence of water in the gas phase or bonded to the support in the sintering of the catalyst during the reduction step. It was stated (101) that water induces sintering due mainly to an increase in the reduction rates resulting from water promoting hydrogen spillover. This results in high localized temperatures, which are responsible for the sintering. However, Geus (102) claimed that the probable cause of the enhancement of sintering by water is that water increases the surface mobility of the metal or metal complexes.

## D. Influence of the Presence of Chlorine

As naphtha reforming catalysts are prepared from chlorine-containing precursors, it is interesting to explore the effect of chlorine compounds on the sintering process. It is well known that on $Pt/Al_2O_3$ catalysts after calcination and reduction, there is appreciable retention of chlorine by the alumina (103). As discussed before (90), the chloride present on the catalyst slows down the sintering of the platinum and acts as a redispersion agent in an oxidizing atmosphere. Lietz et al. (93) confirmed that a high chloride content on the alumina reduces sintering in oxygen atmospheres. Straguzzi et al. (104) also found that hydrogen-sintered $Pt/Al_2O_3$ samples were easily redispersed by air treatment at 500°C, the increase being dependent on the sintering time and the residual level of chloride. Successive sintering ($H_2$)-redispersion (air) cycles with a given sample indicated a slow decrease in the final dispersion, particularly for heavily sintered samples. This trend was reversed by restoring the original level of chloride. The authors give the following explanation for the results: the support provides sites where the Pt atoms can be localized, and the concentration of those sites is related to the level of chloride. When chloride removal occurs, due to the presence of water in the gas atmosphere, the sites are destroyed and consequently the redispersion capacity is affected. During sintering, the metal crystallites increase in size and at the same time decrease in number, so a situation is reached in which the concentration of sites in the vicinity of large metal particles becomes insufficient to locate all the platinum atoms on the support. When more chloride is added to the catalyst, new sites are created and the conditions for total redispersion are reestablished.

Lieske et al. (105) found that platinum sintering and redispersion in the presence of oxygen and chlorine are connected with the formation of surface complexes $[PtO_2]_s$ and $[Pt^{IV}O_xCl_y]_s$. The authors summarized all the observed surface species in a comprehensive scheme of reaction paths, indicat-

## Deactivation by Poisoning and Sintering

ing the conditions of formation and transformation of the species. The reaction model is shown in Figure 22. It can be seen that the redispersion of platinum crystallites starts by oxidative chlorine attack on the platinum oxides, resulting in formation of the complex $[Pt^{IV}O_xCl_y]_s$. Chlorine is formed from chloride ions on the alumina surface. The complex is a platinate ion that can migrate on the surface of alumina in the same way as the chloride ion does, being dispersed and trapped on appropriate centers of the alumina surface. On reduction by hydrogen, the complex produces a well-dispersed metallic platinum. On the other hand, platinum sintering is due to the high-temperature decomposition of the surface complexes $[Pt^{IV}O_xCl_y]_s$ and $[PtO_2]_s$ producing platinum atoms or small clusters. These atomic entities then migrate rapidly on the surface of alumina and form large platinum crystals by nucleation.

One of the most comprehensive studies of the influence of chlorine on sintering is that of Wanke et al. (91). Results presented in Table 13 for a set of chloride-free and chlorinated catalysts show that the presence of chlorine is required for redispersion of platinum supported on alumina. It was also agreed that oxygen alone does not result in redispersion of $Pt/Al_2O_3$.

From the published results, Butt and Petersen (96) concluded the following regarding the sintering/redispersion of $Pt/Al_2O_3$ catalysts:

1. Chlorine present either on the support or in the gas phase is required for redispersion, and oxygen alone cannot accomplish this, as reported by Lieske et al. (105).

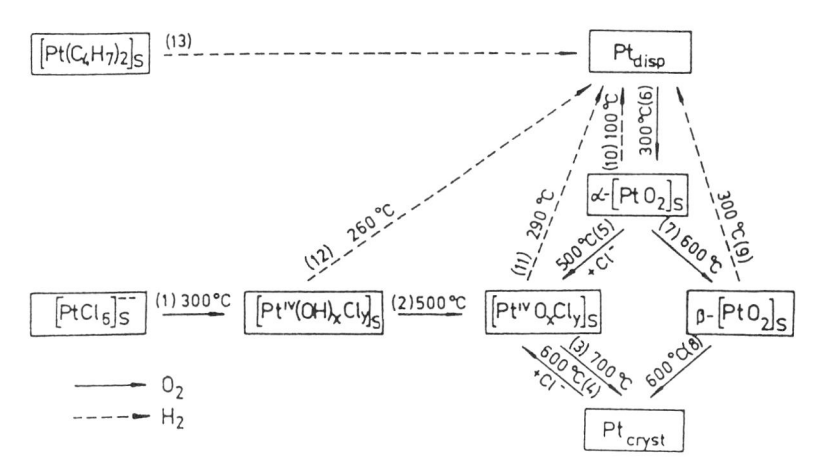

**Figure 22** Reaction pathways of supported species in oxygen (—) and hydrogen (----) at different temperatures. $+Cl^-$ indicates reaction with chloride present in the catalyst or added. From Ref. 105.

2. Chlorine treatment in the absence of oxygen may cause redispersion with chlorine in the gas phase even at temperatures around 350°C, as reported by Bournonville and Martino (90).
3. The appropriate chlorine treatment can retard sintering, but the extent of such retardation is very much dependent on conditions of pretreatment. At lower temperatures, sintering may be retarded, whereas at higher temperatures, the chlorine obviously leads to increased rates of redispersion.

### E. Influence of Sintering on Catalytic Reactions

Very few works have been reported on the effect of sintering on the catalytic reforming reactions. One of the first works on $Pt/Al_2O_3$ catalysts dealing with Pt sintering kinetics was reported by Herrman et al. (106), who correlated the rate of decrease in surface area of a $Pt/Al_2O_3$ catalyst as second order in the remaining surface area. This approach formed the basis of another study by Maat and Moscou (107), who investigated the sintering effect on both activity and selectivity in naphtha reforming catalysts. The experiment consisted of heating the original sample for an indicated period of time in an inert atmosphere, making the determination of the surface area of the Pt, and then correlating with results from the study of the activity in normal heptane reforming at 200 psia, 500°C, 2.44 g/h-g catalyst, and an $H_2/n\text{-}C_7$ molar ratio of 5.3. Although some effects on activity were found, they were not as significant as expected. For example, a 98% reduction of Pt surface area resulted in a 25% reduction in conversion. This result was attributed to the intrinsic activity associated with different crystallite sizes. From the selectivity results, as shown in Figure 23, it can be seen that selectivity between the various normal heptane reforming products changes markedly with different degrees of sintering. The changes in activity are reflected primarily in decreased cyclization reactions. Because the aromatics produced in dehydrocyclization reactions would contribute more to the octane rating of the product than any of the other components, this decrease represents a real change in product quality resulting from a selectivity alteration caused by deactivation. The important conclusion of this work is that catalyst deactivation by sintering influences not only the activity but also the selectivity properties. As stated later by Butt and Petersen (96), Maat and Moscou did not report any notable sintering of the support, so presumably the acidic function is relatively unchanged and the apparent increase in isomerization activity on sintering is the result of increased reactant availability in comparison with the metallic function.

The sintering of metal particles that results in loss of metal surface area can affect the activity and selectivity balance of the reforming reactions. For $n$-hexane reforming, it was found that an increase in the platinum particle size leads to a slight increase in dehydrocyclization and a decrease in isomerization (108). This effect is more pronounced when the catalyst is also poisoned by coke

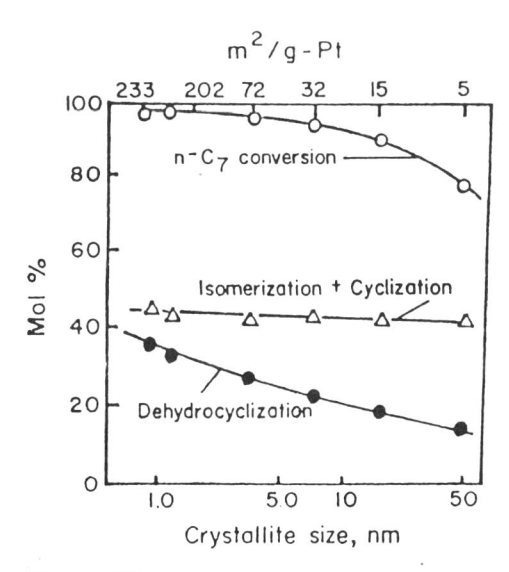

**Figure 23** Reforming of normal heptane as a function of the extent of sintering. From Ref. 107.

deposition. As stated by Franck and Martino (86), the effects of sintering appear most clearly in an industrial unit after catalyst regeneration in which the platinum has not been correctly redispersed. The low initial activity is often noted as a consequence of low dispersion that affects hydrogenation, dehydrogenation, dehydrocyclization, and hydrogenolysis reactions.

The effect of iridium agglomeration on the performance of platinum-iridium catalysts was studied by McVicker and Ziemiak (109). The authors found that an inverse relationship exists between either the rate of dehydrocyclization of normal heptane or naphtha reforming activity and the degree of iridium agglomeration.

The previous examples indicate that sintering of platinum catalysts can produce significant changes in the activity and selectivity of the reforming reactions. Notable is the fact that when smaller metal crystallites are present (highly dispersed metal catalysts), the effect of sintering become more important.

## F. Metal Crystal Redispersion

The opposite process to sintering is redispersion. This process is of tremendous industrial importance in catalytic reforming and is commonly practiced because of the cost of the Pt/Al$_2$O$_3$. As a result, it is still the subject of many studies (110,111) and patents (112,113). Redispersion results in a decrease in particle size accompanied by a net increase in specific surface area. Two models have been proposed to explain redispersion (114). The so-called thermodynamics redispersion model hypothesizes the formation of metal oxides that detach from the crystallites and migrate to active sites. The crystallite splitting model hypothesizes the formation of an oxide scale, which stresses the particle and cracks it open, creating new surface area.

In general, the process used for redispersion involves treating a thermally deactivated catalyst with a strong gaseous oxidant such as chlorine at elevated temperatures. The reagent readily attacks the metal particles to form chlorides or volatile carbonyl chlorides. Redispersion results from the movement of these oxidation products over the support, either by gas-phase transport or by surface migration. Subsequent decomposition of the dispersed oxidation product at a relatively low temperature leads to metal in a more dispersed state.

McHenry et al. (115) noted an increase in the dehydrocyclization rate of normal heptane at 490°C using Pt/Al$_2$O$_3$ catalyst. The rate increase was attributed to an increase in the amount of soluble platinum, which was believed to result from the formation of a Pt/Al$_2$O$_3$-Cl complex that increased the active surface area of the catalyst. Since then, other studies have observed an increase in the platinum surface area under different thermal treatment and atmosphere conditions (116,117). Figure 24 (118) shows that for two Pt/Al$_2$O$_3$ catalysts, the extent of redispersion depends on the treatment temperature and type of

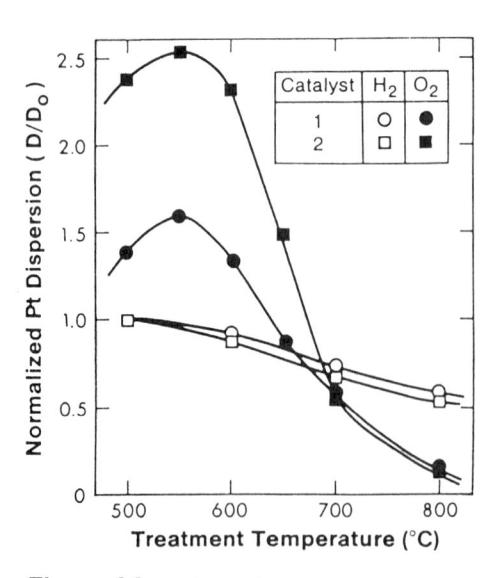

**Figure 24** Effects of 1 h treatment in oxygen and hydrogen on the dispersion of Pt/Al$_2$O$_3$ catalysts. From Ref. 118.

atmosphere, with the process not working in hydrogen. To the contrary, Weller and Montagna (119) found that cyclic treatment in hydrogen and oxygen at 550°C produced redispersion of the metal under reducing conditions, while sintering was observed during the oxygen part of the operation. It was suggested (94) that an oxygen treatment could recover only the fraction of platinum complexed with alumina during prior high-temperature $H_2$ treatment and that it could not actually redisperse the sintered large platinum crystallites. Lieske et al. (105) have stated that redispersion is not possible using oxygen itself and that the presence of chlorine is needed. Bournonville and Martino (90) also agree that chlorine plays an important role in the redispersion of platinum alumina catalysts. The effect of chlorine treatment on the dispersion of platinum metal particles supported on alumina was studied by Foger and Jaeger (120) using XRD, TEM, ultraviolet diffuse reflectance, and temperature-programmed reduction (TPR). They concluded that successful redispersion of platinum by chlorine treatment is achieved in a four-step process:

1. Attack of platinum particles to form volatile $\beta$-$PtCl_2$
2. Vapor-phase transport of the chloride
3. Adsorption of the $PtCl_2$ molecule
4. Further chlorination to form complex chlorides of $Pt^{4+}$ strongly bonded to the surface

The authors found that with $Pt/Al_2O_3$ the formation of anionic complexes of Pt(IV) is favored due to the relatively strong interaction of such compounds with the alumina surface, and an increase in metal dispersion is generally observed after chlorine treatment followed by reduction.

Differences in the size of the particles upon alternate heating in $H_2$ and $O_2$ during redispersion of $Pt/Al_2O_3$ catalysts was reported in a study by Sushumna and Ruckenstein (121). The catalyst particles, studied using electron microscopy and electron diffraction, appear smaller on heating in oxygen and larger on heating in hydrogen. This behavior is apparently in contrast to the normally observed expansion of particles on oxidation and contraction on reduction. The alternation in size is explained on the basis of the formation of an undetectable film around each particle due to the strong interaction between crystallites and support in an oxygen atmosphere and its withdrawal to merge with the respective particle in a hydrogen atmosphere. The film remains in the immediate vicinity and in contact with the parent particle and does not interconnect with another to form a contiguous surface film. The authors also suggested that one of the possible mechanisms of redispersion observed in oxidizing atmospheres involves extension and film formation.

In summary, it can be concluded that platinum on alumina reforming catalysts can be redispersed using high-temperature treatment in oxygen with the presence of chlorine. This process is also called oxychlorination. Industrial practice involves exposure of the catalyst to HCl or $CCl_4$, at 450–500°C in an atmosphere containing 2–10% oxygen diluted with nitrogen for a period of 1–4 h (122). One possible mechanism for oxychlorination is shown in Figure 25 (66), which involves the adsorption of $O_2$ and $Cl_2$ on the platinum crystallites. $AlCl_3$ is formed and then complexed into $PtCl_2(AlCl_3)_2$. The complex dissociatively adsorbs on alumina to form $PtCl_2O_2$ complexes, which can be reduced to monodisperse active platinum clusters (66,122).

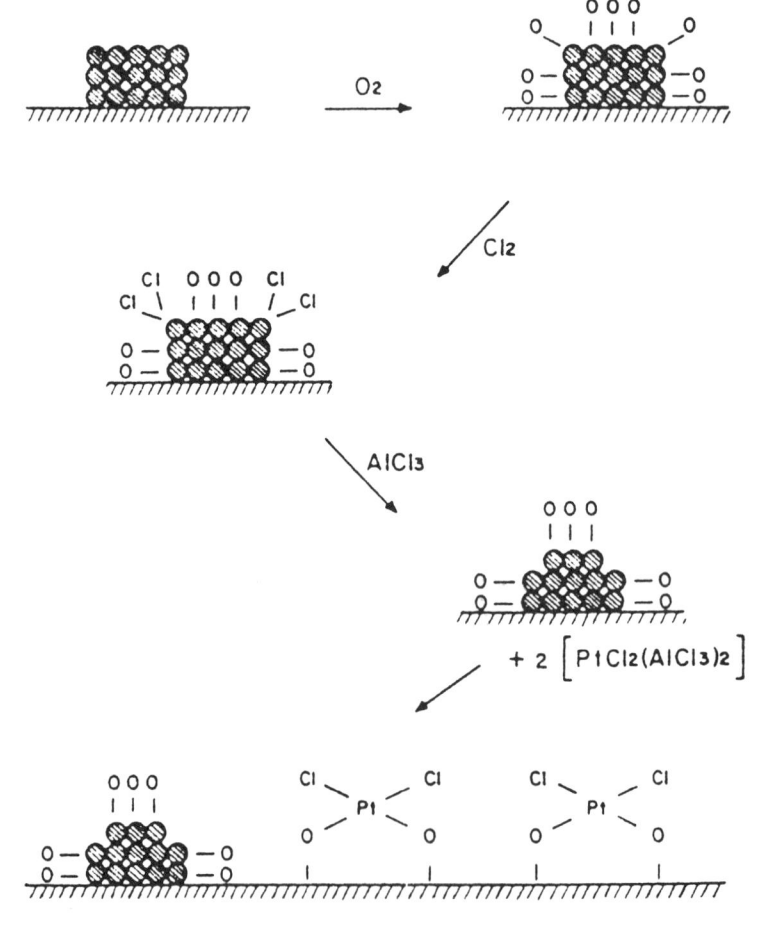

**Figure 25** Mechanism for redispersion by oxychlorination of $Pt/Al_2O_3$ catalysts. From Ref. 67.

## G. Mechanistic Model for Sintering-Redispersion of Pt/Al$_2$O$_3$ Reforming Catalysts

An interpretation of the results published on sintering and redispersion of Pt/Al$_2$O$_3$ catalysts was proposed by Wanke et al. (91) in a comprehensive mechanistic model that involves all the chemical species present during the sintering and redispersion process. This model is summarized in Figure 26. In developing the model, it was assumed that sintering in hydrogen and inert atmospheres occurs by migration of Pt atoms or small Pt clusters, and, as the authors mentioned, there is little evidence to support this assumption. On the other hand, in oxidizing atmospheres the intermediate species are reasonably well established. Wanke and co-workers (91) explained the model in the following way:

Changes in Pt dispersion during treatment in oxidizing atmospheres are governed by the formation of Pt$^{4+}$ species (105,123,124). The stability and mobility of the support-Pt$^{4+}$ complexes determine whether sintering or redispersion occurs. During low-temperature oxygen treatment of well-dispersed Pt/Al$_2$O$_3$ catalysts, Pt oxides are formed (125) and subsequent reduction restores the platinum to a well-dispersed state. Bulk platinum oxides are unstable at high temperatures. The rate of sintering at these conditions is rapid

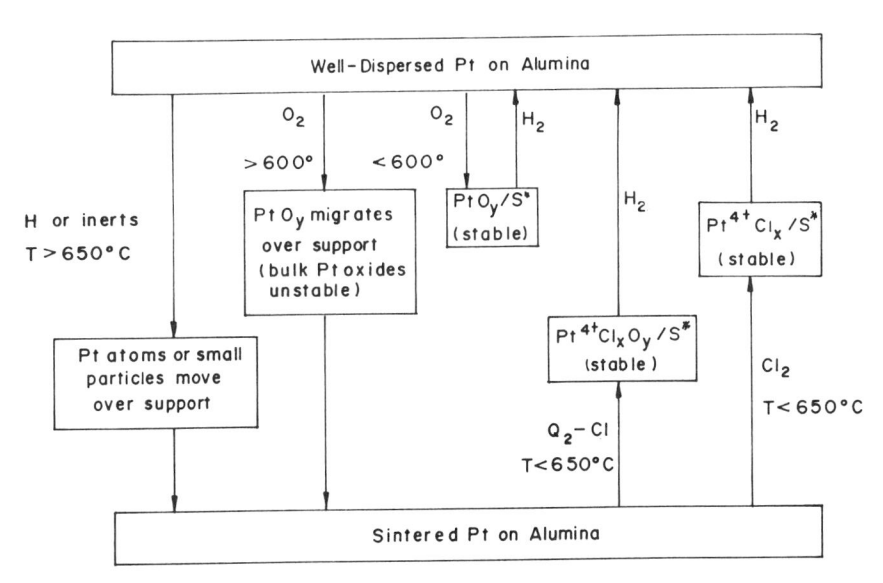

**Figure 26** Schematic representation of sintering and redispersion for Pt/Al$_2$O$_3$ in different atmospheres. From Ref. 91.

because oxygen facilitates the removal of platinum (as $PtO_y$) from the metal particles. It is likely that some of the $PtO_y$ is trapped on specific sites on the alumina surface while the remainder of the $PtO_y$ is captured by metal crystallites. Reduction of $Pt/Al_2O_3$ which has been treated in oxygen at high temperature results in a catalyst with a bimodal Pt particle size distribution. The small Pt clusters come from the $PtO_y$ which was trapped at special sites on the alumina. The temperature at which sintering starts in oxygen depends on the chloride content of the alumina. Higher temperatures are required to sinter chloride-containing catalysts. Sintering in oxygen usually starts at 600 to 650°C.

Chloride not only stabilizes $Pt/Al_2O_3$ catalysts but also is required for redispersion of sintered catalysts. Surface species such as $Pt^{4+}O_yCl_x \cdot Al_2O_3$ (105) and $Pt^{4+}Cl_x \cdot Al_2O_3$ (120) have been identified as the agents responsible for redispersion. These species are well dispersed over the surface of the alumina and are formed during treatment in $O_2$-$Cl_2$ or $Cl_2$ at temperatures below about 650°C. At higher temperatures, these species are unstable and sintering results. Although the crucial role of chlorine in the redispersion of $Pt/Al_2O_3$ has been well established, the detailed reaction mechanism for the formation of these surface complexes is not known. For example, it is not known whether chlorine is necessary for the detachment of Pt from the Pt crystallites or whether chlorine is required only for stabilization of the Pt-surface complexes, since redispersion of sintered $Pt/Al_2O_3$ occurs in the absence of chlorine in the gas phase. Chloride in the support is sufficient for redispersion to occur. Wanke et al. (91) also mentioned the possibility that chloride on the alumina mobilizes protons on the support which attack the Pt particles, resulting in migration of molecular Pt species onto the support surface.

## IV. CONCLUDING REMARKS

As demonstrated in this chapter, a large variety of effects have been observed in the poisoning and sintering of bifunctional naphtha reforming catalysts. On the poisoning side, it was mainly suggested that the poison molecule can have an electronic or geometric effect on the nonpoisoned catalyst surface which enables it to modify the intrinsic catalytic properties of the surface. Any explanation about the change of activity, selectivity, and stability due to the adsorption of the poison molecule will have to take into consideration the interaction between the metal, the hydrocarbon, and the poison-containing compound.

Regarding the sintering-redispersion of platinum catalysts, it was shown that these processes are strong functions of temperature, atmosphere, composition, and characteristics of the support. Furthermore, the rates of sintering are much lower in inert and reducing atmospheres than in oxygen atmospheres at temperatures at which sintering occurs in oxygen. A mechanistic model for sintering-redispersion stressed the importance of the presence of chlorine in these processes.

## ACKNOWLEDGMENT

The author wishes to acknowledge the support provided by King Fahd University of Petroleum and Minerals during the preparation of the manuscript.

## REFERENCES

1. Chapter 9, this volume.
2. C. H. Bartholomew, P. K. Agrawal, and J. R. Katzer, *Adv. Catal. 31*, 135 (1982).
3. J. Barbier, E. Lamy-Patara, P. Marecot, J. P. Bitiaux, J. Cosyns, and F. Verna, *Adv. Catal. 37*, 279 (1990).
4. E. B. Maxted, *Adv. Catal. 3*, 129 (1951).
5. J. Barbier, in *Deactivation and Poisoning of Catalysts* (J. Oudar and H. Wise, eds.). Marcel Dekker, New York, 1985, p. 121.
6. B. C. Gates, J. R. Katzer, and C. A. Schuit, *Chemistry of Catalytic Processes*. McGraw-Hill, New York, 1979, p. 184.
7. J. W. Jenkins, U.S. Patent 3,449,461 (1961).
8. H. E. Kluksdahl, U.S. Patent 3,325,737 (1968).
9. J. Oudar, *Mater. Sci. Eng. 42*, 101 (1980).
10. J. W. Schultze, *Croat. Chem. Acta 48*, 643 (1976).
11. H. P. Bonzel and R. Ku, *J. Chem. Phys. 58*, 4617 (1973).
12. S. R. Keleman, T. E. Fisher, and J. A. Schwarz, *Surf. Sci. 81*, 440 (1979).
13. J. G. McCarty and H. Wise, *J. Catal. 94*, 543 (1985).
14. J. Oudar, S. Pinol, C. M. Pradier, and Y. Berthier, *J. Catal. 107*, 445 (1987).
15. J. Oudar and H. Wise, *Deactivation and Poisoning of Catalysts*. Marcel Dekker, New York, 1985, p. 51.
16. J. Oudar, *Catal. Rev. Sci. Eng. 22*(1), 171 (1980).
17. T. E. Fisher and S. R. Keleman, *J. Catal. 53*, 24 (1978).
18. J. A. Schwarz and S. R. Keleman, *Surf. Sci. 87*, 510 (1979).
19. J. Biswas, G. M. Bickle, P. G. Gray, D. D. Do, and J. Barbier, *Catal. Rev. Sci. Eng. 30*, 161 (1988).
20. P. G. Menon and J. Prasad, 6th International Congress on Catalysis, London, 1977, p. 1077.
21. J. Barbier, P. Marecot, L. Tifouti, M. Guenin, and R. Frety, *Appl. Catal. 19*, 375 (1985).
22. J. M. Parera, C. R. Apesteguia, J. F. Plaza de los Reyes, and T. F. Garetto, *React Kinet. Catal. Lett. 15*, 167 (1980).
23. C. Apesteguia, J. Barbier, J. F. Plaza de los Reyes, T. F. Garetto, and J. M. Parera, *Appl. Catal. 1*, 159 (1981).
24. C. Apesteguia and J. Barbier, *React. Kinet. Catal. Lett. 19*(3-4), 351 (1982).
25. P. A. Van Trimpont, G. B. Marin, and G. F. Froment, *Appl. Catal. 24*, 53 (1986).

26. G. M. Bickle, J. Biswas, and D. D. Do, *Appl. Catal.* **36**, 259 (1988).
27. C. R. Apesteguia, S. M. Trevizan, T. F. Garetto, J. F. Plaza de los Reyes, and J. M. Parera, *React. Kinet. Catal. Lett.* **20**, 1 (1982).
28. C. R. Apesteguia, and J. Barbier, *Bull. Soc. Chim. (France)*, 5-6, I-165 (1982).
29. C. R. Apesteguia, C. E. Brema, T. F. Garetto, A. Borgna, and J. M. Parera, *J. Catal.* **89**, 52 (1984).
30. J. Barbier, P. Marecot, and L. Tifouti, *Appl. Catal.* **19**, 375 (1985).
31. M. Bernard and F. Busnot, *Usuel de Chimie Generale de Minerale*. Paris, 1984, p. 92.
32. G. Marin, Thesis, Institut Francais du Petrole (1978).
33. J. Barbier, P. Marecot and L. Tifouti, 9th Iberoamerican Symposium on Catalysis, Lisbon, 1984, p. 540.
34. K. E. Foger and J. R. Anderson, *J. Catal.* **54**, 318 (1978).
35. J. C. Vedrine, M. Dufaux, C. Naccache, and B. Imelik, *J. Chem. Soc. Faraday Trans. 1*, **440** (1978).
36. P. Gallezot, J. Datka, J. Massardier, M. Primet, and B. Imelik, 6th International Congress on Catalysis, London, 1977, p. 696.
37. C. R. Apesteguia and J. Barbier, *J. Catal.* **78**, 352 (1982).
38. P. Biloen, J. N. Helle, H. Verbeek, F.M. Dautzenberg, and W. M. Sachtler, *J. Catal.* **63**, 112 (1980).
39. V. K. Shum, J. B. Butt, and W. M. H. Sachtler, *J. Catal.* **96**, 371 (1985).
40. J. Jothimurugesan, A. K. Nayak, G. K. Mehta, K. N. Rai, S. Bhatia, and R. D. Srivastva, *AIChE J.* **31**, 1997 (1985).
41. J. Barbier, G. Corro, Y. Zhang, J. Beournonville, and J. P. Franck, *Appl. Catal.* **13**, 245 (1985).
42. J. Barbier and P. Marecot, *J. Catal.* **102**, 21 (1986).
43. J. M. Parera and J. N. Beltramini, *J. Catal.* **112**, 357 (1988).
44. M. Guenin, M. Breysse, R. Frety, K. Tifouty, P. Marecot, and J. Barbier, *J. Catal.* **105**, 144 (1987).
45. M. J. Sterba, P. C. Weinert, A. G. Lickus, E. L. Pollitzer, and J. C. Hayes, *Oil Gas J.* **66**(53), 140 (1968).
46. J. H. Sinfelt, U.S. Patent 3,835,034 (1974).
47. W. K. Meerbott, A. H. Cherry, B. Chernoff, J. Crocoll, J. D. Heldman, and C. J. Kaemmerlen, *Ind. Eng. Chem.* **46**(10), 2026 (1954).
48. H. Heinemann, H. Shalit, and W. S. Briggs, *Ind. Eng. Chem.* **45**, 800 (1953).
49. A. G. Oblad, H. Shalit, and H. T. Tadd, *Adv. Catal.* **9**, 518 (1957).
50. W. F. Engel, Dutch Patent 84,714 (1957).
51. V. Haensel, U.S. Patent 3,006,841 (1961).
52. W. C. Pfefferle, *Preprints Division of Petroleum Chemistry, American Chemical Society, Houston Meeting, 1970*, p. A21.
53. K. M. Minachev and D. A. Kondratchev, *Izvest. Akad. Nauk. SSSR*, 300 (1960).
54. N. R. Bursian and G. N. Maslyanskii, *Khim. Technol. Topliv Masel.* **6**, 6 (1961).
55. J. C. Hayes, R. T. Mistche, E. L. Pollitzer, and E. H. Homeier, *Preprints 16th National Meeting American Chemical Society, Los Angeles, 1974*, p. 334.
56. R. Maurel, G. Leclerq, and J. Barbier, *J. Catal.* **37**, 324 (1975).
57. M. J. Sterba and V. Haensel, *Ind. Eng. Chem. Prod. Res. Dev.* **15**, 2 (1976).
58. P. B. Weisz, Proceedings, Second International Congress Catalysis, Paris, 1960, p. 937.

59. P. A. Van Trimpont, G. B. Marin, and G. F. Froment, *Appl. Catal. 17*, 161 (1985).
60. R. W. Coughlin, A. Hasan, and K. Kawakami, *J. Catal. 88*, 193 (1984).
61. G. M. Bickle, J. N. Beltramini, and D. D. Do, *Ind. Eng. Chem. Res. 29*, 1801 (1990).
62. C. R. Apesteguia and J. Barbier, *React. Kinet. Catal. Lett. 19*, 351 (1982).
63. W. M. H. Sachtler and P. Biloen, *Division of Petroleum Chemistry, American Chemical Society*, Seattle meeting, 1983.
64. J. Barbier, G. Corro, Y. Zhang, J. P. Bournonville, and J. P. Franck, *Appl. Catal. 16*, 269 (1985).
65. G. B. McVicker, P. J. Collins, and J. J. Ziemiak, *J. Catal. 46*, 438 (1977).
66. P. F. Lovell, U.S. Patent 3,565,789 (1971).
67. J. P. Franck and G. Martino, in *Progress in Catalyst Deactivation* (J. L. Figueiredo, ed.). Martinus Nijhoff, The Hague, 1982, p. 355.
68. W. P. Hettinger, C. D. Keith, J. L. Gring, and J. W. Teter, *Ind. Eng. Chem. 47*(4), 719 (1955).
69. J. Barbier, A. Morales, P. Marecot, and R. Maurel, *Bull. Soc. Chim. Belg. 88*(7–8), 569 (1979).
70. J. B. Butt, C. L. M. Joyal, and C. E. Megiris, in *Catalyst Deactivation* (E. E. Petersen and A. T. Bell, eds.). Marcel Dekker, New York, 1987, p. 17.
71. C. Paal and W. Hartmann, *Bert. 51*, 711 (1918).
72. J. Barbier and R. Maurel, *Chim. Phys. 4*, 75 (1978).
73. E. B. Maxted, *J. Chem. Soc.* 1987 (1949).
74. E. Ruckenstein and D. B. Dadyburjor, *J. Catal. 48*, 73 (1977).
75. H. C. Yao, M. Sieg, and H. K. Plummer, *J. Catal. 68*, 365 (1979).
76. B. Pulvermacher and E. Ruckenstein, *J. Catal. 35*, 115 (1974).
77. D. B. Dadyburjor, in *Catalyst Deactivation* (B. Delmon and G. F. Froment, eds.). Elsevier, Amsterdam, 1987, p. 21.
78. Z. Wagner, *Electrochemistry, 65*, 581 (1961).
79. E. Ruckenstein and B. Pulvermacher, *AIChE J. 19*, 356 (1973).
80. P. Wynblatt and N. A. Gjostein, *Prog. Solid State Chem. 9*, 21 (1975).
81. F. W. Kopf, W. C. Pfefferle, M. H. Dalson, W. A. Decker, and J. A. Nevison, Preprints, API, Division of Refining, 1969, p. 23.
82. A. Ramaswamy, P. Ratnasamy, S. Sivasanacker, and A. J. Leonard, Proceedings, 6th International Congress on Catalysis, London, 1976, Vol. 1, p. 855.
83. H. Charcosset, R. Frety, G. Leclerq, B. Moraweck, L. Tournayan, and J. Varloud, *React. Kinet. Catal. Lett. 10*, 301 (1979).
84. X. Bingyan, C. Hailin, W. Jinrong, and Z. Lixin, *J. Fuel Chem. Technol. 14*(4), 315 (1986).
85. J. P. Bournonville, Thesis, Paris, 1979.
86. J. P. Franck and G. P. Martino, in *Deactivation and Poisoning of Catalysts* (J. Oudar and H. Wise, eds.). Marcel Dekker, New York, 1985, p. 205.
87. T. F. Garetto and C. R. Apesteguia, Actas X Iberoamerican Symposium on Catalysis, Merida, Venezuela, 1986, p. 988.
88. W. L. Callender and J. M. Miller, Proceedings, 8th International Congress on Catalysis, Berlin, 1984, p. 491.
89. C. L. Pieck, E. L. Jablonski, and J. M. Parera, *Appl. Catal. 62*, 47 (1990).

90. J. P. Bournonville and G. Martino, in *Catalyst Deactivation* (B. Delmon and G. F. Froment, eds.). Elsevier, Amsterdam, 1980, p. 159.
91. S. E. Wanke, J. A. Szymura, and T. T. Yu, in *Catalyst Deactivation* (E. E. Petersen and A. T. Bell, eds.). Marcel Dekker, New York, 1986, p. 65.
92. J. A. Szymura and S. E. Wanke, 9th North American Meeting of the Catalysis Society, 1985, paper B2.
93. G. Lietz, H. Lieske, H. Spindler, W. Hanke, and J. Volter, *J. Catal. 81*, 17 (1983).
94. F. M. Dautzenberg and H. B. Volters, *J. Catal. 51*, 26 (1978).
95. J. den Otter and F. M. Dautzenberg, *J. Catal. 53*, 116 (1978).
96. J. B. Butt and E. E. Petersen, *Activation, Deactivation and Poisoning of Catalysts*. Academic Press, New York, 1988, p. 171.
97. S. A. Hasan, F. H. Khalil, and F. G. El-Gamal, *J. Catal. 44*, 5 (1976).
98. I. Shushamna and E. Ruckenstein, *J. Catal. 109*, 433 (1988).
99. P. J. F. Harris, *J. Catal. 97*, 527 (1986).
100. M. Uchida, H. Arai, and H. Tominaga, *Shokubai (Catalyst) 22*(4), 310 (1980).
101. T. Inui and T. Miyake, *J. Catal. 86*, 446 (1984).
102. J. W. Geus, in *Chemisorption and Reaction on Metal Films* (J. R. Anderson, ed.). Academic Press, New York, 1971.
103. A. A. Castro, O. A. Scelza, E. R. Benvenuto, G. T. Baronetti, and J. M. Parera, *J. Catal. 69*, 222 (1981).
104. G. I. Straguzzi, H. R. Aduriz, and C. E. Gigola, *J. Catal. 66*, 171 (1983).
105. H. Lieske, G. Lietz, H. Spindler, and J. Volter, *J. Catal. 81*, 8 (1983).
106. R. H. Herrman, S. F. Adler, M. S. Goldstein, and R. M. deBaun, *J. Phys. Chem. 65*, 2189 (1961).
107. H. J. Maat and L. Moscou, Proceedings, 3rd International Congress on Catalysis 1965, p. 1277.
108. P. P. Lankhorst, H. C. de Jongste, and V. Ponec, *Catalyst Deactivation*. Stud. Surf. Sci. Catal., vol. 6. Elsevier, Amsterdam, 1980, p. 43.
109. G. B. McVicker and J. J. Ziemiak, *Appl. Catal. 14*, 229 (1985).
110. P. C. Flynn and S. E. Wanke, *J. Catal. 34*, 390 (1974).
111. R. M. Fiedorow and S. E. Wanke, *J. Catal. 43*, 34 (1976).
112. D. R. Hogin, R. C. Morberck, and H. R. Sanders, U.S. Patent 2,916,440 (1959).
113. R. Coe and H. Randlett, U.S. Patent 3,278,419 (1966).
114. Y. F. Chu and E. Ruckenstein, *J. Catal. 55*, 287 (1978).
115. K. W. McHenry, R. J. Bertolacini, H. M. Brennan, J. L. Wilson, and H. S. Seelig, Proceedings, International Congress on Catalysis, 2nd, 1960, vol. 2, p. 2295.
116. M. F. L. Johnson and C. D. Keith, *J. Phys. Chem. 67*, 200 (1963).
117. M. Kraft and H. Spindler, Proceedings, International Congress on Catalysis, 4th, 1971, p. 1252.
118. S. E. Wanke, in *Progress in Catalyst Deactivation* (J. L. Figueiredo, ed.). Martinus Nijhoff, Boston, 1982, p. 315.
119. S. W. Weller and A. A. Montagna, *J. Catal. 20*, 394 (1971).
120. K. Foger and H. Jaeger, *J. Catal. 92*, 64 (1985).
121. I. Sushumna and E. Ruckenstein, *J. Catal. 108*, 77 (1987).
122. C. H. Bartolomew, *Chem. Eng. 91*(23), 96 (1984).

123. P. Birke, S. Engels, K. Becker, and D. Neubauer, *Chem. Technol. 31*, 473 (1979).
124. P. Birke, S. Engels, K. Becker, and D. Neubauer, *Chem. Technol. 32*, 253 (1979).
125. T. Fukushima and J. R. Katzer, Proceedings, 7th International Congress on Catalysis, Tokyo, 1981, p. 79.

# 11
## Regeneration of Reforming Catalysts

### Jorge Norberto Beltramini
*King Fahd University of Petroleum and Minerals, Dhahran, Saudi Arabia*

## I. INTRODUCTION

The formation of carbonaceous deposits on a catalyst surface is an undesirable reaction that takes place during naphtha reforming processes. This deposit gradually deactivates the catalyst by blocking its active sites. To compensate for this decay of activity, increasing the operation temperature is the common industrial practice. However, after a certain time, it becomes necessary to regenerate the catalyst to restore its original catalyst properties. The operation of burning carbon from the catalyst surface is the most time-consuming part of the whole regeneration process.

Weisz and Goodwin (1) examined the burnoff of carbonaceous deposits of porous catalysts and observed that the presence of oxides of transition metals increased coke burnoff rates by several orders of magnitude. Work done by Baker et al. (2) extended this idea, showing the power of platinum to catalyze the oxidation of graphite at 500°C. This is of considerable importance for reforming catalysts, because at temperatures above 500°C sintering of the platinum becomes a problem. Thus, reforming catalysts can theoretically be regenerated successfully at temperatures below the sintering temperature.

The catalyst regeneration process consists of the following steps:

1. Elimination of coke by controlled burning.
2. Catalyst rejuvenation, which comprises:
   (a) Restoration of the acid function and metallic dispersion by chlorination.
   (b) Reduction with hydrogen.
   (c) Passivation by sulfiding.

The reactivity of the coked catalyst during regeneration depends on the coke structure, which is dependent on the nature of the catalyst and the reaction conditions. The chemistry and nature of carbon deposition and the characterization of coked catalysts are reviewed first, followed by a discussion of the behavior toward gasification using hydrogen and oxygen. Additional results on regeneration of coked-sulfided catalysts are also presented. Finally, the rejuvenation procedure for the catalyst and a typical industrial regeneration procedure are outlined.

## II. CHEMISTRY AND NATURE OF CARBON DEPOSITION

The reforming process requires the catalyst to be bifunctional, consisting of a metallic function and an acidic function. The metallic function of the reforming catalyst is represented by Pt, or its bimetallic successors containing Re, Ir, Sn, Ge, etc., as the second component, dispersed on a chlorinated alumina support that also provides the acid function. The metallic component mainly provides hydrogenation-dehydrogenation and hydrogenolysis activity, whereas the chlorinated alumina provides the cracking and isomerization activity. Dehydrocyclization, ring enlargement, and the bifunctional isomerization reactions require the participation of both functions. Side reactions occurring during reforming also lead to the formation of carbonaceous residue or coke on the surface of the catalyst. This causes a reduction in the activity and selectivity for the desired reforming products and consequently a reduction in the product octane number. Deactivation due to coke formation is a complex process that afflicts the majority of organic gas-phase catalytic reactions. In addition, depending on the severity of the deactivation by coke, periodic shutdowns are necessary for regeneration of the coked catalyst. The conditions for regeneration of the catalyst depend on the reactivity of the coke. The nature of the coke deposits, their mechanisms of formation, and their behavior toward gasification are important variables that need to be considered when a regeneration procedure is established. The variation in the nature of coke deposits has a significant effect on the burning conditions and, as a result, graphite is often used as a model compound to establish the mechanism of coke burning (2). Excellent reviews in this area are available in the literature (3,4).

Carbonaceous deposits on reforming catalysts tend to be pseudographitic or aromatic in nature and the hydrogen-to-carbon ratio of these deposits depends on the specific conditions of their formation. A typical model for production of coke on platinum is shown in Figure 1. A series of fragmentation reactions and successive dehydrogenation reactions lead to the formation of carbon atoms, which may combine to form more toxic deposits. The coke deposits initially have a hydrogen content of about 1.0 to 1.5 hydrogen atoms per surface carbon atom (5). Removal of this hydrogen by successive fragmentation can also be achieved by increasing temperature. The concept of the formation of a layer of graphitic coke on the metal sites and its removal by hydrogen was introduced by Biswas et al. (6). Other mechanisms reported in the literature for metal site coking indicate the presence of different types of coke on the surface (7,8). The routes are based on polymerization reactions occurring on metal surfaces.

Experimental results suggest that the initial deposition of coke is on the metal crystallites and their surroundings and continues on the support (9). A mechanism of coke formation involving both functions that has been accepted involves dehydrogenation activity on the metal to form unsaturated compounds, which are the highly reactive coke precursors, and condensation activity on the acid sites where the highly reactive coke precursors react to produce coke (10). As a result, for this type of catalyst there are at least two different types of coke, one on the metallic sites and surrounding area and another on the support (11,12). It was also noted that the coke on the metal is more readily hydrogenated and is eliminated first during the regeneration process, whereas the coke on the support requires a longer period of time and higher temperatures to be removed (13).

Coking on reforming catalysts is a dynamic process in which partially hydrogenated carbonaceous species are produced. The species migrate across the surface to a nucleation site, at which they grow to form a deposit. This deposit is dehydrogenated and undergoes reordering as a function of time and high temperatures, producing a more graphitic coke (14). Gasification may, however, be influenced if other metallic components are added to the system (16). Addition of iridium to Pt/Al$_2$O$_3$ catalyst results in the formation of less polymerized coke structures which provide an excellent metal-carbon contact.

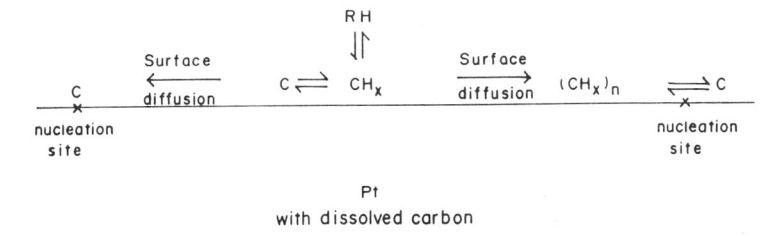

**Figure 1**   Model for the production of carbon on platinum. (From Ref. 5.)

The texture of these deposits facilitates the access of the gasifying agent principally by the spillover phenomenon (17), resulting in a decrease in the burning temperature. On the other hand, addition of Re, Ge, and Sn to platinum leads to a more orderly coke structure deposited mainly on the alumina support. As a consequence, a longer regeneration time is needed for these catalysts. Another reason for this longer regeneration time is that added Ge or Sn decreases the spillover effect of the metal and hence decreases the rate of coke combustion (18).

## III. CHARACTERIZATION OF COKE DEPOSITS

Coke deposited on the surface of naphtha reforming catalysts can be characterized by combustion using temperature-programmed oxidation (TPO) (19,20). As shown in Figure 2 by Beltramini and Datta (21), TPO of coked $Pt/Al_2O_3$ catalysts was found to give a twin-peaked combustion pattern. One peak (maximum at 475°C) occurs at nearly the same temperature as the peak produced by combusting coke on alumina (480°C), and the other peak occurs at 350°C. This second peak has been observed to increase when platinum loading is

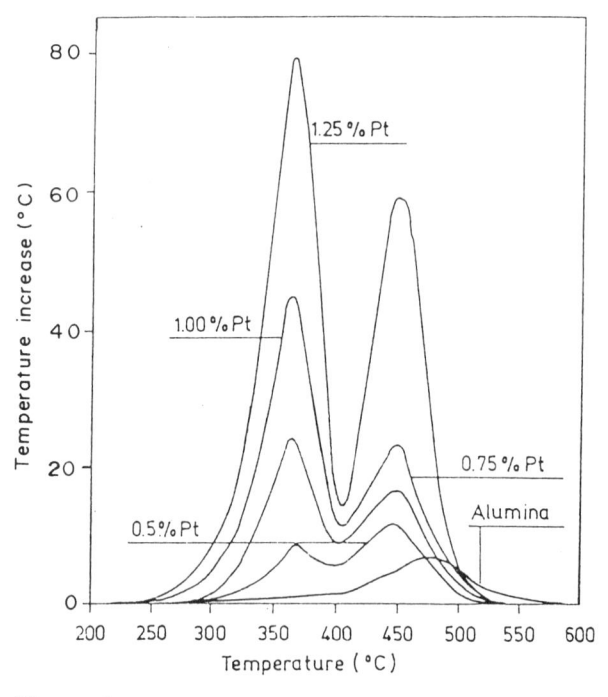

**Figure 2** TPO of $Pt/Al_2O_3$ coked catalysts. (From Ref. 21.)

increased. These results have corroborated the assumption that coke deposited on the metal is oxidizable at lower temperatures than that on the support (22).

Coke components can be analyzed after partial solvent extraction of the coke using a 4:1 molar ratio benzene-methanol solution (22). The soluble part of the coke can be characterized by gas chromatography–mass spectrometry (GC-MS) and proton nuclear magnetic resonance (H-NMR). The non-soluble coke fraction and the total coke can be analyzed by X-ray diffraction (XRD) after dissolving the $Al_2O_3$ with concentrated HCl. Figure 3 shows the chromatogram and the structures found by GC-MS analysis of the extract from a Pt/$Al_2O_3$ catalyst coked in the presence of methylcyclopentane (23). The H-NMR spectra for the same sample are presented in Fig. 4, which shows two zones, one at low shifts that corresponds to aliphatic protons and the other, starting at 6.5 ppm, corresponding to protons of condensed aromatic rings. By integrating the area corresponding to aliphatic protons, it was found that the alkyl chains

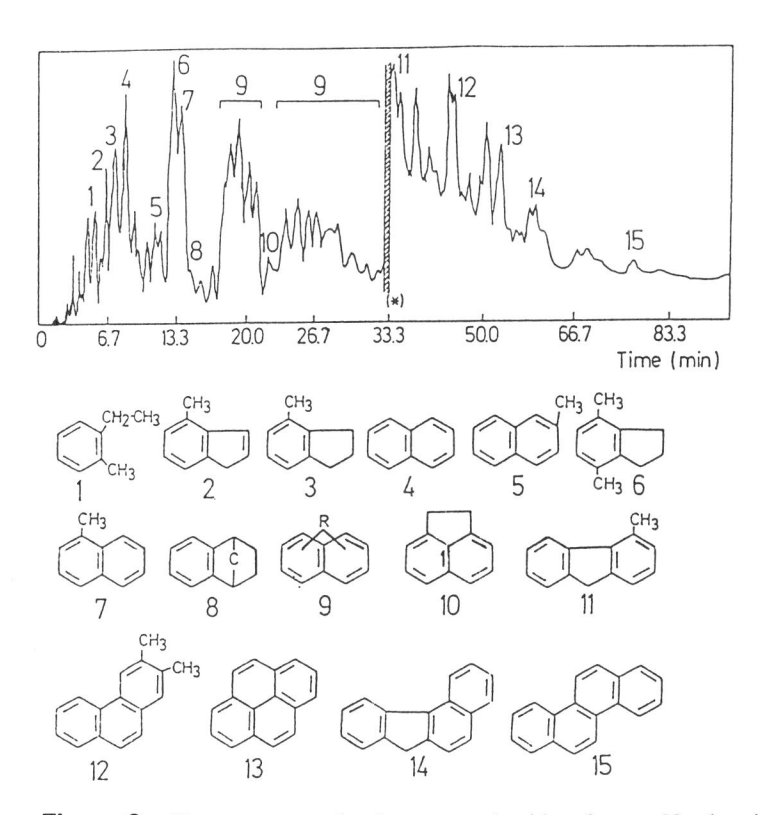

**Figure 3** Chromatogram of coke extracted with solvents. Number indicates the structure of the most important compounds recognized by MS. (*) Change of attenuation. R = two or three methyl groups attached to any of the two rings. (From Ref. 23.)

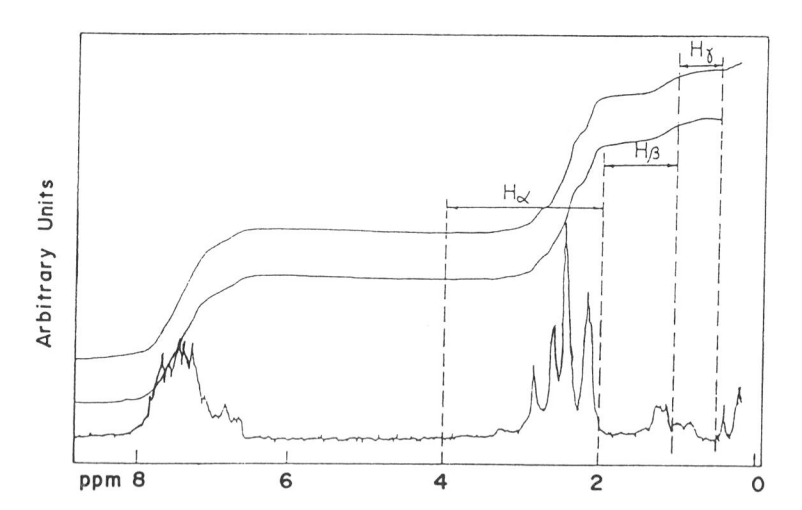

**Figure 4** H-NMR spectra of coke extracted with solvents. $H_\gamma$, Hydrogen in saturated groups $\gamma$ to aromatic rings; $H_\beta$, hydrogen in saturated groups $\beta$ to aromatic rings; $H_\alpha$, hydrogen in saturated groups $\alpha$ to aromatic rings. The last band was taken broadly enough to include saturated carbons in acenaphthene or indene-type structures. (From Ref. 23.)

joined to the aromatics were methyl groups. In addition, protons were found which correspond to saturated carbons in acenaphthene or indene-type structures. The region corresponding to aromatic proton signals included hydrogen attached to nonsaturated five-carbon-atom rings (23).

Studies of hydrocarbon deposits carried out using infrared (IR) spectroscopy (24) show that presence of aromatic C−H bonds, methylene groups, and aromatic rings (25). XRD analysis of coke performed after destruction of alumina showed the existence of poorly organized structure separated by 3.4 Å. This distance is greater than the spacing between graphite layers (3.34 Å). This greater separation between planes found in coke may be due to the presence of five-carbon-atom rings or alkyl chains joined to the rings (26).

Characterization of coked catalysts using electron probe microanalysis shows that coke is distributed homogeneously within the catalyst pellet. On the other hand, a microexamination revealed variations in local concentrations, suggesting the existence of three-dimensional deposits and hence indicating that the existence of a carbon monolayer is impossible (27).

## IV. REGENERATION OF COKED CATALYSTS

### A. Regeneration with Hydrogen

Hydrogen is found to have a significant impact on hydrocarbon reactions over the naphtha reforming process. It is already known that high hydrogen pressures lower the concentration of coke precursors, thereby keeping the coking rates low. This is because the strong dehydrogenation reactions involved in coking are restricted by the presence of hydrogen (28). On the metal surfaces, coke levels are controlled by hydrogen cleaning (29). It was postulated (5) that the cleaning action of hydrogen occurs by catalytic hydrogenation, whereas the slow removal of coke occurs by catalytic hydrogasification. Biswas and coworkers (6) postulated the mechanism shown in Figure 5 for the hydrogen cleaning of platinum metal surfaces. This mechanism is based on the distinction of two types of coke on metal sites, one being easily removed by hydrogen (reversible coke) and the other less readily removed (irreversible coke).

Trimm (30) suggested that the coke cleaning mechanism is due to reaction with dissociated hydrogen. Somorjai (31) showed that the hydrogen dissociation reaction has a high activation energy on crystallite faces but occurs readily on corner Pt atoms. It is known that, on bifunctional catalysts, a small amount of the coke is deposited on the metal, whereas the majority is accumulated on the alumina support (32). Hence a large part of the coke which is destroyed by hydrogen treatment is deposited on the support. These phenomena can be explained by considering either the mobility of carbonaceous deposits on the catalyst surface or the hydrogen mobility. Parera et al. (33) proposed an increase in coke removal by hydrogen spilled over from metal to alumina support.

The presence of an optimum surface chloride concentration in the regeneration of $Pt/Al_2O_3$ catalysts by hydrogen was found to be an important parameter by Parmeliana et al. (34). A catalyst with about 0.5 wt % $Cl^-$ shows complete regeneration during static and dynamic hydrogen spillover from platinum to alumina, which restores the catalyst surface, freeing it from hydrocarbon residues (35).

On platinum particles, addition of iridium or rhenium is known to promote the methanation of coke deposits (36) by increasing the hydrogenolysis activity of the metal. However, in the case of $Pt-Ir/Al_2O_3$, iridium promotes hydrogen

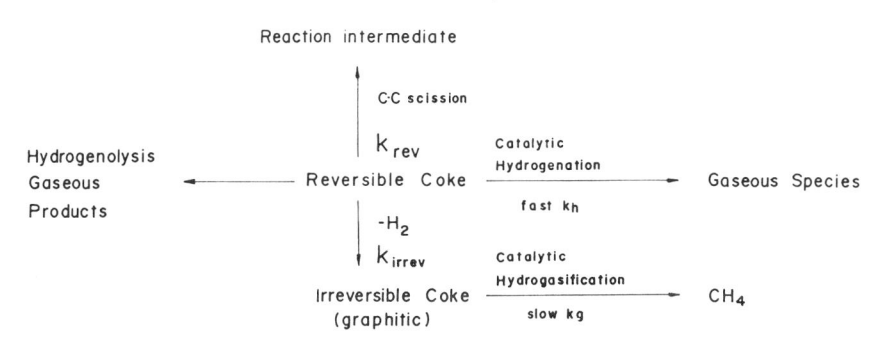

**Figure 5** Model of carbon removal by hydrogen. (From Ref. 6.)

**Table 1** Effect of the Amount of Coke and the Temperature of Hydrogen Treatment on the Regeneration of a Coked 1% Pt/Al$_2$O$_3$ Catalyst

| Catalyst | Carbon (wt %) | $a/a_0$[a] |
|---|---|---|
| Pt/Al$_2$O$_3$ | 0 | 1 |
| Pt/Al$_2$O$_3$ coked 1 h | 2.61 | 0.25 |
| Pt/Al$_2$O$_3$ coked 1 h, regenerated 1 h, 400°C | 2.32 | 0.29 |
| Pt/Al$_2$O$_3$ coked 1 h, regenerated 1 h, 430°C | 1.69 | 0.69 |
| Pt/Al$_2$O$_3$ coked 0.5 h | 1.30 | 0.26 |
| Pt/Al$_2$O$_3$ coked 0.5 h, regenerated 1 h, 400°C | 1.00 | 0.32 |
| Pt/Al$_2$O$_3$ coked 0.5 h, regenerated 1 h, 430°C | 0.65 | 0.70 |

[a] $a/a_0$, relative activity for benzene hydrogeneration.
*Source*: Ref. 39.

spillover, which can increase coke removal from the support (37). On the other hand, this explanation cannot be supported in the case of Pt-Re/Al$_2$O$_3$, because addition of rhenium to platinum is found to have no effect on the spillover effect of the metal (38).

The effect of the nature of bifunctional catalysts on their regeneration by hydrogen treatment was studied for precoked catalysts by Marécot et al. (39). Table 1 shows that carbon remaining decreases and, consequently, the activity of the catalyst for benzene hydrogenation increases with the temperature of hydrogen treatment. Table 2 shows the effect of the time of hydrogen treatment at 400°C and 430°C on the amount of coke remaining on the catalyst, the relative activity in benzene hydrogenation, and the hydrogen content of the coke. It was found that increasing the time of treatment results in decreasing coke con-

**Table 2** Effect of Regeneration Time Under Hydrogen Flow on the Carbon Deposit of the Coked 1% Pt/Al$_2$O$_3$ Catalyst

| Regeneration time (h) | Carbon C$_T$ (wt %) | Regeneration 400°C Carbon C$_H$ (wt %) | H/C | $a/a_0$ | Regeneration 430°C Carbon C$_H$ (wt %) | H/C | $a/a_0$ |
|---|---|---|---|---|---|---|---|
| 0 | 2.61 | 2.61 | 0.48 | 0.25 | 2.61 | 0.48 | 0.25 |
| 1 | 2.61 | 2.32 | 0.44 | 0.29 | 1.69 | 0.42 | 0.69 |
| 5 | 2.61 | 1.84 | 0.41 | 0.43 | 1.25 | 0.40 | 0.92 |
| 10 | 2.61 | 1.83 | 0.40 | 0.44 | 0.97 | 0.33 | 0.92 |
| 20 | 2.61 | 1.69 | 0.39 | 0.45 | 0.92 | 0.29 | 1.00 |

[a] $a/a_0$, relative activity for benzene hydrogenation.
*Source*: Ref. 39.

tent but carbonaceous deposits become more and more dehydrogenated, showing that the more hydrogenated compounds are eliminated first. On the other hand, the metallic activity is totally regenerated by a hydrogen treatment of 20 h at 430°C, although a large amount of coke still remains on the catalyst. This regeneration of the metallic activity and the lower hydrogen-to-carbon ratio of the residual coke after hydrogen treatment agree with previous work showing that coke deposited on the metal is more hydrogenated than coke deposited on the support (40). Recently, this was confirmed from activation energy values of coke removal from metal and acid sites using hydrogen at different treatment time on coked $Pt/Al_2O_3$ catalyst (41). The results presented in Table 3 show that there is a greater increase in activation energy for coke removal from the acid sites than from the metal sites.

The reactivity of coke depends on its chemical nature and particularly on its degree of graphitization. Marécot et al. (39) found that the greater regeneration ability of well-dispersed platinum catalysts can be explained as follows: on large metallic particles, coke would be more dehydrogenated to a pregraphitic state and so would be less reactive in the presence of hydrogen. These results are in good agreement with those of Sachtler (42), who confirmed the high sensitivity of the poorly dispersed platinum catalysts toward the coking reaction. Later, Marécot et al. (43) confirmed these results in a regeneration study of reforming catalysts impregnated with coronene.

A kinetic reaction model for platinum-catalyzed hydrogen gasification under typical reformer operating conditions was proposed by Biswas et al. (6). The model is shown in Table 4. It has been proposed that at reforming temperatures of 450–550°C, the rate-limiting step for catalyzed hydrogen gasification is controlled by the gasification kinetic reaction. This is justified by the following two reasons:

1. The surface diffusion across the metal-coke interface is fast, thus interfacial equilibrium can be assumed (44).
2. The strong dependence of the hydrogasification rate on the amount of deposited coke.

**Table 3** Kinetic Models Parameters

| Hydrogen treatment time (h) | Activation energy (kcal/mol) | |
| --- | --- | --- |
| | Metal sites | Acid Sites |
| 1 | 2.75 | 12.86 |
| 3 | 3.37 | 24.2 |

*Source*: Ref. 41.

**Table 4** Proposed Mechanisms for Metal Site Reforming, Catalytic Gasification, Hydrogenation, and Coking Reactions

| | | |
|---|---|---|
| Main reforming reaction: | $A \Longleftrightarrow B$ | |
| | $A + M \Longleftrightarrow A-M$ | $K_A = \dfrac{[A-M]}{[M][A]}$ |
| | $A-M \Longleftrightarrow B-M$ | |
| | $B-M \Longleftrightarrow B + M$ | $K_B = \dfrac{[B-M]}{[M][B]}$ |
| Reversible coking: | $[PM] \xrightarrow{k_{rev}} C_{rev}$ | |
| Irreversible coking: | $C_{rev} \xrightarrow{k_{irr}} C_{irr}$ | |
| Gasification mechanism: | $\tfrac{1}{2}H_2 + M \Longleftrightarrow M-H$ | $K_H = \dfrac{[M-H]}{[M][H_2]^{1/2}}$ |
| | $M-H \Longleftrightarrow H(ads)$ | $K_1 = \dfrac{[H(ads)]}{[M-H]}$ |
| | $C_{irr} + H(ads) \Longleftrightarrow CH(ads) + M$ | $K_2 = \dfrac{[CH(ads)][M]}{[C_{irr}][H(ads)]}$ |
| | $CH(ads) + H(ads) \Longleftrightarrow CH_2(ads)$ | $K_3 = \dfrac{[CH_2(ads)]}{[CH(ads)][H(ads)]}$ |
| | $CH_2(ads) + H(ads) \xrightarrow{k_g} CH_3(ads)$ | |
| | $CH_3(ads) + H(ads) \xrightarrow{fast} CH_4$ | |
| Hydrogenation mechanism: | | |
| I. Dissociation control | $\tfrac{1}{2}H_2 + M \xrightarrow{k_{hd}} M-H$ | |
| | $M-H + C_{rev} \xrightarrow{k_h} M + C-H$ | |
| II. Hydrogen transfer | Unknown | |

As shown in Table 4, the addition of adsorbed atomic hydrogen on the coke to the $(CH_2)$ intermediate was found to be slow step (45), with all other steps being in equilibrium. From a site balance for all major species on platinum, the expressions for the catalyzed hydrogasification rate $(r_g)$ and catalytic hydrogenation rate $(r_h)$ are the following, respectively:

$$r_g = \frac{k[M_0]^3(1-a)a^2 K_H^3 [H_2]^{3/2}}{1 + K_H[H_2]^{1/2} + K_A[A] + K_B[B] + K_P[P]^2} \quad (1)$$

$$r_h = \frac{k_{hd}[H_2]^{1/2}[M_0]a}{1 + K_A[A] + K_B[B] + K_P[P] + [M-H]} \quad (2)$$

where

$$\frac{[M]}{[M_0]} = \frac{a}{[1] + K_H[H_2]^{1/2} + K_A[A] + K_B[B] + K_P[P]} \tag{3}$$

and A, B, and P are the reactants, product, and coke precursor species, respectively. The local metal activity is related to coke content by

$$a = 1 - \frac{[C_{rev}] + [C_{irrev}]}{[M_0]} \tag{4}$$

where $C_{rev}$ and $C_{irrev}$ are the reversible and irreversible amounts of deposited coke.

In the case of catalytic hydrogenation of coke, the rate of coke removal is affected by the mode of hydrogen supply, i.e., dissociation of molecular hydrogen or hydrogen transfer from adsorbed coke and associated species (46). The hydrogen transfer reaction is 5–10 times slower than molecular hydrogen dissociation at 300–400°C, but its importance increases at higher temperatures. The notion that the hydrogen transfer reaction may be important in reforming reactions is not new. Figure 6 (6) shows the effect of hydrogen partial pressure on hydrogenation rate, indicating that the hydrogen dissociation is more important than hydrogen transfer. The experimentally observed effect of hydrogen partial pressure to decrease hydrogenation rate, compared to hydrogasification, is due to the reaction order of 0.5 compared to 1.5 for hydrogasification (6). From the detailed information presented, it seems that the role of platinum

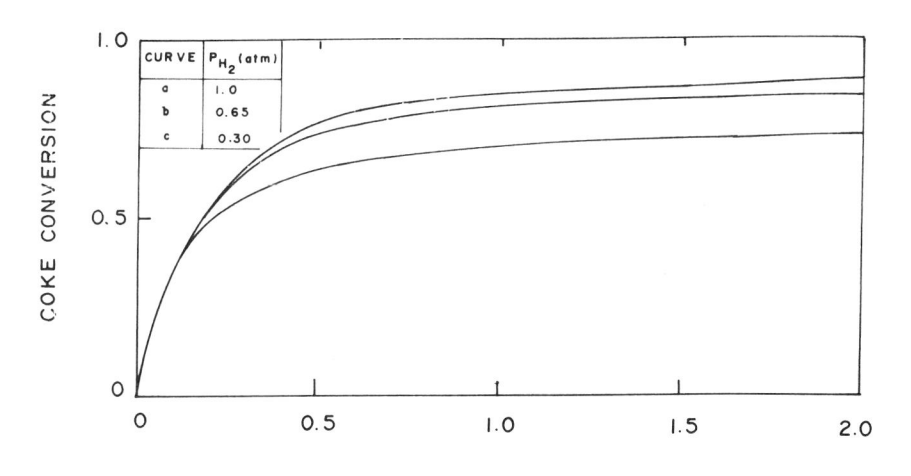

**Figure 6** Effect of pressure on hydrogenation rate of reversible coke at 500°C. (From Ref. 6.)

catalysts in carbon gastification using hydrogen is still not well established and further work on model systems is required.

## B. Regeneration with Oxygen

Regeneration in the presence of oxygen can be complicated by the fact that metal sintering can be promoted at high temperatures (47). The burning of coke starts with the coke deposited on the platinum because of the catalytic action of platinum and because the deposit is richer in hydrogen. The burning then spreads to the carbon on the support by the action of oxygen spillover, with burning at lower temperatures when more platinum is present. The last part of the coke to be burned is the most distant from the metal and is the coke poorest in hydrogen (48).

Comparing the TPO of $Pt/Al_2O_3$ catalysts with different metal-to-acid ratios, it is accepted that Pt catalyzes the burning of the coke. The higher the metal-to-acid ratio, the lower the burning temperature (11). Similar results were reported by Baccaud et al. (49), who found that Pt decreases the oxidation temperature of graphite. There is an additional factor that can reduce the coke burning temperature. When greater amounts of Pt are present, the coke is less polymerized because of the Pt-catalyzed hydrogenation of coke precursors occurring during the reaction. The less polymerized deposit would oxidize at a lower temperature.

Spillover of oxygen was also found to be important during coke removal (33). TPO results for samples of $Pt/Al_2O_3$ and $Al_2O_3$ coked using naphtha and mixed together (Figures 7 and 8) show that when Pt is present, the oxidation starts at lower temperatures and the temperature peak is also at a lower temperature. Although the amount of carbon is the same, the area of the TPO ther-

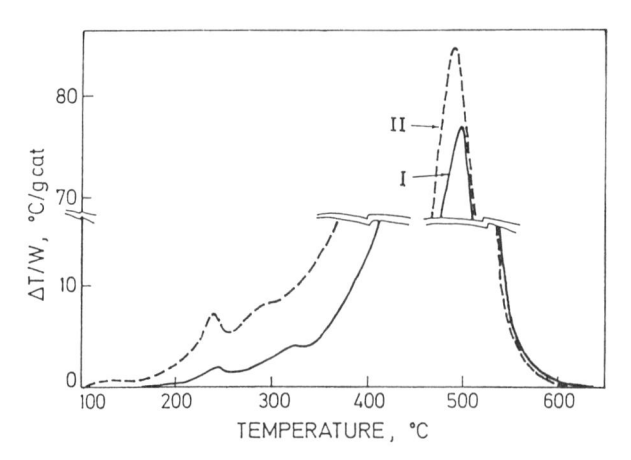

**Figure 7** TPO of alumina samples mixed with platinum-alumina after coking with naphtha during 6 h. (I and II refer to the degree of mixing.) (From Ref. 33.)

*Regeneration of Reforming Catalysts* 377

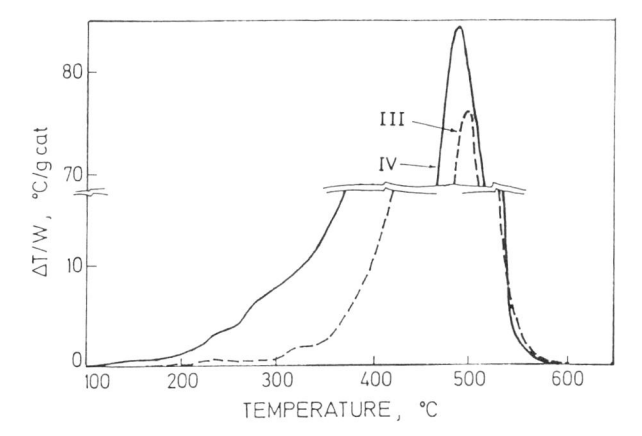

**Figure 8** TPO of alumina samples mixed with platinum-alumina after coking with naphtha during 6 h. (III and IV refers to the degree of mixing.) (From Ref. 33.)

mogram in the case of Pt/Al$_2$O$_3$ is greater than the one found for Al$_2$O$_3$, demonstrating that the carbon oxidation in the presence of platinum produces CO$_2$ whereas in the absence of platinum it produces a mixture of CO + CO$_2$, and releases less heat. The acceleration of coke burning in the presence of platinum can be explained by considering that oxygen is adsorbed on platinum and, in an activated form, travels by spillover along the support to the coked alumina particles. The activated oxygen spills along the pores of the coked alumina, eliminating the coke by burning at lower temperatures than that of the gaseous oxygen.

Traffano and Parera (38) found that the concentration of rhenium in Pt-Re/Al$_2$O$_3$ catalysts significantly influences the regenerative behavior. For bimetallic catalysts with less than 50% rhenium in the total metal concentration, regeneration of the catalyst occurred at the same temperature as for Pt/Al$_2$O$_3$ (around 500°C). When the rhenium fraction was above 50%, the behavior was similar to that of Re/Al$_2$O$_3$ or pure Al$_2$O$_3$, in that coke would not burn off until much higher temperatures were reached. Rhenium is not capable of promoting spillover, so when the rhenium concentration exceeds that of platinum, significant changes in the geometric and electronic effects between the two metals occur, resulting in a decrease of spillover capacity.

The influences of the operating temperature, pressure, and flow rate on the CO$_2$ concentration in the outlet gases were studied to monitor the evolution of coke burning on a commercially coked Pt-Re/Al$_2$O$_3$ catalyst (50). Monitoring the radial coke distribution during the burning of the catalyst pellet led to a model in which two types of coke with different degrees of polymerization were present. It was found that the exothermicity of combustion of the hydrogen adsorbed on the metal is important as the ignitor of the coke burning reaction.

The coke deposited on the metal can be eliminated by treatment with oxygen at low temperature (51). Table 5 shows that this treatment causes the oxidation of a small amount of carbon, but can regenerate the accessibility and the activity of metal atoms (52). Studies conducted using electron energy loss spectroscopy (53) showed that after combustion at nearly 300°C, partial regeneration of the catalyst is achieved, and the region up to a radius of 20 Å surrounding a 10-Å metal crystallite is free of carbon.

## C. Kinetics of Carbon Combustion

Knowledge of coke burning kinetics is essential in the simulation and design of the regeneration process. In contrast to catalytic cracking, the kinetics of the carbon combustion reaction on naphtha reforming catalysts have been scarcely studied. In general, the kinetics of carbon burning by oxygen can be represented by

$$-r_c = kP y_{O_2} C_C \qquad (5)$$

where $P$ is the total pressure, $y_{O_2}$ the mole fraction of oxygen, and $C_C$ the weight of coke on the catalyst. The validity of this correlation is certainly dependent on the amount of coke present initially (54). Studies by Walker et al. (55) supported the idea that the rate with respect to oxygen partial pressure is first order. However, a second-order dependence on carbon content has been reported in another study (56).

For bimetallic naphtha reforming catalysts, a modified model was proposed by Beltramini and Datta (57) in which the carbon on the catalyst is considered to be associated with two types of coke. Carbon deposited on metal sites and

**Table 5** Influence of Temperature of Oxidation of the Coke Deposited on Pt/Al$_2$O$_3$ on Regeneration of Accessibility of the Metal

| Catalyst | Oxidation temperature (°C) | C (%) | C% on the metal | Metal accessibility (atoms of Pt per g catalyst) | Activity in benzene hydrogenation/ mol h$^{-1}$g$^{-1}$ |
|---|---|---|---|---|---|
| Pt/Al$_2$O$_3$ (3.6%) noncoked |  |  |  | 20 × 10$^{18}$ | 0.02 |
| Pt/Al$_2$O$_3$ (3.6% coked) | 250 | 1.9 | 0.20 | 2 × 10$^{18}$ | 0.002 |
|  | 250 | 1.7 | 0.03 | 15 × 10$^{18}$ | 0.015 |
|  | 300 | 1.7 | 0.00 | 19 × 10$^{18}$ | 0.02 |
|  | 350 | 1.7 |  | 19 × 10$^{18}$ | 0.02 |

*Source*: Ref. 52.

## Regeneration of Reforming Catalysts

surroundings is designated $C_m$, while the carbon associated with the acid sites is designated $C_a$. The difference between $C_m$ and $C_a$ lies in their reactivity toward oxidation.

The two independent reactions for modeling carbon removal kinetics may be represented by the following equations:

$$C_m \text{(carbon on metal sites)} \xrightarrow{O_2} CO, CO_2 \tag{6}$$

$$C_a \text{(carbon on acid sites)} \xrightarrow{O_2} CO, CO_2 \tag{7}$$

The kinetics of the carbon removal are considered to be first order with respect to carbon concentration at constant oxygen partial pressure:

$$\frac{dC_m}{dt} = -k_m C_m \tag{8}$$

$$\frac{dC_a}{dt} = -k_a C_a \tag{9}$$

The concentration of total carbon ($C_t$) is related to the concentrations of the two types of carbon by the following expression:

$$C_t = C_m + C_a \tag{10}$$

Integrating Eqs. (8) and (9) with the appropriate initial conditions gives

$$C_m = C_m^0 \exp(-k_m t) \tag{11}$$

$$C_a = C_a^0 \exp(-k_a t) \tag{12}$$

where $C_m^0$ and $C_a^0$ represent the initial carbon associated with metal and acid sites, respectively.

Substitution of Eqs. (11) and (12) into (10) and dividing by the initial concentration of carbon leads to

$$\frac{C_t}{C_t^0} = (1 - \beta) \exp(-k_m t) + \beta \exp(-k_a t) \tag{13}$$

where $\beta$ is the fraction of the initial carbon associated with carbon on acid sites at zero time. Equation (13) was evaluated using TPO data for a Pt/Al$_2$O$_3$ catalyst coked with methylcyclopentane at different times. A good correlation with the data was observed, confirming the first-order reaction assumption. Combustion of coke on metal sites has an activation energy of 8.4 kcal mol$^{-1}$ and for coke on acid sites the activation energy is 18.8 kcal mol$^{-1}$. In the case of a coked alumina sample, the activation energy was found to be 22.6 kcal mol$^{-1}$. The lower value of activation energy for acid sites in comparison with alumina alone confirms the previous finding on the catalytic effect of platinum on the burning of the coke from acid sites.

Activation energy and reaction order with respect to oxygen for the coke burning of a Pt-Re/Al$_2$O$_3$ catalyst were also evaluated by Pieck et al. (58). Results confirm that on a bimetallic naphtha reforming catalyst, coke is composed of two main types of carbonaceous deposits. For the more hydrogenated coke fraction located on the metal, an activation energy for oxidation of 8.9 kcal mol$^{-1}$ was determined (Figure 9). Regarding the more polymerized coke fraction located on the support, the activation energy was found to be 24.2–26.2 kcal mol$^{-1}$. The greater activation energy obtained may be due to the fact that it corresponds to burning of the highly polymerized coke that burns at higher temperatures. It was also noted that when the catalyst was previously hydrogenated there was a considerable decrease in the activation energy (3.2 kcal mol$^{-1}$). This decrease is attributed to the favorable effect of the hydrogen adsorbed on the metal, which plays an important role as the ignitor of the combustion (50). The reaction order with respect to oxygen was found to be closer to 0.5.

To assess the recovery of the catalytic functions after coke burning, benzene hydrogenation and *n*-pentane isomerization were used as a test reaction, for the metallic function and acid function respectively (59). Figs. 10 and 11 show the conversion of benzene to cyclohexane and normal pentane to isopentane, respectively, on partially decoked sulfided and unsulfided Pt/Al$_2$O$_3$ catalysts as a function of burning temperature after reduction at 260°C and 500°C. The behavior of sulfided and unsulfided catalysts is similar: when reduced at 260°C, the metallic activity is recovered as coke is eliminated from the metallic function and the sulfided catalyst shows a more rapid recovery because the amount of coke on the metallic function is smaller. However, when reducing at 500°C, there is a migration of coke from the support to the metallic function, thus promoting its deactivation. When coke on the support is removed by oxidation, there is no migration, but there is a deactivating action from the sulfur compounds migrating back to the metal. It was also found that Pt/Al$_2$O$_3$, regenerated catalysts are more active for benzene hydrogenation than the fresh one. This was explained by the fact that oxidation-reduction treatments improved the catalyst activity. The authors also noticed that regenerated Pt-S/Al$_2$O$_3$ catalyst was more active for hydrogenation than the fresh catalyst. Perhaps, during coking, burning and reduction steps, some sulfur is lost from the metallic function, thus producing an increase in activity. On acid sites, normal pentane isomerization shows that in the range of coke burning temperature of 280–450°C, the normal pentane conversion is low and constant and similar to that of the coked catalysts. This temperature range coincides with the range in which coke is eliminated from the support, and thus the acidic and metallic catalyst activities are fully recovered. Since isopentane is the main product and its formation is related to coke removal, the elimination of coke from the support restores the activity of the acid sites.

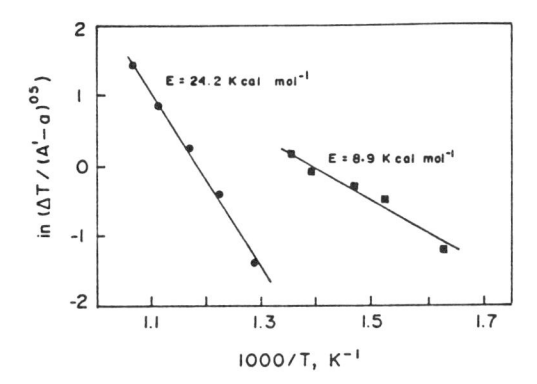

**Figure 9** Experimental data for the determination of activation energy. (■) Less polymerized coke; (●) more polymerized coke. (From Ref. 58).

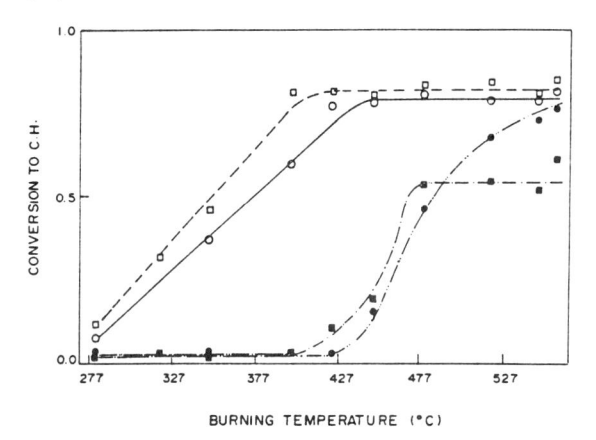

**Figure 10** Conversion of benzene to cyclohexane as a function of coke burning and for two reduction temperatures. (□) Pt/Al$_2$O$_3$ reduced at 260°C before catalytic test, (■) Pt/Al$_2$O$_3$ reduced at 500°C before test. (○) Pt-S/Al$_2$O$_3$ reduced at 260°C before test. (●) Pt-S/Al$_2$O$_3$ reduced at 500°C before test. (From Ref. 59.)

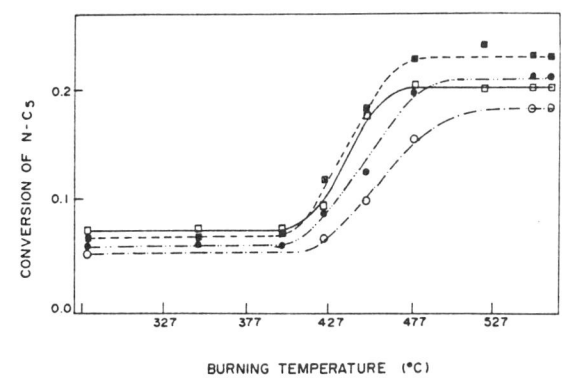

**Figure 11** Conversion of normal pentane as a function of coke burning temperature. Symbols as in Figure 10. (From Ref. 59.)

## V. REGENERATION OF SULFUR-CONTAMINATED CATALYSTS

When naphtha reforming catalysts are contaminated with sulfur, a special procedure has to be employed during the regeneration to diminish the sulfation of the support. Oxidation and reduction studies of the formation and elimination of sulfates on $Pt/Al_2O_3$ catalysts indicate that oxidation eliminated the sulfur deposited on the metal; however, the oxidized sulfur compounds were strongly held on the support (60). Industrially, the situation is more complex because sulfur cannot be completely eliminated during normal operation of a reformer.

In an industrial unit, sulfur causes the formation of iron sulfide from the reactor materials. When carbon removal takes place, the iron sulfide is oxidized into metal oxide, $Fe_2O_3$, and sulfur trioxide, $SO_3$. The $SO_3$ thus formed reacts with the alumina to form the thermally very stable aluminum sulfate, diminishing the surface hydroxyl concentration. The consequences of this sulfating of the alumina are as follows (61):

1. At the time of oxychlorination, redispersion of the metallic phase is more difficult.
2. At the time of reduction, there is a serious sintering risk due to the presence of water and hydrogen sulfide resulting from the reduction of the aluminum sulfate.

When the sulfur content of the catalyst exceeds 1000 ppm, a special regeneration procedure to remove the sulfur is followed before any attempt to burn the coke (61). Sulfur can be removed either after oxychlorination or before catalyst regeneration. The general procedure consists of reduction of the sulfur species under hydrogen at about 500°C. The formed hydrogen sulfide is eliminated from the gas by washing with an aqueous sodium carbonate solution. It is recommended that the procedure be continued until the exit hydrogen sulfide concentration is lower than 2 ppm volume for at least 1 h. When performed after oxychlorination, in order to increase the aluminum sulfate reduction and limit the loss of chloride from the catalyst, a chlorinated compound is injected during the operation. Figure 12 presents results obtained after the conventional regeneration procedure is applied to sulfided catalysts (62). It is clear that in the presence of sulfate ions, operating temperatures have to be increased more rapidly to maintain the octane number constant unless the sulfate removal step is employed.

Bonzel and Ku (63) have shown that, on a clean platinum surface, sulfur can be oxidized to gaseous sulfur dioxide at temperatures of 165–400°C. For a $Pt/Al_2O_3$ catalyst, the $SO_2$ occurs above 400°C, creating an aluminum sulfate species (64–66) which remains strongly bound to the alumina during oxidation and redispersion. Two different behaviors of the sulfur species when the catalyst is reduced in hydrogen have been proposed in the literature. Franck and Martino (61) stated that during reduction at 500°C, the sulfate is reduced

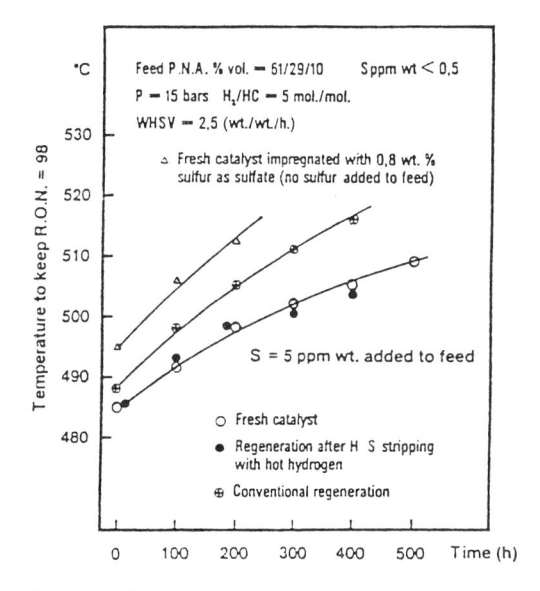

Figure 12   Regeneration of sulfided catalysts. (From Ref. 62.)

to hydrogen sulfide, escaping via the exit gases of the reformer. An interesting point to note is that reduction at 300°C cannot convert sulfates to hydrogen sulfide. Mathieu and Primet (67) Melchor et al. (68), and Apesteguia et al. (69) have shown that the platinum is irreversibly poisoned by sulfur from hydrogen sulfide dissociation during reduction at 500°C.

The regeneration of a coked-sulfided Pt-Re/Al$_2$O$_3$ catlyst has been examined by Schay et al. (70) and Apesteguia et al. (69). Schay et al. (70) argued that only a fraction of the sulfur can be removed from Pt-Re/Al$_2$O$_3$. Because the Re−S bond is strong, they proposed, based on their temperature-programmed reduction results, that the removed sulfur came from the platinum that was segregated from the Pt-Re alloy phase. However, whether this displaced sulfur adsorbed on the alumina or vented with the outlet gases was not addressed. Apesteguia et al. (69) noted that the level of sulfates on the alumina during regeneration can be decreased by 60% if the chlorine content at 510°C is controlled at optimum levels.

Garetto et al. (71) found that if the sulfate is incorporated into the catalyst during coke burning at levels higher than 0.08–0.1% S, it will enhance the coke formation on the catalyst and increase the rate of catalyst deactivation during the following operation cycle. These phenomena result from the difficulty of restoring the chloride level on the support as a result of the blockage of the Lewis acid sites of the alumina by sulfate ions, and from the excessive sulfidation of the metallic fraction by H$_2$S produced during the reduction of the sulfated catalyst with hydrogen.

Experiments on Pt/Al$_2$O$_3$ and Pt-Re/Al$_2$O$_3$ by Bickle et al. (72) indicated that sulfur did not leave the catalyst system during oxidative regeneration but remained bound to the alumina. During the reduction of regenerated Pt-S/Al$_2$O$_3$, the sulfate species present were converted to hydrogen sulfide, which dissociatively readsorbed on the platinum. Pt-Re/Al$_2$O$_3$ regeneration was found to be more complicated if sulfate and sulfite species were present on the alumina surface during calcination. These sulfite species are not reconverted to hydrogen sulfide during reduction. Quite surprisingly, the regeneration of a coked sulfided, Pt-Re/Al$_2$O$_3$ produced a lined-out catalyst whose product distribution was comparable to that of a lined-out Pt-Re/Al$_2$O$_3$ catalyst (Table 6). This result may indicate that sulfur has been vented from the system.

Figure 13 shows a surface science model explaining the effect of the presence of sulfur during oxidative regeneration which was proposed by Bickle et al. (72). In this model, it is hypothesized that the introduction of oxygen at 250°C oxidizes the irreversible sulfur to sulfur dioxide. This SO$_2$ is then adsorbed on the alumina support as either sulfite or sulfate species. As the temperature continues to increase, any sulfite or sulfate species present react on the alumina, forming sulfates (73). The free platinum surface revealed after sulfur removal is immediately available for combustion of the pseudographitic coke in the localized area, as was found by CO$_2$ evolution at 250°C on the TPO analysis. After this initial evolution, combustion of the remaining coke on the metal follows the same pattern as that of coked Pt/Al$_2$O$_3$, where the rate of atomic oxygen spillover to the graphitic coke surface dictates the rate of combustion. Simultaneously with coke combustion on the metal, some atomic

**Table 6** Product Distribution for Lined-out Unregenerated and Regenerated Pt-Re/Al$_2$O$_3$-Cl[a]

| Catalyst history | Group products yield (mol/100 mol of *n*-heptane) ||||
|---|---|---|---|---|
| | *n*-heptane | Isomers | Toluene | Cracking |
| Lined-out Pt-Re/Al$_2$O$_3$ | 29.7 | 26.6 | 16.0 | 59.6 |
| Regenerated and lined-out Pt-Re/Al$_2$O$_3$ | 27.1 | 33.5 | 11.3 | 55.9 |
| Lined-out and presulfided Pt-Re/Al$_2$O$_3$ | 39.8 | 33.6 | 7.2 | 35.1 |
| Regenerated, lined-out, and presulfided (previously) Pt-Re(S)/Al$_2$O$_3$ | 20.0 | 36.5 | 12.9 | 59.0 |

[a] At 480°C, 10 atm, WHSV = 5 h$^{-1}$, H$_2$/HC = 10, feed *n*-heptane.
*Source*: Ref. 72.

# Regeneration of Reforming Catalysts

(1)

IRREVERSIBLE SULFUR
IRREVERSIBLE COKE
PLATINUM
ACIDIC COKE

Original "sulfided-coked" Pt/Al$_2$O$_3$

(2)

SO$_2$ ADSORPTION
PLATINUM
ACIDIC COKE

Initial oxygen introduction removes S$_I$ as SO$_2$, which adsorbs to the alumina

(3)

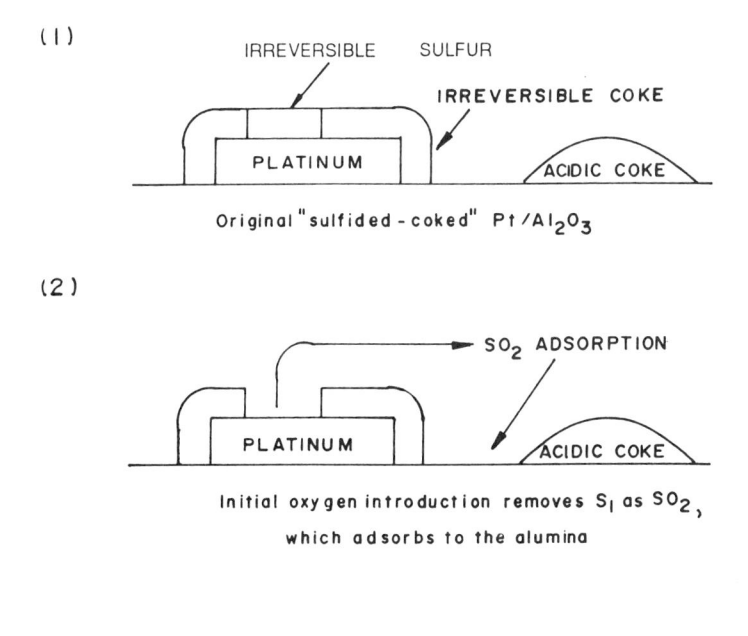

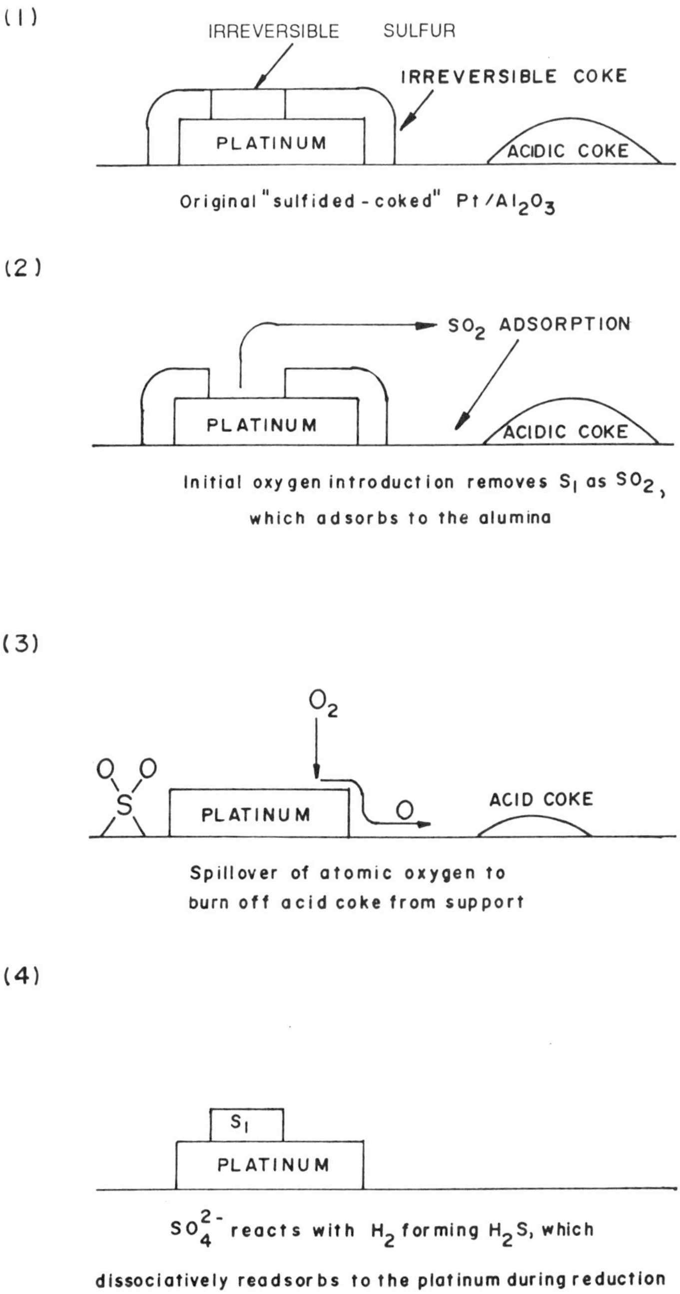

O$_2$
PLATINUM
ACID COKE

Spillover of atomic oxygen to burn off acid coke from support

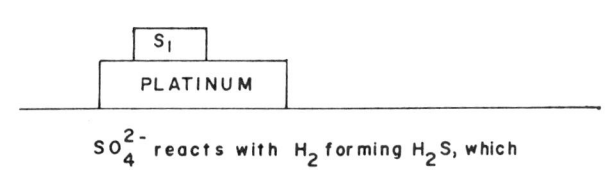

(4)

S$_I$
PLATINUM

SO$_4^{2-}$ reacts with H$_2$ forming H$_2$S, which dissociatively readsorbs to the platinum during reduction

**Figure 13** Model of the regeneration of sulfided coked Pt/Al$_2$O$_3$ catalysts. (From Ref. 72.)

oxygen spills over to the alumina and diffuses along the surface to the acidic coke, which is then combusted. Because it is not in intimate contact with platinum, the combined effect of surface diffusion of oxygen atoms and uncatalyzed acid coke combustion requires high temperatures (as indicated by the higher temperature TPO peak) for reaction.

Partial burning of the coke deposited on commercial Pt-Re(S)/Al$_2$O$_3$ was also studied using a mixture containing 1.9 vol % oxygen in nitrogen at different temperatures and regeneration times (74). Figure 14 show the TPO of original and partially regenerated samples at 350°C and 520°C for 2 h. Figure 15 shows the TPO when the regeneration conditions were 425°C for 2 and 9 h, respectively. It is seen that the more hydrogenated coke, which burns at a low temperature, may be removed by mild treatment (2 h at 350°C or 425°C) without any noticeable modifications to the less hydrogenated coke. When the burning severity is increased, part of the less hydrogenated coke is eliminated simultaneously with the more hydrogenated coke. The catalytic activity and selectivity of the partially regenerated catalysts were evaluated for benzene

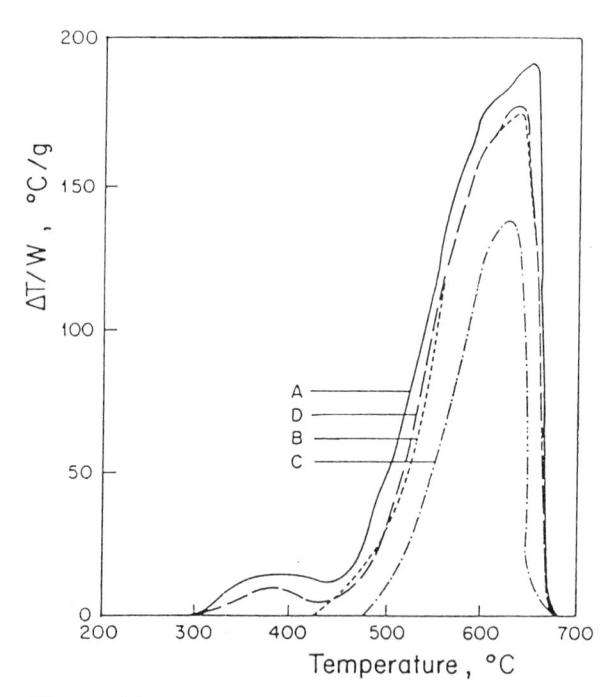

**Figure 14** TPO of the original and partially regenerated catalysts. Effect of temperature. (A) Original coked catalyst; (B) 350°C; (C) 520°C; (D) as B plus 2 h at 500°C in a hydrogen atmosphere. (From Ref. 73.)

*Regeneration of Reforming Catalysts* 387

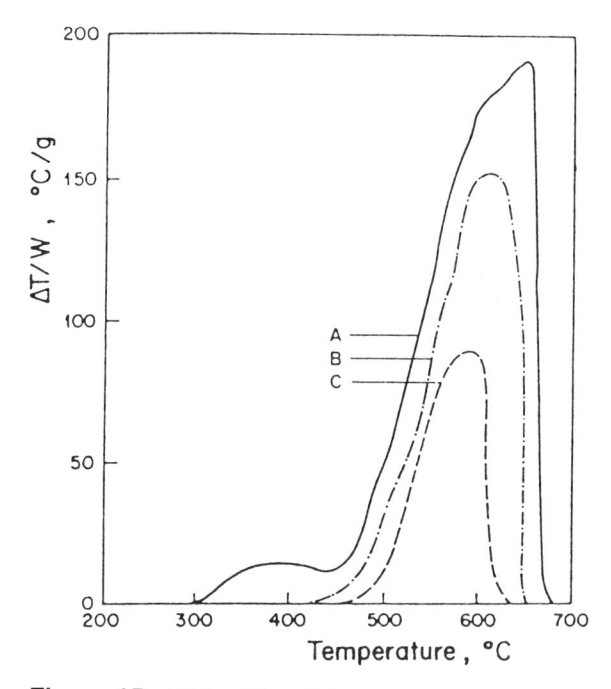

**Figure 15** TPO of the original and partially regenerated catalysts at 425°C. Effect of time. (A) Original coked catalyst; (B) 2 h; (C) 9 h. Conditions as in Figure 14. (From Ref. 73.)

hydrogenation and *n*-pentane isomerization. Figure 16 shows that the activity for isomerization increases as the amount of coke removed increases. This was explained by considering that the increase in carbon removal makes the support surface cleaner, and therefore more active sites become available for that reaction. On the other hand, the benzene hydrogenation results seem to depend on both carbon deposition and sulfur adsorption. When a partial regeneration is carried out, the coke on the metal is easily eliminated; hence, a mild treatment yields the same effect as a severe treatment. This was found to be true at temperatures below 480°C. Above this temperature, some modifications of the sulfur on the metal occur. Sulfur on the metal is oxidized to $SO_3$, which then migrates to the support, producing aluminum sulfate. As sulfur is removed from the metallic function, the hydrogenating activity increases. At 500°C, under a hydrogen atmosphere, the sulfate on the alumina is reduced to sulfite and migrates to the metal. Simultaneously, part of the coke on the support also migrates to the metal, thus reducing its activity below that of the lower reduction temperature catalyst (69).

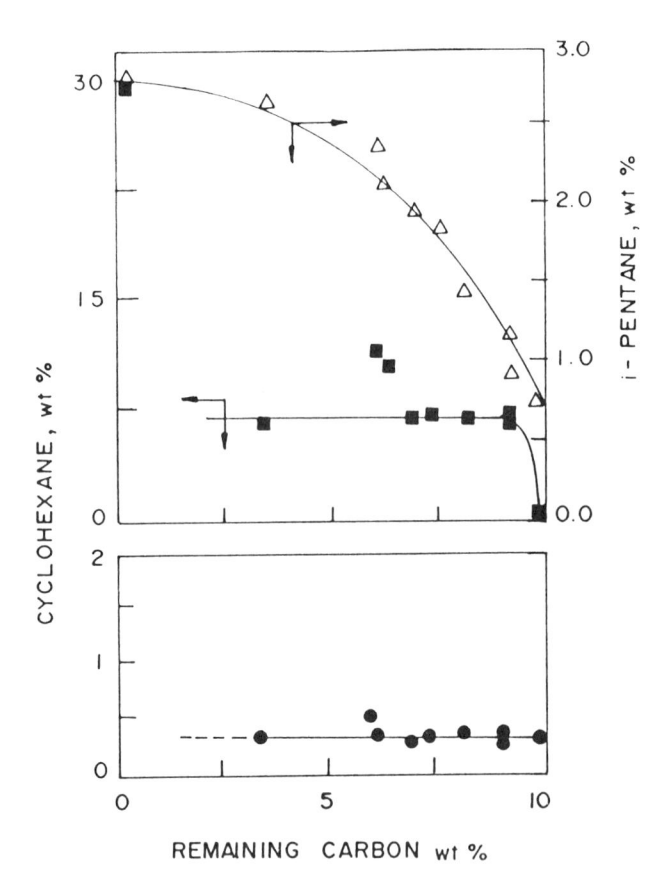

**Figure 16** Percentage of cyclohexane and isopentane in the test reactions as a function of the percentage of carbon remaining on the catalyst. Temperature of reduction (■) 260°C; (●) 500°C. (From Ref. 73.)

## VI. CATALYST REJUVENATION

The object of the rejuvenation step is to return the reforming catalyst to a state as close as possible to the properties of the fresh catalyst. This means that after burning the carbon from the catalyst it will be necessary to restore metals to their reduced state, high metal dispersion levels, and the proper level of acid function by controlled injection of chloride compounds. To achieve these goals, the general steps to follow are redispersion, reduction, and sulfidation.

## A. Redispersion

The aim of the redispersion operation is to restore the metal crystallite distribution to its fresh initial conditions. As explained in a previous chapter (47), reforming catalysts are susceptible to sintering. As the metal crystallites increase in diameter (greater than 10 Å) the activity of the naphtha reforming catalyst decreases (75). The agglomeration of metal crystals is caused mainly by high temperatures, the presence of water, additives, coke, and sulfur (76).

In conjunction with the sintering process, the redispersion process was extensively discussed in Chapter 10. In general, the redispersion can be accomplished by treating the catalyst at high temperature with a gaseous mixture of oxygen and chloride compounds. The technique used in an industrial plant requires continuous circulation of the regeneration gas, which normally contains 5 to 6 mol % oxygen, with injection of chloride compounds during a period of approximately 4 h maintaining the catalyst temperature at 480–510°C.

## B. Reduction

After redispersion, the metals on the catalyst surface are normally in an oxidized state, and to activate the catalyst for the reforming operation they must be reduced to the metallic state. The reduction operation is conducted in the presence of hydrogen at temperatures ranging from 450 to 500°C (77). To conduct this operation, normally the first step is to cool down the reactor and remove all traces of oxygen. For this purpose, the unit is depressured and the oxygen removal is conducted by purging with nitrogen. There still remains some dispute concerning the quality of the hydrogen used for the reduction step. Some operators believe that high-purity hydrogen is needed, but there is a growing use of streams with 90–95% hydrogen purity to serve the same purpose.

The effect of direct reduction by hydrogen without preoxidation was studied by Barbier et al. (78) using monometallic $Pt/Al_2O_3$ catalysts. They found that in the presence of water there is an inhibition effect on the reduction reaction, which can also induce platinum sintering. On the other hand, HCl has no kinetic effect on the reduction but is able to increase the metallic dispersion. At the end, increasing hydrogen pressure during reduction implies decreasing platinum accessibility.

## C. Sulfidation

The purpose of the sulfidation step is to deactivate the superactive catalyst sites that can produce excessive hydrogenolysis and hydrocracking when the hydrocarbon feed is restarted. The operation can be conducted by addition of hydrogen sulfide or mercaptan compounds to the circulating hydrogen. The effect of the sulfidation step on the activity, selectivity, and deactivation of the catalyst was largely discussed in previous chapters of this book.

## VII. TYPICAL INDUSTRIAL REGENERATION PROCEDURE

Regeneration of $Pt/Al_2O_3$ catalysts was successfully carried out industrially well before the first laboratory results of Baker et al. (2). Franck and Martino (61) outlined a typical procedure for regeneration of a coked platinum reforming catalyst. The procedure consisted of three major steps:

1. Nitrogen carrier gas is passed over the catalyst up to 200°C to remove any adsorbed hydrocarbons.
2. A well-controlled concentration between 0.5 and 2.0% oxygen in nitrogen carrier gas is passed over the bed with ramped heating to 500°C. Ramping should not be performed too quickly, according to Satterfield (79), because localized hot spots with temperature up to 200°C above the average bed temperature can occur during the exothermic coke combustion. Care must be taken to avoid this temperature runaway, which leads to sintering of the catalyst.
3. Chlorine addition to the system under an air atmosphere at 500°C during oxidation is practiced to redisperse the platinum crystallites (80).

After these steps, the standard oxidation and reduction pretreatment must be performed before the catalyst is ready for operation again.

An industrial patent (81) recommends the following procedure for catalyst regeneration: after the feed is stopped to the reformer unit, a recycle gas is introduced for a period of time to remove residual heavy hydrocarbons. The heaters, reactors, and recycle gas system are isolated from the remainder of the unit. Hydrogen is replaced with nitrogen and at nearly 400°C a small concentration of oxygen is introduced to the system to initiate the burning of the coke. Temperatures within the reactor must be carefully monitored during this operation to avoid an excessive increase in temperature which could result in damage of the catalyst (metal and support sintering). The carbon is then removed in a series of steps in which either the temperature or the oxygen concentration of the regeneration gas is increased until there is no further evidence of coke combustion.

Various regeneration procedures are used in industrial plants. Most refineries introduce oxygen in the first reactor and then continue the regeneration process to the last reactor. In other cases, oxygen is introduced into the first and last reactors simultaneously. This saves time because the last reactor is the one that contains the largest amount of coke and hence requires the longest time for carbon burn. The process of regeneration is normally conducted at atmospheric pressure using air as a source of oxygen. However, the process can also be conducted at high pressure with the use of available pure oxygen in a nitrogen atmosphere. The advantage of this procedure is that it reduces the time required for complete burning of the coke.

Another variant of the process involves removing the coked catalyst and sending it to the catalyst manufacturer, where the carbon is removed in a more controlled environment (off-site regeneration). The regenerated catalyst is sent back to the plant and reloaded into the reactors.

During the past two decades efforts have been directed toward the design of a continuous catalytic reforming process. The development of such a process arose from the anticipated need to produce unleaded gasoline, which requires operation at high severity (higher octane number and high-purity hydrogen) (82). A low-pressure process with continuous regeneration was commercialized by UOP. In this process, small quantities of catalyst are continuously withdrawn from an operating reactor and then transported to a regeneration unit, where the carbon is burned off and the catalyst chloride level is adjusted. The regenerated catalyst is then returned to the top of the reactor system (83).

## VIII. CONCLUSIONS

Regeneration of bifunctional naphtha reforming catalysts is a very complex procedure. Knowledge of the coke and catalyst structure as well as of coke burnoff kinetics is essential for the design of the regeneration process. The balance between metal function and acid function in bifunctional reforming catalysts is rather delicate. Hence, regeneration of the catalyst by burning off the coke, redispersing the metal, and restoring the acidity of the support is complicated.

There is still a wide scope for research to investigate the use of hydrogen as a gasifying agent for the deposited coke on the surface of the catalyst. At present, the principal role of hydrogen is to inhibit the coking reaction. In the case of coked sulfided reforming catalysts, it is clear that the regeneration has not been studied in great detail despite its considerable importance to industry. The incentive for a better process efficiency continues to exist in catalytic reforming. Improvement in the catalyst regeneration process will be an economic advantage for the whole process.

## ACKNOWLEDGMENT

The author wishes to acknowledge the support provided by King Fahd University of Petroleum and Minerals during the preparation of the manuscript.

## REFERENCES

1. P. B. Weisz and R. B. Goodwin, *J. Catal.* 6, 227 (1966).
2. R. T. K. Baker, J. A. France, L. Rouse, and R. J. Waite, *J. Catal.* 41, 22 (1976).
3. J. M. Thomas, *Chemistry and Physics of Carbon*, Vol. 1 (P. L. Walker, ed.). Marcel Dekker, New York, 1965, p. 122.
4. R. T. K. Baker, *CRC Crit. Rev. Solid State Sci.* 375 (1976).
5. S. M. Davis, F. Zaera, M. Salmeron, B. E. Gordon, and G. A. Somorjai, *Preprint Amer. Chem. Soc.*, LbL 14611 (1982).
6. J. Biswas, P. G. Gray, and D. D. Do, *Appl. Catal.* 32, 249 (1987).
7. A. Sarkany, T. Lieske, T. Szilagyi, and L. Toth, *Proceedings, 8th International Congress on Catalysis*, Berlin, 1984.
8. D. L. Trimm, *Appl. Catal.* 5, 263 (1983).
9. J. M. Parera, N. S. Figoli, E. M. Traffano, J. N. Beltramini, and E. E. Martinelli, *Appl. Catal.* 5, 33 (1983).
10. R. J. Verderone, C. L. Pieck, M. R. Sad, and J. M. Parera, *Appl. Catal.* 21, 239 (1986).
11. J. M. Parera, N. S. Figoli, and E. M. Traffano, *J. Catal.* 79, 484 (1983).
12. C. L. Pieck, E. L. Jablonski, R. J. Verderone, J. M. Grau, and J. M. Parera, *Catal. Today* 5, 463 (1989).
13. J. Barbier, in *Catalyst Deactivation* (B. Delmon and G. F. Froment, eds.). Elsevier, Amsterdam, 1987, p. 1.
14. J. Biswas, G. M. Bickle, P. G. Gray, and D. D. Do, in *Catalyst Deactivation* (B. Delmon and G. F. Froment, eds.). Elsevier, Amsterdam, 1987, p. 553.
15. R. T. Rewick, P. R. Wentreck, and H. Wise, *Fuel*, 53, 274 (1974).
16. R. T. Baker and R. D. Sherwood, *J. Catal.* 61, 378 (1980).
17. P. A. Sermon and G. C. Bond, *Catal. Rev. Sci. Eng.* 8, 211 (1973).
18. J. N. Beltramini and R. Datta, Proceedings, Chemeca 88, Sydney, Australia, 1988, p. 555.
19. J. Barbier, P. Marecot, N. Martin, L. Elassal, and R. Maurel, in *Catalyst Deactivation* (B. Delmon and G. F. Froment, eds.). Elsevier, Amsterdam, 1980, p. 53.
20. J. Barbier, E. Churin, J. M. Parera, and J. Riviere, *React. Kinet. Catal. Lett.* 29(2), 323 (1985).
21. J. N. Beltramini and R. Datta, *React. Kinet. Catal. Lett.* 44(2), 345 (1991).
22. J. N. Beltramini, E. J. Churin, and J. M. Parera, *Appl. Catal.* 19, 203 (1985).
23. J. M. Parera, N. S. Figoli, J. N. Beltramini, E. J. Churin, and R. A. Cabrol, Proceedings, 8th International Congress on Catalysis, Berlin, 1984, p. 593.
24. D. A. Eisenbach and E. J. Gallei, *J. Catal.* 56, 377 (1977).
25. P. E. Eberly, C. N. Kimberley, W. H. Miller, and H. V. Drushel, *Ind. Eng. Chem.* 5, 193 (1966).
26. N. S. Figoli, J. N. Beltramini, C. A. Querini, and J. M. Parera, *Appl. Catal.* 26, 39 (1986).
27. D. Espinat, Thesis, Institute Francaise du Petrole (1982).
28. V. K. Shum, J. B. Butt, and W. M. H. Sachtler, *Appl. Catal.* 11, 184 (1984).
29. E. H. van Broekhoven, J. W. Schoonhoven, and F. M. Ponec, *Surf. Sci.* 156, 899 (1985).

30. D. L. Trimm, in *Deactivation and Poisoning of Catalysts* (J. Oudar and H. Wise, eds.). Marcel Dekker, New York, 1985.
31. G. A. Somorjai, *Chemistry in Two Dimensions: Surfaces*. Cornell University Press, Ithaca, 1981.
32. J. N. Beltramini, T. J. Wessel, and R. Datta, *AIChE J.* 37(6), 845 (1991).
33. J. M. Parera, E. M. Traffano, J. C. Musso, and C. L. Pieck, in *Spillover of Adsorbed Species* (G. M. Pajonk, S. J. Teichner, and J. E. Germain, eds.). Elsevier, Amsterdam, 1983, p. 101.
34. A. Parmeliana, F. Frusteri, A. Mezzaopica, and N. Giordano, *J. Catal. 111*, 235 (1988).
35. J. M. Parera, N. S. Figoli, E. L. Jablonski, M. R. Sad, and J. N. Beltramini, in *Catalyst Deactivation* (B. Delmon and G. F. Froment, eds.). Elsevier, Amsterdam, 1980, p. 571.
36. J. L. Carter, G. C. McVicker, N. Weisshan, W. S. Kmak, and J. Sinfelt, *Appl. Catal. 3*, 327 (1982).
37. J. N. Beltramini and D. L. Trimm, *Erdol Kohle Ergas Petrochemie, (Hydeocarbon Technology) 9*, 400 (1987).
38. E. M. Traffano and J. M. Parera, *Appl. Catal. 28*, 193 (1986).
39. P. Marécot, S. Peyrovi, D. Bahloul, and J. Barbier, *Appl. Catal. 66*, 181 (1990).
40. J. Barbier, E. J. Churin, and P. Marecot, *Bull. Soc. Chim. France 6*, 910 (1987).
41. J. N. Beltramini and K. Dasgupta, private communication (1992).
42. W. M. H. Sachtler, *J. Mol. Catal. 25*, 1 (1984).
43. P. Marécot, S. Peyrovi, D. Bahloul, and J. Barbier, *Appl. Catal. 66*, 191 (1990).
44. D. R. Olander and M. Balloch, *J. Catal. 60*, 41 (1979).
45. B. J. Wood and H. Wise, *J. Chem. Phys. 63*, 4772 (1975).
46. S. M. Davis and G. A. Somorjai, *J. Phys. Chem. 87*, 1545 (1983).
47. J. N. Beltramini, *Reforming Catalysts*, Marcel Dekker, New York, 1994.
48. B. B. Zharkov, L. B. Galperin, V. L. Medzhinskii, L. F. Butochnikova, A. N. Krasilnikov, and I. D. Yakoleva, *React. Kinet. Catal. Lett. 32*(2), 457 (1986).
49. R. Baccaud, H. Charcosste, M. Guenin, R. Torres-Hidalgo, and L. Tournayan, *Appl. Catal. 1*, 81 (1981).
50. C. L. Pieck, E. L. Jablonski, R. J. Verderone, J. M. Grau, and J. M. Parera, 4th Symposio de Catalise, San Pablo, Brazil, 1987.
51. J. Barbier, *Appl. Catal. 23*, 225 (1986).
52. L. Elassal, Thesis, Poitiers, France (1983).
53. E. Freud, J. P. Franck, and J. P. Bournonville, Institut Francaise du Petrole, private communication (1981).
54. A. Bondi, R. S. Miller, and W. G. Schlaffer, *Ind. Eng. Chem. Proc. Des. Dev. 1*, 196 (1962).
55. P. L. Walker, F. Rusinko, and L. G. Austin, *Adv. Catal. 11*, 34 (1959).
56. J. C. Dart, R. T. Savage, and C. G. Kirkbride, *Chem. Eng. Prog. 45*, 102 (1949).
57. J. N. Beltramini and R. Datta, Catalyst Deactivation Symposium, AIChE Meeting, Los Angeles, 1991.
58. C. L. Pieck, R. J. Verderone, E. L. Jablonski, and J. M. Parera, *Appl. Catal. 55*, 1 (1989).
59. C. L. Pieck, E. L. Jablonski, and J. M. Parera, *Appl. Catal. 70*, 19 (1991).
60. C. Apesteguia, A. Borgna, T. Garetto, C. Brema, and J. M. Parera, Proceedings, 8th Ibreoamerican Symposium in Catalysis, Spain, 1982, p. 551.

61. J. P. Franck and G. Martino, in *Progress in Catalyst Deactivation* (J. L. Figueriedo, ed.). Martinus Nijoff, The Hague, 1982, p. 355.
62. M. Berthelin, Institut Francaise du Petrole, private communication (1981).
63. H. P. Bonzel and R. Ku, *J. Chem. Phys. 59*, 1641 (1973).
64. M. P. Rosynek and F. L. Stry, *J. Catal. 41*, 312 (1976).
65. C. C. Chang, *J. Catal. 53*, 374 (1978).
66. A. Datta, R. G. Cavell, R. W. Tower, and Z. M. George, *J. Phys. Chem. 89*, 443 (1985).
67. M. V. Mathieu and M. Primet, *Appl. Catal. 9*, 361 (1984).
68. A. Melchor, E. Garbowski, M. V. Mathieu, and M. Primet, *React. Kinet. Catal. Lett. 29*, 371 (1985).
69. C. R. Apesteguia, T. F. Garetto, and A. Borgna, *J. Catal. 106*, 73 (1987).
70. Z. Schay, K. Matusek, and L. Guezi, *Appl. Catal. 10*, 173 (1984).
71. T. Garetto, A. Borgna, and C. R. Apesteguia, *Ind. Eng. Chem. Res. 31*, 1283 (1992).
72. G. M. Bickle, J. N. Beltramini, and D. D. Do, Proceedings, Chemeca 88, Syndney, Australia, 1988.
73. H. C. Yao, H. K. Stepien, and H. S. Gandhi, *J. Catal. 67*, 231 (1981).
74. C. L. Pieck, E. L. Jablonski, R. J. Verderone, and J. M. Parera, *Appl. Catal. 56*, 1 (1989).
75. R. H. Herrman, S. F. Adler, M. S. Goldstein, and R. M. deBaun, *J. Phys. Chem. 65*, 2189 (1965).
76. H. Charcosset, R. Frety, G. Leclerq, B. Moraweck, L. Tournayan, and J. Varloud, *React. Kinet. Catal. Lett. 10*, 301 (1979).
77. J. Benson and M. Boudart, *J. Catal. 4*, 704 (1965).
78. J. Barbier, D. Bahloul, and P. Marecot, *J. Catal. 137*, 377 (1992).
79. C. N. Satterfield, *Mass Transfer in Heterogeneous Catalysts*. MIT Press, Cambridge, MA, 1970, p. 208.
80. H. Lieske, G. LItz, H. Spindler, and J. Volter, *J. Catal. 81*, 8 (1983).
81. A. R. Greenwood and K. D. Vesely, U.S. Patent 3,647,680 (1972).
82. M. H. Fromager and J. J. Patouillet, *Hydroc. Proc. 58*(4), 171 (1979).
83. J. A. Weiszmann, in *Handbook of Petroleum Refining Process* (R. Meyers, ed.). McGraw-Hill, New York, 1986, p. 3.1.

# 12
## Recovery of Pt and Re from Spent Reforming Catalysts

**J. P. Rosso and Mahmoud I. El Guindy**
*Gemini Industries, Inc., Santa Ana, California*

## I. INTRODUCTION

Usage of precious metal–bearing catalysts has increased significantly over the last 10 years as the demand for fuels has increased. Each precious metal catalyst changeout represents millions of dollars in costs to the petroleum or petrochemical company: fresh catalyst fabrication costs, the costs of the precious metal being placed on the catalyst, the contractor's costs to change the catalyst out, the loss of production while the unit is shutdown, the spent catalyst reclaim costs, and so forth. Costs must be controlled and recovery, where possible, must be maximized. Petroleum refining companies must pay strict attention to the proper handling of their spent reforming catalyst. Thousands, even millions of dollars may be at stake. In addition, there may be potential long-term environmental liability if disposal is not properly handled (1–5).

This chapter addresses in a practical fashion some of the different areas in which a petroleum refining company can effect procedural and attitudinal changes which can save tens of thousands of dollars. The chapter also contains a discussion of the material management procedures followed at the spent catalyst processing facility and an overall presentation of physical sampling and chemical processing of the catalyst.

## II. SPENT CATALYST HANDLING AT THE PETROLEUM REFINERY

### A. Catalyst Values

#### 1. The Basic Assumptions

300,000 lb net dry weight of spent catalyst × 14.583 conversion from avoirdupois lb to troy oz × 0.350 wt % of platinum content when catalyst was new

= 15,312.15 troy oz platinum content × $350.00 Jan. 1992 average platinum price per troy oz

= $5,359,252 total gross platinum value of the catalyst or $17.86 per pound of spent catalyst

A simple 0.5% error in the weight of the catalyst as calculated in the example results in a difference of 1500 lb of catalyst. That "simple" tolerance error has a value of $26,790.00. Or, looking at it another way, assuming 350 lb of catalyst in each 55-gallon drum being shipped, then each drum is worth approximately $6,251.00.

It can be seen by examining these values that a catalyst reclaim project must be well thought out and executed. A few safeguards and procedures established at the plant level can help ensure that the values of the metals are protected.

#### 2. Records

The situation arises in most petroleum companies that the catalyst often lasts longer in service than the people who were originally responsible for it. So, when the time comes to change out the unit, there has been no continuity of data, nor have catalyst handling procedures been established. With an average life span of anywhere from 2 years for some isomerization catalysts to as long as 10 or 12 years for some reforming catalysts, the personnel who originally purchased and/or installed the catalyst are rarely the people who are responsible for the reclaim project (4).

As a result, the process engineers currently running the unit may know only that the unit was "supposed to be" loaded with brand X catalyst. They "believe" the catalyst to contain X amount of platinum. That is as far as their knowledge of the catalyst in the unit goes. Even their records on unit operation may be incomplete. How many times was the unit regenerated/screened/reloaded? Was the unit ever "topped off"? Was an additional layer of support material added to stabilize the catalyst bed? They often have no record of these matters.

These are all critical factors which can and do occur during the life cycle of the catalyst. These situations can affect the accuracy of the estimates projected for values sent to reclaim as well as the actual reclaim results themselves. An error made in the projection of catalyst values at this step can result in many hours of fruitless search for precious metal that was never there to begin with.

a. *Example.* A petroleum refinery did not have complete and accurate information on their units when they began their catalyst changeout.

1. They had no verification, confirmed or recorded, in either their purchasing or accounting records of the exact amount of precious metal catalyst assigned to the units.
2. They had incomplete or nonexistent records of what specifically had happened to the catalyst while in use.
3. They did not supervise their reactor servicing company when the units were dumped. The service company was told to screen the material, but no one specified the screen sizes to be used.
4. They did not examine the shipment before it left. They assumed that their "net" weights, provided by the catalyst servicing company, consisted solely of catalyst. The result was that the company made some erroneous calculations:
    (a) First, they assumed that all 175,000 pounds shipped was actually catalyst. As it turned out, only 155,000 pounds of the material shipped was actually catalyst. The other 20,000 pounds consisted of inert/ceramic support media which had not been removed during screening by the reactor servicing company. It turned out that both 1-inch and 1/4-inch support slugs were used in the reactors. The catalyst had been screened, but the service company had used only a number 1 or number 2 screen. They had completely removed all of the 1-inch support slugs but all the 1/4-inch slugs were still commingled with the catalyst. The error was discovered at the reclaim facility. The reclaimer then had to screen off and remove the 20,000 pounds of 1/4-inch support material.
    (b) Because the oil company had used their own overstated erroneous weights to make their platinum recovery estimates, they had overstated their catalyst weight by the 20,000 pounds of 1/4-inch support. They had to go back to their management and readjust for approximately 1020 oz of platinum that had never had been there to begin with.

## B. Material Handling

The first thing a catalyst user must do is set up an inventory or stores procedure to track and account for every pound and every drum of catalyst purchased. This should start with the issuing of the purchase order (PO) to the catalyst manufacturer. The PO should specify the number of total pounds of catalyst to be manufactured from the platinum provided by the petroleum company. A procedure for verifying the precious metal content of the fresh catalyst after manufacture by representative sampling and exchange of assays should be included in the original agreement.

Next, the exact total pounds of manufactured catalyst should be provided to the petroleum company by the catalyst manufacturer. A complete shipping manifest specifying the total number of pounds of catalyst, broken down drum by drum, should be received. The drum count and weights should be verified at the petroleum refinery immediately upon delivery of the catalyst. All the drums should be shipped from the manufacturer with locking rings on the lids and the rings should be sealed with numbered wire or plastic seals.

Once the weights and drum count have been verified at the petroleum refinery, the drums of catalyst should be stored in a secure warehouse. They should be segregated from any other types of catalyst that might be in use in other areas of the plant. A simple double-signature, double-check system should be initiated for material entry and exit. No catalyst should ever leave the warehouse without it. In addition, the drums should be clearly marked in such a way as to make them easy to distinguish from any other catalyst used (7).

These steps avoid confusion and intermixing of catalyst types. Often, a reactor needs to be "topped off" after a regeneration or dump/screen/reload operation. If there are no labels on the drums or if an efficient double-check system is not in place in the warehouse, it is quite easy for warehouse operators inadvertently to grab pallets of base metal catalyst, not realizing that they have done so. The wrong catalyst is then used for topping off and is loaded into the reactor.

Depending on the size of the reactor, a few drums of base metal catalyst will not necessarily affect reactor performance to a measurable degree. However, someone trying to estimate the precious metal content of the reactor will now be off in the calculation because of the weight of the base metal catalyst which was added.

The reverse scenario can (and has) also occurred. Drums of precious metal catalyst are inadvertently loaded into a unit which is supposed to use base metal catalyst. In this event the platinum catalyst will be lost when the spent base metal catalyst is dumped from the unit and sent for disposal.

*Example*: A refiner did not follow a strict stores or warehouse procedure as has been recommended. Two days before approximately 150,000 lb of platinum/rhenium catalyst was supposed to be shipped for reclaim, a warehouse operator inadvertently loaded the wrong drums of catalyst onto trucks for use as roadbed material at the refinery. The error was not discovered until 2 days later, when the precious metal catalyst was to be shipped. By then the platinum catalyst had been used as roadbed over the entire refinery. The unused balance had been thrown in the on-site landfill. Using the values established in Section A, they laid down a $2,679,000 roadbed!

## C. Regeneration: On-site and Off-site

Proper catalyst regeneration is another key step. Catalyst that is contaminated with wet hydrocarbon presents a serious material safety and handling problem. Material sent off to be reclaimed in this condition will have to undergo extensive labeling, special packaging, special shipping, and special manifesting. When it arrives at the reclaim facility, it will still have to undergo treatment before sampling and recovery can begin. Catalyst cannot be accurately sampled if it is covered by heavy amounts of carbon or if it is wet with residual material. This situation invariably results in much higher costs for the reclaim project than would otherwise have been necessary. Therefore it is highly recommended that the petroleum company have the catalyst regenerated before sending it off to the reclaim facility. On-site regeneration is sometimes more cost-efficient; however, it is not always a viable option. In those cases it is recommended that an off-site regenerator be chosen to perform this service. Most off-site regenerators welcome the presence of a company representative or a contract agent during this operation.

When the material has been fully burned, it is reloaded into drums or bins and sent directly to the reclaimer. Whether through on-site or off-site regeneration, the reclaimer should now have clean, uncontaminated catalyst to work with. Clean catalyst allows for more accurate sampling. A more accurate assay is the direct result of a more accurate sample, which again maximizes the return of recovered precious metal from the spent catalyst.

## D. Screening and Separation

In most cases the petroleum company will utilize a reactor or catalyst service contractor to dump/screen/reload catalyst into a unit. When going through this procedure there are usually separate groups of whole catalyst, "fines" catalyst, and support medium. This presegregation is very important to maintain, as these materials can contain different amounts of precious metals.

The whole catalyst usually contains something near the original precious metal specification content. The "fines," however, can be as much as 50% higher or lower in precious metal content. This depends on how the catalyst was manufactured originally. The support medium has no precious metal content. The actual volume from any given changeout will vary. This must all be taken into account when trying to make any precious metal content projections.

Caution is advised in overseeing all outside contact with the catalyst. It is possible for material to be left inadvertently in the bottom of screening or dust collecting equipment. All equipment should be checked before it leaves the plant. It should be remembered that each pound of catalyst is worth approximately $17.86.

## E. Labels and Seals

The vast majority (+80%) of all spent precious metal catalyst shipped does not have proper labels. Each drum should have:

1. The name and address of the petroleum company shipping the material
2. A simple description of the type of catalyst, e.g., R-32/KX-120/E-301
3. The destination name and address

In addition to lacking labels, most shipments are sent without seals on the individual drums. The petroleum plant should seal all drums of spent catalyst with numbered wire or plastic seals. Sealing drums is good, inexpensive insurance to prevent curious people from "inspecting" the material. It also forces a double check of the drums to ensure that all lids and rings are locked onto the drums in the proper fashion.

*Example*:  A refiner decided to use steel flow bins to ship catalyst. This is normally a very good idea. Flow bins have many advantages over drums: no disposal when through, time and labor savings in both loading and unloading, increased safety for workers because they can use a forklift to move them, etc. However, flow bins must be inspected to ensure that the locking lugs on the bottom slide gates are in the fully upright and locked position.

In this instance, no one supervised the closing of the slide gates, nor did anyone double-check when loading the bins onto a flat-bed trailer. Two bins did not have their gates properly locked. All the catalyst in those two bins eventually leaked out during transit to the reclaimer. In all, approximately 10,000 lb of catalyst with an estimated value of $178,600.00 was lost along the highway.

## F. Transport and Insurance

A large, reputable freight carrier should always be used. The truck should be engaged for exclusive use even if your material is only a partial load. The difference in freight rate for exclusive use versus a commingled shipment is only a few hundred dollars at most. Considering the value of the catalyst ($6251 per drum), this is a small premium to pay for the added security. The use of enclosed vans is strongly encouraged. Although flat-bed trucks are potentially easier to load and unload, they present a serious security risk. An enclosed van utilizes "container within a container" practice for more secure shipping. Should the truck break down or any other incident occur, your material is safely locked inside. If a drum or bin spills inside the van, it is easily swept up upon receipt and unloading at the reclaim facility. A spill on a flat bed is invariably lost forever.

The shipment should be insured for its full, estimated platinum value. Unless your company routinely self-insures, be sure to request full value insurance from the freight carrier. A certificate of insurance from the carrier's broker should be requested for your review. The policy should be currently in force with a large, financially sound company. Most carriers routinely maintain $250,000 of insurance for materials in transit in any single 40-foot van. Most catalyst shipments of 30,000 lb or more per van will require insurance of $1 million per van.

## III. SPENT CATALYST HANDLING AT THE PROCESSING FACILITY

This section is devoted to discussion of the various physical and chemical actions performed at the precious metals processing facility.

### A. Physical Inspection and Material Accounting

Reforming and similar alumina-supported catalyst are manufactured in the form of extrudates or spheres made from high-surface-area alumina ($\gamma$-alumina) as the carrier for Pt or Pd and a promoter such as Re. The catalysts are identified by the producers using a combination of letters and numerals, e.g., R-32, PHF-4, or E-601.

Upon receiving the material at the processor, an examination of the transport vehicle bed and the catalyst containers is performed. Abnormalities are noted and recorded on the accompanying shipping documents. Spilled catalyst and open or damaged containers are reported to management and documented, including photographs of the damage or spill. The seals on sealed containers are also examined and documented. The drum or container count is verified and compared with information supplied by the shipper. Any differences are noted and verified and, if beyond normal, variances are reported immediately to the shipper. A receiving document is then prepared and sent to the shipper. Copies are distributed to production personnel. All materials are kept intact pending confirmation of sampling arrangements with the customer.

### B. Weighing and Sampling

It is inherent in the nature of the catalyst and the techniques used for impregnation of the support with the active metals that variations in precious metal loading from one particle to another could occur. These variations do not affect the performance of the product as a catalyst but require careful consideration when sampling.

Inert ceramic balls are occasionally added to the catalyst before loading in the reactors to increase space utilization of a given amount of catalyst and to stabilize the catalyst bed. Some of these inert materials remain with the catalyst and have to be removed before sampling. Fines and smaller particles are generated during the use of the catalyst through regeneration cycles and loading or unloading operations. These abnormal constituents may have a different metal content and would have to be removed and sampled separately.

It is thus clear that the sampling of spent catalyst is not a simple task. Accordingly, specially designed and engineered sampling systems and devices are needed to obtain properly representative samples of the spent catalyst and any other fractions containing valuable metals, e.g., the fines fraction.

One example of such a system is shown in Figure 1. The spent catalyst containers are first opened, examined for uniformity and for conformity to shipping documents, and then gross weighed. They are elevated and emptied into the feed hopper. Tare weights of the empty containers are then obtained.

In the screening operation, the first set of screens effects the removal of large particles and extraneous materials including inert support medium and debris. This material is inspected and disposed of.

The second screen is sized to separate the smaller fraction and direct this material to a proportional rotary sampler, where a 5% cut is taken as a sample.

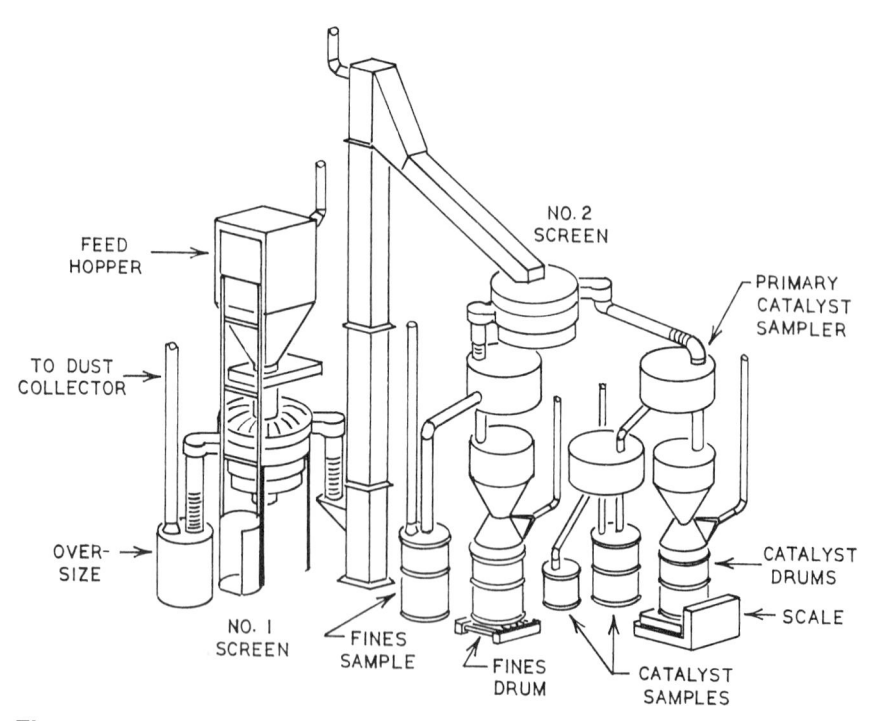

**Figure 1** Catalyst sampling system.

The whole catalyst representing the bulk of the material is directed to a separate rotary proportional sampler, where a 1–2% sample is automatically taken and directed to a secondary sample cutter to produce a final representative sample weighing 15–30 lb.

Individual samples of the as-received catalyst are then taken from this bulk sample to be used for determination of solids content, or "loss on ignition." Furthermore, about half the bulk sample is taken using a riffle mechanism and ground in a suitable grinder, blended, and split to produce ground, homogeneous samples for the analysis of Pt, Re, and any other contained metals. A set of samples is sealed and retained to be used for referee analysis, if needed (8). Compositional factors affecting the processing of the spent catalyst, such as amount of insoluble alumina, carbon, iron, and base metals contamination, are also determined using portions from the original bulk sample.

Following the completion of sampling of a given material, all weights of the various fractions are added and compared with receiving weights and customer-advised weights to ensure complete accounting. Any discrepancies are reported to management immediately.

## C. Chemical Processing

The most commonly employed techniques for recovery of both platinum and rhenium (if present) from the catalysts mentioned above make use of the reactivity of the alumina with both acids and caustic to form soluble salts, according to the following reactions:

$$Al_2O_3 + 3H_2SO_4 \longrightarrow Al_2(SO_4)_3 + 3H_2O \qquad (1)$$

$$Al_2O_3 + 2NaOH \longrightarrow 2NaAlO_2 + H_2O \qquad (2)$$

In the acid medium the platinum is not affected, while the rhenium is solubilized producing a perrhenic acid solution as follows:

$$Re_2O_7 + H_2O \longrightarrow 2HReO_4 \qquad (3)$$

After completion of the dissolution reactions, solid-liquid separation produces the insoluble platinum as a residual material to be processed. The processing involves oxidizing chloride media reactions as in Eq. (4) to produce chloroplatinic acid, which is then purified to platinum metal (10).

$$Pt + 6HCl + O_2 \text{ (oxidizer)} \longrightarrow H_2PtCl_6 + 2H_2O \qquad (4)$$

The soluble rhenium is removed from the solution by liquid or solid ion exchange and purified by chemical procedures to produce pure crystalline ammonium perrhenate.

The caustic system of processing is presently employed at one facility for a small amount of their processing requirements, specifically the catalysts containing Ir as a promoter. The sulfuric acid digestion system is by far the most commonly employed because of its ease of operation and versatility. Accordingly, only the latter system will be discussed in detail.

## D. Sulfuric Acid Processing of Spent Catalyst

Although the reaction as represented by Eq. (1) is simple, many factors influence the start of the reaction, maintenance of reactivity, and quality of both Pt insolubles and aluminum sulfate product. Thus, a laboratory evaluation of each shipment of catalyst is necessary in order that operations be kept at good productivity and high quality of products. The generalized steps involved in this recovery and purification operation, the factors affecting the reactions, and the necessary preprocess evaluation are discussed in the following.

### 1. Sulfuric Acid Digestion

This reaction is exothermic although it requires initial heating to the boiling point. The reaction mixture should have enough water to maintain the alumina in solution at all times. Reaction temperature should be maintained close to boiling for the duration.

### 2. Solid-Liquid Separation

Good settling and coagulation of the insoluble mixture of alumina, carbon, platinum, and other constituents is essential to quantitative separation and recovery of the precious metals and to the production of high-grade aluminum sulfate.

### 3. Rhenium Removal and Purification

The fraction of contained Re which is soluble in sulfuric acid is recovered by contacting the filtered aluminum sulfate solution with a liquid or solid ion exchange medium capable of removing the Re species (11). The Re is then eluted from the exchange medium and purified by chemical separations and crystallization. Quality is determined by analysis. The product has to meet the required grade, particularly for reuse in catalyst production.

### 4. Processing Pt Residues

The filtered insolubles (residues) are calcined and ground before chloride medium dissolution of the contained Pt. The product chloroplatinic acid solution is chemically processed to remove impurities and the Pt is precipitated as ammonium chloroplatinate (1). The latter salt is then reduced to fine Pt sponge, which is calcined for further purification. Purity of the product Pt is determined by optical emission spectroscopy or ICP. Product meeting a minimum purity of 99.95% is shipped to catalyst manufacturers or depositories as requested by the customer.

## E. Laboratory Process Evaluation of Catalyst Samples

A complete pilot plant scale run to study the reaction characteristics of each customer lot and determine material behavior during reaction, filtration, and calcination is usually carried out. Analysis of the products gives information on the following.

### 1. Solubility of the Alumina Substrate

Lower alumina solubility could be the result of activities during the use, regeneration, or carbon burn-off of the catalyst. Side reactions or extra heat generation could convert some of the soluble alumina to lower-surface-area types, e.g., $\delta$ or $\alpha$ alumina. In addition, carbon present on the surface of the catalyst could act as a barrier to the diffusion of the reactants, thus causing lower reaction rates. Lower solubility results in a lower-grade insoluble (% Pt) requiring more processing activity (manpower, residence time, or chemicals) for the same amount of precious metals produced. These factors could also adversely influence Re solubility and in certain cases prevent all the Re from dissolving. This results in a search for alternative and more costly processing techniques.

### 2. Base Metals and Iron

Base metals and iron influence the quality of the aluminum sulfate. The precious metals processor markets the product aluminum sulfate, resulting in a reduction of the costs incurred in processing. If this alum product is out of specification and cannot be sold, it results in additional costs for alternative disposal in addition to lost revenue.

### 3. Carbon Content in Catalyst and Leach Products

The presence of carbon on the catalyst surface affects the solubility of all components and results in lower yields and higher costs for all process steps. Furthermore, a long calcination period at a higher temperature may be required to expose and activate the surfaces. In high-contamination situations (known as cokeballs) the amount of economically recoverable Pt decreases by as much as 10% and no Re can be recovered. Thus, analysis of carbon content in the incoming material and in leach insolubles is made to guide the choice of downstream processing steps.

### 4. Metal Return and Accounting

Following the sampling of the catalyst as outlined, both processors and customers analyze their samples for the metal content. An assay exchange is then carried out and the metals contents are compared. If the numbers fall within acceptable limits, a content is agreed upon and becomes due on settlement date. An umpire assay could be necessary if the parties do not agree.

Most processors of spent catalyst do their work on a toll refining basis, assuming no ownership of the contained metals. Upon completion of work, the metals are shipped to customers' depository accounts at catalyst manufacturers.

## IV. CONCLUDING REMARKS

The examples presented in this chapter are all true. They have all occurred during the past 5 years. They have been encountered by both large and small petroleum companies all over the world. The reason they occurred is that the refiners continued to treat their catalyst as an "orphan" and did not take the time to establish some minimal controls and procedures.

If one remembers the calculations cited at the beginning, each 0.5% error will cost $26,796.00 in lost, unrecovered platinum. As the price of platinum rises, so do these values. The expense of training personnel and educating them in the proper way to handle this material is easily justified in the prevention of lost revenues.

Proper regeneration of the catalyst charge prior to unloading could result in large savings on cost reclamation and possible elimination of hazardous waste labeling and future environmental liabilities and costs.

## REFERENCES

1. E. W. Hobart and C. L. Maul, Consideration in utilizing independent laboratories for PGM analysis. In *Precious Metals Recovery and Refining, Proceedings of a Seminar of the International Precious Metals Institute*, November 12-14, 1989, Scottsdale, Arizona (L. Manziek, ed.). Historical Publications, Austin, 1989, pp. 1-5.
2. J. P. Rosso, Maximizing the return of precious metals from spent petrochemical catalysts. In *Precious Metals Recovery and Refining, Proceedings of a Seminar of the International Precious Metals Institute*, November 12-14, 1989, Scottsdale, Arizona (L. Manziek, ed.). Historical Publications, Austin, 1989, pp. 41-47.
3. J. C. Bullock, Superfund liability of precious metal refinery customers. In *Precious Metals Recovery and Refining, Proceedings of a Seminar of the International Precious Metals Institute*, November 12-14, 1989, Scottsdale, Arizona (L. Manziek, ed.). Historical Publications, Austin, 1989, pp. 51-58.
4. D. G. Luning, Platinum group metals available for reclamation from the petroleum industry. In *The Platinum Group Metals. An In-depth View of the Industry, Proceedings of a Seminar of the International Precious Metals Institute*, April 10-13, 1983, Colonial Williamsburg, Virginia (D. E. Lundy and E. D. Zysk, eds.). I.P.M.I. Publishing, 1983, pp. 123-126.
5. J. A. Alvarez and R. J. Parker, Uses of platinum in the petroleum refining industry. In *The Platinum Group Metals. An In-depth View of the Industry, Proceedings of a Seminar of the International Precious Metals Institute*, April 10-13, 1983, Colonial Williamsburg, Virginia (D. E. Lundy and E. D. Zysk, eds.). I.P.M.I. Publishing, 1983, pp. 299-313.

6. S. Kallman, Precious metal bearing catalysts. In *Sampling and Assaying of Precious Metals II, Proceedings of a Seminar of the International Precious Metals Institute*, March 18-19, 1980, San Francisco (M. E. Browning and D. A. Corrigan, eds.). I.P.M.I. Publishing, 1980, pp. 173-189.
7. E. Levendosky, Sampling of spent catalyst. In *Sampling and Assaying of Precious Metals, Proceedings of a Seminar of the International Precious Metals Institute*, April 13-14, 1978, Morristown, New Jersey (M. E. Browning and D. A. Corrigan, eds.). I.P.M.I. Publishing, 1978, pp. 65-82.
8. E. B. Mayo, The role of representatives and umpires. In *Sampling and Assaying of Precious Metals, Proceedings of a Seminar of the International Precious Metals Institute*, April 13-14, 1978, Morristown, New Jersey (M. E. Browning and D. A. Corrigan, eds.). I.P.M.I. Publishing, 1978, pp. 83-96.
9. M. L. Batchelder, ed., *1986 N.P.R.A. Q & A Session on Refining and Petrochemical Technology*. Farrar, Tulsa, Oklahoma, 1986, pp. 118-122.
10. *Kirk-Othmer: Encyclopedia of Chemical Technology*, 3rd ed., Vol. 18. Wiley, New York, 1982, pp. 236-238.
11. E. E. Malouf et al., in *Rhenium* (B. W. Gonser, Ed.) Elsevier, New York, 1960, pp. 7-9.

# 13
# Catalytic Reforming Processes

**Abdullah M. Aitani**

*King Fahd University of Petroleum and Minerals, Dhahran, Saudi Arabia*

## I. INTRODUCTION

Catalytic reforming of naphtha has evolved rapidly during the past four decades to become one of the most advanced processes available to the refining industry. The main objective of this process is to upgrade low-octane naphtha to high-octane reformate for use as high-performance gasoline fuel. The process may be alternatively operated to produce high yields of aromatics for petrochemical feedstocks. Catalytic reforming combines catalyst, hardware, and process technology to produce optimal results. At present, several processes are being used in the refining industry; the main differences between these processes are (1) nature of catalysts, (2) the catalyst regeneration procedures, and (3) the conformation of the equipment. Reforming technology, catalysts and processes continue to lead petroleum refining in innovation, safety, and reliability as well as positive environmental impact and profitability. The main elements of the reforming process technology include:

- Operating strategy and steady-state optimization
- Process control
- Catalyst selectivity and stability
- Environmental control
- Integration with other refinery units and the refinery energy balance.

The reforming process continues to satisfy changing gasoline pool requirements. Refiners and catalyst manufacturers are examining means for catalyst performance improvement and process design improvement to enhance the selectivity to aromatics and high-octane reformate. The changes in process design are frequently accompanied by modification of the catalyst for improved performance to achieve one of the following objectives:

- Higher reformate octane yields
- More efficient catalyst regeneration
- Longer catalyst life and enhanced surface stability
- Lower-pressure operation and less hydrogen recycle

In industrial practice, reformers are typically operated to produce a product with a constant octane number. The operating conditions of the reforming unit are usually described in terms of severity. The most widely used measure of reforming severity has been the research octane number (RON). As the catalyst deactivates, the severity (reactor temperatures) of the process may be increased to compensate for the lower activity. In this way the octane number of the product can be maintained at the desired level. A reformer can meet many product demands through its wide range of design and operating flexibility. Commercial reformers process virgin and cracked naphthas of varying characteristics over operating pressures of 3.5–35 bar and temperatures of 725–800 K in the presence of added hydrogen to suppress both coking and excessive degrees of aromatization. However, the need for high-octane reformate requires the operation of existing reformer units at the lowest possible operating pressures and the use of high-stability reforming catalysts. The lower operating pressures are applied mainly in the fully and continuously regenerative types of processes.

## II. PROCESS DESCRIPTION

The industrial reforming unit comprises three sections: feedstock pretreatment section, reaction section, and product separation/stabilization section.

### A. Feedstock Pretreatment

The feedstock, which is generally a straight-run or synthetic naphtha, is pretreated before it enters the reaction section of the reformer. Chapter 8 of this book presents a comprehensive review of naphtha pretreatment. Feedstocks, such as FCC naphtha, hydrocracked naphtha, and raffinate, have more $C_5$ and $C_6$ naphthenes than do most straight-run naphthas, which makes them more difficult to reform. Stocks such as coker naphthas tend to carry with them polycyclic material for a given boiling range. For this reason, these synthetic naphthas should not be compared directly with straight-run naphthas (3). The objective of the feed pretreatment is to remove permanent reforming catalyst poisons (such as arsenic, lead, and copper) and to reduce the temporary catalyst poisons (such as sulfur, oxygen, and nitrogen compounds) to low levels (2). The catalyst can tolerate only a few parts per billion of these permanent poisons. Suggested feedstock specifications for reformers using monometallic catalysts are the following (1):

S $\leqslant$ 10 ppm or 5 ppm at high severity
N $\leqslant$ 1ppm
$H_2O$ $\leqslant$ 4 ppm
Pb + As + Cu < 20 ppb

Hydrotreatment involves vapor-phase reaction with hydrogen over a sulfur-resistant catalyst such as $Co\text{-}Mo/Al_2O_3$ followed by cooling, phase separation, and efficient stripping of all $H_2S$ and $NH_3$ out of the treated naphtha. Water is also removed during the stripping because its presence leads to modification of the acidity of the reforming catalysts. A sulfur guard bed may be used to supplement feed hydrotreating. It further reduces sulfur to about 0.2 ppm in the reformer feed and generally lengthens the catalyst life between regenerations. On the other hand, excessive chloride on the catalyst shifts yield selectivity and frequently results in excessive hydrocracking. The effects of high levels of chloride in the feed can be partially offset by adding water or alcohol to the feed to wash off the excess chloride. However, it is desirable to maintain a chloride level on the catalyst at a predetermined value (2).

## B. Reaction Section

The reaction section of the conventional reformer consists of multibed reactors in series, with pre- and reheaters between the reactors. Three or four reactors are frequently used to obtain a product with high octane number. The naphtha feedstock is mixed with recycle hydrogen and heated to the desired temperature before it enters the first reactor. The overall reactions are endothermic and best results can be obtained by using isothermal reaction zones placed directly in a furnace. However, this operation is difficult to achieve because of the large number of reactions taking place. Instead, the reaction beds are separated into a number of adiabatic zones operating at about 755 K with heaters between stages to supply the necessary heat of reaction and hold the overall train near a constant temperature. Various reaction sections have been developed and the main difference is related to the type of catalyst and its regeneration.

The selection of an appropriate reactor system is based on the required performance of the reforming unit. Initially, downflow (axial) reactors were used because of the more economic reactor internals. However, due to the large pressure drop in these reactors, radial-downflow reactors are being used for fixed-bed systems. Reforming units may have radial or axial flow reactors or both. Catalysts in the form of extrudates or spheres with optimized size are used in fixed-bed systems in order to avoid diffusion limitations and achieve acceptable pressure drops (1). Adiabatic reactors are used in all reforming units with heat being supplied by intermediate furnaces placed between the fixed-bed reactors operating in series.

The type of catalyst (monometallic or bimetallic) used has a profound effect on the overall performance of the reaction section (4-7). In general, the catalyst is made of metallic phase dispersed on a solid acidic support. The metallic phase, which is platinum promoted with other metals such as Re, Ir, Sn, or Ge, is responsible for hydrogenation-dehydrogenation reactions. The balance between the metallic phase and the acidic support is important for smooth operation. Both selectivity and activity of the catalyst are key factors of the reforming process. The advantage of bimetallic over monometallic catalysts is their higher stability under reforming conditions. A disadvantage is the higher sensitivity toward poisons, process upsets, and susceptibility to nonoptimum regenerations. The present-day reforming catalysts have been improved to withstand the high severity of operation. High surface area stability for repeated regeneration, use of a multimetallic phase, better chloride retention, and reduced attrition losses are characteristics of continuous-regeneration catalysts.

The reaction temperature in the first reactor decreases rapidly due to the dehydrogenation of cycloalkanes to aromatics. The effluent from the first reactor is reheated and fed to the second reactor, where it undergoes dehydroisomerization of cyclopentanic naphthenes at a lower rate than that of the first reactor. As reaction occurs, the temperature drop in the second reactor is in the range 295-305 K. In the last stage of reactors, reactions such as dehydrocyclization and cracking occur; however, slight variations in temperature are generally observed. The small change in temperature is attributed to the exothermic hydrocracking reaction that takes place in this reactor. Because some of the preceding reactions are very fast, the first reactor is always smaller than the second, but the third and fourth are even larger. The amount of platinum-containing catalyst in the first reactor is usually about 10-20% of the total catalyst charge (1).

Process variables of the reforming unit are determined based on the thermodynamics and kinetics of the various endothermic reactions taking place as well as the characteristics of the feedstock. Usual feedstocks for the reforming units are naphthas with a boiling range of 365-500 K. The composition of the feedstock has a major effect on the reactor temperature and product yield. Paraffin-rich feeds are more difficult to reform. The main purpose of the reforming process is to produce aromatics from naphthenes and paraffins. Major operating variables that affect the performance of the reforming unit are reviewed in Chapter 7 of this book.

## C. Product Separation and Stabilization

The effluent from the reaction section is cooled and separated into gaseous and liquid products. The gaseous product consists of hydrogen and $C_1$-$C_4$ hydrocarbons, with 60-90 mol % as hydrogen. A portion of this gas is recycled to the inlet of the reaction section, where it is mixed with naphtha feed at a ratio of 5-10 moles of recycle gas per mole of naphtha. The net hydrogen production can be purified depending on its use or sent to hydrogen users in the refinery (1).

The liquid product, commonly known as reformate, consists of $C_5$ through $C_{10}$ hydrocarbons. Aromatic hydrocarbons are present in the range of 60-70 wt %. The crude reformate is sent to a product stabilizer, where volatile and light hydrocarbons are removed from the high-octane liquid product. Stabilization is generally carried out in a single-column operation under sufficient pressure to permit condensation of reflux from the overhead vapor. Part or all of the net overhead product is taken off as a vapor stream.

## III. PROCESS CLASSIFICATION

Reforming processes are generally classified into three types: semiregenerative, cyclic (fully regenerative), or continuous regenerative process (moving bed). This classification is based on the frequency and mode of regeneration. Table 1 presents a worldwide distribution of catalytic reformers by process design (8). There are more than 800 commercial installations of catalytic reforming worldwide with a total capacity of 9.5 million barrels a day (9). Currently, most grassroots reformers are equipped with continuous catalyst regeneration.

### A. Semiregenerative Process

The semiregenerative process is characterized by continuous operation over long periods, with decreasing catalyst activity. Eventually the reformers are

**Table 1** Regional Distribution of Catalytic Reformers by Process Design in 1993

| Region | % of Reformers | | | |
|---|---|---|---|---|
| | Semiregenerative | Cyclic | Continuous | Other[a] |
| United States | 57 | 13 | 14 | 16 |
| Europe | 66 | 11 | 17 | 6 |
| Japan | 70 | 10 | 15 | 5 |

[a] Other includes nonregenerative catalyst and moving bed systems.
*Source*: From Ref. 8.

shut down periodically as a result of coke deposition to regenerate the catalyst in situ. Regeneration is carried out at low pressure (approximately 8 bar) with air as the source of oxygen. The development of bimetallic and multimetallic reforming catalysts with the ability to tolerate high coke levels has allowed the semiregenerative units to operate at 14–17 bar with similar cycle lengths obtained at higher pressures. It is believed that all reforming licensors have semiregenerative process design options. The schematic flow diagram of such a process is shown in Figure 1 (10).

The semiregenerative process is a conventional reforming process which operates continuously over a period of up to 1 year. As the catalytic activity decreases, the yield of aromatics and the purity of the by-product hydrogen drop because of increased hydrocracking. Semiregenerative reformers are generally built with three to four catalyst beds in series (11). The fourth reactor is

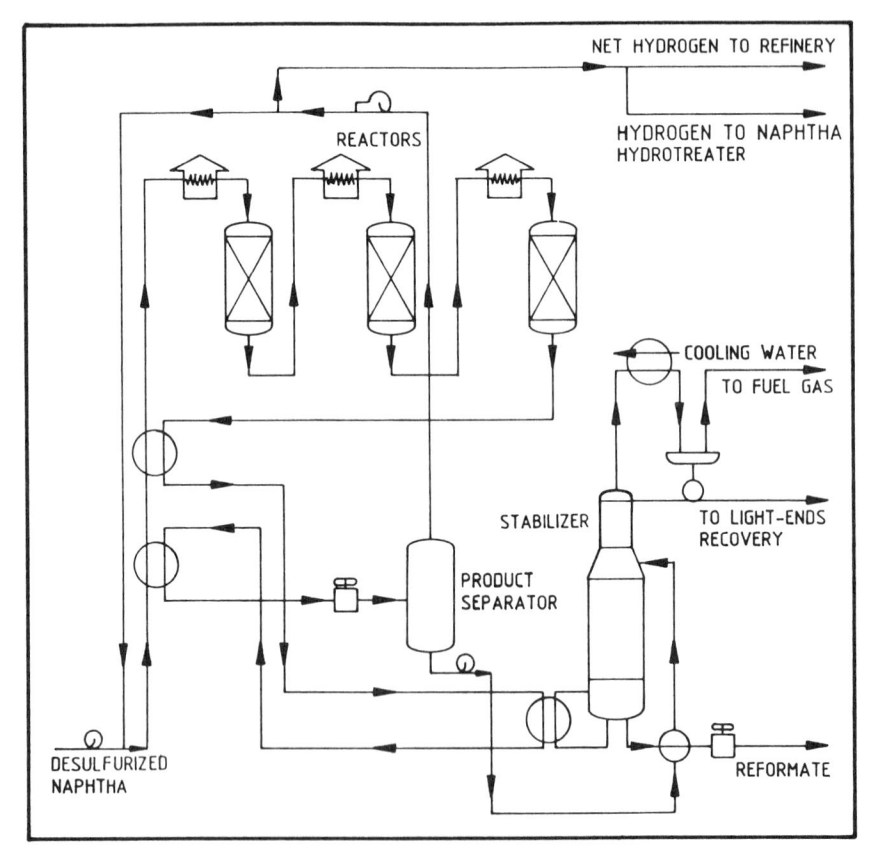

**Figure 1** Schematic flow diagram of a semiregenerative reforming process. From Ref. 10.

usually added to some units to allow an increase in either severity or throughput while maintaining the same cycle length. The longer the required cycle length, the greater the required amount of catalyst. Conversion is maintained more or less constant by raising the reactor temperatures as catalyst activity declines. Sometimes, when the capacity of a semiregenerative reformer is expanded, two existing reactors are placed in parallel, and a new, usually smaller, reactor is added. Frequently, the parallel reactors are placed in the terminal position. When evaluating unit performance, these reactors are treated as though they are a single reactor of equivalent volume (2).

RON that can be achieved in this process is usually in the range of 85–100, depending on an optimization between feedstock quality, gasoline qualities, and quantities required as well as the operating conditions required to achieve a certain planned cycle length (6 months to 1 year). The catalyst can be regenerated in situ at the end of an operating cycle. Often the catalyst inventory can be regenerated 5–10 times before its activity falls below the economic minimum, whereupon it is removed and replaced (10).

## B. Cyclic (Fully Regenerative) Process

The cyclic process typically uses five or six fixed catalyst beds, similar to the semiregenerative process, with one additional swing reactor, which is a spare reactor. It can substitute any of the regular reactors in a train while the regular reactor is being regenerated. In this way, only one reactor at a time has to be taken out of operation for regeneration, while the process continues in operation. Usually all the reactors are of the same size. In this case, the catalyst in the early stages is less utilized; therefore, it will be regenerated at much longer intervals than the later stages. The cyclic process may be operated at a low pressure, a wide-boiling-range feed, and a low hydrogen-to-feed ratio. Coke laydown rates at these low pressures and high octane severity (RON of 100–104 range) are so high that the catalyst in individual reactors becomes exhausted in time intervals of less than 1 week to 1 month. The schematic flow diagram of the cyclic process and its regeneration system is shown in Figure 2 (10).

The process design of the cyclic process takes advantage of low unit pressures to gain a higher $C_{5+}$ reformate yield and hydrogen production. The overall catalyst activity, conversions, and hydrogen purity vary much less with time than in the semiregenerative process. However, a drawback of this process is that all reactors alternate frequently between a reducing atmosphere during normal operation and an oxidizing atmosphere during regeneration. This switching policy needs a complex process layout with high safety precautions and requires that all the reactors be of the same maximum size to make switches between them possible. Process licensors such as Exxon, Engelhard, and Amoco offer cyclic reforming processes with swing reactors.

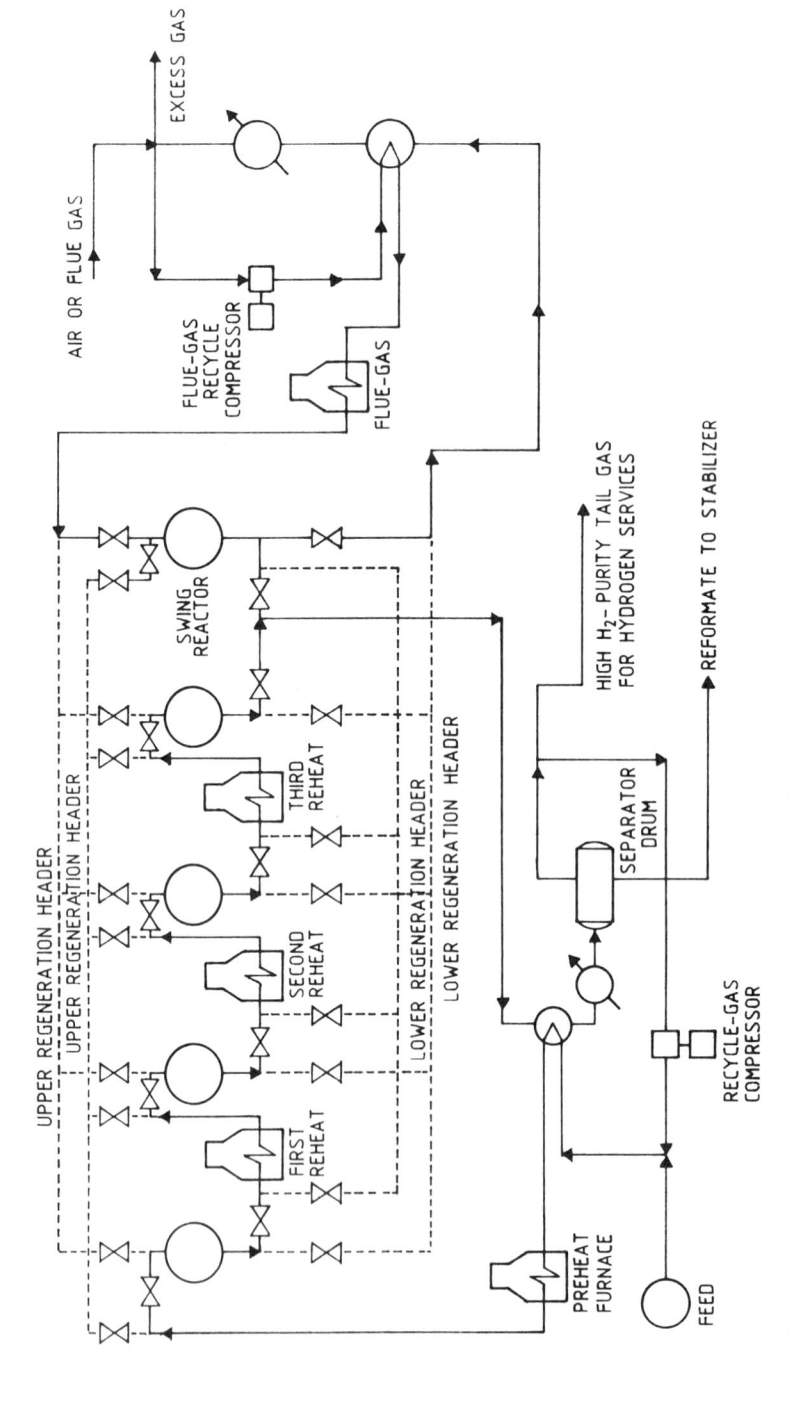

**Figure 2** Schematic flow diagram of a cyclic reforming process. From Ref. 10.

## C. Continuous Regenerative (Moving Bed) Process

The continuous reforming process was developed by UOP in the late 1960s to produce large quantities of high-octane reformate and high-purity hydrogen on a continuous basis (3). IFP also licenses a moving-bed system for continuous regeneration of the catalyst (12).

The process is characterized by high catalyst activity with reduced catalyst requirements, more uniform reformate of higher aromatic content, and high hydrogen purity. The process can achieve and surpass reforming severities as applied in the cyclic process but avoids the drawbacks of the cyclic process. In this process, small quantities of catalyst are continuously withdrawn from an operating reactor, transported to a regeneration unit, regenerated, and returned to the reactor system. In the most common moving-bed design, all the reactors are stacked on top of one other. The fourth (last) reactor may be set beside the other stacked reactors. The reactor system has a common catalyst bed that moves as a column of particles from top to bottom of the reactor section. A schematic diagram is shown in Figure 3 (10). Coked catalyst is withdrawn from the last reactor and sent to the regeneration reactor, where the catalyst is regenerated on a continuous basis. However, the final step of the regeneration, i.e., reduction of the oxidized platinum and second metal, takes place in the top

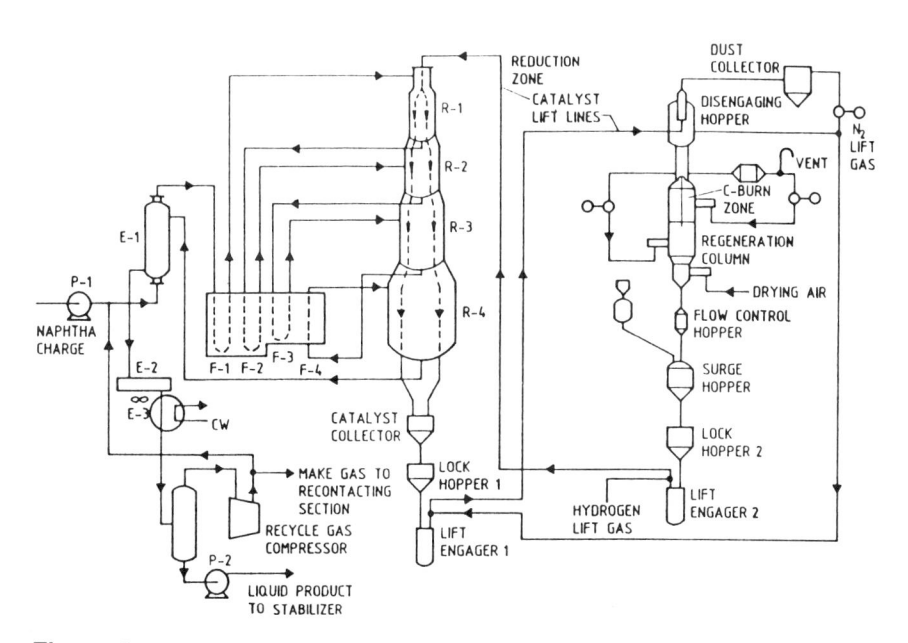

**Figure 3** Flow diagram of UOP-Platforming process with continuous catalyst regeneration. From Ref. 10.

of the first reactor or at the bottom of the regeneration train (10). Fresh or regenerated catalyst is added to the top of the first reactor to maintain a constant quantity of catalyst. Catalyst transport through the reactors and the regenerator is by gravity flow, whereas the transport of catalyst from the last reactor to the top of the regenerator and back to the first reactor is by the gas-lift method (13). Catalyst circulation rate is controlled to prevent any decline in reformate yield or hydrogen production over time on stream.

In another design, the individual reactors are placed separately, as in the semiregenerative process, with modifications for moving the catalyst from the bottom of one reactor to the top of the next reactor in line. The regenerated catalyst is added to the first reactor and the spent catalyst is withdrawn from the last reactor and transported back to the regenerator. The continuous reforming process is capable of operation at low pressures and high severity by managing the rapid coke deposition on the catalyst at an acceptable level. Additional benefits include elimination of downtime for catalyst regeneration and steady production of hydrogen of constant purity. Operating pressures are in the 3.5–17 bar range and design reformate octane number is in the 95–108 range.

## IV. COMMERCIAL REFORMING PROCESSES

Several commercial reforming processes are available for license worldwide. These processes differ in the type of operation (semiregenerative, cyclic, or continuous), catalyst type, and process engineering design. Table 2 presents a brief description of the major commercial naphtha reforming processes. All licensors agree on the necessity of hydrotreating the feed to remove permanent reforming catalyst poisons and to reduce the temporary catalyst poisons to low levels. The first U.S. commercial catalytic reforming process was introduced in 1939 as a result of large demand for high-octane gasoline and aromatics during World War II.

### A. UOP-Platforming

The UOP-Platforming process is used to upgrade low-octane naphthas to high-octane motor fuel blending components or for the production of aromatics for petrochemical feedstocks. Platforming was the first process to use platinum-on-alumina catalysts. The process has been adapted to bimetallic catalyst and to both semiregenerative and continuous operation (3,14,15).

To meet the demand of increased severity, UOP has improved the performance of the Platforming process by incorporating a continuous catalyst regeneration (CCR) system. This process uses stacked radial-flow reactors and a CCR section to maintain a steady-state reforming operation at optimum process conditions: fresh catalyst performance, low reactor pressure, and minimum

**Table 2** Commercial Catalytic Reforming Processes

| Licensor | Process name | Commercial installations | Application (upgrade low-octane naphthas) | Feedstock (naphtha) | Process type | References |
|---|---|---|---|---|---|---|
| Universal Oil Products (UOP) Process Division | Platforming | Over 700 units | •High-octane motor fuel blending<br>•Aromatics | •Straight run<br>•Hydrocracked, FCC, Thermal cracked | •Semiregenerative<br>•Continuous catalyst regeneration | 3, 14–16 |
| Institut Français du Pétrole (IFP) | Catalytic reforming | 90 units licensed (30 units continuous regeneration) | •High-octane reformate<br>•BTX and LPG | •Straight run<br>•Catalytic and thermal cracked | •Semiregenerative<br>•Continuous catalyst regeneration | 12, 16–18 |
| Chevron Research Co. | Rheniforming | 73 units | •High-octane gasoline blend stock<br>•Aromatics | •Straight run<br>•FCC<br>•Hydrocracked | •Semiregenerative | 16,23 |
| Engelhard Corp. | Magnaforming | 150 units | •High-octane reformate | •Straight run<br>•Hydrocracked<br>•Coker naphtha | •Semiregenerative<br>•Semicyclic | 16,20 |
| Exxon Research and Engineering Co. | Powerforming | 1.4 million BPSD | •Gasoline blending stocks, or<br>•Aromatics | •Straight run | •Semiregenerative or cyclic | 6, 21 |
| Amoco Oil Co. | Ultraforming | 150 units | •High-octane reformate<br>•Aromatics | •Straight run<br>•Catalytic and thermal cracked | Cyclic or semiregenerative | 6,22 |
| Houdry Division of Air Products and Chemicals, Inc. | Houdriforming | 0.25 million BPSD | •Aviation blending stocks<br>•Aromatics<br>•Motor gasoline | •Variety of feedstocks having end boiling points up to 200°C | •Semiregenerative | 22 |

recycle gas circulation. A typical Platforming diagram with CCR section is presented in Figure 3. The basic flow pattern through the Platforming unit with CCR is essentially the same as with conventional fixed-bed units. However, the catalyst is continuously withdrawn from the last reactor and transferred to the regenerator. The withdrawn catalyst is regenerated and returned to the top of the reactor stack, maintaining nearly fresh catalyst quality. The typical operating conditions for the current design of the UOP CCR process are: reactor pressure, 6.8 bar; LHSV, $1.6\,h^{-1}$; $H_2/HC$, molar, 2-3; and RON, clear, 100-107 (14). The simultaneous use of CCR technology and bimetallic catalysts has given UOP a unique position in the field of catalytic reformer process licensing (3).

One recent development in CCR technology is second-generation CCR Platforming with several modifications in the reactor and regenerator sections. The high-efficiency regenerator design resulted in an increased coke-burning capacity with reduced regeneration severity and complexity (14). The operation at ultralow pressure (3.4 bar) and the use of low-platinum R-34 catalyst ensured the highest yield of the reformate and aromatic product with more cost-effective process operation. The net gas recovery schemes maximized the yields of reformate and hydrogen. Moreover, the new regenerator allowed higher regeneration rates to support the coke generation of the low-pressure operation and high conversion levels. Table 3 presents typical yields of the Platforming process on a somewhat unusual feed (16).

It is believed that the UOP technology is used in over 40% of all reforming installations with more than 700 units in service worldwide. The individual capacities of these units range from 150 to 63,000 BPSD. A total of 113

**Table 3** Specific Case of UOP Semiregenerative and Continuous Platforming Processes

| Feed | Middle East Naphtha | | |
|---|---|---|---|
| Full boiling range, K | 365-385 | | |
| P/N/A, LV % | 69.4/21.4/9.2 | | |
| Gravity, °API | 59 | | |
| Operating conditions | Semiregenerative | Continuous | |
| Pressure, bar | 20.4 | 8.5 | 3.4 |
| Product, RON clear | 100 | 100 | 100 |
| Catalyst | R-62 | R-34 | R-34 |
| Yields on feed | | | |
| Hydrogen, wt % | 1.9 | 3.0 | 3.6 |
| $C_{5+}$, LV % | 70.4 | 78.1 | 81.8 |
| Aromatics, LV % | 46.0 | 54.5 | 57.6 |

*Source*: From Ref. 16.

stacked-reactor Platforming units are operating with continuous regeneration facilities. As of September 1993, the total capacity of the CCR units is 2.3 million BPSD with 16 units operating at ultralow (3.4 bar) reactor pressure. Forty-five additional CCR platforming units are in design and construction.

## B. IFP Reforming

The IFP catalytic reforming process is used to upgrade various types of naphthas to produce high-octane reformate, BTX, and LPG. The reforming process can be supplied in either semiregenerative or continuous operation. The semiregenerative version is a conventional reforming process in which the catalyst is regenerated in situ at the end of each cycle. The operating pressure of this process is in the range of 11.5–24.0 bar with low pressure drop in the hydrogen loop. The product RON, clear, is in the range 90–100 (16).

The decrease in the reformate yield during the run of the semiregenerative version (in spite of improvement in catalyst stability) has led IFP to develop a catalyst moving-bed system that allows continuous regeneration of the catalyst. This version, the Octanizing process, is an advanced design that reflects the results of several decades of research and development efforts carried out at IFP (19). The heart of the Octanizing technology lies in its original catalyst circulation and continuous catalyst regeneration systems. A general flow sheet of the IFP regenerative reforming process is presented in Figure 4. The flow

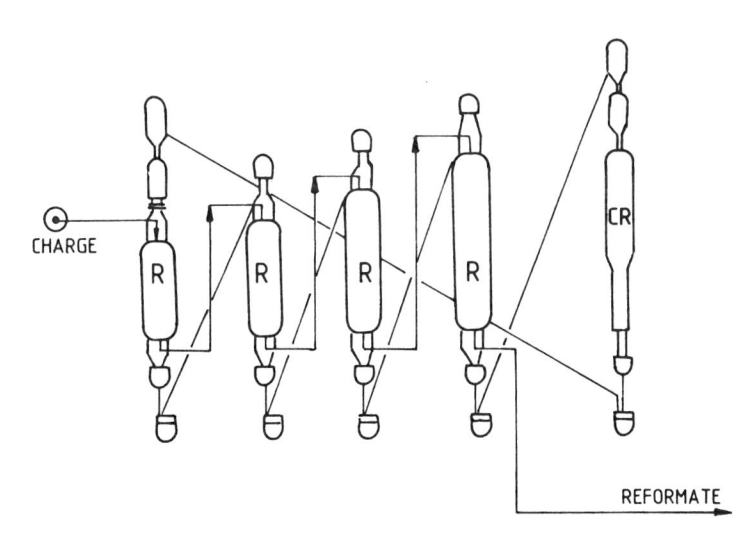

**Figure 4** Flow diagram of IFP regenerative reforming process. R, reactor; CR, continuous regenerator. Reprinted by permission of *Hydrocarbon Processing*, November 1990. Copyright 1990 by Gulf Publishing Co., all rights reserved.

scheme is similar in principle to that of any reforming process. The overall process comprises the following (1):

- A conventional reaction system consisting of a series of four radial-flow reactors which use a highly stable and selective catalyst suitable for continuous regeneration
- A catalyst transfer system using gas lift to carry the catalyst from one reactor to the next and finally to the regenerator
- A catalyst regeneration section which includes a purge to remove combustible gases, followed by catalyst regeneration

IFP regenerative technology has been improved to allow faster circulation of the catalyst and, as a consequence, increased regeneration frequency as required by the more severe operating conditions (low pressure, low $H_2$-to-hydrocarbon ratio). The Octanizing process features high on-stream efficiency, flexibility, and reliability. Major improvements, compared with previous designs, are the development of catalysts of increased activity, selectivity, and hydrothermal stability along with substantial increases in the yields of $C_{5+}$ reformate and hydrogen. Table 4 presents typical yields of the IFP conventional and regenerative process (16). In 1990, the total number of installations using the IFP technology was 90 licensed units; of these 30 units were designed for continuous regenerative technology (16).

## C. Engelhard-Magnaforming

The Magnaforming process, which is licensed by Engelhard Minerals and Chemicals Corporation, is used to upgrade low-octane naphtha to high-octane reformate. The product is a premium blending component or aromatic hydrocarbon source. A typical Magnaforming process diagram is presented in Figure 5 (6,16,20).

**Table 4** Typical Yields of IFP-Reforming Process

| Feed (naphtha) | Cut from Light Arabian Feedstock | |
|---|---|---|
| Full boiling range, K | 365–445 | |
| Operating conditions | Conventional process | Octanizing process |
| Stream factor, h/y | 8000 | 8400 |
| Operating pressure, bar | 9.5–14.5 | <5 |
| RON clear | 100 | 102 |
| MON, clear | 89 | 90.5 |
| Yields on feed | | |
| Hydrogen, wt % | 2.8 | 3.8 |
| $C_{5+}$, wt % | 83 | 88 |

*Source*: From Ref. 16.

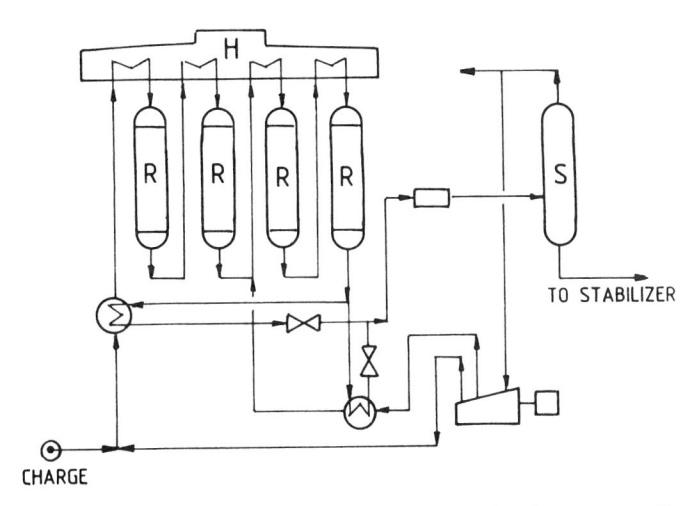

**Figure 5** Flow diagram of Engelhard-Magnaforming process. R, reactor; H, heater; S, separator. Reprinted from *Hydrocarbon Processing,* November 1990. Copyright 1990 by Gulf Publishing Co., all rights reserved.

The feature that most distinguishes the Magnaforming process from other reforming processes is its use of a split hydrogen recycle stream to increase liquid yields and improve operating performance. About half of the recycled gas is compressed and recycled to the first two reactors, which are operated at mild conditions. The greater portion of the recycle gas is returned to the terminal reactors, which operate under severe conditions. It is believed that substantial compressor power saving can be achieved by splitting the recycle.

Engelhard does not offer a continuous design, but it does offer a semiregenerative design and a combination of semiregenerative and cyclic regeneration. The combination design is made by supplementing the terminal reactors by a swing reactor which can alternate with the terminal reactors. It can also operate in parallel with the terminal reactors, permitting these reactors to be regenerated without unit shutdown. Smaller Magnaforming units use a conventional three-reactor system, compared with four reactors in large units.

The process was initially designed to operate with established monometallic platinum catalysts but was adapted to include the newer platinum-rhenium-based catalyst of the E600 and E800 series. The catalysts provide greater activity and stability, enabling use in units where high-severity operation and long cycle lengths are required. A wide range of catalysts have been used in the Magnaforming process in order to optimize operating performance and to produce desired product specifications. Many units incorporate Sulfur Guard Technology to reduce sulfur in reformer feed to ultralow levels. Table 5 presents examples of product yields of the Magnaforming process as a function

**Table 5** Yields of Engelhard Magnaforming Process

| Feed (naphtha) | | | |
|---|---|---|---|
| Full boiling range, K | 340–480 | | |
| P/N/A, LV% | 55.0/34.4/10.6 | | |
| Operating conditions | | | |
| Reactor pressure, bar | 24 | 17 | 10 |
| Product, RON clear | 100 | 100 | 100 |
| Yields on feed | | | |
| $H_2$, wt % | 2.5 | 2.8 | 3.1 |
| $C_1$–$C_3$, wt % | 8.5 | 6.1 | 4.0 |
| $iC_4 + nC_4$, LV % | 7.1 | 5.2 | 3.4 |
| $C_{5+}$, LV % | 78.9 | 81.5 | 84.0 |

*Source*: From Ref. 16.

of average reactor pressure. In 1990 there were approximately 150 units totaling 1.8 million BPSD using the Engelhard reforming technology (16).

## D. Exxon-Powerforming

The Powerforming process is used to produce gasoline blending stocks from low-octane naphthas. Alternatively, the process may be operated to give high yields of benzene or other aromatics or to produce aviation blending stocks. The process also produces large quantities of hydrogen, which can be used to desulfurize or improve other products. The Powerforming process may be designed to operate as either a semiregenerative or a cyclic unit to permit selection of the best operation for a specific reforming application. The reactor system of the semiregenerative units is designed in such a way that it can be readily converted into the corresponding cyclic operation with swing reactor. Both systems use high-activity bimetallic catalysts of Exxon's KX series in order to minimize catalyst and investment costs while maximizing product yield (21). Figure 6 presents a flow diagram of the Powerforming process (6).

Cyclic Powerformers are designed to operate at low pressure, with a wide range of feed boiling point and low hydrogen-to-feed ratio. The unit has four reactors in series plus a swing reactor. The use of the swing reactor allows any of the on-stream reactors to be taken out of service for regeneration while maintaining continuous operation of the unit. The frequency of regeneration can be varied to meet changing process objectives as well as operating at high-severity, low-pressure conditions, since coke deposition is maintained at low levels. Regeneration is generally performed on a predetermined schedule to avoid having two or more reactors regenerated at the same time. Usually the terminal reactors are scheduled for more frequent regenerations than the early-stage reactors. Run lengths of up to 6 years between shutdowns could be

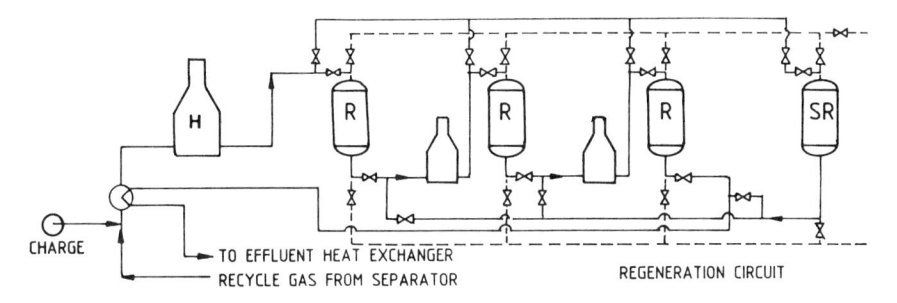

**Figure 6** Flow diagram of Exxon-Powerforming process. H, heater; SR, swing reactor; R, reactor. From Ref. 16.

achieved. Typical yields of both semiregenerative and cyclic Powerformers are presented in Table 6. In 1984, the total capacity of the Powerformer units was more than 1.4 million BPSD.

### E. Chevron-Rheniforming

The Rheniforming process is used to convert naphthas to high-octane gasoline blendstock or aromatics plant feedstock. The process has gained wide acceptance since Chevron patented the bimetallic Pt/Re catalyst in 1968. A typical Rheniformer process in presented in Figure 7 (16,23). Rheniforming is basically a semiregenerative process which comprises a sulfur sorber, three radial-flow reactors in series, a separator, and a stabilizer. The process is characterized by the sulfur control sorb, which reduces sulfur to 0.2 ppm in the reformer feed. A new Rheniforming F/H catalyst system has been used, which permits low-pressure operation. The high resistance to fouling of the catalyst

**Table 6** Yields of the Powerforming Process Using Semiregenerative and Cyclic Operation

| Feed (naphtha) | | |
|---|---|---|
| Gravity, °API | 57.2 | |
| P/N/A, LV % | 57/30/13 | |
| | Type of operation | |
| Yields on feed | Semiregenerative | Cyclic |
| $H_2$, wt % | 2.3 | 2.6 |
| $C_1$-$C_4$, wt % | 13.1 | 11.2 |
| $C_{5+}$, LV % | 78.5 | 79.1 |
| RON, clear | 99 | 101 |

*Source*: From Ref. 22.

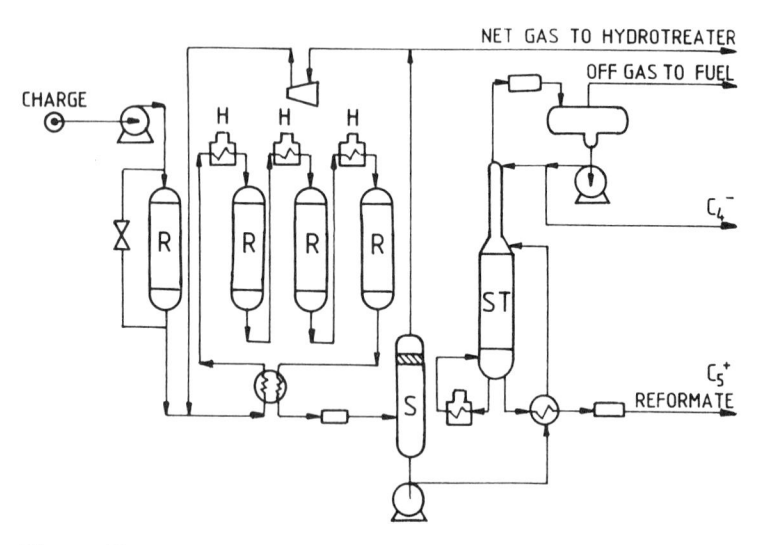

**Figure 7** Flow diagram of Chevron-Rheniforming process. R, reactor; S, separator; H, heater; ST, stabilizer. From Ref. 16.

system increases the yields of aromatic naphtha product and hydrogen due to the long cycle lengths, which reach 6 months or more. Optimized operating techniques permit maintenance of high catalyst activity throughout each cycle and return to fresh activity after each regeneration.

The increased resistance to fouling also provides for expansion of existing plants by using higher space velocities, lower recycle ratios, or increased product octanes. Converted units are operating with $H_2/HC$ ratios of 2.5–3.5 and long cycles between regeneration. Table 7 presents some characteristics of the Rheniformer process and typical yields for two different naphtha feeds. It is believed that a total of 73 Rheniformers are on stream with a total capacity of more than 1 million BPSD (16).

### F. Amoco-Ultraforming

The Ultraforming process is used to upgrade low-octane naphthas to high-octane blending stocks and aromatics. The process is a fixed-bed cyclic system with a swing reactor incorporated in the reaction section which is usually specified for aromatic (BTX) production. The system can be adapted to semiregenerative operation with the conventional three radial-flow reactors in series (6,22).

The process uses rugged, proprietary catalysts permitting frequent regeneration and high-severity operations at low pressures. The catalyst system has relatively low precious metals content, and its estimated life is perhaps 4 years

*Catalytic Reforming Processes*

**Table 7** Yields of the Rheniforming Process Using Different Naphtha Feeds

| Feed (naphtha) | Hydrotreated | | Hydrocracked |
|---|---|---|---|
| Type | Paraffinic | | Naphthenic |
| Full boiling range, K | 365–390 | | 365–475 |
| P/N/A, LV % | 69.6/23.4/8.0 | | 32.6/55.5/11.9 |
| Sulfur, ppm | <0.2 | | <0.2 |
| Nitrogen, ppm | <0.5 | | <0.5 |
| Operating conditions | | | |
| Outlet pressure, bar | 6.0 | 13.6 | 13.6 |
| Product, RON clear | 98 | 99 | 100 |
| Yields on feed | | | |
| Hydrogen, wt % | 3.1 | 2.5 | 2.9 |
| $C_{5+}$, LV % | 80.1 | 73.5 | 84.7 |
| Aromatics, LV % | 66.5 | 67.9 | 69.9 |
| Paraffins, LV % | 32.4 | 31.2 | 27.5 |
| Naphthenes, LV % | 1.1 | 0.9 | 2.6 |

*Source*: From Ref. 16.

for cyclic operation versus 8 years for semiregenerative operation (22). The swing reactor in cyclic operation replaces any reactor while the catalyst bed in this reactor is being regenerated. Normally, the reactors are all the same size; however, the first reactor is loaded with half the usual amount of catalyst.

Ultraformers may be designed to produce high-purity xylene and toluene, which can be separated by straight distillation before the extraction step. The benzene fraction can be recovered by extractive distillation. High yields of $C_{5+}$ reformate and hydrogen have been reported for the Ultraforming process. Table 8 presents typical yields for various naphtha feeds. The total capacity of

**Table 8** Yields of the Ultraforming Process Using Different Naphtha Feeds

| Feed (naphtha) | Mid-east | Mid-continent | Heavy hydrocracked |
|---|---|---|---|
| Full boiling range, K | 375–460 | 350–455 | 380–465 |
| P/N/A, LV % | 68/19/13 | 52/35/13 | 25/36/39 |
| Yields on feed | | | |
| $H_2$, scf/bbl | 1,280 | 1,370 | 950 |
| $C_1$–$C_4$, wt % | 13.3 | 12.3 | 9.8 |
| $C_{5+}$, LV % | 78.0 | 77.6 | 83.6 |
| Aromatics, LV % | 68 | 78 | 84 |
| RON, clear | 99 | 103 | 106 |

*Source*: From Ref. 22.

Ultraforming is over 530,000 BPSD for 39 commercial units worldwide. However, no new Ultraformers have been licensed in recent years (22).

### G. Houdriforming

The Houdriforming process is licensed by the Houdry Division of Air Products and Chemicals, Inc. The process is used to upgrade various naphthas to aviation blending stocks, aromatics, and high-octane gasoline in the range of 80–100 RON, clear. The process operates in a conventional semiregenerative mode with four reactors in series for BTX production, compared with three reactors for gasoline (22).

The catalyst used is usually $Pt/Al_2O_3$ or may be bimetallic. A small "guard case" hydrogenation pretreater can be used to prevent catalyst poisons in the naphtha feedstock from reaching the catalyst in the reforming reactors. The guard case is filled with the usual reforming catalyst but operated at a lower temperature. It is constructed as an integral stage of the Houdriforming operation when required for the feedstock.

At moderate severity, the process may be operated continuously for either high-octane gasoline or aromatics, without provision for catalyst regeneration. However, operation at high severity requires frequent in situ catalyst regeneration. Typical operating conditions are: temperature 755–810 K, pressure 10–27 bar, LHSV $1-4\,h^{-1}$, and $H_2/HC$ ratio 3–6. Table 9 presents yields of the Houdriforming process. It has been reported that the total capacity of Houdriforming units is about 250,000 BPSD (22).

## V. COMMERCIAL REFORMING CATALYSTS

Since the 1950s commercial reforming catalysts have been essentially platinum metal supported on chlorided alumina. Currently, platinum and another

**Table 9** Yields of the Houdriforming Process

| | |
|---|---|
| Feed (naphtha) | |
| Gravity, °API | 52.6 |
| Full boiling range, K | 365–465 |
| P/N/A, LV % | 43/38/9 |
| Yields on feed | |
| RON, clear | 100 |
| Paraffins, LV % | 21 |
| Naphthenes, LV % | 2 |
| Aromatics, LV % | 77 |

*Source*: From Ref. 22.

metal—often rhenium, tin, or iridium—account for most commercial bimetallic reforming catalysts. The catalyst is most often presented as 1/16, 1/8, or 1/4-inch $Al_2O_3$ cyclindrical extrudates or beads into which Pt has been deposited. In commercial catalysts, platinum concentration ranges between 0.3 and 0.7% and chloride is added (0.1-1.0%) to the alumina support (eta or gamma) to provide acidity. At present, there are 10 international manufacturers of reforming catalysts producing more than 80 different types of catalysts suitable for different applications and for a variety of feedstocks (24). The current demand for reforming catalysts is mainly a replacement market with about 75 to 80% of it as bimetallics.

Reforming catalyst manufacturers continue to develop new catalyst formulations designed to meet a wide array of challenges. Many of these challenges involve environmental regulations that refiners have been, and will be, required to meet during the 1990s (24). Table 10 presents a recent compilation of all commercial reforming catalysts that are available by sale or license to refiners. This compilation is based on information presented in the Oil and Gas Journal—Worldwide Catalyst Report published in 1993 (24). The list is designed to provide a ready reference for both refiners and manufacturers and to sort out the sometimes confusing nonmenclature used to designate or label catalysts (24).

## VI. CONCLUSIONS

Catalytic reforming is a well-established process in the refining industry for the production of high-octane gasoline. The process also enjoys a virtual monopoly in terms of supplying aromatics to the petrochemical industry. Reforming process licensors are responding rapidly to the growing pressure of quality standards instigated by environmental protection legislation. The growing demand for octane (and for unleaded gasoline) has necessitated operation at very low pressure and recycle ratios. The use of bimetallic catalysts has allowed revamping of high-pressure reformers to low-pressure operation, resulting in higher reformate and hydrogen yields. Moreover, the successful operation of the continuous catalyst regeneration process has provided a considerable octane enhancement due to the ability to operate at very low pressure with maximum time on stream and catalyst activity. The anticipated future reduction of volatility, aromatics, and benzene contents of gasoline will require improving reforming catalysts to produce less benzene in the BTX composition and raising the initial boiling point of the reformer feed. In addition to severity reduction, operational changes can substantially reduce reformate benzene content. Benzene may also be removed either by prefractionating naphtha feed or by post-treating the reformate product.

The longer-term trend for catalytic reforming shows increased interest in the development of more selective and stable catalysts capable of entirely con-

**Table 10** International Reforming Catalyst Compilation[a]

| Catalyst designation | Primary differentiating characteristics | Application (feedstock) | Application (product) | Form | Carrier support | Active agents |
|---|---|---|---|---|---|---|
| **AKZO Chemicals** | | | | | | |
| CK-300 series | All monometallics, diff. active metal contents | Naphtha | Gaso. or BTX | Cyl. | $Al_2O_3$ | PtCl |
| CK-433 | Bimetallic | Naphtha | Gaso. or BTX | Cyl. | $Al_2O_3$ | PtReCl |
| CK-522 TRILOBE | Bimetallic, high part. dens. | Naphtha | Gaso. or BTX | Sh. ext. | $Al_2O_3$ | Pt/ReCl |
| CK-542 TRILOBE | High Re, high part. dens. shaped catalyst | Low-sulfur naphtha | Gaso. or BTX | Sh. ext. | $Al_2O_3$ | Pt/ReCl |
| ARC-555 | Bimetallic/economic | Naphtha | Gaso. or BTX | Cyl/Sh | $Al_2O_3$ | Pt/ReCl |
| ARC-111E | Low-pressure fixed bed | Naphtha | Gaso. or BTX | Ext. | $Al_2O_3$ | Pt/Sn |
| ARC-111B | Continuous operation | Naphtha | Gaso. or BTX | Sph. | $Al_2O_3$ | Pt/Sn |
| **Chevron Research & Technology Co.** | | | | | | |
| Rheniforming | Latest version, long cycle life, stable product yields | Naphthas, SR and cracked-coal naphthas | Gaso. & BTX | Cyl. | "p" | Pt/Re |
| Type F | | | | | | |
| Type H | Used with type F for higher operations | | Gaso. & BTX | Cyl. | "p" | Pt/Re |
| **Criterion Catalyst Co. LP** | | | | | | |
| AROFORM PHF-5 | Monometallic | Naphtha | Gaso. or BTX | Cyl. | $Al_2O_3$ | "p" |
| PRHF-30 | Bimetallic, different amounts of active metals | Naphtha | Gaso. or BTX | Cyl. | $Al_2O_3$ | "p" |
| PRHF-37 | | | | | | |
| PRHF-50 | | | | | | |
| PRHF-58 | | | | | | |
| PRHF-33 | | | | | | |
| PR-28 | Bimetallic | Low-sulfur naphtha | Gaso. or BTX | Sh. ext. | $Al_2O_3$ | PrReCl |
| AEROFORM PFH-4 | Monometallic | Naphtha | Gaso. or BTX | Cyl. | $Al_2O_3$ | PtCl |
| TRILOBE P-8 | Monometallic | Naphtha | Gaso. or BTX | Sh. ext. | $Al_2O_3$ | PtCl |
| AEROFORM PR-6&6A | Bimetallic | Naphtha | Gaso. or BTX | Cyl. | $Al_2O_3$ | PtReCl |
| TRILOBE PR-8 8A & 8E | Bimetallic | Naphtha | Gaso. or BTX | Sh. ext. | $Al_2O_3$ | PtReCl |
| TRILOBE PR-18 & 18A | Bimetallic | Low sulfur naphtha | Gaso. or BTX | Sh. ext. | $Al_2O_3$ | PtReCl |
| PS-7 | Multimetallic | Naphtha | Gaso. or BTX | Cyl. or sh. ext. | $Al_2O_3$ | PtSnCl |
| PS-10 | Multimetallic | Naphtha | Gaso. or BTX | Sph. | $Al_2O_3$ | PtSnCl |

## Engelhard Corporation (Acreon)

| | | | | | |
|---|---|---|---|---|---|
| E-301[b] | Moderate stability | Naphtha, S < 5 ppmw | Gaso. or aro. | Cyl. | Al$_2$O$_3$ | Pt |
| E-302[b] | More stable than E-301 | Naphtha, S < 10 ppmw | Gaso. or aro. | Cyl. | Al$_2$O$_3$ | Pt |
| E-311[b] | Maximize LPG production | Naphtha, S < 5 ppmw | LPG gaso. | Cyl. | Al$_2$O$_3$ | Pt |
| E-601[b] | High stability | Naphtha, S < 1 ppmw | Gaso. or aro. | Cyl. | Al$_2$O$_3$ | Rt, Re |
| E-603[b] | High stab. lower metal load | Naphtha, S < 1 ppmw | Gaso. or aro. | Cyl. | Al$_2$O$_3$ | Rt, Re |
| E-611[b] | Very high stability | Naphthas, S = 1 ppmw | Gaso. or aro. | Cyl. | Al$_2$O$_3$ | Rt, Re |
| E-801 | Stab. or E-603 at higher act. | Naphthas, S < 1 ppmw | Gaso. or aro. | Cyl. | Al$_2$O$_3$ | Pt/Re |
| E-802 | Stab. or E-611 at higher act. | Naphthas, S < 0.5 ppmw | Gaso. or aro. | Cyl. | Al$_2$O$_3$ | Pt/Re |
| E-803 | Stab. or E-611 at higher act. opt. metals loading | Naphthas, S < 0.5 ppmw | Gaso. or aro. | Cyl. | Al$_2$O$_3$ | Pt/Re |
| E-804 | Stab. and act. of E-801 opt. metals loading | Naphthas, S < 0.5 ppmw | Gaso. or aro. | Cyl. | Al$_2$O$_3$ | Pt/Re |
| E-1000 | High stab. for continuous units | Naphtha | GO or aromatics | Sph. | Al$_2$O$_3$ | PtSn |

## Exxon Research & Engineering Co.

| | | | | | |
|---|---|---|---|---|---|
| KX-120 | Multimetallic semiregen. or cyclic units, diff. promotor | Virgin/cracked naphthas | High oct. gaso. & aro. | | | Pt, etc. |
| KX-130 | Multimetallic semiregen. or cyclic units, diff. promotor | Virgin/cracked naphthas | High oct. gaso. & aro. | | | Pt, etc. |
| KX-160 | Multimetallic semiregen. or cyclic units, diff. promotor | Virgin/cracked naphthas | High oct. gaso. & aro. | | | Pt, etc. |

## Instituto Mexicano del Petroleo

| | | | | | |
|---|---|---|---|---|---|
| IMP-RNA-1 | High activity bimetallic | Desulfurized naphtha, aro. production | BTX, high-oct. gaso. | Cyl. ext. | Al$_2$O$_3$ | Pt/Re |
| IMP-RNA-2 | High activity bimetallic | | BTX, high-oct. gaso. | Trilobe ext. | Al$_2$O$_3$ | Pt/Re |
| IMP-RNA-4, continuous process | High activity bimetallic | Desulfurized naphtha, aro. production | High-oct. gaso. | Sph. | Al$_2$O$_3$ | Pt/Sn |

## Katalysatorenwerke Hüls GmbH

| | | | | | |
|---|---|---|---|---|---|
| H-2440 | | Fraction 35–195°C | Gaso. or BTX | Cyl. ext. | Al$_2$O$_3$ | Pt, 0.3–0.8% monometallic |

(*continued*)

**Table 10** Continued

| Catalyst designation | Primary differentiating characteristics | Application (feedstock) | Application (product) | Form | Carrier support | Active agents |
|---|---|---|---|---|---|---|
| **Leuna-Werke AG** | | | | | | |
| 8815/03 | Monometallic, different amounts of active metal | Naphtha | Gaso. or BTX | Ext. | $Al_2O_3$ | Pt |
| 8815/05 | Monometallic, different amounts of active metal | Naphtha | Gaso. or BTX | Ext. | $Al_2O_3$ | Pt |
| 8819 | All bimetallics, different amounts of active metal | Naphtha, S < 1 ppm | Gaso. or BTX | Ext. | $Al_2O_3$ | Pt/Re Cl |
| 8819 WS | All bimetallics, different amounts of active metal | Naphtha, S < 1 ppm | Gaso. or BTX | Vortex | $Al_2O_3$ | Pt/Re Cl |
| 8819B | All bimetallics, different amounts of active metal | Naphtha, S < 1 ppm | Gaso. or aromatics | Vortex | $Al_2O_3$ | Pt/Re |
| 8819C | All bimetallics, different amounts of active metal | Naphtha, S < 1 ppm | Gaso. or aromatics | Vortex | $Al_2O_3$ | Pt/Re |
| 8821C | All bimetallics, different amounts of active metal | Naphtha, S < 1 ppm | Gaso. or aromatics | Sph. | $Al_2O_3$ | Pt/Re |
| 8841 | Cont. regen. reformer, high mechanical stability | Naphtha, S < 1 ppm | Gaso. or aromatics | Sph. | $Al_2O_3$ | Pt/Sn |
| 8821 | Cont. regen. reformer, high mechanical stability | Naphtha, S < 1 ppm | Gaso. or aromatics | Sph. | $Al_2O_3$ | Pt/Sn |
| 8822 | Cont. regen. reformer, high mechanical stability | Naphtha, S < 0.5 ppm | Gaso. or aromatics | Vortex | $Al_2O_3$ | Pt/Sn |
| 8823 | Cont. regen. reformer, high mechanical stability | Naphtha, S < 0.5 ppm | Gaso. or aromatics | Sph. | $Al_2O_3$ | Pt/Sn |

| | | | | | | |
|---|---|---|---|---|---|---|
| 8840 | Cont. regen. reformer, high mechanical stability | Naphtha, S < 1 ppm | Gaso. | Sph. | Al$_2$O$_3$ | Pt + prom. |
| 8860 | Selct. HC or n-paraffins | Naphtha | Gaso. | Vortex | Al$_2$O$_3$/zeo. | Pt |
| **Procatalyse (Acreon)** | | | | | | |
| RG-412 | H.P. reformer | Naphtha | Gaso. | Cyl. | Al$_2$O$_3$ | Pt 0.35% |
| RG-482 | M.P. & L.P. reformer | Naphtha | Gaso. | Cyl. | Al$_2$O$_3$ | Pt 0.3% + Re |
| RG-492 | M.P. & L.P. reformer | Naphtha | Gaso. | Cyl. | Al$_2$O$_3$ | Pt/Re |
| RG-442 | Reformer | Naphtha | Gaso. and LPG | Cyl. | Al$_2$O$_3$ | Pt + IR + promo. |
| CR-201 | Cont. regen. reformer | Naphtha | Gaso. | Sph. | Al$_2$O$_3$ | PtSn |
| AR-403 | Cont. regen. reformer (IFP aromizing) | Naphtha | Aro. | Sph. | Al$_2$O$_3$ | PtSn |
| AR-405 | CCR (IFP aromizing) | Naphtha | Aro. | Sph. | Al$_2$O$_3$ | PtSn |
| MD-101 | Sulfur removal | Naphtha | Naphtha | Sph. | Al$_2$O$_3$ | PtSn |
| RG-582 | M.P. & L. P. reformer | Naphtha | Gaso. | Cyl. | Al$_2$O$_3$ | PtRe promoter |
| RG 592 | M.P. & L. P. reformer | Naphtha | Gaso. | Cyl. | Al$_2$O$_3$ | PtRe promoter |
| **UOP** | | | | | | |
| R-11 | Higher act. monometallic | Semi-regen. or swing | Gaso. aro. | Sph. | Al$_2$O$_3$ | Pt |
| R-12 | Monometallic, higher stab. than R-11 | Semi-regen. | Gaso. aro. | Sph. | Al$_2$O$_3$ | Pt |
| R-55 | Monometallic, higher stab. than R-11 | Semi-regen. or swing | Gaso. aro./ | Ext. | Al$_2$O$_3$ | Pt |
| R-15 | Monometallic | Semi-regen. | LPG gaso. | Sph. | SiO$_2$/Al$_2$O$_3$ | Pt |
| R-16H | Bimetallic, higher stab. than monometallics | High severity, semi-regen. | Gaso. aro. | Sph. | Al$_2$O$_3$ | Pt/Re |
| R-16F | Bimetallic, lower Pt. higher stab. than monometallics | High severity, semi-regen. | Gaso. aro. | Sph. | Al$_2$O$_3$ | Pt/Re |
| R-16G | Bimetallic, higher act. and stab. than 16F and H | High severity, semi-regen. | Gaso. aro. | Sph. | Al$_2$O$_3$ | Pt/Re |
| R-18 | Bimetallic, higher act. and stab. than R-16 series | High severity, semi-regen. | Gaso. aro. | Sph. | Al$_2$O$_3$ | Pt/Re |
| R-22 | Bimetallic, higher yields, reduced state | CCR or semiregen. | Higher pro. of oct. bbl. and aro. | Sph. | Al$_2$O$_3$ | "P" |

(*continued*)

**Table 10** Continued

| Catalyst designation | Primary differentiating characteristics | Application (feedstock) | Application (product) | Form | Carrier support | Active agents |
|---|---|---|---|---|---|---|
| **UOP** *(continued)* | | | | | | |
| R-30 | Bimetallic, higher yields, reduced state | CCR or swing CCR | Higher pro. of oct. bbl. and aro. | Sph. | $Al_2O_3$ | "P" |
| R-32 | Bimetallic, higher yields, reduced state | CCR or swing CCR | Higher pro. of oct. bbl. and aro. | Sph. | $Al_2O_3$ | P |
| R-50 | Bimetallic, higher act. and stab. than R-16 series | High severity, semi-regen. | Gaso. or aro. | Ext. | $Al_2O_3$ | Pt/Re |
| R-51 | Bimetallic, higher act. and stab. than R-16 series | High severity, semi-regen. | Gaso. or aro. | Ext. | $Al_2O_3$ | Pt/Re |
| R-56 | Bimetallic, higher act. and stab. than R-16 series | High severity, semi-regen. | Gaso. or aro. | Ext. | $Al_2O_3$ | Pt/Re |
| R-60 | Bimetallic, higher stab. than R-50, high act. | High severity, semi-regen. | Gaso. or aro. | Sph. | $Al_2O_3$ | Pt/Re |
| R-62 | Bimetallic, lower Pt than R-60, higher stab. than R-50, high act. | High severity, semi-regen. | Gaso. or aro. | Sph. | $Al_2O_3$ | Pt/Re |
| R-132 | Bimetallic, higher yields, reduced state | CCR or swing CCR | Higher prod. of oct. bbl & aro. | Sph. | $Al_2O_3$ | "P" |

[a] Abbreviations: aro - Aromatic or aromatics; BTX - Benzene, toluene, xylenes (aromatics); Cyl. - Cylinder; Ext. - Extrudate; GO - Gas oil; Gaso. - Gasoline; LPG - Liquified petroleum gas; Oct. - Octane; "P" - Proprietary; Prom. - Promoted or promoter; SA - Surface area; SR - Straight run; Stab. - Stable or stability; Sph. - Sphere; Sh. ext. - Shaped extrudate; VGO - Vacuum gas oil; Zeo. - Zeolite; x - Unrestricted availability; L - Available under license only.
[b] Catalysts available in higher density at 49.5 lb/cu.ft. They are designated HD, i.e., E-301HD.
*Source:* From Ref. 24.

verting naphtha feed into aromatics and hydrogen alone. Moreover, there is a good incentive to replace fixed-bed reformers with reformers designed for continuous catalyst regeneration, since increased reforming severity in fixed-bed units is limited by yield loss. It is believed that reforming catalyst manufacturers and process licensors will continue to improve the performance of the catalytic reforming process in order to meet the environmental regulations of the 1990s.

## ACKNOWLEDGMENT

The author acknowledges the support of the Research Institute, King Fahd University of Petroleum and Minerals, Dhahran, during the preparation of the manuscript.

## REFERENCES

1. J. F. Le Page, *Applied Heterogeneous Catalysis*. IFP Publications, Editions Technip, Paris, 1987, p. 467
2. M. D. Edgar, Catalytic reforming of naphtha in petroleum refineries. In *Applied Industrial Catalysis* (B. E. Leach, ed.). Academic Press, New York, 1983, vol. 1, p. 123.
3. J. A. Weiszmann, UOP Platforming process. In *Handbook of Petroleum Refining Processes* (R. Meyers, ed.). McGraw-Hill, New York, 1986, p. 3.1.
4. J. H. Sinfelt, *Bimetallic Catalysts*. Wiley, New York, 1983, p. 131.
5. D. M. Little, *Catalytic Reforming*. PennWell, Tulsa, OK, 1985, p. 40.
6. F. G. Ciapetta and D. N. Wallace, Catalytic naphtha reforming. *Catal. Rev. 5*, 67 (1971).
7. R. D. Srivastava, *Heterogeneous Catalytic Science*. CRC Press, Boca Raton, FL, 1988, p. 125.
8. L. Bell, Worldwide refining. *Oil Gas J.*, Dec. 20, 1993, p. 46.
9. *International Petroleum Encyclopedia*, Vol. 25, PennWell Publishing Co., Tulsa, OK, 1992, p. 321.
10. *The Petroleum Handbook*, Elsevier, Amsterdam, 1983, p. 268.
11. B. C. Gates, J. R. Katzer, and G. C. Schuit, *Chemistry of Catalytic Processes*. McGraw-Hill, New York, 1979, p. 236.
12. *IFP—Technology Symposium: Oil Refining and Petrochemicals Production*. IFP Publications, Paris, 1981, p. 3.
13. R. Prins, Modern processes for the catalytic reforming of hydrocarbons. In *Chemistry and Chemical Engineering of Catalytic Processes* (R. Prins and G. Schuit, eds.). Sijthoff and Noordhoff, Alphen aan den Rijn, The Netherlands, 1980, p. 389.
14. R. Peer, R. Bennet, and S. Bakas, Continuous reformer catalyst regeneration technology improved. *Oil Gas J.*, May 30, 1988, p. 52.
15. R. Peer, R. Bennett, D. Felch, and R. G. Kabza, Platforming leading octane technology in the 1990's. UOP Publications, Des Plaines, IL, 1990.

16. Refining Handbook '90, *Hydrocarbon Process.*, November 1990, p. 118.
17. P. Bonnifay, B. Cha, J. Barbier, A. Vidal, and R. Huin, Maximizing aromatics production goals of IFP process. *Oil Gas J.*, Jan. 19, 1976, p. 48.
18. M. Berthelin, J. Bournonville, E. Diab, and J. Frank, *Catalytic Reforming Catalysts*. Procatlyse Publications, Paris, 1986.
19. A. Hennico, L. Mank, C. Mansuy, and D. H. Smith, Texas reformer designed for two-step expansion. *Oil Gas J.*, June 8, 1992, p. 54.
20. J. Nevison, C. Obaditch, and M. Dalson, Magnaforming units compared. *Hydrocarbon Process.*, June 1974, p. 110.
21. R. Cecil, W. Kmak, J. Sinfelt, and L. Chambers, New reforming catalyst highly active. *What's New in Cat Reforming*. Petroleum Publishing Co., Tulsa, OK, 1973, p. 29.
22. Refining Handbook '80, *Hydrocarbon Process.*, September 1980, p. 115.
23. M. Freiburger, W. C. Buss, and A. G. Bridge, Recent catalyst and process improvement in commercial rheniforming. Presented at the 1980 NPRA Meeting, March 23–25, 1980, New Orleans.
24. A. K. Rhodes, Worldwide catalyst report. *Oil Gas J.*, Oct. 11, 1993, p. 41.

# 14
# Modeling Commercial Reformers

**Lee E. Turpin**

*Profimatics, Inc., Thousand Oaks, California*

## I. INTRODUCTION

Business decisions on commercial catalytic reformers are based on information provided from process data and laboratory analyses. To enable the refiner to maximize the effectiveness of readily available process data, techniques are presented for collection of raw process data, determining the validity of the data, and conversion of the data into usable information via a process simulation model. Once a catalytic reformer is accurately modeled, business decisions can be made with respect to catalyst performance, product yields, mechanical or process changes in the system, and control and optimization of the catalytic reforming process.

The chapter also contains examples of typical refinery problems that can be easily and accurately solved using a kinetic model of a catalytic reformer. The solution techniques are described for calculating the value of incremental feedstock, replacing the feed/effluent exchanger, and calculating the benefit of reducing octane through advanced process control.

The final section discusses the application of a kinetic model to advanced process control for on-line optimization. Typical advanced controls are presented along with methods for calculating their economic benefits.

## II. TEST RUNS

The objective of a test run is to measure the performance of a process unit. In the case of a catalytic reformer, a test run can be used to measure the performance of both mechanical equipment and the reforming catalyst. A successful test run is predicated on six basic issues:

1. Identification of test run objective(s)
2. Proper selection of streams to be crossed by the material and energy balance envelope
3. Steady-state operation
4. Accurate process measurements of flows, temperatures, and pressures
5. Correct sample collection
6. Coordination between process engineering, unit operations, refinery planning, and the laboratory

The cost of performing routine test runs is insignificant in comparison with the lost opportunity cost of not having valid information for making business decisions.

### A. Test Run Objectives

Test run objectives must be identified prior to defining data collection requirements and specifying test run operating conditions. Identifying the test run objectives and designing a test run to meet those objectives will minimize the data collection and laboratory test requirements and the disruption of refinery operations.

### B. Material and Energy Balance Envelope

Once the test run objective has been identified, the test run material and energy balance envelope (1) can be defined. This definition includes identifying the process measurements to be recorded and the samples to be collected for the test run. This envelope varies from one catalytic reformer unit to another but normally includes the feed preheat section, reactor section, product separator, and fractionator. The material and energy balance envelope should include as few pieces of process equipment as needed to meet the test run objectives, should cross as few process streams as possible, and must cross only process and utility lines that can be accurately measured and sampled using the instrumentation and sample points available. A line containing two-phase flow cannot be crossed by the material and energy balance envelope; neither can a line without a flowmeter with corresponding temperature and pressure indications and a sample collection point. Care should be taken to avoid process conditions in a line that will affect the measuring instrumentation or the sample collection, such as using an orifice plate as the primary flow measuring element when the process stream being measured is at its bubble point. A product stream that is used for fuel within the process should be included twice in the material and energy balance: once as a product stream and once as an energy source. It is critical to properly test, measure, and account for all feed and product streams that cross the material and energy balance envelope.

## C. Steady-State Operation

It is important to have the unit in steady-state operation at the time of the test run. Steady-state operation can be determined by monitoring the standard deviation of various process variables on a once-per-minute basis for 30 minutes. Table 1 shows the process variables to be checked and the desired allowable standard deviation for each variable. If any variable exceeds the standard deviation specification, the unit is not in steady state. A program can easily be installed in a distributed control system (DCS) to generate the data required to determine a steady-state operation.

## D. Precise Flow Measurement

The major sources of errors in most refinery test runs are feed and product flow measurements. Commercial catalytic reformer units uses orifice plates as the primary flow measurement elements. To minimize flow errors it is recommended that the orifice diameter of each orifice plate be verified at each unit turnaround and any orifice plates that exhibit wear be replaced at that time. Approximately 2 weeks before the test run, the differential pressure element for each flowmeter should be zeroed and spanned. During the week preceding the test run, daily material balances should be performed to verify the flowmeter calibrations before doing the test run. If feed and/or product tanks are available, meter readings should be compared to changes in tank inventories.

**Table 1** Steady-State Specifications

| Process variable | Allowable standard deviation |
|---|---|
| Feed rates | 0.1% of flow |
| Product rates | |
|   On flow control | 0.1% of flow |
|   On level control | 0.5% of flow |
| Recycle rate | 0.2% of flow |
| Reactor inlet temperatures | 0.2°C |
| Reactor delta temperatures | 1.5°C |

### E. Correct Sample Collection

Refinery laboratories are often criticized unfairly for reporting erroneous results, when in fact problems are often not in the analysis performed but rather in the technique and/or equipment used for the sample collection. For example, one common problem is caused when reformate is sampled from the reformate run-down line. The same is collected at flowing temperature in a glass sample bottle and taken to the laboratory to be tested for research octane number clear (RONC), Reid vapor pressure (RVP), gravity, and component analysis by gas chromatography (GC). If the sample is drawn directly from the run-down line into a sample container, a small amount of the butane (maybe 0.5% of the total sample) can flash while the sample container is being filled. Because the blending octane number of the flashed sample is very close to that of the reformate and the RONC test procedure is known to have a wide variation in reported results, the reported RONC is not questioned. Likewise, the gravity analysis is not questioned because the small quantity of butane flashed does not change the gravity of the reformate beyond what is expected. The GC analysis reports 50 to 75% of the expected butane concentration, and the fault is placed on the laboratory for a "bad" GC analysis when in reality it was a sample collection error.

The following guidelines are provided to minimize sample collection errors. Note that the guidelines apply to sample collection in a catalytic reformer process and are not inclusive of guidelines for other process areas. These guidelines are for ensuring that the samples are representative of the streams being sampled. Safety and environmental impact considerations are not included.

1. All samples should come from the center of the process line being sampled. This can easily be achieved by cutting the tip off of a thermowell, installing the thermowell into the process line, and taking the sample through the modified thermowell.

2. Samples being drawn into an open container must be cooled below the temperature which will result in part of the sample flashing. If the sample cannot be cooled to a point at which flashing does not occur, a sample bomb will be used to collect the sample.

3. Samples must not only be collected in one phase, they must also be analyzed in the laboratory in one phase. For example, the reformer product separator liquid can be sampled as a single-phase stream in the field. But when the sample is depressured in the laboratory at atmospheric pressure, the laboratory must deal with two phases. Proprietary laboratory techniques have been developed to handle this problem and must be used if a sample is to be measured in two phases in the laboratory. It is common practice to collect a sample in the liquid phase and then vaporize the sample in the laboratory for analysis.

4. Extreme care must be taken to ensure that vapor sample bombs are not leaking. This requires routine maintenance on the sample bombs. It is recommended that the same bomb always be used for a given sample. This avoids contamination problems and possible valve seat failures. The bomb used to collect the hydrogen net gas should have brass valves with Teflon seats to minimize hydrogen leaks. If the hydrogen net gas sample bomb is used in a corrosive gas service, e.g., hydrogen sulfide absorber–rich gas, there is a high probability that the valves will eventually leak.

5. Liquid samples taken in bombs should be taken with the bomb in the vertical position and the sample entering from the bottom and vented at the top. The inlet valve should be wide open and the outlet valve used to control the purging of the bomb.

6. Vapor samples should be taken into bombs held in the vertical position with the sample entering the top of the bomb and purged out the bottom. The inlet valve should be wide open and the discharge valve should be used to control the purge.

7. Sample bombs should be purged at a rate that does not cool the bomb (due to the Joule-Thomson cooling effect of expansion of the purge material) below the process stream temperature.

It is recommended that three sets of samples be taken at 1-h intervals for each set of test run conditions. Samples should be taken as close to the reactor section as practically possible. Line/vessel fill calculations need to be made to ensure that the sample collections are coordinated with the recording of unit operating data. When on-line analyzers are used, data collection must be coordinated with the sample injection into the process analyzer. This is particularly true in the case of on-line analyzers with long cycle periods.

## F. Planning Coordination

Coordination between process engineering, unit operations, refinery planning, and the laboratory is critical to a successful test run. Because a test run is performed for process engineering, the process engineering group has the responsibility for the coordination. The key issue in coordinating a test run is finding a point in time at which the feedstock needed for the test run will be in adequate supply to allow steady-state operations and the unit operating conditions can be changed to meet the test run requirements. If the required unit operating conditions include a specific octane severity or a number of severities, this must be coordinated with the refinery planning department to avoid causing problems in product blending and shipping. Once a time has been found that will meet feed and severity requirements, coordination must be arranged with unit operations for proper unit staffing and to minimize disturbances from outside sources. For example, maintenance work should be minimized during the test run. The net effect of properly planning and executing a test run is the development of reliable data for refinery decision making at a minimum cost for data collection and analysis.

## III. DATA RELIABILITY

In many areas of business and commerce, subjective units of measure such as good, fair, and bad are used to describe the quality of products, systems, and measurements. Refinery engineers require a much more objective set of standards to define the quality of data used in process modeling. Precisely defined terms of accuracy and precision are used to describe the quality of data.

### A. Accuracy and Precision

Accuracy is defined as the closeness between a reported measured variable and the variable's true value or accepted standard. For example, if the true specific gravity of a stream is 0.750 and the gravity measured and reported by a laboratory is 0.754, the accuracy would be reported as +0.004. The reproducibility of a measurement under controlled conditions is referred to as precision (2). With sufficient testing, using the same test procedure, statistical precision terms of repeatability and reproducibility can be established. These terms define the probability of a measured value being within a prescribed accuracy. In commercial operations such as refineries, engineers work with reported values based on standard tests with defined precision.

The accuracy and precision of data are based on the equipment used to measure a process variable and the test procedure followed by the technician administering the test. In most instances, the accuracy of a reported value is not readily available because the property being measured is an indication of the desired value and not the value itself. In the case of a temperature, the data point collected is the measurement of the EMF of a thermocouple. The data point representing a stream's gravity is the measurement of how far a float sinks into a sample collected from the stream. The data collected in most instances are a representation of a process variable. A significant point to remember about the data collected is that the device used to measure the process variable is usually the most cost-effective device and not necessarily the device that will yield the greatest precision or accuracy.

The precision concepts of repeatability and reproducibility magnify the inaccuracy of data measurement. A very simple way to test equipment calibration would be to introduce a standard into the test equipment, execute the test procedure, record the reported test result, and then adjust the indicator on a scale by the delta between the recorded test value and the standard's value. The accuracy of the equipment calibration is compromised by the repeatability of the test procedure (3). Calibration techniques vary from one type of measuring device to another. Some equipment cannot be calibrated once it has left the manufacturer.

### B. Laboratory Data

Laboratory data are the results of tests on the sample provided, not on the process stream. Any error caused by sample collection cannot be reported. Most laboratory tests are performed in compliance with a standardized test procedure defined by the American Society for Testing and Materials (ASTM) or some other recognized testing organization. Most of the standardized test procedures include a section on the precision of the test results. One term used is the *repeatability* of the test technique; a second is the *reproducibility*. The repeatability value is the maximum expected difference between the true and measured values, 95% of the time, if the same test technician repeats the procedure on the same sample using the same equipment. The reproducibility value is the maximum expected difference, 95% of the time, if a second test technician repeats the procedure on the same sample using a second set of equipment. Note that the definitions of reproducibility and repeatability do not address the accuracy of the reported value but rather provide guidance on the expected validity of the test procedure.

Table 2 shows the ASTM procedure number, repeatability, and reproducibility of many tests commonly performed on catalytic reformer feed and reformate streams (4). Accuracy is a function of sampling technique, equipment calibration, and reproducibility of the test technique.

### C. On-Line Analyzers

On-line analyzers involve different issues when discussing data validity. The magnitude of the accuracy error is reduced in the areas of sampling technique and reproducibility of the test technique. In a properly designed and installed

**Table 2** ASTM Procedures Commonly Used on Catalytic Reformer Feed and Reformate Streams

| Property | ASTM procedure | Repeatability | Reproducibility |
|---|---|---|---|
| Gravity | D287 | 0.2° API | 0.5° API |
| PONA | D1319 | | |
|   35% paraffins | | 1.7 vol % | 5.3 vol% |
|   65% aromatics | | 1.5 vol % | 3.3 vol% |
| Octane | D2699 | Not given | 0.7 |
| Distillation | D86 | See ASTM procedure | |
| Components | D323 | 1.7 kPa | 3.8 kPa |

sampling system, the variability of the samples collected is much lower in an automated system than in a manual system. The possibilities for sample contamination and flashing are very small. The test technique being used is automated and the test repeatability is much smaller than for manual systems. Equipment calibration is, however, a major source of concern. Many analyzers display drift, which is a progressive deviation between reported and actual values. On-line analyzers are particularly susceptible to drift. In addition, many of the on-line analyzers do not use the same test technique as the ASTM (or similar standard test procedure), yet results are reported as an ASTM value (5).

## D. Process Measurements of Flow, Temperature, and Pressure

Process measurements of flow, temperature, and pressure exhibit measured value validity problems similar to those of laboratory data. Each process value has a primary measuring element (sometimes referred to as the sensor), a transmitter, and an indicator.

### 1. Flow Measurement

The most commonly used device for flow measurement in a catalytic reformer is the square-edged orifice plate. In rare instances positive-displacement meters and pitot tube meters have been installed. Commonly accepted values for the accuracy of flow measurement via an orifice plate are 10% of flow at 25% of full meter range, 2.5% of flow at 50% of full meter range, and 1.1% of flow at 75% of full meter range. These values represent the cumulative probable error of all of the elements in the measuring device: the orifice plate, differential pressure transmitter, and indicator. It is assumed that flowmeters are being compensated for stream pressure, temperature, and density for each reading. Using average or typical stream pressure, temperature, and density can lead to significant errors in flow measurement (6).

### 2. Temperature Measurement

It is common practice with a catalytic reformer to measure temperatures in the range of 425 to 550°C in the reactor section with type K thermocouples. Type J thermocouples are used for temperatures below 425°C. Switching thermocouples for the different temperature ranges does improve accuracy, but neither type J nor type K thermocouples yield the accuracy of a resistance bulb. The thermocouples are much less expensive and provide sufficient accuracy for operating the unit, which is the intent of the engineers designing the plant.

The calibration tolerances for both type J and type K thermocouples are the greater of 2.2°C or ±0.75% (7). These values represent the cumulative probable error of all of the elements in the measuring device: the thermocouple, extension wire, bridge, and indicator. Thermocouples are normally checked at the time of installation but are rarely calibrated during plant operation unless a problem is suspected by plant personnel. Thermocouple accuracy is also very dependent on proper installation of the thermocouple within a thermowell and proper design and installation of the thermowell assembly in the process hardware.

### 3. Pressure Measurement

Pressure measurement systems normally consist of a pressure sensor, transmitter, and indicator. In some instances, strain gauges are used in the pressure sensor element to convert pressure to a signal that can be easily transmitted. For the most part, the pressure-sensing element is a bellows with integrated capacitance plates. The bellows/capacitance-type sensors are configured as a differential pressure sensor combined with a transmitter. In pressure measurement service, one leg of the differential transmitter is connected to the process and the second leg references the atmosphere. The accuracy of the strain gauge–type pressure indicator is approximately 2% of the scaled value. The integrated bellows/capacitance-type pressure measurement has an accuracy of 0.2 to 0.3% of the full-scale value.

## IV. DATA VERIFICATION

Refiners must make the best use of the data available when making business decisions. Because the data values available from process measurements and laboratory tests are not going to be true values, techniques must be developed to determine the overall validity of the reported data. The following procedures outline 10 tests that can be made on catalytic reformer stream flow and composition measurements to verify the overall data accuracy (8). Some of the tests are subjective and are meant only to identify obvious problems with data presented. Other tests are objective and must meet specific closures. If all tests pass the requirements, it can be concluded that the measured stream flow and composition data are valid. The 10 test techniques will be demonstrated using the test run data summarized in Table 3 (8).

### A. Material Balance

The material balance on reformer test run data should close within ±1.0 wt %. A spreadsheet can be used to simplify calculations. It is recommended that all extraneous streams entering the material balance that do not go through the reactor section be treated as negative products. This will result in the individual component yields being calculated on a reactor system outlet basis. The material balance error is defined by the following equation:

$$\text{Material balance error} = \frac{\text{weight feed} - \text{weight products}}{\text{weight feed}} \times 100 \tag{1}$$

**Table 3** Test Run Sample Case: Reformer Stream Flow and Composition Data

| Stream | Feed | Net gas | Stabilizer O/H gas | Stabilizer O/H liquid | Reformate |
|---|---|---|---|---|---|
| Flow | | | | | |
| KL/d | 1,1518 | | | 170.9 | 1,224 |
| mNM$^3$/d | | 314.5 | 5.68 | | |
| Volume % | | | | | |
| H2 | | 92.45 | 67.22 | | |
| C1 | | 3.24 | 7.05 | 0.12 | |
| C2 | | 1.99 | 9.47 | 5.24 | |
| C3 | | 1.43 | 9.32 | 20.55 | |
| i-C4 | | 0.35 | 2.37 | 12.14 | 0.06 |
| n-C4 | | 0.27 | 0.27 | 17.34 | 0.29 |
| P5 | 1.93 | 0.22 | 2.16 | 38.18 | 1.40 |
| P6 | 9.57 | 0.05 | 0.20 | 5.38 | 13.89 |
| P7 | 12.16 | | | 0.12 | 9.89 |
| P8 | 13.07 | | | | 7.85 |
| P9 | 5.23 | | | | 1.95 |
| P10 | 1.18 | | | | 0.18 |
| P11 | 0.13 | | | | |
| MCP | 3.66 | | | | 1.78 |
| CH | 3.69 | | | | 0.20 |
| N7 | 13.13 | | | | 1.71 |
| N8 | 11.95 | | | | 0.90 |
| N9 | 10.08 | | | | 0.25 |
| N10 | 2.40 | | | | 0.04 |
| N11 | 0.32 | | | | |
| A6 | 0.59 | | | | 5.31 |
| A7 | 2.37 | | | | 18.13 |
| A8 | 4.31 | | | | 21.80 |
| A9 | 2.80 | | | | 11.31 |
| A10 | 0.90 | | | | 0.24 |
| A11 | 0.01 | | | | 0.24 |
| Specific gravity | | | | | |
| | 0.7543 | | | | 0.8012 |
| D86 distillation | | | | | |
| I.B.P. | 71.0 | | | | 68.5 |
| 10% pt. | 93.5 | | | | 89.5 |
| 30% pt. | 109.0 | | | | 99.5 |
| 50% pt. | 121.5 | | | | 112.0 |
| 70% pt. | 132.0 | | | | 135.0 |
| 90% pt. | 153.0 | | | | 164.5 |
| F.B.P. | 189.0 | | | | 213.5 |

Table 4 shows the material balance for the sample case data to be 0.017%, which is within the ±1% closure specification. The values for all streams crossing the material balance are provided. Energy balance streams are not shown.

A common mistake when evaluating test run data is to normalize reformer product flows to close the material balance. This practice skews all data and serves to mask a real data problem. When the material balance data are not normalized, quite often it is simple to identify the source of the error.

## B. Hydrogen Balance

The hydrogen balance is calculated by adding the weight of hydrogen in each stream to determine the total amount of hydrogen in the feed and products. The hydrogen balance error should not exceed ±0.5%. The hydrogen balance error is defined as

$$\text{Hydrogen error} = \frac{\text{wt. feed hydrogen} - \text{wt. product hydrogen}}{\text{wt. feed hydrogen}} \times 100 \quad (2)$$

The hydrogen balance can be calculated quite quickly in a spreadsheet. The weight of hydrogen in a stream can be calculated by summing the weight of hydrogen contributed by each component using the following: The weight of hydrogen contributed by component CiHj is equal to the weight of the stream multiplied by the weight fraction of CiHj times a factor HFACTOR.

$$\text{HFACTOR} = \frac{i \times \text{mole weight of hydrogen}}{j \times \text{mole wt. of carbon} + i \times \text{mole wt. of hydrogen}} \quad (3)$$

In the sample test run data the hydrogen balance was 0.06%. This is well within the 0.5% criterion for acceptable data.

**Table 4** Sample Case Material Balance

| Feed/product streams | kg/h |
|---|---|
| Feed | 47,636 |
| Products | |
| Net gas | 2,444 |
| Stabilizer O/H liquid | 4,276 |
| Stabilizer O/H gas | 137 |
| Reformate | 40,771 |
| Total products | 47,628 |
| Difference | 8 |
| Material balance error | 0.017% |

## C. Cyclic Ring Balance

The knowledge that the relative rate of cyclization of paraffins to naphthenes and aromatics increases approximately linearly with carbon number and that the relative rate of paraffin cracking increases nearly linearly with carbon number allows the use of increases in ring compounds to check for errors in the feed and reformate analyses (9–11). The number of rings in the feed and reformate for each carbon number is calculated by adding the moles of naphthenes and aromatics by carbon number. The relative increase can then be calculated by calculating a percentage change in the rings for each carbon number. This is simply the difference in feed and product rings divided by the feed rings by carbon number and multiplied by 100. If the cracking reactions did not occur, a plot of the percent increase in rings versus carbon number would yield a nearly straight line. But, because the reformer catalyst is bifunctional and cracking does occur and increases significantly with $C_{9+}$ compounds, the theoretical linear plot shifts to a function that is increasing through $C_8$ or $C_9$ followed by a drastic change in slope. Table 5 shows the calculated rings for the sample case feed and reformate streams. Figure 1 is a typical plot of the percent increase in rings. The loss of $C_6$ rings is typical at the sample unit operating pressure and severity. If either the pressure is decreased or the severity is increased, the position of the plot in Figure 1 will be higher on the $Y$ axis. The plot will maintain the same general shape. By changing the boiling range of the feed, the shape of the plot can be altered slightly. For example, a heavier feed will result in a percentage increase in $C_9$ and heavier rings to the point where the increase in $C_9$ rings will be in line with the points for $C_7$ and $C_8$.

## D. Butane Ratio

When virgin naphtha is charged through a catalytic reformer unit, the mole ratio of $i$-butane to $n$-butane in the reactor product is approximately 0.8. Pro-

**Table 5** Calculated Percent Increase in Ring Compounds

| Carbon number | Percent increase in rings |
|---|---|
| 6 | −15 |
| 7 | +18 |
| 8 | +25 |
| 9 | −19 |
| 10 | −28 |
| 11 | −37 |

## Modeling Commercial Reformers

% RING INCREASE

| CARBON # | % INCREASE |
|---|---|
| 6 | −15 |
| 7 | 18 |
| 8 | 25 |
| 9 | −19 |
| 10 | −28 |
| 11 | −37 |

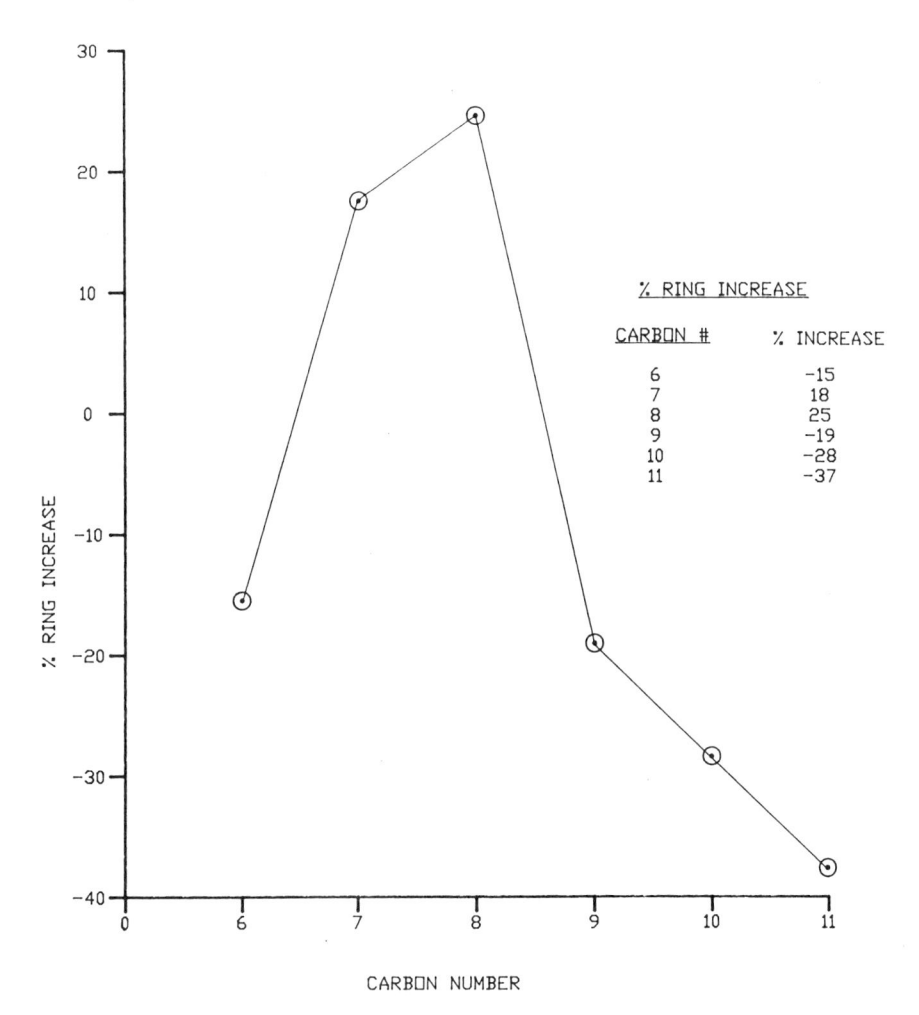

**Figure 1** Percentage increase in rings by carbon number.

cessing cracked naphthas through a reformer results in a slightly higher $i$-butane/$n$-butane ratio. This check is extremely useful in identifying problems with reformate samples. If the reformate has been allowed to weather in the laboratory, or if the sample was taken hot, some of the butane will be missing from the sample. Because the concentration of normal butane in reformate is typically four times that of $i$-butane, more normal butane will be unaccounted for. The result is that the $i$-butane/$n$-butane mole ratio will be out of the range 0.8 ±0.15. The $i$-butane/$n$-butane ratio for the sample case is 0.7, which is within the acceptable range.

## E. Light Ends Ratios

Because the stabilizer overhead streams are relatively small in comparison to the reformate flow, large errors in these streams will not be identified by the material and hydrogen balances. Comparing light ends ratios is extremely useful in identifying significant errors in light ends stream flow rates that may not create a significant deviation in either the material or hydrogen balance. The quantities of methane, ethane, propane, and mixed butanes produced in the catalytic reforming process are approximately equimolar when the chloride on the catalyst is about 1% (12, 13). The exact distribution of the light ends (methane, ethane, propane, and mixed butanes) is dependent on reactor conditions, operating conditions, and especially catalyst chloride level (14). For each of the light ends components, a ratio is calculated whereby the moles of each component are divided by the sum of the moles of the light ends components. Table 6 contains light ends molar yields and resulting ratios calculated from the sample data.

The objective of the procedure for reviewing the light ends distribution is not to verify exactly the equimolar light ends distribution but rather to determine whether unusual data have been reported. It must be stressed that the chloride level on the catalyst will significantly affect the light ends ratios (15, 16). If the feed composition and operating severity for a unit are kept constant, calculating light ends ratios on a routine basis will provide an excellent method for monitoring the chloride level on the catalyst.

## F. Stream Component Sums

Summing the components of each stream GC analysis will aid in resolving data problems. Quite often laboratory personnel simply report a sample analysis of less than 100% rather than report a percentage of the sample unidentified. Transcription errors between laboratory reports and engineering calculations also occur. To identify possible sources of problems, each stream GC analysis should be summed to ensure that the data being used are complete. The sum of the components for each stream in the sample data is shown in Table 7.

**Table 6** Light Ends Yields and Ratios

| Component | Molar yield | Moles/ sum of moles |
| --- | --- | --- |
| Methane | 482.2 | 0.21 |
| Ethane | 552.9 | 0.24 |
| Propane | 627.3 | 0.27 |
| Mixed butanes | 638.7 | 0.28 |

**Table 7** Stream Component Sums

| Stream | Sum of components (%) |
|---|---|
| Feed | 99.48 |
| Reformate | 99.92 |
| Net gas | 100.00 |
| Stabilizer gas | 98.06 |
| Stabilizer liquid | 99.07 |

## G. Feed and Reformate Probability Plots

Gas chromatographic analyses of the feed and reformate stream are subject to numerous sources of error. One way to detect major errors in the analyses of the feed and reformate is to plot the cumulative sum of the percent components for each of paraffins, naphthenes, and aromatics against carbon number. When the cumulative values are plotted on probability paper by carbon number, the lines for the feed paraffins, naphthenes, and aromatics and the reformate paraffins and aromatics should be nearly straight lines. It is impractical to use this data validation technique on the reformate naphthenes because of the small concentrations. Figures 2 and 3 demonstrate the plots for the feed paraffins and the reformate aromatics. If the laboratory has reported all components heavier than "x" as Cx+ the plots end abruptly at carbon number x. If a blended feed, such as might be produced by mixing coker naphtha and crude still naphtha, is plotted, quite often a distinctive kink will be noticed in all five plots.

The purpose of making these plots is to identify potential problems with the reported stream analyses. If the plots are not straight lines, it does not mean that the data are bad. Nonlinear plots should serve as a flag to the process engineer that there is a potential problem with the reported data. The results of this procedure, when combined with the calculated versus measured density comparisons and with the calculated versus measured distillations, serve as a basis for determining the validity of the feed and reformate sample analyses.

## H. Net Gas Component Plot

One of the more common problems with reported reformer test run laboratory data is the net gas analysis. An error in the net gas composition is not easily identified from either the hydrogen balance or the material balance. Plotting the component composition by carbon number as shown in Figure 4 allows a simple visual screen of the stream analysis. If the produced light ends distribution is nearly equimolar, the resulting curve should be descending with increasing

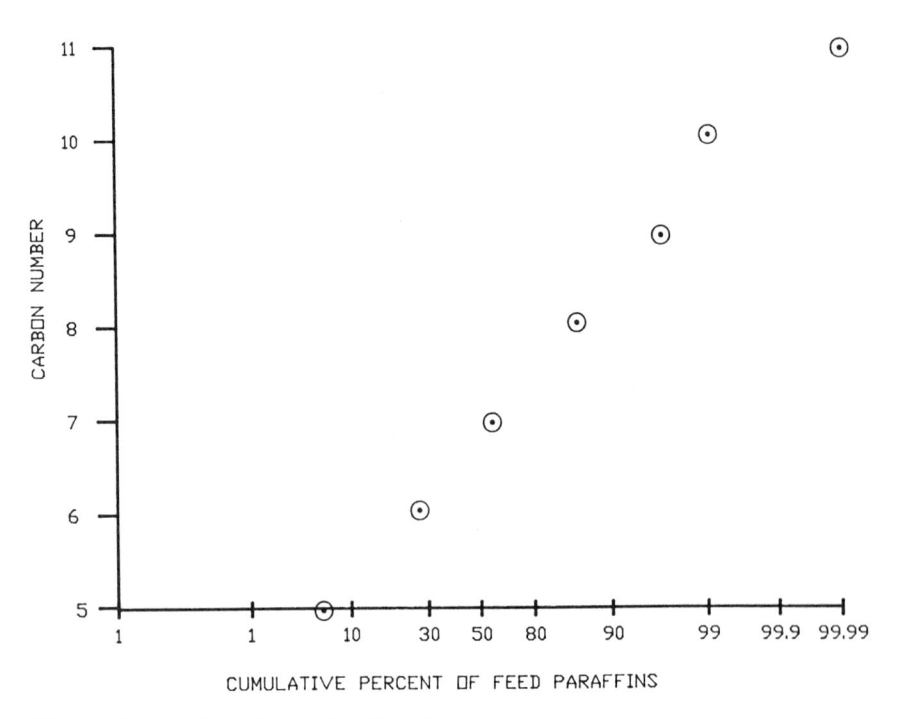

**Figure 2** Feed paraffins probability plot.

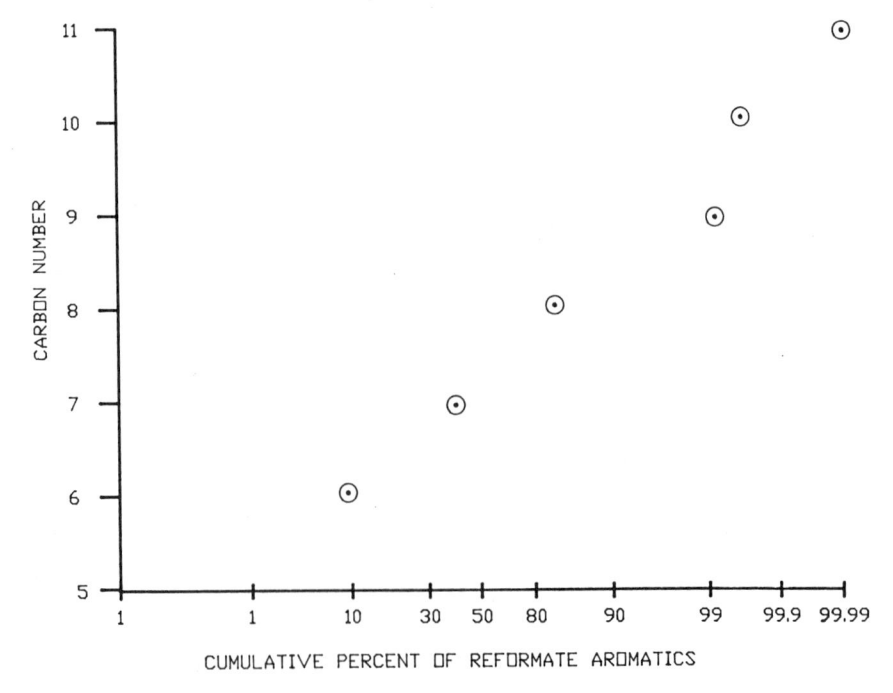

**Figure 3** Reformate aromatics probability plot.

## Modeling Commercial Reformers

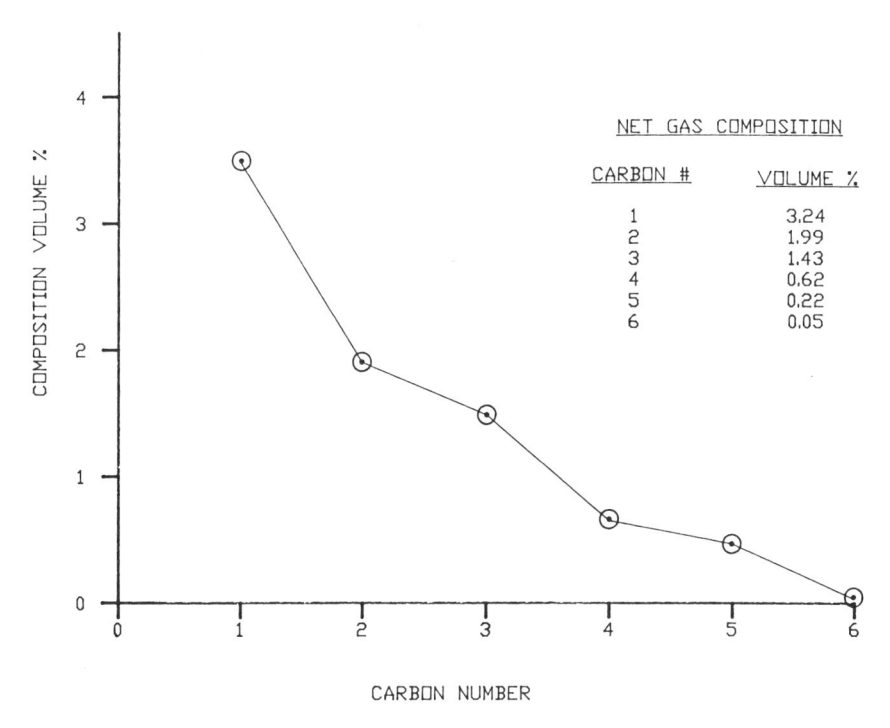

**Figure 4** Recycle gas composition versus carbon number.

carbon number and asymptotic to zero concentration because of the relative volatilities of the light ends.

If the unit is running heavily chlorided, the propane and butane production may be significantly higher than the methane production because of the increased hydrocracking. In this case, the methane concentration in the net gas may actually be lower than the ethane concentration.

The results of this data analysis procedure, combined with the results of the light ends distribution analysis and fundamental knowledge of the catalyst chloride level, allow the engineer to judge the accuracy of the net gas sample analysis.

### I. Feed and Reformate Gravities

The reformer feed and reformate gravities can be calculated from the GC analyses. The calculated gravities should match the laboratory measured data within the ASTM (D287) test reproducibility limit of 0.5 API (4). At 50.0 API, 0.5 API is approximately 0.002 specific gravity. The measured and calculated gravities of the sample data reformate were both 45.1 API. The measured feed gravity was 56.1 API and the calculated feed gravity was 56.2 API. Both pairs of calculated and measured gravities are within the recommended tolerance.

## J. Feed and Reformate Distillations

D86 distillations can also be calculated from the feed and reformate component analyses. The differences between the calculated and measured 10% through 90% points should not exceed the ASTM (D86) reproducibility limits. The ASTM reproducibility limits for each point are dependent on the curvature of the particular distillation. See the ASTM test procedure for details on calculating these values. Several procedures have been prepared to convert a stream composition to a distillation curve. Most of these conversion techniques are valid between the 10% and 90% points but calculate questionable results for the IBP and end point. Table 8 is a summary of the measured and calculated distillation points, the difference between the measured and calculated distillation points, and the ASTM reproducibility limits for both the feed and reformate streams. The differences between calculated and measured values are much larger than anticipated.

## V. MODELING TECHNIQUES

Data concerning how a commercial reformer operated yesterday, or what the yields were on last week's feedstock, are in and of themselves worthless as information for making business and engineering decisions in the future. It is common practice in the refining industry to use mathematical models to turn

**Table 8** Measured and Calculated Feed and Reformate Distillations

| Distillation point (%) | Measured value | Calculated value | Delta | Reproducibility limit |
|---|---|---|---|---|
| | | Feed data | | |
| 10 | 83.5 | 91.0 | 2.5 | 3.0 |
| 30 | 109.0 | 102.0 | 7.0 | 3.0 |
| 50 | 121.5 | 108.0 | 13.5 | 3.0 |
| 70 | 132.0 | 120.5 | 11.5 | 3.0 |
| 90 | 153.0 | 142.5 | 10.5 | 3.5 |
| | | Reformate data | | |
| 10 | 89.5 | 85.0 | 4.5 | 3.0 |
| 30 | 99.5 | 102.0 | 2.5 | 3.0 |
| 50 | 112.0 | 105.0 | 7.0 | 3.0 |
| 70 | 135.0 | 127.5 | 7.5 | 3.0 |
| 90 | 164.5 | 160.0 | 4.5 | 3.5 |

the data collected from past operations into useful information on how the process will perform in the future (17). Some of the more common questions requiring predictions of reformer performance include:

1. What is the effect of changing a unit operation such as recycle rate on catalyst life?
2. What is the value of an incremental feedstock?
3. What are the magnitudes of the linear programming vectors used in the refinery economic planning model?
4. What is the economic benefit of replacing a feed effluent exchanger system?
5. What is the optimum operation on the unit?
6. How can benzene production be minimized?
7. What is the optimum temperature profile on the reactor system?

Three techniques are commonly used for modeling commercial catalytic reforming units: correlation models, pseudokinetic models, and kinetic models (8). Correlation models consist of a set of equations based on regressions of observed operating data. Pseudokinetic models are an improvement over the correlation-type models, and they calculate bulk reactions such as the gross amount of naphthenes converted to aromatics. Kinetic models use detailed reaction kinetics and fundamental engineering equations to calculate the change of the naphtha feed into the various reformer products.

### A. Correlation Models

Conceptually, correlation models are relatively easy to develop because they are simple regressions of observed data. There are three methods for creating correlation models: correlating massive quantities of data with unknown accuracy, performing a series of planned test runs on a specific reformer, and performing a series of pilot plant studies. All correlation models have some common advantages and disadvantages. The biggest single advantage of correlation models is they offer quick solutions that can be obtained by hand calculation or with a simple computer program. Correlation models have six major disadvantages:

1. Correlation models are not in heat and material balance.
2. Developing a correlation model requires a substantial amount of test run data to be collected at various feed rates, operation conditions, and with multiple feedstocks.
3. Changes in catalyst or catalyst performance may negate the validity of the model.
4. The model application is limited to the range over which the test run data were collected.

5. Changes in individual reactor operation cannot be modeled.
6. Individual component yields such as benzene cannot be accurately predicted (18).

### 1. Correlation Models from a Large Data Base

The major advantage of correlating a model from a large data base is that information pertaining to various catalyst types, activity decline, reactor loadings, and operating pressures can be included in the correlations. Another advantage of this type of modeling system is that the data base can be maintained and expanded with time, allowing recorrelation of the data for improved model performance. The weakness in this approach to modeling is the limited validity of the data in the data base. When collecting large amounts of data over a period of time, it is difficult to maintain high standards of data retrieval and preparation.

Development of a data base with operating data from many reformers under a large number of operating conditions is limited to large oil refineries with numerous reformers from which data can be drawn or to a catalyst vendor that can collect data from a large client base. An oil company attempting to develop this type of model must ensure that catalyst license agreements are not violated by putting the operating data of particular units into a model.

### 2. Correlation Models from Test Runs

A second approach to creating a correlation model of a catalytic reformer is to run a series of test runs on a commercial reformer. This requires careful planning to ensure that all of the independent variables to be used in the model are perturbed over a sufficient operating range to create a complete model. The main advantage of creating a model using this technique is that the model can be created in a relatively short period of time. It is possible to control the data collection to minimize extraneous information from entering the data base. Data errors can be minimized. The disadvantages, however, are quite significant. It is very expensive to move a commercial unit from its optimum operating point to a series of nonoptimum operations just to collect data. A tremendous work load is placed on the laboratory support resources. A model created using this technique is limited in application to the commercial unit from which the data was collected. Because only one catalyst loading is represented in the data base, the model will be limited to that specific catalyst loading. When the catalyst deteriorates with time due to changes in surface area or metals contamination, the model becomes invalid.

A correlation model based on test runs on a commercial unit can be developed by any company operating a commercial catalytic reformer. Before proceeding with such a program, the refiner should verify that developing a model will not be in violation of the catalyst license agreement.

## 3. Correlation Models from Pilot Plant Data

Correlating data from adiabatic pilot plant runs offers many of the advantages of correlating a model from large data bases and from specific test runs. First, pilot plant operating conditions can be varied over a wide range of operations, the catalyst can be changed, and the data base can be set up to allow regular model enhancements by updating the correlations on a routine basis. These similarities to the large data base type of correlation make the model much more versatile. As with the correlation model based on test run data, the model can be developed in a relatively short period of time and information collection can be controlled to ensure data validity. Another advantage of using a pilot plant over a commercial unit for making the test runs is the reduced cost.

The two main disadvantages of this type of model formation are that a pilot plant must be available to run the tests and the test results must be scaled up for commercial operation. Development of correlation models based on pilot plant data is limited to companies that have access to an adiabatic reformer pilot plant.

## B. Pseudokinetic Models

The development of pseudokinetic models is considerably more difficult. The basic reaction chemistry must be included in the calculations. Both forward and reverse reactions of the gross cracking, cyclization, and dehydrogenation reactions can be included (13). The calculations in a pseudokinetic model are much more complex than in a correlation model and need to be automated through a computer program. The pseudokinetic model of the reforming process does offer the advantages of (1) being in material balance, (2) requiring fewer test runs for implementation, (3) allowing for changes in catalyst or catalyst performance, and (4) predicting operations over an extended operating range. The pseudokinetic model has the disadvantages of (1) not being in heat balance, (2) needing several sets of test run data with multiple feedstocks to match the operation of the model to a particular commercial unit, and (3) not being able to predict yields of individual components such as benzene. This type of model has limited application at present. When this type of model was first conceived, computers were limited and the simplified approach to kinetics offered an attractive compromise between a correlation model and a full kinetic model.

## C. Kinetic Models

A kinetic model defines the rates of various reactions and the associated heat and material balance using a system of equations. Developing a kinetic model from basic reaction kinetics and fundamental engineering relationships is a major investment in engineering resources. The quantity and quality of information from a kinetic model are far superior to that of a correlation model or a pseudokinetic model. A kinetic model is in both heat and material balance. Individual reactor calculations can be made with great accuracy. A kinetic model is not limited to the range over which the test run data were collected, but rather to the validity range of the fundamental equations used in the kinetic model development. Yields of individual components such as benzene can be accurately predicted. The major problems with kinetic models are the engineering support needed for development and the time needed to make calculations. A kinetic model must be computer program based. The characteristics of a kinetic model are (8, 19)

1. Paraffin isomerization is included in the reactor calculations because of its importance in calculating octane from components.
2. Feeds are represented by components. Options to use a distillation curve for feed characterization often exist, but the software converts the distillation curve to components for internal use.
3. A choice is made of either reactor inlet temperatures or reformate octane as the independent variable for the reactor solution.
4. Individual reactors are modeled in adiabatic operation.
5. Cracking, cyclization, and dehydrogenation reactions are component based. The cyclization and dehydrogenation reactions are modeled as equilibrium reactions.
6. Cracking, cyclization, and dehydrogenation activities are included and are adjustable to account for catalyst decay.
7. System pressures are adjusted for changes in operating parameters such as feed rate, recycle rate, product separator pressure, and unit operating severity.

Two types of kinetic models are in common use. A catalyst-based kinetic model back-calculates catalyst activities from observations of the yields and operations. An activity-based kinetic model uses known activities of the reforming catalyst to set the kinetic rate equations (19).

## 1. Catalyst-Based Kinetic Models

Although the development effort is quite substantial on a catalyst-based kinetic model, only minimal operating data from a process unit are needed to calibrate the activity terms in the reaction rate equations. To match the catalyst-based kinetic model to the process unit, information is collected from one to three test runs performed during routine operation. Special test runs outside normal operation are not required. There is a slight cost for added laboratory tests during calibration, but there is no lost opportunity cost as is the case when developing a correlation model. The test run data allow not only the kinetic model activity terms to be updated but also individual rate constants to be adjusted if necessary. Update factors are also calculated for pressure drops through the system, flash calculations, heater duties, compressor power requirements, octane predictions, and RVP calculations.

The catalyst-based kinetic model is particularly well suited for estimating yields and operations based on present catalyst performance. This is the type of model that best serves the general needs of a refinery engineering and planning staff. Refinery linear optimization program vectors can easily be updated using this type of model.

As the catalyst deteriorates over time due to surface area loss or metals contamination, a new set of operating data is collected and the model updated. The characteristic of updating the model to observed catalyst activities is particularly useful in evaluating the optimum operation under the present operating conditions. The updates can be done quickly and easily and allow a refiner to optimize profits with the present catalyst performance. The catalyst-based kinetic models are not well suited to performing unit designs or for catalyst selection studies.

## 2. Activity-Based Kinetic Models

Reformer catalyst activities can be calculated from pilot plant data. Activity-based kinetic models include a catalyst activity term that can easily be adjusted for changes in catalyst activity. This type of kinetic model is excellent for reformer design studies and for catalyst selection. The yield predictions from an activity-based model are predicated on normal catalyst performance and thus are excellent for long-term planning studies.

The characteristics of an activity-based model make it an ideal model for a catalyst company, a design and construction company, or a consulting company involved in long-range planning. Using a kinetic model for these types of activities is preferred over a correlation model because it does not require any information about a specific commercial operation.

## 3. Primary Reactions and Their Kinetic Expressions

Kinetic models can easily handle interdependent and forward/reverse reactions (12, 20). Kinetic models are excellent tools for predicting individual component yields as well as yield patterns such as $C_{5+}$ in a reformer. The basic reforming reactions are the reversible reactions of paraffin cyclization, isomerization of alkylcyclopentanes and alkylcyclohexanes, naphthene dehydrogenation, paraffin hydrocracking, and isomerization of paraffins to iso-paraffins (19–22). Hydrodealkylation of naphthenes and aromatics is a very slow reaction at typical reforming operating conditions and is insignificant compared with the other basic reactions.

The primary reactions involved in the production of benzene in the order of increasing relative rates are cyclization of hexane to methylcyclopentane, the isomerization of methylcyclopentane to cyclohexane, and dehydrogenation of cyclohexane to benzene. The forward reaction of hexane to cyclohexane is much slower than the reverse reaction, resulting in a significant amount of cyclohexane in the reformer feed being saturated to hexane (8, 18).

For most reformer operations, characterization of the feed through $C_{11}$ is satisfactory. There are several thousand paraffin, olefin, naphthene, and aromatic isomers in the typical reformer charge (23). Isomers must be grouped into general types by carbon number to make the feed characterization manageable (8, 19). In addition to the large number of isomers in a reformer process, it is important to note that there may be several reaction mechanisms to reach a particular component. Because of the limited data available, it is necessary to select one mechanism for modeling each type of reaction.

Irreversible reactions are modeled using the following first-order kinetic expression:

$$dC = a \times K \times e^{-E/RT} \times C_i \times PF \times W/F \qquad (4)$$

where  $dC$ = change in concentration of reactant $i$
$a$ = catalyst activity
$K$ = reaction rate constant
$E$ = activation energy
$R$ = gas law constant
$T$ = temperature
$PF$ = pressure factor
$C_i$ = Inlet concentration of reactant $i$
$W$ = catalyst weight
$F$ = flow rate

The reversible reactions use the same basic equation expanded for the reverse reaction to

$$dC = a \times (K_f \times e^{-E_f/RT} \times C_i \times PF^x - K_r \times e^{-E_r/RT} \times PPH^x \times C_p) \times W \qquad (5)$$

where  $K_f$ = forward reaction rate constant
$E_f$ = forward activation energy
$K_r$ = reverse reaction rate constant
$E_r$ = reverse activation energy
$C_p$ = inlet concentration of product
$PPH^x$ = hydrogen partial pressure with order term $x$

These kinetic expressions are simple, and necessary data such as reaction rate constants, activation energies, and equilibrium constants are reported in the literature. Reactant flow and catalyst volume are defined to be independent variables. The hydrogen pressure can easily be calculated from process conditions (8).

Catalyst activity can be either specified by a model user or back-calculated for a specific operation given operating conditions. In reformer operations both metal and acid activities exist (19, 24). The effect of coke laydown on the catalyst can be made by adjusting catalyst activity (25).

Commercial reactors are adiabatic, not isothermal as implied in the conventional kinetic equations. In addition, reactants are being produced as the feed and recycle pass through the reactor system. These problems are solved by treating each reactor as a series of plug flow reactors with discrete temperatures, hydrogen partial pressures, inlet component concentrations, and catalyst activities. Solving heat and material balances around each succeeding plug flow reactor allows the net reactor product and thus succeeding reactor feed, temperature, and hydrogen partial pressure to be calculated. The net effect is an integration of the heat and material balance through the reactor volume.

### 4. Computer Programming Concepts

The hydrogen recycle stream common to all catalytic reformer units, and thus to all kinetic models of the reforming process, requires special handling. Historically, reformer models have been written with closed equations and have utilized internal convergence techniques. Techniques such as the bisection method, the secant method, and the Newton-Raphson method have been used to handle convergence problems such as those raised by the recycle gas stream. In recent years open equation modeling techniques with simultaneous solutions have become available (26). The open equation modeling technology is driven by computer memory size and performance.

### 5. Unit Optimization

All variables in a catalytic reformer can be classified as independent variables or dependent variables. Independent variables are those that are set by engineering in the design and construction of the unit or set by control instruments. Independent variables include reactor sizes, catalyst loadings, feed pumps, heat exchangers, and fired heaters, plus all of the other mechanical aspects of the unit. Independent variables set by process instrumentation include reactor inlet temperatures, feed rate, product separator pressure, and recycle rate. The dependent variables include the variables resulting from the unit operation. These include heater duties, reactor outlet temperatures, product yields, utilities, and system pressures. To optimize the reformer process, the independent variables that can be adjusted are set to maximize a single dependent variable such as profit (called the objective function), subject to limitations of various dependent variables (called constraints). The optimum operation of a catalytic reformer is easy to calculate once the independent variables are identified, unit operating constraints are defined, and economic values are placed on feed, products, and operating costs.

*a. Optimization Software.* A host of optimization software is available for reformer optimization. Models written with open equation technology include an SQP optimizer for simultaneous equation solutions that can be used to calculate optimum operating conditions. Models written with closed equations can be readily integrated with an SLP or SQP optimizer for calculating the optimum operation. In either case, independent variables must be defined along with an acceptable operating range. A single objective function must be specified, and unit constraints must be defined.

*b. Identification of Independent Variables.* It is critical in the identification of independent variables for optimization to consider the effects of movement of the independent variables on the entire refinery operation. If, for example, the product separator is allowed to vary and the hydrogen is at too low a pressure to exit the process unit into the refinery hydrogen pressure, the optimization solution is invalid. Typically, the independent variables in a catalytic reformer unit optimization include reactor inlet temperatures, recycle rate, feed rate, and product separator pressure.

*c. Identification of Objective Function.* The objective function for most catalytic reformer operations is profit. It must be remembered when setting up a single process unit optimization study that the solution must fall within the refinery's overall optimization. Running the reformer at 100 RONC to maximize unit profits may violate the refinery master operating plan, which has the reformer operating at 94 RONC to meet octane blending requirements. When the objective function is profit, the unit operation is subject to the overall refinery objective.

*d. Definition of Constraints.* Constraints are defined by unit process equipment limitations and refinery objectives. In a motor fuels reformer operation, octane is normally constrained. Mechanical constraints typically include heater firing, compressor load, hydrogen header pressure, and feed rates.

*e. Feed and Product Values and Operating Costs.* Feed and product values can normally be taken as the refinery optimization program's intermediate stream values. Operating costs for the first heaters and recycle compressor are also normally available from the refinery operating plan. The most difficult operating cost to determine is the cost of regenerations in a semiregenerative reformer. The cost of lost production may have to be included.

## VI. MODEL APPLICATIONS

A kinetic model of the reforming process is a fundamental engineering tool in a modern refinery. Three engineering problems are presented in this section to demonstrate how a kinetic model can be used to easily solve typical refinery

*Modeling Commercial Reformers*

**Table 9** Product Values and Operating Costs

| Product values | | Operating costs | |
|---|---|---|---|
| Hydrogen | $ 12.00/KNM$^3$ | Fuel gas | $7.80/GCAL |
| Methane/ethane | $ 7.80/gCAL | Compressor | $0.08/KWH |
| Propane | $ 70.00/m$^3$ | Catalyst | $75K/regeneration |
| $i$-Butane | $ 95.00/m$^3$ | | |
| $n$-Butane | $ 80.00/m$^3$ | | |
| C$_{5+}$ reformate | $150.00 at 95 RONC | | |
| Octane value | $ 4.50/OCT-m$^3$ | | |

problems. The base reformer operation (8) used in these demonstration problems has a capacity of 1750 m$^3$/d and is a three-reactor semiregenerative reformer operating at an average reactor pressure of 19 kg/cm$^2$. The typical feed rate is 1500 m$^3$/d of a full-range naphtha with 45 vol % naphthenes and 11 vol % aromatics. Stream analyses and flow rates of a typical operation are given in Section III of this chapter. The average hydrogen-to-hydrocarbon recycle ratio is 3:1. For the three problems presented, the product values and operating costs found in Table 9 will be used.

### A. What Is the Value of a Second Feedstock?

Product economics and environmental regulations are requiring refiners to reevaluate internal stream alignments and the way gasoline is being produced. Naphtha that was once put in gasoline is now being sold. Streams internal to refineries are being modified to meet specific environmental regulations. The need for hydrogen for treating various refined products is creating new objectives for reformer operations.

For this example, a second source of feedstock is assumed, and the composition shown in Table 10 has been identified as the potential feedstock for the reformer. What is its value?

**Table 10** Incremental Feed Stock Composition

| Carbon number | Paraffins | Naphthenes | Aromatics |
|---|---|---|---|
| C5 | | 3.0[a] | |
| C6 | 20.6 | 3.6[b] | 1.0 |
| C7 | 22.6 | 12.3 | 2.6 |
| C8 | 23.0 | 4.9 | 2.0 |
| C9 | 2.2 | 1.9 | 0.3 |

[a] MCP (methylcyclopentane).
[b] CH (cyclohexane).

## 1. Problem Analysis

The question of the value of reformer feed typically has two correct answers. If the stream is going to be a replacement for the present reformer feed, it will have a different value than if it is being evaluated as an incremental feedstock to be added to the present reformer feed. The evaluation, assuming the alternate stream is a replacement for the present reformer feed, can be handled with a correlation-type model. However, evaluating an incremental feed can be done properly only with a kinetic model that has the capability of blending the streams together on a component basis and evaluating the resulting feed.

## 2. Evaluation of a Second Feedstock as a Replacement Feedstock

Evaluating the second feedstock as a replacement feedstock is done by running a kinetic model simulation of the base operation with the second feedstock composition specified. The model used for this demonstration allows the user to specify not only the unit operating conditions and feedstock data but also product values and operating costs.

A summary of product and utility prices, product yields, and utilities consumption is given in the first two columns of Table 11. The third column gives the product value and operating cost contributions expressed on a per unit of feed basis (unit price times yield divided by feed rate). The sum of the product and utility contributions is the feed value, in this case $119.95 per cubic meter.

With the knowledge of the calculated value of various naphthas as reformer feed and their values in alternate processes, a refiner can select the optimum plant operating strategy. The technique presented in this example is particularly useful to refiners who process purchased naphtha and can select one of several naphthas available. The problem is reduced to subtracting the calculated feed value from the purchase price for each naphtha available and then selecting the naphtha with the highest delta.

## 3. Evaluation of a Second Feedstock as an Incremental Feedstock

Assuming an additional 240 m$^3$/d can be processed in the reformer while maintaining a constant hydrogen recycle rate and constant octane, the solution to this problem requires two model runs. First, the base case operation is modeled to establish base case yields and operating costs. A second simulation is made with 240 m$^3$/d of the incremental feed added to the base feed. The simulation model used for this problem allows multiple feeds streams to be defined by the user, and the feed streams are summed by molar flow rate within the software.

Table 12 shows the yields, operating costs, and profit contributions of the reformer products and utilities for the present base operation. Note that the net profit per day is $193,281. Table 13 shows the same information for the second case at the base feed rate plus 240 m$^3$/d. The net profit is $222,606 per day. This is an incremental profit of $29,325, or $122.19 per cubic meter of incre-

**Table 11** Sample Replacement Feed Case

| | Unit price | Yield/consumption | Per unit of feed ($/m$^3$) |
|---|---|---|---|
| Hydrogen value | 12.000 $/kNm$^3$ | 236.63 kNm$^3$/d | $ 1.870 |
| Fuel value | 7.800 $/Gcal | 58.69 kNm$^3$/d | 3.647 |
| Propane value | 70.000 $m$^3$ | 136.92 m$^3$/d | 6.312 |
| i-Butane value | 95.000 $/m$^3$ | 60.94 m$^3$/d | 3.813 |
| n-Butane value | 80.000 $/m$^3$ | 83.57 m$^3$/d | 4.403 |
| Gasoline value | 150.000 $/m$^3$ | 1111.40 m$^3$/d | 109.791 |
| Octane value | 4.500 $/RONC-m$^3$ | | −5.636 |
| Catalyst regen. cost | (75.000)m$/regen | | (0.09967) |
| Fired fuel cost | (7.800)$/Gcal | 633.65 Gcal | (3.255) |
| Recycle compressor | (0.040)$/kWh | 33988.34 kWh | (0.895) |
| Net profit per unit of feed | | | $119.950 |

At 1518 cubic meters per day, the total feed value is $182,136 per day.

**Table 12** Sample Base Case Operation

|  | Unit price | Yield/consumption | Per unit of feed ($/m$^3$) |
|---|---|---|---|
| Hydrogen value | 12.000 $/kNm$^3$ | 294.26 kNm$^3$/d | $ 2.325 |
| Fuel value | 7.800 $/Gcal | 23.28 kNm$^3$/d | 1.446 |
| Propane value | 70.000 $m$^3$ | 54.45 m$^3$/d | 2.510 |
| $i$-Butane value | 95.000 $/m$^3$ | 23.88 m$^3$/d | 1.494 |
| $n$-Butane value | 80.000 $/m$^3$ | 83.57 m$^3$/d | 1.970 |
| Gasoline value | 150.000 $/m$^3$ | 1301.03 m$^3$/d | 128.523 |
| Octane value | 4.500 $/RONC-m$^3$ |  | −6.632 |
| Catalyst regen. cost | (75.000)m$/regen |  | (0.03589) |
| Fired fuel cost | (7.800)$/Gcal | 663.72 Gcal | (3.409) |
| Recycle compressor | (0.040)$/kWh | 34224.10 kWh | (0.902) |
| Net profit per unit of feed |  |  | $127.290 |

At 1518 cubic meters per day, the total feed value is $192,281 per day.

mental feedstock. Knowing that the incremental feed is worth $122.19 per cubic meter allows the refiner to compare running this naphtha through the reformer or some alternative process. For example, the value of the incremental naphtha as gasoline blend stock may be $100 per cubic meter. In this case the refiner can improve profits by reforming the naphtha instead of blending it.

**Table 13** Incremental Feed Case

|  | Unit price | Yield/consumption | Per unit of feed ($/m$^3$) |
|---|---|---|---|
| Hydrogen value | 12.000 $/kNm$^3$ | 320.71 kNm$^3$/d | $ 2.189 |
| Fuel value | 7.800 $/Gcal | 33.71 kNm$^3$/d | 1.809 |
| Propane value | 70.000 $m$^3$ | 78.82 m$^3$/d | 3.138 |
| $i$-Butane value | 95.000 $/m$^3$ | 34.68 m$^3$/d | 1.874 |
| $n$-Butane value | 80.000 $/m$^3$ | 52.21 m$^3$/d | 2.376 |
| Gasoline value | 150.000 $/m$^3$ | 1457.57 m$^3$/d | 125.878 |
| Octane value | 4.500 $/RONC-m$^3$ |  | −6.501 |
| Catalyst regen. cost | (75.000)m$/regen |  | (0.05484) |
| Fired fuel cost | (7.800)$/Gcal | 738.82 Gcal | (3.277) |
| Recycle compressor | (0.040)$/kWh | 36432.11 kWh | (0.829) |
| Net profit per unit of feed |  |  | $126.600 |

At 1518 cubic meters per day, the total feed value is $222,606 per day.

## 4. Solution Analysis

A kinetic model allows the calculation of both base feed stock operations and the effect of incremental feeds. In this example, the intended use of a feedstock will create different answers to a simple question. The techniques demonstrated here can also be used to evaluate the incremental value of only a portion of feedstock. As an example, if the question is "what is the incremental value of the $C_6$ portion of the feedstock?" the incremental value can be calculated by first simulating the operation with the base feedstock. Then a second simulation is made using the feedstock less the $C_6$ portion.

## B. What Is the Economic Benefit of Replacing a Feed/Effluent Exchanger?

Many of the reformer units built in the 1960s are still in operation with hot side approach temperatures on the feed/effluent exchangers of 100 to 150°C and high pressure drops on both the shell and tube side. In many cases it is not economically justifiable to replace the whole unit, but revamps with plate exchangers being utilized in modern reformers are quite common.

### 1. Problem Analysis

In this example, an engineering study has been performed, and it is estimated that the pressure drop on the feed side of the feed/effluent exchanger can be reduced by 60 kg/cm$^2$, that the effluent side pressure drop can be reduced by 30 kg/cm$^2$, and that the hot side approach temperature can be reduced by 85°C.

### 2. Evaluation of a New Exchanger

The problem is broken into two parts. The first part requires simulating the present operation with a model that allows the user to vary the exchanger approach temperatures and pressure drops. The base case yields, product values, and utility costs are shown in Table 12 along with a net profit of $127.29 per cubic meter of feed.

The second phase of the problem is to simulate the operation with the reduced approach temperature and reduced pressure drops. Table 14 shows the new yields, product values, and operating costs. The net profit in this case if $128.47 per cubic meter. This is an incremental profit improvement of $1.18 per cubic meter or, at a charge rate of 1500 cubic meters per day, a profit improvement of $646,000 per year.

### 3. Solution Analysis

It is noted that the improvement in net profit came from three areas. The change in pressure drop reduced the compressor cost by $0.21 per cubic meter of feed; the lower approach temperature reduced the utility cost by $0.93 per cubic meter of feed; the lower reactor operating pressure increased the $C_{5+}$

**Table 14** New Feed/Effluent Exchanger

|  | Unit price | Yield/consumption | Per unit of feed ($/m$^3$) |
|---|---|---|---|
| Hydrogen value | 12.000 $/kNm$^3$ | 304.97 kNm$^3$/d | $ 2.410 |
| Fuel value | 7.800 $/Gcal | 21.32 kNm$^3$/d | 1.325 |
| Propane value | 70.000 $m$^3$ | 49.87 m$^3$/d | 2.299 |
| $i$-Butane value | 95.000 $/m$^3$ | 21.87 m$^3$/d | 1.368 |
| $n$-Butane value | 80.000 $/m$^3$ | 34.60 m$^3$/d | 1.823 |
| Gasoline value | 150.000 $/m$^3$ | 1306.67 m$^3$/d | 129.081 |
| Octane value | 4.500 $/RONC-m$^3$ |  | −6.630 |
| Catalyst regen. cost | (75.000)m$/regen |  | (0.03738) |
| Fired fuel cost | (7.800)$/Gcal | 482.29 Gcal | (2.477) |
| Recycle compressor | (0.040)$/kWh | 26318.58 kWh | (0.693) |
| Net profit per unit of feed |  |  | $128.468 |

At 1518 cubic meters per day, the total feed value is $195,070 per day.

yield, and thus the product value, by $0.04 per cubic meter of feed. Had the effect of the change in reactor pressure not been considered in the calculations, the estimated benefit would have been low by nearly 4%. Using a model to do the calculations for process modification or equipment constraint problems allows variability studies to be computed easily for detailed economic evaluations (27).

## C. What Is the Benefit of Reducing Octane Variability with Advanced Process Controls?

One of the benefits normally calculated as a justification for installing advanced process controls on a catalytic reformer unit is the benefit derived from reducing octane variability (28).

### 1. Problem Analysis

The value of installing an on-line octane analyzer and then coupling the analyzer with an advanced control function is highly dependent on the unit product prices and operating costs. The answer lies in the curvature of the octane versus profit curve. Figure 5 is a plot of octane versus profit. It can be seen that the plot is roughly a second-order relationship. If the relationship was first order, there would be no benefit from reducing variability. Using Figure 5 for an example, the unit is operated at an average of 96 octane, and the octane varies 2 numbers positive resulting in a change in profit of $5328 per day ($214,147 − $208,819 = $5328). This does not have as big an effect as

*Modeling Commercial Reformers* 469

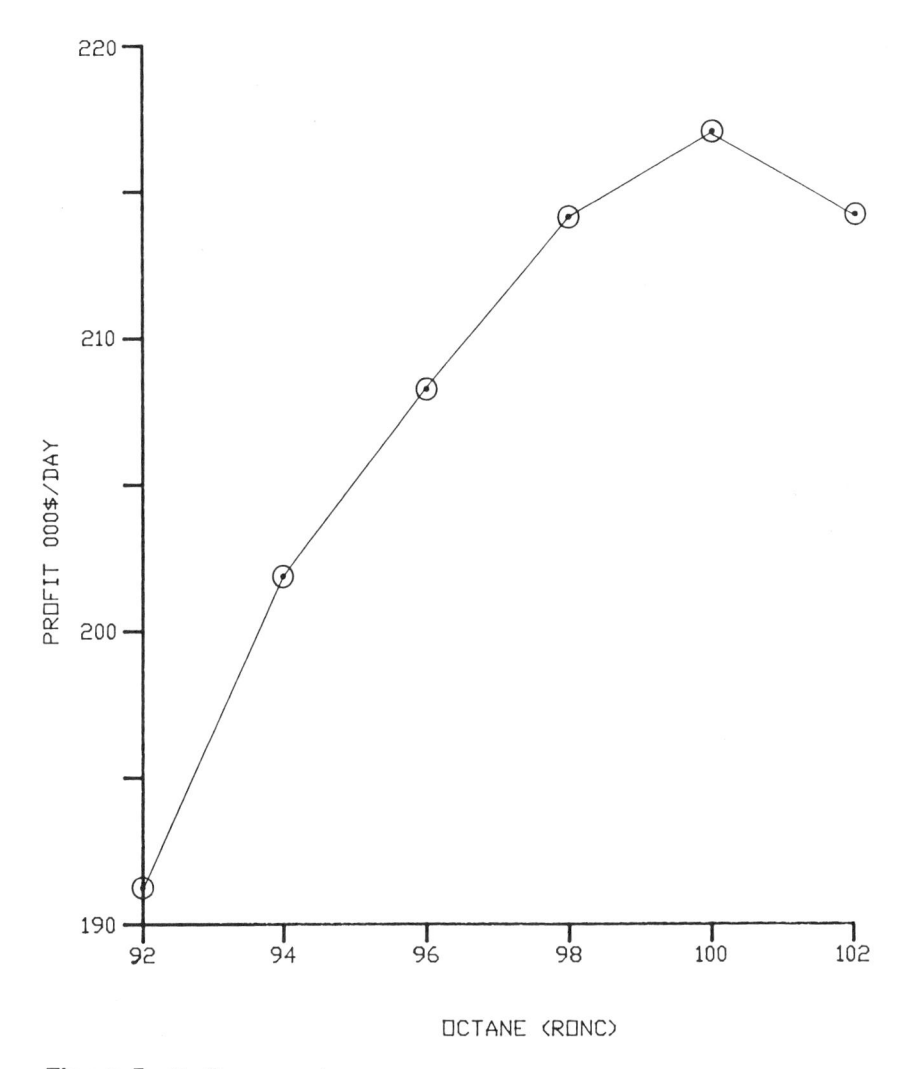

**Figure 5** Profit versus octane.

decreasing the octane 2 numbers for a change in profit of $7680 per day ($208,819 − $201,139 = $7680). The average effect is then ($7680 − $5328)/2 or $1176 per day. If the variability in change is only 1 octane number, the net average change is $578 per day. As the relationship between profit and octane changes due to changes in product values and utility costs, the effect of octane variability changes (29).

For this sample problem, it is assumed the measured standard deviation of the octane is 1.25.

## 2. Evaluation of Benefit from Reduced Octane Variability

The problem is evaluated by first simulating the reformer unit over a series of octanes to create an octane versus profit function and generating a second-order equation in the form

$$\text{Profit} = A + B \times \text{RONC} + C \times \text{RONC}^2 \tag{6}$$

Profit is in dollars per day. Figure 5 shows the relationship for the sample case. A second-order regression of profit as a function of octane yields the following relationship:

$$\text{Profit}(\$000) = -3284. + 69.75 \times \text{RONC} - 0.3475 \times \text{RONC}^2 \tag{7}$$

The effect of reducing octane variability can then be calculated by using the equation (30)

$$\Delta \text{Profit} = C(s_c^2 - s^2) \tag{8}$$

where  $\Delta$Profit = change in profit from reduction in variability
  C = quadratic term in curvature equation
  $s_c$ = standard deviation with advanced process control
  s = standard deviation with only regulatory control

It is common practice in advanced process control benefit studies to assume that the standard deviation of process variables will be decreased by 50% with the installation of advanced process controls. This reduces the delta profit equation to

$$\Delta \text{Profit} = -0.75 \times C \times s^2 \tag{9}$$

$$\Delta \text{Profit} = -0.75 \times (347.5) \times (1.25)^2 = \$407 \text{ per day} \tag{10}$$

## 3. Solution Analysis

The change in profit due to reduced variability is significantly affected by the curvature of the octane versus profit relationship and the measurement of present octane variability. The possibility of error due to curvature can be addressed by a detailed economic variability analysis. Measuring octane variability is quite difficult without an on-line analyzer. One technique is to use a process model to calculate the octane at regular intervals of 1 to 2 h.

## VII. ADVANCED PROCESS CONTROL AND OPTIMIZATION

Three levels of control functions exist on a modern catalytic reformer. At the lowest level of controls are the conventional temperature, pressure, and flow regulators. Controls at the second level in a reformer are commonly referred to as advanced process controls. The final level of control is overall process unit optimization.

## A. Conventional Process Controls

Single-input, single-output controllers using standard PID (proportional/integral/derivative) controllers are located in a DCS. The feedback to the controllers consists of process measurements such as temperature, pressure, or flow. Figure 6 is a simple flow diagram showing the typical location of controllers in a conventional semiregenerative reformer unit. These controllers execute in the frequency range of 3 to 60 times a minute. The output of the conventional process controllers is normally used to set valve openings either directly or through a valve positioner.

## B. Advanced Process Controls

The advanced process controllers typically include multivariable control functions using calculated variables for feedback or long-lag-time analyzer feedback. Quite often the functions will include several input variables and several output variables and can handle controlling the process against one or more constraints. Figure 7 shows the relationship of the advanced process controllers relative to the conventional process controls. The advanced control functions are located in either the DCS or an associated advanced control computer. Frequency of execution varies from once per minute to once every 5 minutes. Advanced process controllers are normally used to adjust the setpoints of conventional process controllers.

The advanced process control functions typically found on a catalytic reformer are for (1) feed maximization, (2) octane control, (3) reactor temperature control, (4) recycle gas control, (5) pressure minimization, (6) stabilizer reflux control, (7) reformate RVP control, and (8) fired heater excess oxygen control (31).

### 1. Feed Maximization

The feed maximization control function typically adjusts the feed flow controller setpoint, supplemental feed controller setpoint, and feed from storage controller setpoint to maintain a maximum feed rate to the reformer up to the limit imposed by one or more constraints. This type of controller is normally a multivariable controller and can handle multiple feed streams and multiple constraints. Typical process unit constraints include maximum heater absorbed duty, minimum hydrogen-to-hydrocarbon ratio, octane requirement, maximum rate of catalyst deactivation, and maximum measured heater tube skin temperature (27).

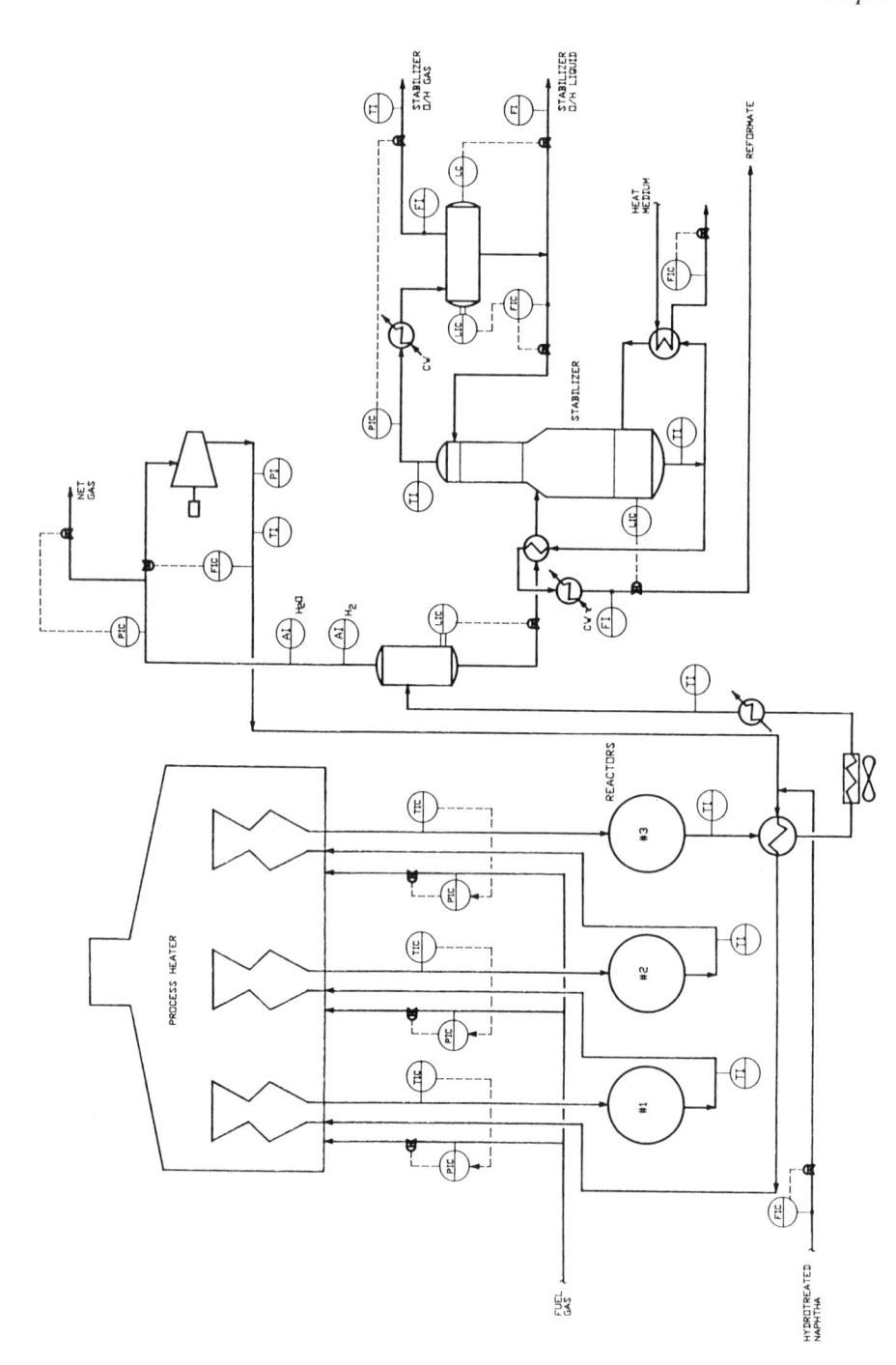

**Figure 6** Typical catalytic reformer unit regulatory controls.

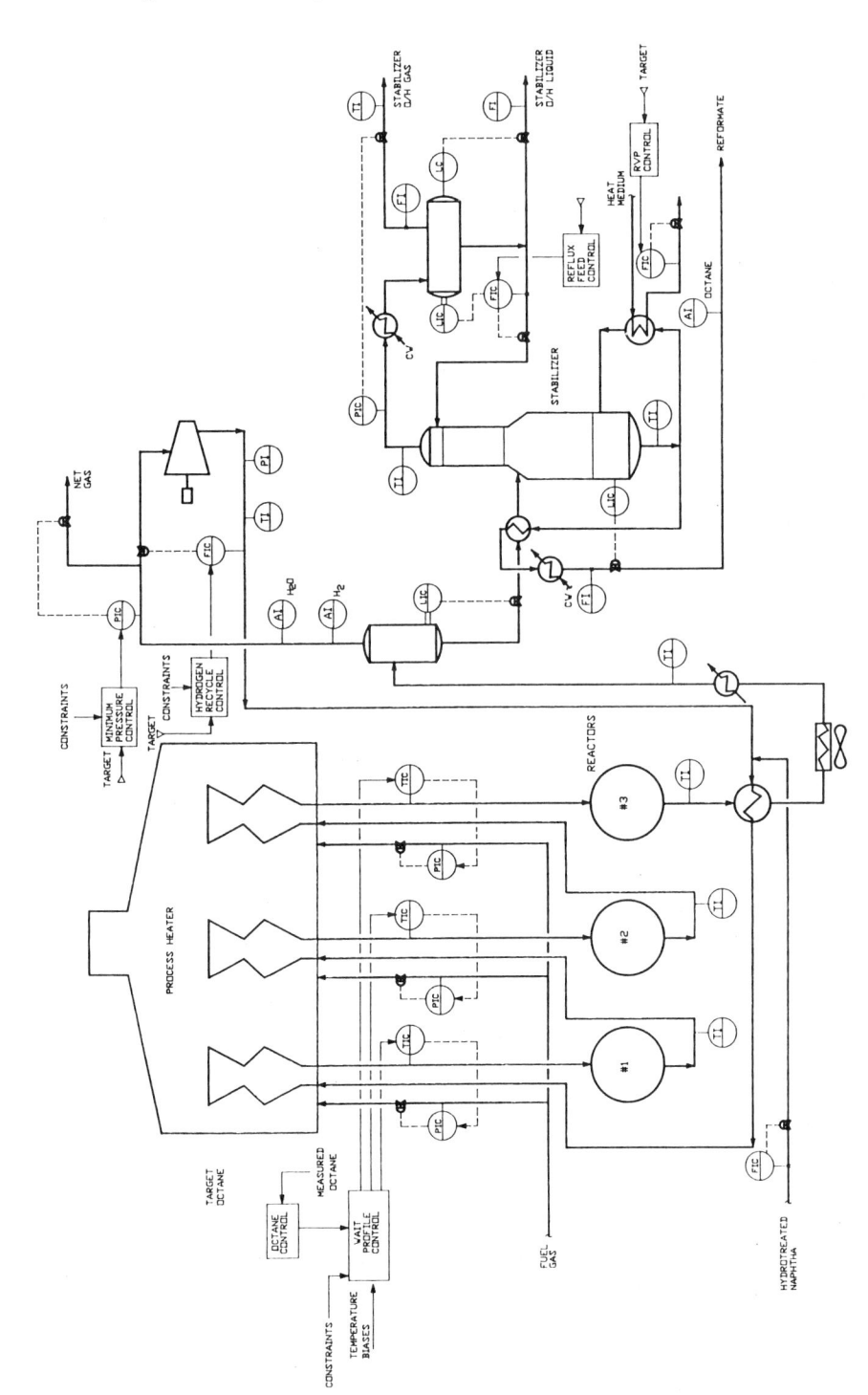

**Figure 7** Typical catalytic reformer unit advanced process controls.

## 2. Octane Control

The octane control adjusts the target temperature for the reactor temperature controller to control the reformer gasoline octane at an operator-specified target. The type of control algorithm used for octane control will be a function of the source of feedback.

The most common octane feedbacks include calculated values, direct measurement of octane from an on-line ASTM-CFR analyzer, and octane inferred from an alternate analyzer. Octane can be calculated by a process model from GC analysis or from a simple correlation of feed properties and measured operating data.

Use of an on-line ASTM-CFR engine is the only way to measure octane directly, but it is the most expensive to install, requires the most maintenance, has the highest down time, and has the lowest overall reliability. Octane can be inferred from several indirect analyzer measurements such as those provided by the Monirex Octane Analyzer, the Foxboro "Cool Flame" Analyzer, refractive index, or a NIR analyzer (5, 32).

## 3. Reactor Temperature Control

The reactor temperature control function adjusts reactor inlet temperature controller setpoints to maintain a target weighted average inlet temperature (WAIT) and the designated inlet temperature profile. The reactor temperature controller uses multivariable control techniques to enforce a WAIT target specified from the octane control module while ensuring that various constraints are not exceeded. Typical constraints enforced by the controller are maximum furnace absorbed duty, maximum furnace tube skin temperature, and maximum furnace outlet temperature.

## 4. Recycle Gas Control

The recycle gas control adjusts the hydrogen recycle rate to maintain a selected target (e.g., hydrogen-to-hydrocarbon ratio or hydrogen partial pressure). An on-line analyzer and/or laboratory analyses determine hydrogen purity of the recycle stream. The rate can be constrained to maximum or minimum rates or recycle compressor capacity.

## 5. Pressure Minimization

The pressure minimization control function is used to adjust the reactor system pressure via the product separator pressure controller until a system constraint or optimum pressure is attained. System constraints may include maximum recycle compressor horsepower, maximum pressure drop, downstream hydrogen minimum pressure requirements, and control valve positions.

## 6. Stabilizer Reflux Control

Stabilizer reflux control is typically a simple PI controller used to maintain a calculated reflux-to-feed ratio, an internal reflux rate, internal reflux ratio, or reflux-to-distillate ratio.

## 7. Reformate RVP Control

Reformate RVP control adjusts the bottoms temperature controller setpoint in order to keep the reformate RVP at a specified target. RVP feedback is frequently calculated but can also be measured using a fast, accurate, and reliable analyzer. It is quite common to use this control function to control the RVP of the light reformate when the reformate is sent directly to a splitter. Because of the long lag time involved, a multivariable control function must be implemented.

## 8. Fired Heater Excess Oxygen Control

The fired heater oxygen control function is used to adjust the stack damper positioner setpoint to maintain flue gas oxygen at a specified target. The function has the capability to handle constraints such as minimum stack temperature and minimum damper position.

## C. On-Line Optimization

On-line optimization includes a complex model plus optimizer for optimization of the process unit. Optimization of the unit corresponds to finding the most profitable operating targets, i.e., feed rate, reactor temperatures, system pressure, hydrogen recycle rate, and octane target. Figure 8 shows the configuration of an on-line optimizer with the advanced process controls in a catalytic reformer unit.

### 1. On-Line Optimization Software

The most common optimization technique uses a sequential linear programming (LP) method for constrained or unconstrained optimization for determination of the optimal operating targets for the reforming process. Derivatives for the LP are generated by an on-line rigorous simulation package. The catalyst activities in the process model used to develop the LP derivatives must be updated on-line and frequently. The required frequency for updating is going to vary from unit to unit. The catalyst activity changes very slowly on a semiregenerative unit, and the activity factors may need to be updated only every 2 weeks. On a reformer with catalyst circulation and constant regeneration, the activity factors may need to be updated prior to each optimization.

The frequency of optimization is also unit dependent. The unit should be optimized every time there is an operator-initiated major change in unit opera-

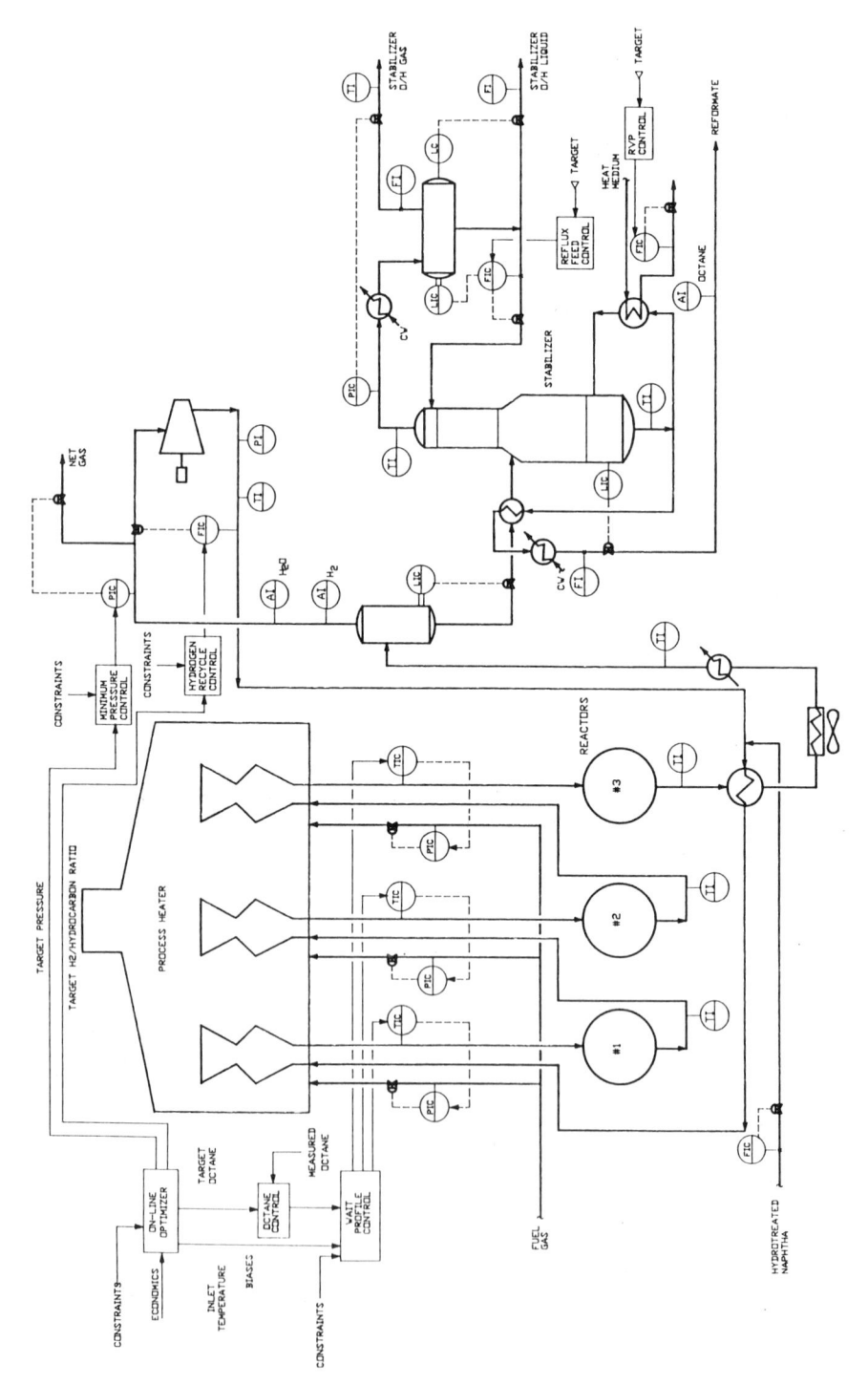

**Figure 8** Typical on-line optimization of a catalytic reformer unit.

tion. A catalytic reformer should also be optimized every time there is a significant change in feed composition. A regular optimization schedule of two or three times a day is adequate for most operations.

## 2. Independent Variables, Objective, and Constraints

The independent variables adjusted by the optimizer may include feed rate, reactor temperatures, system pressure, and hydrogen recycle rate. In almost all cases, the objective of optimization is maximization of profits. Occasionally, other objectives such as maximization of octane, feed rate, and hydrogen production are specified as the optimizer objective. Typical reformer optimization constraints include heater duties, compressor horsepower, catalyst life, minimum hydrogen-to-hydrocarbon ratio, minimum product separator pressure, and minimum/maximum feed rate.

## 3. Constraint Handling

Quite often there are small differences between the measured dependent variable that is being used as a constraint and the mathematically modeled dependent variable. To account for these small differences, a bias equal to the difference between the measured and calculated variables can be applied to the optimizer constraint.

It is not uncommon to have a constraint specified in the unit advanced control strategy that is not modeled. A fired heater tube wall temperature is a typical example. In this type of situation, a relationship must be developed between the actual unit constraint and a dependent variable in the model. To continue the example, tube wall temperature may be the advanced control constraint, and fired heater duty may be the process model constraint.

## 4. Data Validation and Model Updating

In both conventional controls and advanced controls, the objective is to minimize the deviation between a setpoint and the controller's feedback signal. In both situations feedback signals are selected primarily for precision and sensitivity (the rate of change of the feedback signal with respect to the process variable) and secondarily for accuracy. The objective is to keep the unit in a safe steady mode of operation while operating as close to the desired process variable as possible. The objective of the on-line optimizer is to determine the optimum operating conditions for the unit. The data used for these calculations as well as for model updating must be as accurate as possible, and the unit must be at steady state. To ensure that data being used for model updating and on-line optimization are in steady state, it is common practice to screen the data through a steady-state detection routine. This eliminates the possibility of transient data entering the optimization software. After passing through the steady-state screen, a second software screen is used to verify the data validity. A simple example of the steady-state screen would be to check the standard deviation of the feed and product flow rates. If the standard deviation is below a prescribed quantity, the streams are in steady state. The second screen would be a material balance calculation. If the material balance closes within a specified tolerance, the flow rates would be considered valid and acceptable for use in updating the process model or for use in the unit optimization software.

### D. Benefits from Advanced Process Control

The benefits attributable to advanced process control projects can be estimated based on experience or literature values, audits of operations before and after advanced process controls are installed on a unit, or statistical estimates based on the unit's operation prior to installing advanced process controls (30, 33, 34). The objective of each estimation method is to determine how much closer advanced controls can move the average value of a process variable to meet a desired operating target and then apply an appropriate economic factor. Table 15 lists the five most commonly used statistical techniques for calculating benefits from installing advanced process controls (30). Also shown are the general equations used with each calculation method.

#### 1. Same Limit Rule

The same limit rule is used when the present operation exceeds the target operation an unreasonable percentage of the time. In this case it is assumed that after process controls are installed, the process variable will violate the

**Table 15** Benefits Calculation Equations

| Benefits rule | Standard Equation |
|---|---|
| Same limit rule | $\Delta x = P(m)(s - s_c)$ |
| Same percentage rule | $\Delta x = (x_1 - x)(1 - s_c/s)$ |
| Final percentage rule | $\Delta x = x_1 - (x + P(m)s_c)$ |
| Achievable operation rule | $\Delta x = x_1 - x$ |
| Curvature rule | $\Delta \text{profit} = C(s_c^2 - s^2)$ |

| | Nomenclature |
|---|---|
| $\Delta x$ | Change in average value of process variable |
| $x_1$ | Specification limit |
| $x$ | Mean value of process variable |
| $s_c$ | Standard deviation with advanced process controls |
| $s$ | Standard deviation with only regulatory controls |
| $P(m)$ | Probability distribution at percent $m$ |
| $\Delta \text{profit}$ | Change in profit due to reduced variability |
| $C$ | quadratic term in profit $= A + Bx + Cx^2$ |

specified target some limited percentage of the time. For example, if it is assumed that the process variable will violate the target 5% percent of the time, a value for $P(m)$ of 1.65 can be extracted from a normal distribution table. This rule is particularly useful when a mechanical modification is going to be made to a unit that will relieve a constraint to operation.

2. Same Percentage Rule

If during normal operation, a process variable violates the target only an acceptable percentage of the time, the same percentage rule is used to calculate the effect of installing advanced process control. This rule is the most commonly used technique of calculating benefits. It is based on the assumption that the operators are performing their tasks as well as can be expected with the instrumentation provided.

3. Final Percentage Rule

If the standard deviation after advanced process control can be estimated and the acceptable percentage of violations of a process variable is known, the final percentage rule can be applied to calculate advanced control benefits. This rule is particularly useful when estimating the benefits if a DCS has not been installed on the process unit.

4. Achievable Operation Rule

Quite often the work load on human operators exceeds their capability. At this time, the operators prioritize their time and some process variables do not receive the required attention for consistent control. The achievable operation rule is used to calculate the advantage of having the advanced process controls maintain a process variable at its target value when it was being ignored with only normal regulatory controls.

5. Curvature Rule

The curvature rule can be used to estimate the increase in profit of an operation if the standard deviation of a process variable is reduced. This rule is applicable only if the process variable $(x)$ significantly affects profit and with a second-order relation defined by

$$\text{Profit} = A + Bx + Cx^2 \tag{11}$$

6. Benefits from On-Line Optimization

Benefits from advanced process control for a catalytic reformer unit vary from $0.30 and $2.00 per cubic meter of reformer charge with typical benefits in the range of $0.60 to $1.00 per cubic meter of charge (28, 31). Benefits from on-line advanced control typically add an additional 10% to the benefits from advanced process control.

## REFERENCES

1. V. H. Agreda and R. C. Schad, Heat and material balances, *Today's Chemist at Work*, May/June, 24–29 (1993).
2. P. W. Murrill, *Fundamentals of Process Control Therapy*. Instrument Society of America, Research Triangle Park, NC, 1987, p.39.
3. O. Muller-Girard, Do calibration instruments meet your specs? *Intech*, March, 37–39 (1993).
4. *ASTM Manual on Hydrocarbon Analysis*, 4th ed. ASTM, Philadelphia, 1989, pp.46–59, 90–92, 93–99, and 281–286.
5. K. J. Clevett, *Process Analyzer Technology*. Wiley-Interscience, New York, 1986, pp. 659–690.
6. D. W. Spitzer, What affects flowmeter performance? *Intech*, February, 24–27 (1993).
7. *Temperature Measurement Thermocouples*. Instrument Society of America, Research Triangle Park, NC, 1982, p.15.
8. L. E. Turpin, Cut benzene out of reformate. *Hydrocarbon Process.* June, 81–92 (1992).
9. E. H. van Broekhoven, F. Bahlen, and H. Hallie, On the reduction of benzene in reformate. *AIChE National Meeting*, March 1990.
10. G. W. G. McDonald, To judge reformer performance. *Hydrocarbon Process.* June, 1977, pp. 147–150.
11. P. Bonnifay and J. Barbier, Maximizing aromatics production goal of IFP process. *Oil Gas J.*, January 19, 48–52 (1976).
12. F. G. Ciapetta, Catalytic Reforming. *Petro/Chem Eng.* May, C-19–C-31 (1961).
13. R. B. Smith, Kinetic analysis of naphtha reforming With platinum catalyst. *Chem. Eng. Prog.* 55(6), 76–80 (1959).
14. J. H. Sinfelt and J. C. Rohrer, Reactivities of some C6–C8 paraffins over Pt-Al2O3. *J. Chem. Eng. Data* 8(1) 109–111 (1963).
15. J. T. Pistorius, *Troubleshooting catalytic reforming units*. NPRA Annual Meeting, March 1985, AM-85-55.
16. F. G. Ciapetta and D. N. Wallace, Catalytic naphtha reforming. *Catal. Rev.* 5(1), 67–158 (1971).
17. R. W. Wansbrough, Modeling chemical reactors. *Chem. Eng.*, August, 95–102 (1985).
18. G. N. Maslyanskii, A. A. Potapova, and S. A. Barkan, Calculation of benzene yield in catalytic reforming. *Khim. Tekhnol. Topl. Masel*, No. 2, 6–8 (Feb. 1978).
19. J. H. Jenkins and T. W. Stephens, Kinetics of cat reforming. *Hydrocarbon Process.*, November, 163–167 (1980).
20. A. M. Klugman, What affects cat reformer yield. *Hydrocarbon Process.*, January, 95–102 (1976).
21. J. Henningsen and M. Bundaard-Nielson, Catalytic reforming. *Br. Chem. Eng.* 15(11), 1433–1436 (1979).
22. H. G. Krane, A. B. Groth, B. L. Shulman, and J. H. Sinfelt, Reactions in catalytic reforming of naphthas. Fifth World Petroleum Congress, Section III, Paper 4, 1959.

23. P. Van Arkel, J. Beens, H. Spaans, D. Grutterlink, and R. Verbeek, Automated PNA analysis of naphthas and other hydrocarbon samples. *J. Chromatogr. Sci. 25*, 141-148 (April 1987).
24. A. E. Eleazar, G. W. Roberts, H. F. Tse, and R. M. Yarrington, Mass velocity effects in an adiabatic pilot plant simulation of commercial catalytic reforming. *Ind. Eng. Chem. Process Des. Dev. 17(4), 393-400 (1978).*
25. A. Romero, J. Bibao, and J. R. Gonzales-Velasco, Calculation of kinetic parameters for the deactivation of heterogeneous catalyst. *Ind. Eng. Chem. Process Des. Dev.* 20(3), 570-575 (1981).
26. S. E. Gallun, R. H. Luecke, D. E. Scott, and A. M. Morshedi, Use open equations for better models, *Hydrocarbon Process.*, July, 78-90 (1992).
27. R. Lee, R. Rossi, and L. E. Turpin, *Maximizing reformer profitability against mechanical constraints using a kinetic reaction model.* NPRA Computer Conference, CC-91-129, November 1991.
28. H. Besl, T. Cusworth, and E. A. Livingston, Advanced controls improve profitabiity of UOP CCR platforming. *Fuel Reformulation*, November/December, 51-56 (1992).
29. T. M. Stout and R. P. Cline, Control system justification. *Instrumentation Technol.* 23(9), 51-58 (1976).
30. L. E. Turpin, R. P. Cline, and G. D. Martin, Estimating control function benefits. *Hydrocarbon Process.*, June, 68-73 (1991).
31. Advanced process control handbook VII. *Hydrocarbon Process.*, September, 114-115 (1992).
32. M. S. Zetter and B. A. Politzer, On-line octane control with NIR analyzers. *Hydrocarbon Process.*, March, 103-106 (1993).
33. P. R. Latour, Quantify quality control's intangible benefits. *Hydrocarbon Process.*, May, 61-68 (1992).
34. B. D. Stanton, Using historical data to justify controls. *Hydrocarbon Process.*, June, 57-60 (1990).

# 15
# Reforming for Gasoline and Aromatics
## Recent Developments

**Subramanian Sivasanker and Paul Ratnasamy**

*National Chemical Laboratory, Pune, India*

## I. INTRODUCTION

Catalytic reforming is a major petroleum refining process and owes its growth to the ever-increasing demand for high-octane gasoline, the major automotive fuel, and for aromatics, the important building blocks of the petrochemical industry. Ever since the introduction of the UOP Platforming process for octane enhancement of naphtha fractions in the late 1940s (1), both the catalyst and the process have undergone major improvements (2-6). Originally, catalytic reforming was developed to produce high-octane gasoline from straight-run naphthas. Subsequently, applications have extended to the production of aromatics, LPG, and hydrogen and to the upgrading of olefinic stocks and raffinates. Today, about 75 different reforming catalysts are available in the market that are suitable for different applications involving a variety of feedstocks (7). The numerous catalysts and processes available today have been tailored to achieve specific goals using specific feedstocks in the most economic manner.

In this chapter, the various improvements that have taken place over the years with regard to the catalyst formulations and the processes are reviewed. The bulk of all reforming processes are designed for two major applications, gasoline and aromatics production. Even though the two applications are not dissimilar, the discussion is separated in order to differentiate the earlier developments from the later ones and to incorporate the additional processes designed specifically for octane improvement and aromatics production. The developments in the reforming scene that have taken place over the years will be reviewed briefly before descriptions of a number of recent and novel processes used for enhanced octane and aromatic yields are presented, even though many of the novel processes are yet to be commercialized.

## II. REFORMING FOR GASOLINE PRODUCTION

The incentive for the development of naphtha reforming came from the need for higher-octane gasoline with improvements in internal combustion engines used for automotive and aviation purposes. In this regard, World War II was an important motivating force for the development of catalytic reforming. In the 1930s and early 1940s, thermal methods were used for the enhancement of gasoline octane. A breakthrough was the discovery of catalytic reforming using chromia-alumina catalysts and, subsequently, molybdena catalysts. The major breakthrough in catalytic reforming, however, was the commercialization by UOP of the Platforming process based on Pt-alumina catalysts (1). The Pt-alumina catalysts were more active by an order of magnitude or more than the molybdena-alumina and chromia-alumina catalysts. The early developments in the area of catalytic reforming have been reviewed by Ciapetta et al. (2).

The early Pt-alumina catalysts had about 0.3 to 0.6% Pt supported on a fluorided alumina. Subsequently, chlorided $\gamma$-alumina became the support of choice. During the years 1950 to 1969, a large number of monometallic Pt-alumina catalysts came into the market. The early catalysts had to be operated at relatively high pressures (above 20 bars) and lower temperatures (below 520°C), at which conditions the formation of aromatics became limited by thermodynamic constraints (4). Furthermore, high pressures lead to decreased $C_{5+}$ yield due to increased hydrocracking reactions. Reduction of pressures or increasing the temperature of operation decreased the life of the catalyst dramatically, resulting in short cycle lengths. A number of process innovations which extracted the maximum benefits of Pt-alumina catalysts were introduced. Brief accounts of the different processes introduced through the late 1960s have been presented by Ciapetta et al. (2,3) and by Aitani in Chapter 13 of this book.

### A. Catalyst Innovations

During the 1960s, higher-octane gasolines were necessitated by improvements in automobile engines, and in the 1970s and 1980s, environmental concerns, especially the phasing out of lead (8), were the main driving forces for high-octane gasoline production. In 1969, Chevron introduced the Pt-Re-alumina (bimetallic) catalyst (9,10). The Pt-Re-catalyst could operate at lower pressures and higher temperatures, thereby avoiding the thermodynamic limitations on aromatics formation. Subsequently, other bimetallic catalysts containing other metals such as Ir, Sn, and Ge (11–13) and also multimetallics incorporating additional elements (14,15) (usually in ppm levels) came into the market. The bi- and multimetallics have longer cycle lengths at high-severity operations for the following reasons:

1. Lower rate of coke deposition
2. Ability to perform even at high coke levels
3. Lower sintering rate of the metallic function
4. Better activity recovery after regeneration

Comparisons of the relative stabilities of mono-, bi-, and multimetallic catalysts have been reported in a number of publications (16,17). Laboratory studies (17) of coke deposition rates over different bimetallic catalysts during the reforming of $n$-heptane have revealed the following decreasing order in coke pick-up: Pt-Ge > Pt > Pt-Re > Pt-Ir. The lower coke deposition rate observed over Pt-Ir and Pt-Re could be due to the better self-cleaning ability of these catalysts through either hydrogenation or hydrogenolysis of the coke precursors and "soft" coke (18–21). Again, the ability of the bimetallics to operate at higher coke levels than the Pt catalysts is probably due to differences in the type of coke found on the different systems; a "hard" graphitic coke has been reported to be formed on Pt, and a "softer" less refractive coke is present on Pt-Re catalysts (21). An excellent review of coking over Pt and bimetallic catalysts has been presented in Chapter 9 of this book. The lower sintering rates and better activity recovery after regeneration are probably associated with alloy/cluster formation and/or the promoter acting as an anchoring agent for the nearly atomically dispersed Pt (22). The relative sintering rates of Pt and different bimetallic catalysts observed in laboratory studies are presented in Figure 1 (18,24). It is interesting to note that even low-melting metals such as indium, tin, and germanium lower the rate of sintering for platinum. These metals may be acting as anchors of platinum, reducing its mobility on the surface. The increase in cycle lengths achieved through catalyst innovations is presented in Figure 2, for a series of reforming catalysts available from IFP (23). Other innovations in catalyst design have been skewed metal bimetallics and high-density catalysts, some of which may pack better in the reactor because of their special shape (25).

## B. Process Innovations

Catalyst innovations have frequently been accompanied by process innovations. The different reforming processes commercially operational at present are

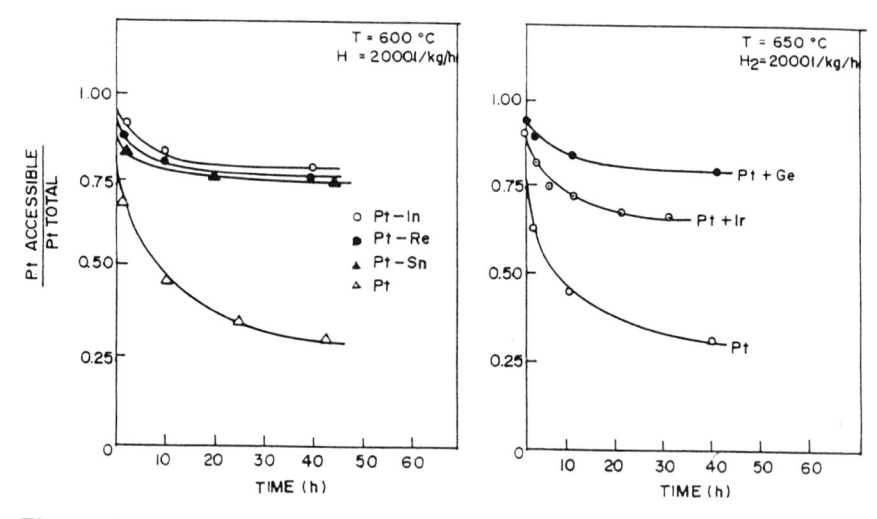

**Figure 1** Relative sintering rates of Pt and bimetallic catalysts. (From Refs. 18 and 24.)

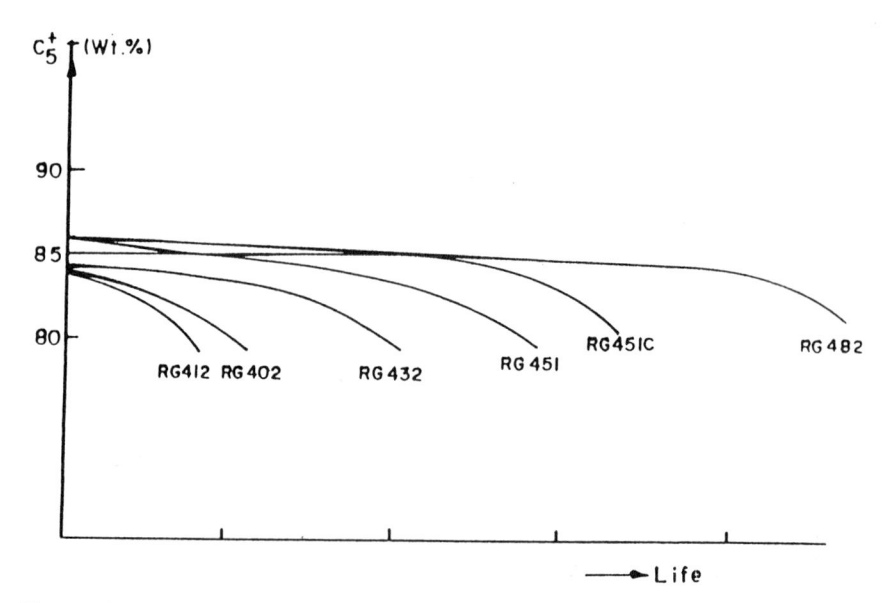

**Figure 2** Evolution of IFP reforming catalysts. RG 412 & 402, monometallic; RG 432, bimetallic; RG 451, multimetallic; RG 451C, improved multimetallic; RG 482, improved bimetallic. [Conditions of testing; pressure = 10 bar; WHSV ($h^{-1}$), 3; $H_2$/HC (mol), 5; RONC, 98.] (From Ref. 23, used with permission.)

*Recent Developments in Reforming* 487

reviewed and discussed in Chapter 13 of this book. Detailed descriptions of the different processes are available in many publications (2,3,26,27). Studies have shown that the dehydrocyclization activities of different bimetallic catalysts peak at different $H_2$ partial pressures (24). Also, the selectivity for the desired products over the various bimetallics depends, to different extents, on the $H_2$ partial pressure. The results of *n*-heptane dehydrocyclization at different $H_2$ partial pressures over various bimetallic catalysts are presented in Figure 3 (24). As a consequence of the improved performance of the bi- and multimetallics, the process severity has increased continuously over the years. This is shown schematically in Figure 4.

Process improvements over the years have also been in the area of moisture and S control of the feedstocks. As chloride is leached out of the catalyst due to

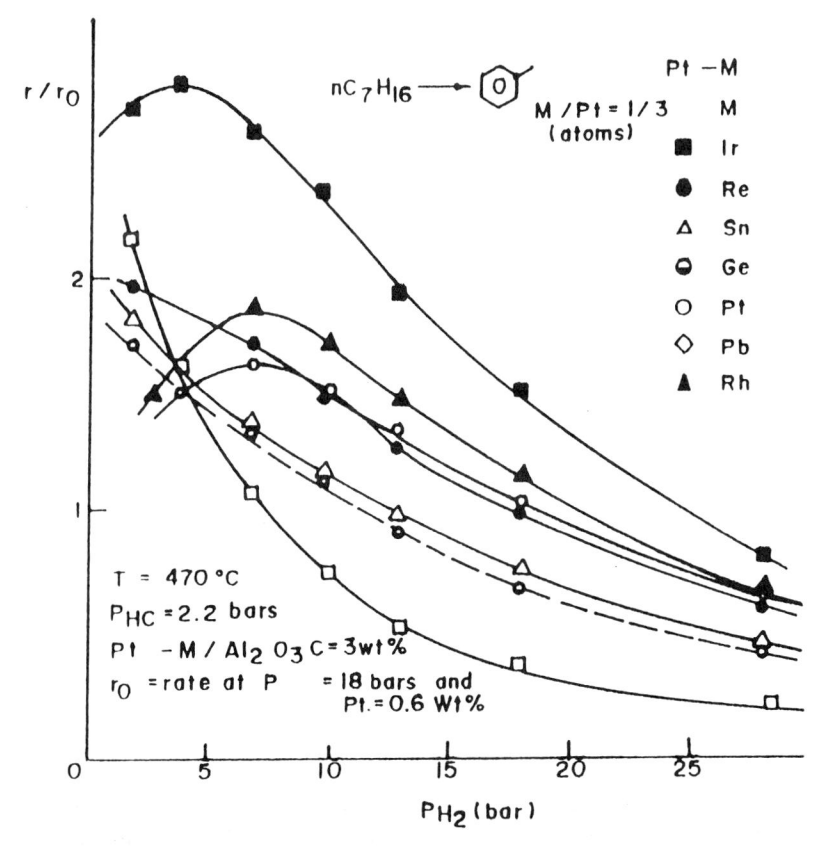

**Figure 3** Rate of dehydrocyclization of *n*-heptane over mono- and bimetallic catalysts: influence of $H_2$ partial pressure. (From Ref. 24.)

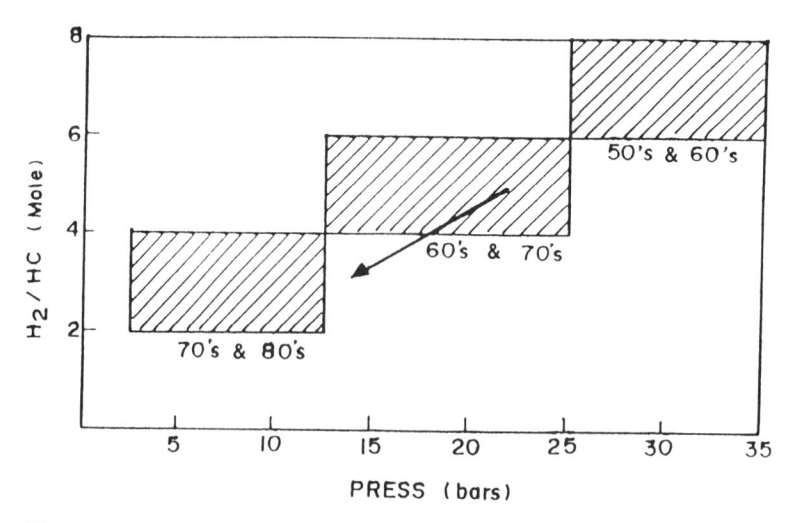

**Figure 4** Increasing trends in the severity of naphtha reforming processes. (Arrow indicates increasing severity of operation.)

the presence of moisture in the feed, process licensors generally recommend injection of ppm levels of chloride along with the feed, normally as an organic chloride, to maintain the optimum chloride level (0.8–1.0 wt %) on the catalyst (4,28). A proper water/chloride balance in the feed is necessary to maintain the required acidity level of the support (4). A small amount of water (about 10–20 ppm) in the feed is usually necessary to sustain the Brönsted acid sites on the support, which are believed to be present in halided alumina (29), and also to prevent excessive coke deposition due to overchlorination. Chlorine has an additional beneficial effect in that it lowers the sintering of the metal (28).

In spite of the many beneficial aspects of the bi- and multimetallic catalysts, they suffer from the disadvantage that they are more prone to poisoning by sulfur (4). Whereas Pt-alumina catalysts operate at S levels as high as 5 ppm, the bi- and multimetallic catalysts require feed S levels < 1 ppm (4,30). Use of a sulfur guard bed to reduce the S level to < .05 ppm (from 1 ppm) has been reported to lead to a cycle length increase by about 25% (30). Lower S levels in the feed permit lower operating pressures, leading to better selectivities and economics. In fact, one of the reasons for the greater tendency of the bimetallic catalysts to be poisoned by S is their operation at low $H_2$ partial pressures, at which the desorption of S from the surface of the metal (as $H_2S$) is less favored. Excellent reviews of the poisoning of metal catalysts by S have been presented in Chapter 10 of this book and in other publications (21,31,32).

## C. Octane boosting of gasoline

### 1. Selectoforming

Erionite is a eight-member ring, small-pore zeolite (0.41 nm diameter) which has shape-selective cracking properties (33); it cracks $n$-paraffins selectively, excluding the other larger molecules which cannot enter its pores. Based on this principle of shape-selective cracking, a process named Selectoforming was introduced by Chen et al. (34,35) in 1968. This was a postreforming process in which the low-octane $n$-paraffins present in the reformate were cracked selectively, thereby boosting its octane number. The catalyst was a nickel-containing, partially $H^+$ exchanged erionite (36). The Selectoforming catalyst is usually loaded in a separate reactor at the tail end and is operated under milder conditions, preferably with an independent gas recycle to enable greater flexibility of operation. Some typical results of Selectoforming are presented in Table 1.

### 2. ZSM-5 as a Promoter for Octane Boosting

ZSM-5 is a medium-pore, high-silica zeolite with excellent cracking properties (37). It has a three-dimensional pore system of two types of pores, an elliptical (0.51 × 0.55 nm) one and a near-circular (0.54 × 0.56 nm) one. Its pore dimensions are of the right size to permit the entry and cracking of the low-octane normal and slightly branched hydrocarbons, making it useful for octane boosting of reformates. The relationship between the relative cracking rates of

**Table 1** Selectoforming for Octane Improvement

| Yields, wt. | Case 1 Charge | Case 1 Products | Case 2 Charge | Case 2 Products |
|---|---|---|---|---|
| $C_1$, $C_2$, $H_2$ | 7.0 | 8.7 | 13.6 | 16.6 |
| $C_3$ | 5.3 | 15.3 | 10.8 | 15.5 |
| $C_4$ | 5.9 | 6.3 | 10.8 | 8.9 |
| $i$-$C_5$ | 6.4 | 7.6 | 8.7 | 8.0 |
| $n$-$C_5$ | 6.1 | 1.6 | 7.0 | 3.9 |
| $C_6$ + $n$-paraffins | 8.9 | 1.8 | 2.6 | 1.3 |
| $C_6$ + isos + aromatics | 60.4 | 58.7 | 46.5 | 45.8 |
| Octane numbers | Reformate | Selecto-formate | Reformate | Selecto-formate |
| $C_{5+}$, RON + 0 | 86.1 | 93.3 | 96.5 | 99.5 |
| $C_{6+}$, RON + 0 | 87.3 | 94.3 | 102.9 | 104.7 |

*Source*: From Ref. 36.

$C_5$–$C_7$ paraffins over ZSM-5 and their octane ratings is presented in Figure 5 (38). Plank et al. (39) have found (Table 2) that addition of 1% ZSM-5 as a promoter to a reforming catalyst increases the octane number of the reformate through selective cracking of the low-octane paraffins.

## 3. M-Forming

M-Forming (40) is another postreforming process introduced by Mobil using ZSM-5 as catalyst. Apart from its shape-selective cracking property, ZSM-5 also possesses other catalytic properties for the oligomerization and aromatization of olefins, alkylation of benzene and toluene, and dealkylation of bulkier alkyl aromatics. All these reactions are involved in M-Forming using an Ni-ZSM-5 catalyst (41). The benefits of M-Forming are more apparent when reforming light naphthas to very high octane levels. A typical case of reforming a $C_6$-80°C Mid-Continent naphtha is presented in Figure 6 (41). The light naphtha was reformed to 84 RONC (research octane number clear) and then

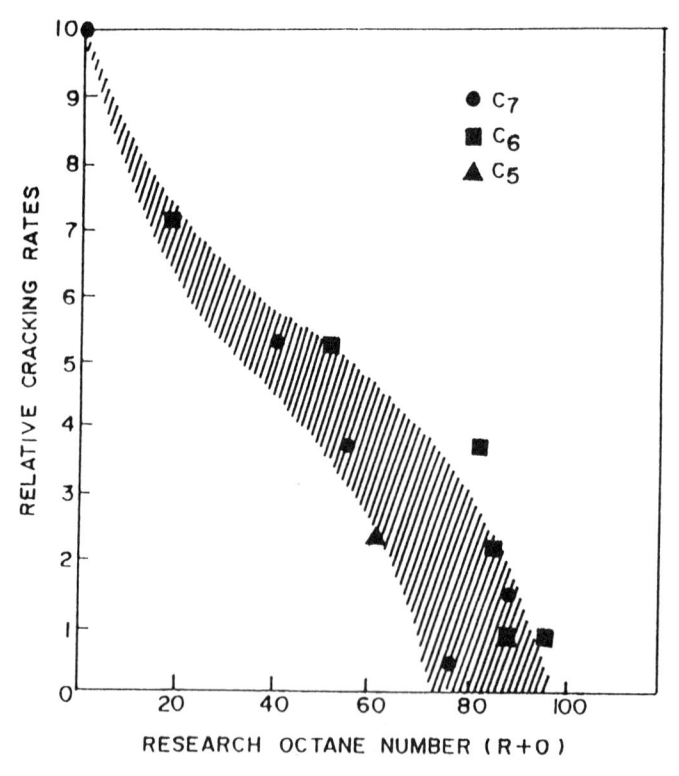

**Figure 5** Relationship between relative cracking rates of $C_5$–$C_7$ paraffins over ZSM-5 and their octane ratings. (From Ref. 38.)

**Table 2** Effect of ZSM-5 on Reforming Catalyst Performance[a]

| Catalyst | Base case | Base catalyst + 1 wt % ZSM-5 | |
|---|---|---|---|
| Temperature, °C | 485 | 485 | 465 |
| $C_{5+}$ octane | 94.5 | 101.8 | 96.4 |
| $C_{5+}$ yield, vol % | 83.4 | 69.7 | 74.4 |
| Gas yield, wt % | | | |
| Methane, ethane | 1.8 | 3.0 | 1.6 |
| Propane | 2.4 | 9.1 | 7.7 |
| Butanes | 2.9 | 9.8 | 9.1 |

[a] Feed: $C_6$-140°C mild continent naphtha. Operating conditions: pressure, 14.6 atm; LHSV ($h^{-1}$), 1.7; hydrogen/hydrocarbon (mol), 9.6.
*Source*: From Ref. 36.

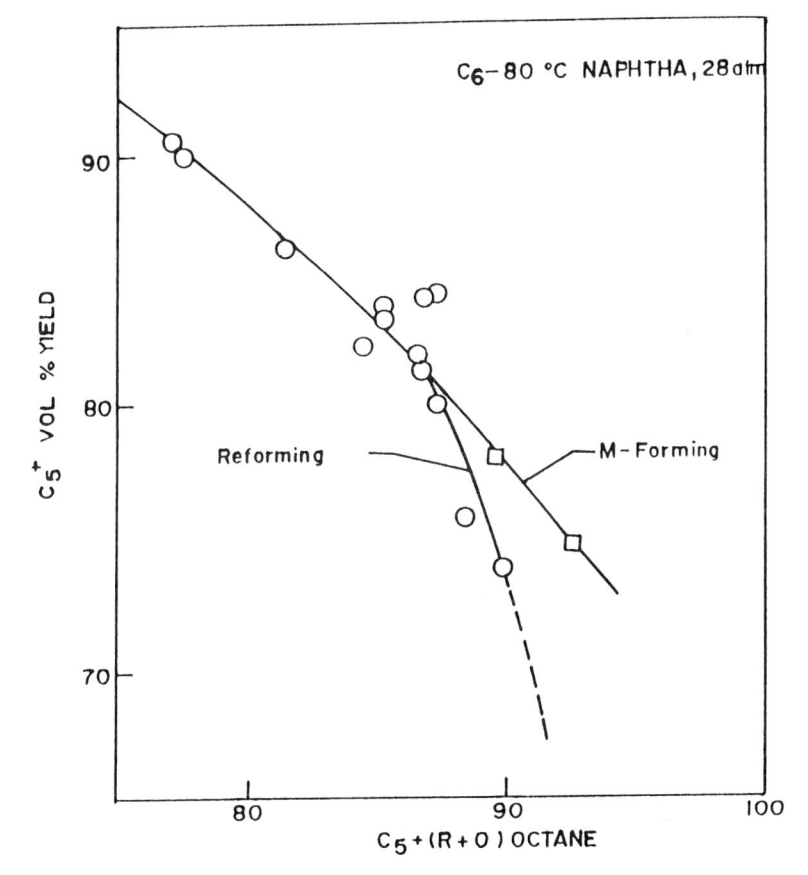

**Figure 6** Comparison of the performance of reforming and M-Forming. (From Ref. 36.)

charged into an M-Forming reactor (pressure 28 bars; 7:1 $H_2$/HC (mole); temperature 315°C). A boost of almost 9 RONC at equivalent liquid yields was reported.

## III. REFORMING FOR AROMATICS PRODUCTION

Reforming for octane improvement of gasoline stocks and reforming for aromatics production are not significantly different processes. One minor difference is that narrower feed cuts are usually used in aromatics production to maximize the yield of the required aromatics. For example, typically a 60–90°C cut is used for benzene production and a 110–140°C cut is used for production of toluene and xylenes. The exact cut points usually depend on the crude source. In gasoline production, a full-range naphtha is often preferred. Again, reforming for aromatics production usually involves operation at the highest possible severities to obtain the maximum possible yield of aromatics, whereas reforming for gasoline, reformer conditions are optimized to produce maximum octane barrels (octane number × $C_{5+}$ yield). Catalysts designed specifically for maximum dehydrocyclization (aromatic production) are commercially available (42).

Besides the conventional reforming technology and catalysts, some novel processes have been introduced for aromatics production. These are the M2-Forming (43), Aroforming (44), and Cyclar (45) processes. Brief descriptions of these processes are presented in the following paragraphs.

### A. M2-Forming Process

This Mobil process (43) is also based on ZSM-5 zeolite. Unlike M-Forming this is not a postreforming process. It also has the ability to handle olefinic feeds. It operates at much higher temperatures (about 538°C) and lower pressures (below 20 bars, preferably close to atmospheric). The reactions involved in M2-Forming are:

1. Cracking of the olefin and paraffin molecules in the feedstock into lighter olefins
2. Oligomerizing and aromatization of the olefins via cyclization and $H_2$ transfer reactions

A major benefit of this process is its ability to transform less valuable olefinic gasolines obtained from FCC and thermal processes into aromatics. The results of the aromatization of FCC gasoline and a SR naphtha are presented in Table 3 (43). It is to be noticed that at a higher temperature of 550°C, a nearly pure aromatic product (99.4 wt % in $C_{5+}$) is obtained.

**Table 3** Aromatization of FCC Gasoline and SR Naphtha

| Feed | Light FCC gasoline | | Virgin naphtha |
|---|---|---|---|
| Boiling range, °C | | | |
| Initial | $C_5$ | | $C_6$ |
| End | 110 | | 110 |
| Elemental analysis, wt % | | | |
| C | 85.4 | | 84.8 |
| H | 14.6 | | 15.2 |
| Type analysis, % | | | |
| Paraffin | 34 | | 47 |
| Olefin | 41 | | — |
| Naphthene | 15 | | 41 |
| Aromatics | 10 | | 12 |
| Reaction conditions | | | |
| Temperature °C | 390 | 550 | 550 |
| WHSV, $h^{-1}$ | 1.2 | 1.3 | 1.3 |
| $C_{5+}$ product, wt % on feed | | | |
| Total yield/53.2 | 54.3 | 47.3 | |
| Aromatics yield | 43.6 | 54.0 | 43.8 |
| Aromatics in $C_{5+}$ wt % | 82.0 | 99.4 | 92.6 |

*Source*: From Ref. 43. Reprinted with permission from *Ind. Eng. Chem. Process Des. Dev. 25*, 151 (1986). Copyright (1986) American Chemical Society.

## B. The Aroforming Process

In this process, a ZSM-5–based catalyst containing $Ga_2O_3$ is used to convert LPG and light naphthas into an aromatic-rich concentrate. An interesting aspect of this process, introduced jointly by Salutec, Australia and IFP, France, is that multiple tubular reactors operating isothermally are used (44). The process uses no $H_2$ recycle and operates cyclically in swing arrangement with 12-h cycle lengths. This is related to the Cyclar and M2-Forming processes, which also use Ga-doped ZSM-5 as the catalyst. However, the lack of $H_2$ recycle precludes the processing of FCC gasolines, unlike the M2-Forming process.

## C. The Cyclar Process

This novel BP-UOP process, based on a Ga-doped ZSM-5 catalyst and employing the UOP continuous catalyst regeneration (CCR) technology, has already been put into commercial practice. It is designed to convert propane and butane (LPG) into aromatics (45,46). The aromatic yields have been claimed to surpass those obtained from the reforming of light naphtha fractions over conventional catalysts (see Figure 7) (45). The yields of benzene and toluene are higher than those from naphtha, while that of $C_{9+}$ aromatics is lower (Figure 7) (45). The product from the Cyclar process can be fractionated directly (to recover the aromatics) as the content of the aliphatics boiling in the BTX range in the reactor effluent is negligible. This is an additional factor in favor of the Cyclar process for BTX production. The possible reaction pathways in Cyclar are illustrated in Figure 8 (45).

## IV. MONOFUNCTIONAL REFORMING CATALYSTS

The conventional reforming catalysts operate primarily by a dual functional mechanism, the acid sites of the support (chlorided-alumina) taking part in iso-

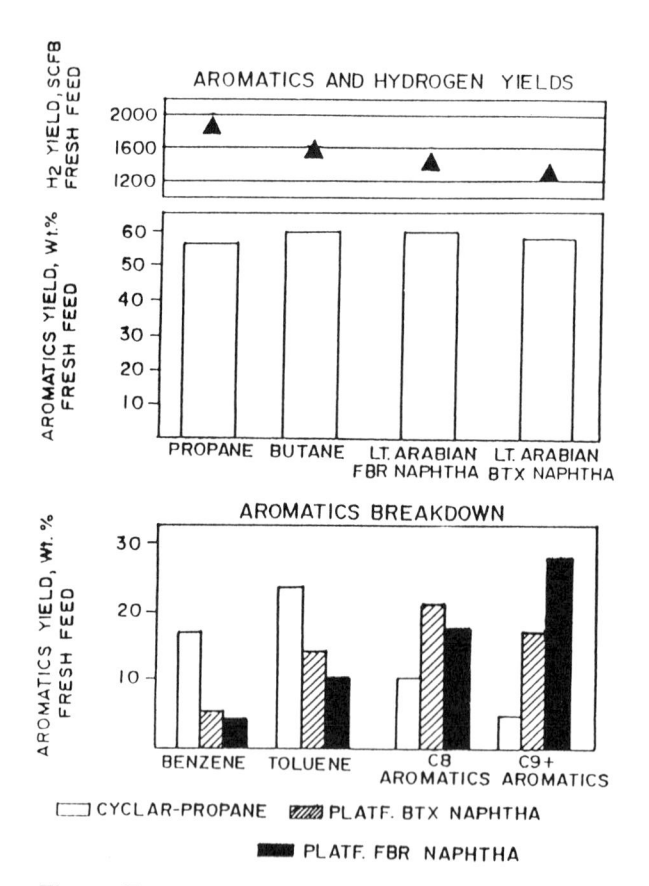

**Figure 7** Aromatics production: comparison of Cyclar with platinum reforming. (Feeds: Lt. Arabian naphtha; FBR, 71–185°C, P/N/A = 68/19/13; BTX, 71–157°C, P/N/A = 70/18/12.) (From Ref. 45, used with permission.)

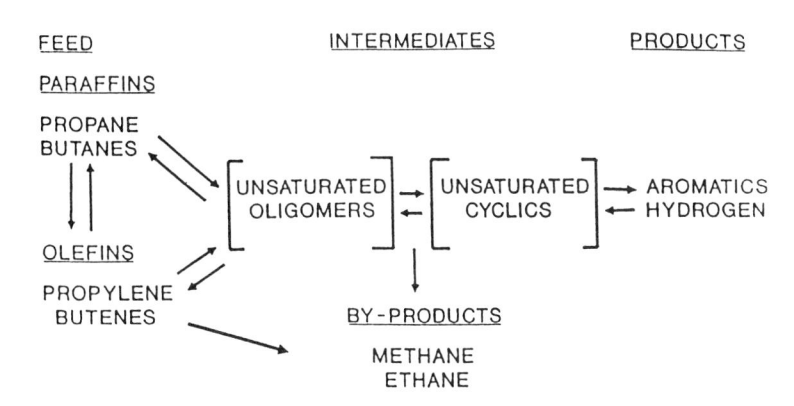

```
FEED                    INTERMEDIATES              PRODUCTS
PARAFFINS
 PROPANE
 BUTANES
    ↕  ↘   ┌──────────┐   ┌──────────┐
    ↕      │UNSATURATED│ → │UNSATURATED│ → AROMATICS
           │ OLIGOMERS │ ← │  CYCLICS  │ ← HYDROGEN
 OLEFINS ↗ └──────────┘   └──────────┘
 PROPYLENE                      ↓
 BUTENES  ↘              BY-PRODUCTS
           →              METHANE
                          ETHANE
```

**Figure 8** Reaction pathways in the Cyclar process. (From Ref. 45, used with permission.)

merization and cyclization reactions and the metallic function acting as the dehydrogenation-hydrogenation agent. For example, the transformation of $n$-hexane to benzene is believed to take place according to the following steps:

1. Dehydrogenation to hexanes (metal catalyzed)
2. Isomerization to cyclohexane (acid catalyzed)
3. Dehydrogenation of cyclohexane to benzene (metal catalyzed) (4)

However, the direct dehydrocyclization of $n$-hexane over the metal (Pt) alone is also possible. This reaction proceeds with high conversions and selectivities when Pt is supported on an alkaline L-zeolite.

## A. Pt-L-Zeolite Catalysts

Zeolite L (47) is a wide-pore zeolite with a typical $SiO_2/Al_2O_3$ ratio of 6. It has a unidimensional pore system consisting of cages of dimensions 0.48 nm (length), 1.24 nm (height) and 1.07 nm (width) interconnected by openings of 0.74 nm diameter. In its basic form, when fully exchanged with alkali ions and incorporated with Pt, it makes an excellent catalyst for the dehydrocyclization of alkanes (48). The preparation of these Pt-L-zeolite catalysts has been described by Hughes et al. (49). Ba-K-L is prepared by exchange of K-L with $Ba^{2+}$ ions and calcined at an optimized temperature (593°C) to avoid blockage of the channel system by the exchanged ions. Pt is then incorporated into the zeolite by an ion-exchange procedure using Pt-amine complexes. Next, the catalyst is dried, oxidized carefully, and reduced. These catalysts possess no acidity and reform hydrocarbons monofunctionally (48). Catalysts prepared from other wide-pore acidic zeolites like the faujasites and mordenites do not possess such excellent activities and selectivities for the aromatization of alkanes (49).

The performance of a Pt/Ba-K-L zeolite in the reforming of $C_6$–$C_9$ n-paraffins is compared to that of a Pt-Re reforming catalyst in Figure 9 (49). Both aromatization activity and selectivity are higher on the monofunctional catalyst than on the conventional bifunctional catalyst for all the hydrocarbons; the advantages become attenuated with increasing carbon number (49). However, one disadvantage of these catalysts is their greater susceptibility to S poisoning, so much so that meaningful cycle lengths are demonstrated typically when the S levels are less than 0.05 ppm in the feed (49). Recent studies (50) show that S in the feed not only lowers the number of available Pt sites for the reaction but also increases the rate of agglomeration of Pt.

Based on the above catalyst, a new reforming process (AROMAX) has been announced by Chevron (51). Patents issued to UOP and Exxon (52,53) indicate that these companies also have interest in aromatics production over Pt-L catalysts. The yield advantage of this process over a conventional bifunctional catalyst has been demonstrated for five different feedstocks boiling in the range 52 to 138°C (51). The data are presented in Figure 10 (51). The properties of the feeds are presented in Table 4 (51). The monofunctional process (AROMAX) yields more octane barrels on all feeds at all severities.

Two different reasons have been advanced for the spectacular activity of Pt-L zeolites. One postulate is that there is an electronic interaction between the zeolite and the Pt metal (54–56); the other is that structural parameters of the zeolite are responsible (57–59). Based on studies of benzene hydrogenation and n-hexane reforming over a series of Pt-L catalysts exchanged with different

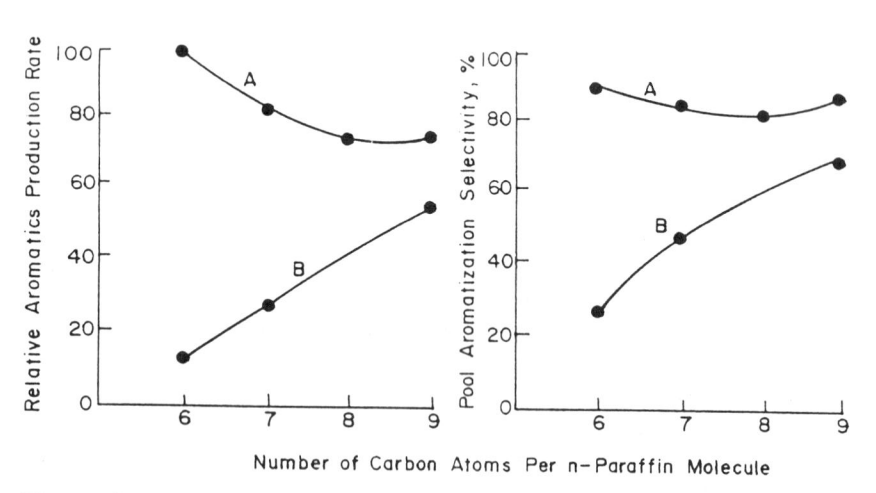

**Figure 9** Comparison of the performance of Pt-Re-$Al_2O_3$ and Pt-BaKL catalysts in the reforming of $C_6$–$C_9$ n-alkanes. [Conditions: 100 psig; 493°C; LHSV ($h^{-1}$), 18. A, Pt-BaKL; B, Pt-Re-$Al_2O_3$.] (From Ref. 49, used with permission.)

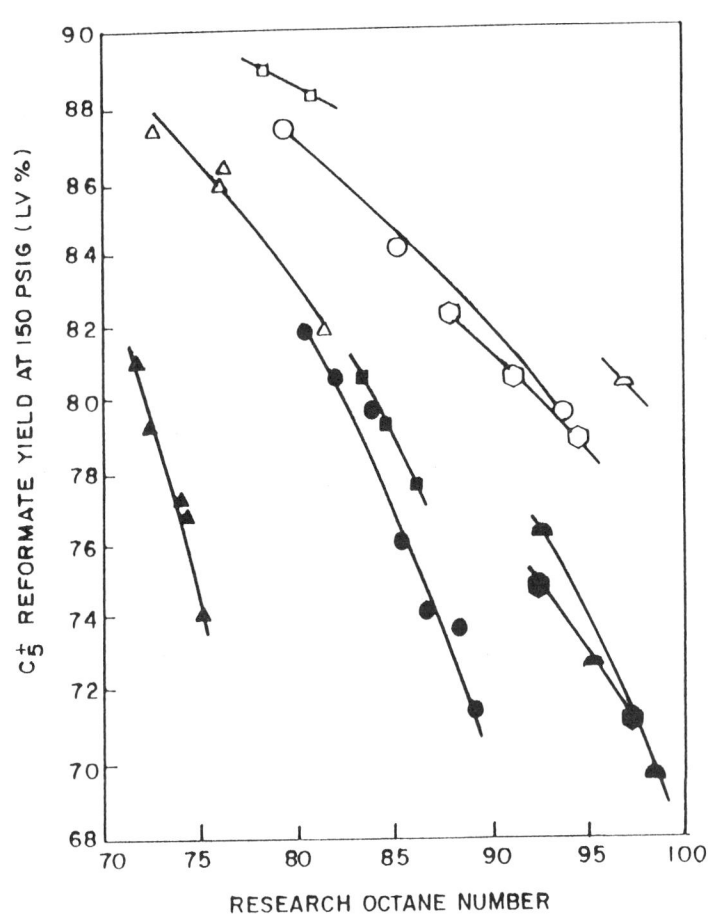

**Figure 10** Comparison of AROMAX and bifunctional catalysts. (△) Mixed hexanes; (○) Raffinate; (⌒) Light Straight Run-1; (□) LSR + Raffinate; (●) LSR-2 (open symbols for AROMAX catalyst and filled symbols for bifunctional catalyst). (From Ref. 51, used with permission.)

**Table 4** Properties of the Light Naphthas Used in the AROMAX Process (Figure 10)

|  | LSR/ Mixed hexanes | Raffinate blend | Raffinate | LSR[a]-1 | LSR[a]-2 |
|---|---|---|---|---|---|
| Boiling range (ASTM D-86) |  |  |  |  |  |
| IBP/FBP (°C) | 65/76 | 67/126 | 52/129 | 77/138 | 88/132 |
| RON | 58.0 | 63.1 | 66.4 | 55.3 | 46.2 |

[a] LSR, light straight run.
*Source*: From Ref. 51.

alkali metals (Li, Na, Rb, or Cs), Besoukhanova et al. (54) have shown that the activity of the catalyst increases with increasing basicity, i.e., in the order Cs > Rb > Na > Li. Furthermore, based on infrared shifts of CO adsorbed on Pt in these catalysts, the authors have suggested that the Pt particles in these catalysts are electron rich, the electron richness of the Pt clusters arising from interaction with the basic $O^{2-}$ ions in the lattice (54). Barthomeuf (55) has reported an inverse relationship between benzene yield in the dehydrocyclization of $n$-hexane and $S_{int}$, the intermediate electronegativity, of a number of Pt-L zeolites exchanged with various cations (Figure 11) (55). $S_{int}$ is inversely related to the negative charge on the oxygen ions. The latter is a measure of the basicity of the zeolite. Larsen and Haller (56) carried out competitive hydrogenation of toluene and benzene over a series of Pt-L catalysts containing Mg, Ca, and Ba ions. They found that the ratio of adsorption constants for toluene and benzene ($K_{t/b}$) increased in the order Mg > Ca > Ba, which is also the order of increasing basicity. They concluded that Pt clusters interact strongly with the zeolite and suggested (56) that the extreme S sensitivity of Pt-L catalyst could be due to the electron-rich nature of the Pt in these catalysts (54), in contrast to the relatively greater S tolerance of Pt in Pt-Y zeolite attributed to the electron deficiency of Pt in acidic zeolites (60).

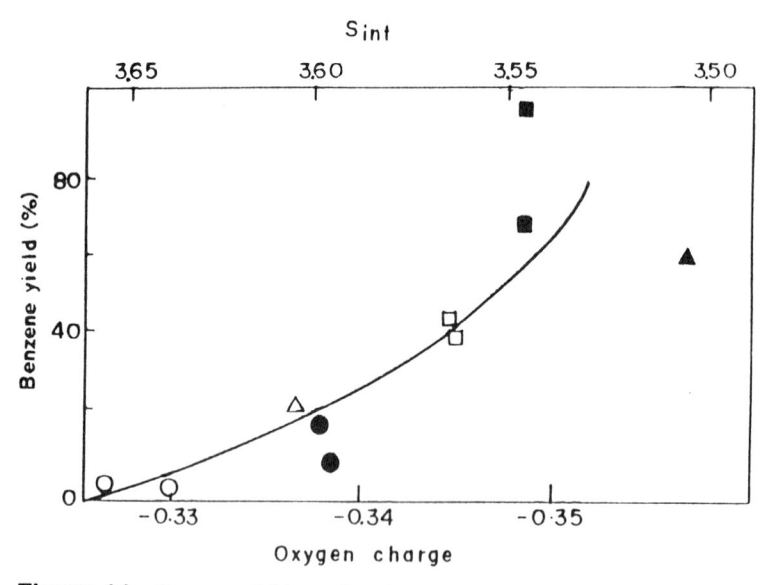

**Figure 11** Benzene yield as a function of relative basicity of Pt-L zeolite catalysts in the dehydrocyclization of $n$-hexane. $S_{int}$ is intermediate electronegativity. Negative oxygen charge is related to basicity. (○) H; (●) Li; (△) Na; (□) K; (■) Rb; (▲) Cs. (From Ref. 55, used with permission.)

Tauster and Steger (57) have, on the other hand, proposed that the pore openings in L-Zeolite collimate diffusing $n$-hexane molecules, leading to end-on adsorption on the Pt clusters situated inside the cancrinite cages. Such end-on adsorption should facilitate 1,6-ring closure leading to aromatization. Their observation that terminal hydrogenolysis activity correlates with aromatization activity for a number of Pt-K-L and Pt-SiO$_2$ catalysts in the conversion of $n$-hexane (57) has been cited as support for their "molecular die" model. Based on structural characterization and molecular modeling studies, Newsam et al. (59) also propose that the steric constraints imposed by the surrounding cage may be responsible for promoting conformations of $n$-hexane molecules (already terminally bonded to Pt atoms) that will lead to 1,6-ring closure. However, Derouane and Vanderveken (58), based on molecular modeling, have proposed a slightly different mechanism (confinement model) in which the $n$-hexane molecules are preorganized in a pseudocycle resembling the transition state of ring closure. The essential pathway for the dehydrocyclization of $n$-hexane according to this mechanism is:

1. The $n$-hexane molecule moves inside the channel, meets a lone Pt atom (or a Pt$_{15}$ cluster in the neighboring cage).
2. It curves itself to form a six-membered ring, the ring formation being assisted by nonbonding interactions with the surface of the channel, forcing the carbon in the 6 position into the vicinity of the Pt atom.
3. A $C_1$–$C_6$ metallocycle is formed on the Pt atom (or cluster), the metallocycle rearranging into cyclohexane, the precursor to benzene.

Based on their studies on the conversion of $n$-hexane over Pt-Y and Pt-L zeolites, Lane et al. (61) have also concluded that the "confinement model" better explains the superior dehydrocyclization activity of Pt-L.

Recently, Katsuno et al. (62) have reported that pretreatment of K-L zeolite with CF$_3$Cl prior to incorporation of Pt enhances the selectivity and life of the catalyst. The authors have shown that coke deposition is decreased when the catalyst is pretreated with CF$_3$Cl (62); the carbon deposition rate on an untreated Pt-K-L zeolite was 88 ppm/h, whereas it was only 6.6 ppm/h on a CF$_3$Cl-treated catalyst. Their studies, using infrared spectroscopy of adsorbed CO, suggest that the Pt in the CF$_3$Cl-treated catalyst is more electron rich than in the untreated catalyst. The Pt particle size was also smaller in the treated catalyst. They have further suggested that AlF$_3$ formed during the halocarbon treatment might narrow the pore dimensions of the zeolite. Whether the slower coke deposition rate of the halogenated catalyst is due to pore size optimization or the electron-rich nature of the Pt is not clear.

Iglesia and Baumgartner (63) have postulated that the good performance of Pt-L zeolites in dehydrocyclization is due to the protection from carbon deposits offered by the LTL pore system to the Pt clusters situated inside them.

Mielczarski et al. (64), however, believe that the uniqueness of Pt-L as a reforming catalyst lies in its ability to stabilize extremely small Pt clusters in a completely nonacidic environment.

In spite of the numerous studies by the different authors, the exact reason for the superior dehydrocyclization activity of Pt-L is not clear. However, certain points emerge:

1. Acidity in the system should be absent.
2. The Pt particles should be well dispersed and be located inside the LTL pore system.
3. The zeolite should be heat treated after exchange with the alkali ions to optimize their locations.
4. The feed should be S-free.

An interesting aspect of Pt-L is its rapid deactivation even if small amounts of acidity are present in it (65). Hence a modified catalyst reactivation procedure in which the chloride deposited on the catalyst during the oxychlorination step (Pt redispersion) is steamed out by wet air has been recommended (66).

### B. Pt/MgO:Al$_2$O$_3$ Catalyst

Davis and Derouane (67) have reported that Pt supported on nonacidic MgO:Al$_2$O$_3$ (Mg/Al 5.1 molar ratio) exhibits unusual aromatization properties. The activity and selectivity of this catalyst were similar to those of Pt-K-L in the reforming of $n$-hexane. For example, the selectivity for benzene observed over Pt/Mg(Al)O was found to be 43.7% at 35% conversion of $n$-hexane and was 43.9% at 39.6% conversion over Pt-K-L at similar operating conditions (67). The reasons for the good performance of this catalyst are not clear. However, an electronic effect of the basic support on the Pt metal as suggested by Barthomeuf et al. (54,55) in the case of Pt-L zeolites has been proposed. Mielczarski et al. (64) believe, on the other hand, that the high aromatization activity of the Pt/Mg(Al)O is due to the support acting as an inert carrier for small Pt clusters. Besides, the Pt-KL used for comparison is also suspected to possess residual acidity created during the reduction of the Pt ions, making it a less than optimum catalyst.

## V. REFORMING OF OLEFINIC FEEDSTOCKS

Naphtha fractions obtained in FCC or in thermal processes such as visbreaking and delayed coking contain substantial amounts of olefins. They also contain high levels of S and N. These naphthas often have to be reformed before blending into the gasoline pool. However, they cannot be reformed directly over the conventional reforming catalysts, which deactivate rapidly in the presence of S, N, and olefinic compounds. The usual procedure is to hydrotreat these feeds before reforming (68). However, patent reports claim (69,70) that FCC gasoline can be successfully reformed over catalysts containing MFI (ZSM-5) isomorphs without any hydrogen pretreatment step.

Pyrolysis gasoline (PG) is obtained during the manufacture of light olefins by the steam cracking of naphtha and other hydrocarbon materials. PG is rich in aromatics, making it a good source of BTX. However, direct extraction of BTX by solvents is difficult because of the presence of olefins and diolefins, which tend to slip into the aromatic fraction and also degrade the solvent. Therefore, pyrolysis gasoline is first hydrogenated (in two stages) and then extracted for aromatics. Extraction of aromatics by solvents, apart from being expensive, often results in the loss of 1–5 wt % aromatics during the extraction process. Usually the raffinate is further reformed to obtain more BTX. The direct reforming of PG over conventional reforming catalysts is not feasible due to rapid catalyst deactivation by the alkenes and the S and N impurities.

Reddy et al. (71) have reported that it is possible to reform pyrolysis gasoline directly over Pt-alumina containing 5–10 wt % ZSM-5. The ZSM-5 component reduces the coke-forming tendency of the catalyst by cracking away the coke precursors. The results of the reforming of pyrolysis gasoline from two different sources are presented in Table 5 (71). Reforming using a ZSM-5 (5 wt %)–containing catalyst leads to yields of additional aromatics. The interesting aspect of the results is the negligible amount of BTX range aliphatics in the products even though the feedstocks themselves had large quantities of these hydrocarbons. In the case of PG, for example, the feed contained 8.18 wt % aliphatics in the BTX boiling range; the product contained only 0.10 wt % of those aliphatics. The small amount of aliphatics present in the product makes it possible to obtain BTX with high purity by a simple fractionation step. Thus, reforming over ZSM-5–containing catalysts could make the production of aromatics from pyrolysis gasoline a much simpler operation than the multistep processes presently in commercial practice. A similar possibility has also been claimed for the M2-Forming process.

The reduced deactivation rate of Pt-alumina in the presence of ZSM-5 is clear from Figure 12 (71), which compares the aging rates of Pt-alumina with and without ZSM-5. The figure shows the plots of total aromatic gain and aliphatic content in the BTX range as a function of time on stream. Pt-alumina deactivates more rapidly than the zeolite catalyst. The reasons for the slower deactivation of the zeolite catalysts are:

1. The resistance of ZSM-5 to accumulate coke inside the pores.
2. The ability of ZSM-5 to crack away the coke precursors, preventing them from building up on the catalyst surface.

**Table 5** Reforming of Pyrolysis Gasoline Fractions[a]

| Conditions: | | | | |
|---|---|---|---|---|
| Temperature, °C | | 500 | | 480 |
| Pressure, bar | | 4.5 | | 4.5 |
| WHSV, h$^{-1}$ | | 2.5 | | 2.5 |
| H$_2$/HC, mol | | 7.0 | | 8.0 |
| TOS, h | | 20 | | 20 |
| Component | Feed 1 (wt %) | Product 1 (wt %) | Feed 2 (wt %) | Product 2 (wt %) |
| C$_1$–C$_6$ | 11.00 | 15.96 | 15.99 | 30.23 |
| Aliphatics in BTX range[b] | 8.18 | 0.10 | 3.67 | 0.11 |
| Benzene | 73.78 | 75.29 | 37.59 | 41.11 |
| Toluene | 7.02 | 8.30 | 15.22 | 17.09 |
| Xylenes | 0.02 | 0.35 | 3.38 | 3.31 |
| Styrene | — | — | 5.61 | — |
| Ethylbenzene | — | — | 1.45 | 5.74 |
| C$_{9+}$ (including dicyclopentadiene) | — | — | 17.09 | 2.41 |
| ΔC$_8$ aromatics | — | +3.12 | — | +4.02 |

[a] Catalyst: ZSM-5 (5%)-Pt(0.6%)-Cl(1.1%)-Al$_2$O$_3$.
[b] Olefins: feed 1, 3.41%; feed 2, 2.34%; no olefins in products.
*Source*: From Ref. 71.

## VI. CONCLUDING REMARKS

The increasing worldwide demand for unleaded and higher-octane gasoline as well as for aromatics will continue to spur the development of more efficient catalysts and processes. Reforming processes have already reached very high severity levels and further increase in severity is also likely. The high-severity processes will continue to replace the low-severity processes all over the world. Again, as always, the search for more stable and poison-tolerant catalysts will continue.

The implementation of clean air regulations, which restrict benzene and aromatics content of gasoline, is bound to have an effect on reforming for gasoline. As a result, the future reforming scenario is likely to be as follows:

In the case of gasoline, processes producing less benzene will be favored. As a corollary, catalysts which alkylate benzene and toluene to xylenes and C$_9$ aromatics will be more in demand. Processing of full-range naphtha may not be desirable, especially to avoid production of benzene. Catalysts which isomerize the lighter C$_6$ and C$_7$ hydrocarbons to the branched isomers with higher octane numbers will be much in demand. Reid vapor pressure constraints will

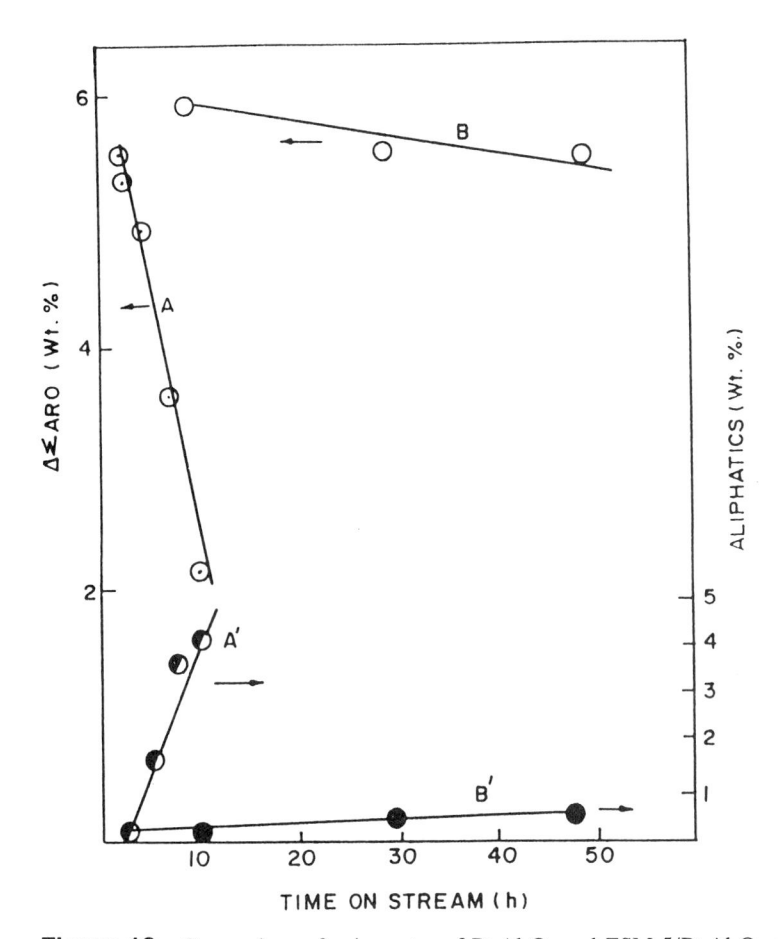

**Figure 12** Comparison of aging rates of Pt-Al$_2$O$_3$ and ZSM-5/Pt-Al$_2$O$_3$ catalysts in the reforming of pyrolysis gasoline. A,A': Pt-Al$_2$O$_3$; B,B': ZSM-5 (5 Wt %)/Pt-Al$_2$O$_3$. Temperature, 500°C; pressure, 4.5 bars; WHSV (h$^{-1}$), 2.25; H$_2$/oil (mol), 6. (From Ref. 71.)

result in eliminating butane from gasoline. The excess butane available may be converted into isobutane and isobutylene or aromatics.

For aromatics, ideally, one would prefer processes which yield only aromatics and hydrogen. As this is not feasible in the near future, processes which yield aromatics and LPG (less C$_1$ and C$_2$) will be desired. Such processes will obviate the need for expensive aromatics extraction steps. Even though the introduction of cleaner alternative fuels and blending components like oxygenates may slowly reduce the importance of reforming for gasoline production, reforming for aromatics production is expected to increase due to the growth of petrochemical industries. In this context, processes designed for aromatics production will become more important in the future.

## REFERENCES

1. V. Haensel, (a) U.S. Patent 2,479,109 (1949); (b) U.S. Patent 2,479,110 (1949).
2. F. G. Ciapetta, R. M. Dobres, and R. W. Baker, Catalytic reforming of pure hydrocarbons and petroleum naphthas. In *Catalysis*, Vol. 6 (P. H. Emmett, ed.). Reinhold, New York, 1958, p. 495.
3. F. G. Ciapetta and D. M. Wallace, Catalytic naphtha reforming. *Catal. Rev. 5*, 67 (1971).
4. B. C. Gates, J. R. Katzer, and G. C. A. Schuit, Reforming, In *Chemistry of Catalytic Processes*. McGraw-Hill, New York, 1979, p. 184.
5. J. H. Sinfelt, Catalytic reforming of hydrocarbons. In *Catalysis Science and Technology*, Vol. 1 (J. R. Anderson and M. Boudart, eds.). Springer-Verlag, Berlin, 1981, p. 257.
6. J. A. Weiszmann, Catalytic re-forming. In *Handbook of Petroleum Refining Processes* (R. A. Meyers, ed.). McGraw-Hill, New York, 1986, p. 3-3.
7. A. K. Rhodes, Survey shows over 1,000 reforming catalysts. *Oil Gas J. 89*(41), 43 (1991).
8. E. F. Schwartzenbek, Catalytic reforming. In *Origin and Refining of Petroleum*, Adv. Chem. Ser. Vol. 103 (R. F. Gould, ed.), 1971, p. 94.
9. H. E. Kluksdahl, U.S. Patent 3,415,737 (1968).
10. R. L. Jacobson, H. E. Kluksdahl, C. S. McCoy, and R. W. Davis, Platinum-rhenium catalysts: major new catalytic reforming development. *Proceedings of the American Petroleum Institute, Division of Refining 49*, 504 (1969).
11. J. H. Sinfelt, U.S. Patent, 3,953,368 (1976).
12. R. E. Rausch, U.S. Patent, 4,016,068 (1977).
13. G. J. Antos, U.S. Patent, 4,101,418 (1978).
14. F. C. Wilhelm, U.S. Patent, 3,972,805 (1976).
15. G. J. Antos, U.S. Patent, 4,165,276 (1981).
16. R. R. Cecil, W. S. Kmak, J. H. Sinfelt, and L. W. Chambers, Developments in Powerforming with advanced catalysts. *Oil Gas J. 70*(32),50 (1972).
17. J. P. Franck and G. Martino, Deactivation and regeneration of catalytic reforming catalysts. In *Progress in Catalyst Deactivation* (J. L. Figueiredo, ed.). Martinus Nijhoff, The Hague, 1982, p. 355.
18. J. P. Franck and G. Martino, Deactivation of reforming catalysts. In *Deactivation and Poisoning of Catalysts* (J. Oudar and H. Wise, eds.). Marcel Dekker, New York, 1985, p. 205.
19. A. V. Ramaswamy, P. Ratnasamy, S. Sivasanker, and A. J. Leonard, Structure and catalytic properties of bimetallic reforming catalysts, In *Proceedings of the 6th International Congress on Catalysis*, London, July 12–16, 1976 (G. C. Bond, P. B. Wells, and F. C. Tompkins, eds.), Vol. 2, 1977, p. 855.
20. J. Biswas, P. G. Gray and D. D. Do, The reformer lineout phenomenon and its fundamental importance to catalyst deactivation, *Appl. Catal. 32*, 249 (1987).
21. J. Biswas, G. M. Bickle, P. G. Gray, D. D. Do, and J. Barbier, The role of depo-

sited poisons and crystallite surface structure in the activity and selectivity of reforming catalysts. *Catal. Rev. Sci. Eng. 30*, 161 (1988).
22. W. M. H. Sachtler, Selectivity and rate of activity decline of bimetallic catalysts. *J. Mol. Catal. 25*, 1 (1984).
23. J. P. Franck and A. Vidal, Catalytic reforming: three decades of development. Paper presented at Technical Seminar, New Delhi, India, Nov. 27, 1991.
24. J. P. Bournonville and J. P. Franck, Hydrogen and catalytic cracking. In *Hydrogen Effects in Catalysis* (Z. Paal and P. G. Menon, eds.). Marcel Dekker, New York, 1988, p. 653.
25. L. A. Gerritsen, Recent developments in catalytic reforming. Ketjen Catalysts Symposium '86, Scheveningen, The Netherlands, May 25–28, 1986, paper H-1, Akzo Chemie Ketjen Catalysts, Amersfoort, The Netherlands, 1986.
26. E. A. Sutton, A. R. Greenwood, and F. H. Adams, A new processing concept for continuous Platforming. *Oil Gas J. 71*(20), 136, (1973).
27. B. J. Cha, R. Huin, H. Van Landeghem, and A. Vidal, New flexible design for IFP high severity reforming technology. *Chem. Ind. (London) (9)*, 373 (1974).
28. J. P. Bournonville and J. P. Franck, Why chlorine monitoring in catalytic reforming is compulsory, Paper presented at the "Workshop on Developments in Catalytic Reforming," Indian Institute of Petroleum, Dehradun, India, March 19–20, 1987.
29. T. V. Antipina, Nature of catalytically active sites in fluorided alumina. *J. Catal. 12*, 108 (1968).
30. R. L. Wiltshire, Improve reformer operation with trace sulfur removal. Paper presented at the "Workshop on Developments in Catalytic Reforming," Indian Institute of Petroleum, Dehradun, India, March 19–20, 1987.
31. J. Oudar, Sulfur adsorption and poisoning of metallic catalysts. *Catal. Rev. Sci. Eng. 22*, 171 (1980).
32. C. H. Bartholomew, P. K. Agrawal, and J. R. Katzer, Sulfur poisoning of metals. *Adv. Catal. 31*, 135 (1982).
33. N. Y. Chen and W. E. Garwood, Some catalytic properties of ZSM-5, a new shape selective zeolite. *J. Catal. 52*, 453 (1978).
34. N. Y. Chen and W. E. Garwood, U.S. Patent 3,379,640 (1968).
35. N. Y. Chen, J. Mazuik, A. B. Schwartz, and P. B. Weisz, Selectoforming, a new process to improve octane and quality of gasoline. *Oil Gas J. 66*(47), 154 (1968).
36. N. Y. Chen, W. E. Garwood, and F. G. Dwyer, *Shape Selective Catalysis in Industrial Applications*. Marcel Dekker, New York, 1989, p. 158.
37. R. J. Argauer and G. R. Landolt, U. S. Patent 3,702,886 (1972).
38. N. Y. Chen and W. O. Haag, Hydrogen transfer in catalysis on zeolites. In *Hydrogen Effects in Catalysis* (Z. Paal and P. G. Menon, eds.). Marcel Dekker, New York, 1988.
39. C. J. Plank, E. J. Rossinski, and E. N. Givens, U.S. Patent 4,141,859 (1979).
40. N. Y. Chen, U.S. Patent 3,729,409 (1973).
41. N. Y. Chen, W. E. Garwood, and R. H. Heck, M-Forming process, *Ind. Eng. Chem. Res. 26*, 706 (1987).
42. Brochure: More octane and more aromatics from naphtha—IFP takes up the challenge with its reformer and aromizer. (Ref. 33004) IFP, Rueil Malmaison Cedex, France.

43. N. Y. Chen and T. Y. Yan, M2-Forming—a process for aromatization of light hydrocarbons. *Ind. Eng. Chem. Process Des. Dev. 25*, 151 (1986).
44. Brochure: Aroforming process: aromatics from paraffins, a new flexible approach to aromatics production. Salutec, Melbourne, Australia and IFP, Rueil Malmaison Cedex, France.
45. J. A. Johnson, J. A. Weiszmann, G. K. Hilder, and A. H. P. Hall, Dehydrocyclodimerization converting LPG to aromatics. Paper presented at the 1984 NPRA Annual Meeting, March 25–27, 1984, San Antonio, TX.
46. C. D. Gosling, F. P. Wilcher, L. Sullivan, and R. A. Mountford, Process LPG to BTX products. *Hydrocarbon Process. 70*(12), 69 (1991).
47. R. M. Barrer and H. Villiger, The crystal structure of synthetic zeolite L. *Zeit. Kristall. 128*, 352 (1969).
48. J. R. Bernard, Hydrocarbon aromatization on platinum alkaline zeolites. In *Proceedings of the 5th International Conference on Zeolites* (L. V. C. Rees, ed.). Heyden, London, 1980, p. 686.
49. T. R. Hughes, W. C. Buss, P. W. Tamm, and R. L. Jacobson, Aromatization of hydrocarbons over platinum alkaline earth zeolites. In *Proceedings of the 7th International Zeolite Conference*. Tokyo, Japan (Y. Murakami, A. Lijima, and J. W. Ward, eds.), Kodansha and Elsevier, Tokyo and Amsterdam, 1986, p. 275.
50. J. L. Kao, G. B. McVicker, M. M. J. Treacy, S. B. Rice, J. L. Robbins, W. E. Gates, J. J. Ziemak, V. R. Cross, and T. H. Vanderspur, Effect of sulfur on the performance of Pt/KL hexane aromatization catalyst. In *Proceedings of the 10th International Congress on Catalysis*, Part B, Budapest, July 19–24, 1992 (L. Guczi, F. Solymosi, and P. Tétényi, eds.). Akadémiai Kiadó, Budapest, 1993, 0. 1019.
51. P. W. Tamm, D. H. Mohr, and C. R. Wilson, Octane enhancement by selective reforming of light paraffins. In *Catalysis 1987*, Studies in Surfaces Science and Catalysis Series, Vol. 38 (J. W. Ward, ed.). Elsevier, Amsterdam, 1987, p. 325.
52. D. L. Ellig and G. J. Antos, U. S. Patent 4,870,223 (1989).
53. S. C. Fung and S. J. Tauster, U. S. Patent 4,595,669 (1986).
54. C. Besoukhanova, J. Guidot, and D. Barthomeuf, Platinum-zeolite interaction in alkaline L zeolites. *J. Chem. Soc. Faraday Trans. I 77*, 1595 (1981).
55. D. Barthomeuf, Acidity and basicity in zeolites. In *Catalysis and Adsorption by Zeolites*. Studies in Surface Science and Catalysis Series, Vol. 65 (G. Ohlmann, H. Pfeifer, and R. Fricke, eds.). Elsevier, Amsterdam, 1991, p. 157.
56. G. Larsen and G. L. Haller, Metal support effects in Pt/L-zeolite catalysts. *Catal. Lett. 3*, 1003 (1989).
57. S. J. Tauster and J. J. Steger, Molecule die catalysis: hexane aromatization over Pt/KL. *J. Catal. 25*, 382 (1990).
58. E. G. Derouane and D. J. Vanderveken, Structural recognition and preorganization in zeolite catalysis: direct aromatization of $n$-hexane on zeolite L–based catalyst. *Appl. Catal. 45*, L15 (1988).
59. J. M. Newsam, B. G. Silbernagel, A. R. Garcia, M. T. Melchior, and S. C. Fung, Fundamental characteristics of the catalyst system platinum loaded zeolite L. In *Structure-Activity Relationships in Heterogeneous Catalysis*, Studies in Surface Science and Catalysis Series, Vol. 67 (R. K. Grasselli and A. W. Sleight, eds.). Elsevier, Amsterdam, 1991, p. 211.

60. R. A. Dalla Betta and M. Boudart, Well dispersed Pt on Y zeolite. Preparation and catalytic activity. In *Proceedings of the 5th International Congress on Catalysis*, Vol. 2 (J. W. Hightower, ed.), North Holland, Amsterdam, 1973, p. 1329.
61. G. S. Lane, F. S. Modica, and J. T. Miller, Platinum/zeolite catalyst for reforming n-hexane: kinetic and mechanistic considerations. *J. Catal. 129*, 145 (1991).
62. H. Katsuno, T. Kukenaga, and M. Sugimoto, New modification method of Pt/L zeolite catalyst for hexanes aromatization. In *Proceedings of the 10th International Congress on Catalysis*, Part C, Budapest, July 19–24, 1992 (L. Guczi, F. Solymosi, and P. Tétényi, eds.). Akadémiai Kiadó, Budapest, 1993, p. 2419.
63. E. Iglesia and J. E. Baumgartner, A mechanistic proposal for alkane dehydrocyclization rates on Pt/L-zeolite. Inhibited deactivation of Pt sites within zeolite channels. In *Proceedings of the 10th International Congress on Catalysis*, Part B, Budapest, July 19–24, 1992 (L. Guczi, F. Solymosi, and P. Tétényi, eds.). Akadémiai Kiadó, Budapest, 1993, p. 993.
64. E. Mielczarski, S. B. Hong, R. J. Davis, and M. E. Davis, Aromatization of n-hexane by platinum-containing molecular sieves. II. n-Hexane reactivity. *J. Catal. 134*, 359 (1992).
65. S. Sivasanker, S. B. Kulkarni, K. R. Murthy, S. R. Padalkar, and P. Ratnasamy, In *Proceedings of the 2nd Indo-Soviet Symposium on Catalysis*, Hyderabad, India (P. K. Rao, ed.). INSDOC, New Delhi, 1986, p. 66.
66. D. H. Mohr, Australian Patent 600,076 (1988).
67. R. J. Davis and E. G. Derouane, A non-porous supported platinum catalyst for aromatization of n-hexane. *Nature 349*, 313 (1991).
68. L. A. Gerritsen, Catalytic reforming of FCC naphtha for production of lead free gasoline. Ketjen Catalysts Symposium '84, Amsterdam, The Netherlands, Akzo Chemie Ketjen Catalysts, Amersfoort, The Netherlands, 1984, p. 199.
69. E. P. Kieffer and S. T. Sie, U. S. Patent 4,512,877 (1985).
70. I. E. Maxwell, F. Muller, F. H. K. Hok, K. K. Heong, and J. Lucien, European Patent EPA 0,430,337 A1 (1990).
71. K. M. Reddy, S. K. Pokhriyal, P. Ratnasamy, and S. Sivasanker, Reforming of pyrolysis gasoline over platinum-alumina catalysts containing MFI type zeolites. *Appl. Catal. A: General 83*, 1 (1992).

# Index

Acid function:
  balance with metal function, 234, 450
  catalysis of paraffin and olefin isomerization, 58, 67
  characterization of, 117–124
  chloride-promoted alumina, 15, 234–237
  and coke production, 288–290, 294, 296, 297, 302, 303
  control of bifunctional mechanisms, 59
  control in reformer, 235–237
  and dehydrocyclization, 235
  excessive hydrocracking, 234, 289, 450
  strength for isomerization, 59
Alumina ($\gamma$- and $\varepsilon$-):
  acidity and basicity of, 83, 84, 117–122

[Alumina ($\gamma$- and $\varepsilon$-)]
  influence of halogens on, 83, 84, 122–124, 209, 234–237
  porosity of, 85, 114–116, 209, 210
  precursors of, 80–82
  strength of, 210
  structure of, 82, 165
  support forming, 84, 85
  surface area of, 81, 85, 114, 115, 165, 209
  surface area retention, 210, 340, 412
  surface hydroxyls of, 83, 87–89, 99, 118, 119, 152
  surface properties of, 81, 83, 88, 89, 118–121, 152
  thermal transformations of, 80–82
Aromatics:
  aroforming, 493

*509*

[Aromatics]
  cyclar, 493, 494
  general properties, 5, 14
  M2-forming, 492
  production from propane, 80, 493
  reduction in fuels, 429

Benzene reduction:
  in gasoline, 16, 429, 502
  through alkylation, 17, 502
Bifunctional catalyst, 15, 27, 113, 186, 207, 208, 211
  reactions on, 49, 54, 58–60, 63–68
Bimetallic catalysts:
  advantages of, 15, 79, 104, 231, 232, 485, 487
  characterization of, 146–160
  electronic effects in, 31, 231, 232
  geometric effects in, 31, 231, 232
  performance evaluation of, 189–192
  platinum/germanium, 15, 72
    catalyst characterization, 159–160, 231
    catalyst preparation, 103
    interaction, 160, 231
    role of germanium in, 231, 288, 341
  platinum/iridium, 15, 71
    catalyst characterization, 157–159, 168
    catalyst preparation, 102, 103, 157–159
    interaction, 102, 141, 157–159, 230
    platinum/rhenium, 15, 96
    platinum/rhenium, alloy formation in, 96, 150–153, 228
    role of iridium in, 230, 232, 287, 288, 341, 367

[Bimetallic catalysts]
  catalyst characterization, 146–153, 168
  catalyst preparation, 96
  rhenium oxidation state in, 96, 150–153, 228
  role of rhenium in, 228, 229, 286, 287, 341, 342, 377
  sulfur interaction with, 150, 151, 153, 184, 232, 233, 316
  platinium/tin, 15, 72
    catalyst preparation, 98–101
    electronic interaction, 31, 100, 101, 154–156, 167, 168, 229, 250
    oxidation state of tin in, 98–101, 153–156, 167, 168, 229
    role of tin in, 229, 288, 341
    presulfiding of, 71, 72, 232–234

Carbonaceous deposits, 49, 50, 72, 217, 279, 366–368
  bimetallic catalysts and, 284–290, 299, 300
  characterization of, 280–283, 368–370
  dehydrogenation of, 218, 286, 303–305, 367
  disordered, 34, 233
  effect of migration of chemisorbed species, 34
  effect of sulfur, 35, 297–302, 332, 333
  effect of surface hydrogen on, 38, 239
  evolution with time on stream, 282–284
  formation by polymerization of surface polyenes, 34–35, 37, 303–305, 367

# Index

[Carbonaceous deposits]
  geometric effects of, 34, 302
  H/C ratio, 233, 280, 284, 286, 293, 301
  as hydrogen transfer agents, 34
  migration of precursors from metal to acid sites, 35, 290, 294, 367
  ordered, 34, 217–219, 233, 282, 284
  reaction conditions and, 291–297
  reversible/irreversible, 33–35
Catalyst deactivation, 75, 182, 208, 302, 305, 306
  by poisoning, 313–340
Catalyst performance evaluation, 182, 243–245
  laboratory tests, 183–188
    of activity with model compounds, 184
    evaluating bimetallic catalysts, 189–192
    isothermal microreactors, 185–188
    isothermal unit with temperature program, 192–199
    miniadiabatic pilot unit, 199–205
    using more practical feedstocks, 185
Catalyst preparation steps:
  drying, 94, 95, 97, 99, 148, 152
  impregnation, 85–87
    competing agents in, 86–88, 90, 93, 94, 96
    diffusion effects on, 86, 89, 90
    germanium precursors, 103
    iridium precursors and species in, 102
    metal profiles, 86, 93, 94, 96, 98, 102, 103, 152, 153

[Catalyst preparation steps]
    platinum precursors and species in, 86, 90–93, 96, 101
    rhenium precursors and species in, 96–98, 152
    support dissolution during, 89, 90
    surface interactions in, 87–94, 97–102, 152
    techniques for, 86, 96–101, 157
    tin precursors and species in, 98–101, 155
  oxidation, 95, 97, 99, 148
  reduction, 95, 98, 100, 101, 141
Catalyst presulfiding, 15, 71, 72, 96, 221, 232–234, 316, 327, 330–333
Catalyst selectivity, 45, 50
Catalytic reforming:
  for aromatics as petrochemicals, 13, 16, 409, 483, 484, 492–494, 501
  cycle length factors, 182, 242, 243, 282, 327–329, 337, 415
  endothermicity, 14, 52, 55, 62, 74, 182
  for high-octane gasoline, 8, 12, 13, 181, 409, 483, 484, 492, 502
  modeling commercial units, 437–461
  monitoring unit performance, 243–245
  operating parameters, 208, 239–243, 412, 413
  process:
    continuous catalyst regeneration, 182, 183, 410, 413, 417, 418, 435
    fully regenerative, 182, 183, 410, 413, 417, 418, 435

[Catalytic reforming]
   semiregenerative, 182, 183, 413–415
   severity, 14, 241–243, 257, 410, 487
   typical feed composition, 13, 239, 240
   unit optimization, 461, 462, 468–479
C–C bond:
   reactant structure sensitivity in, 30
   rupture, 23, 24, 70, 211, 216, 218
$C_5$ cyclization, 19
   metal catalyzed, 20, 23
   surface intermediate, 22
C–H bond:
   dissociative chemisorption, 25
   reactive chemisorption, 25
   rupture, 25, 211, 215, 218
Characterization, 113, 114
   of acidity properties, 166
      by calorimetry, 121, 122, 166
      by Hammett indicators, 117, 118, 124
      by infrared probe techniques, 118–121, 123, 166
   of alumina support:
      acidity, 117–124, 166, 209
      porosity, 115–117, 166, 209, 210
      strength, 114, 115, 210
      surface area, 115, 165
      by gas adsorption, 115, 116
      by mercury penetration, 115, 116
   of metal dispersion and particle size:
      using chemisorption techniques, 127–130, 139, 167
      using nuclear magnetic resonance, 138–140
      by small angle x-ray scattering, 131–133

[Characterization]
   using titration methods, 130, 131, 140
   with transmission electron microscopy, 126, 127, 139, 167
   with x-ray adsorption fine structure analysis, 134–138
   with x-ray diffraction, 125, 126
   of metal properties:
      by Auger electron spectroscopy, 148
      with calorimetry, 144
      using infrared spectroscopy, 144
      with ion scattering spectroscopy, 143, 144
      with temperature programmed reduction, 141
      by x-ray photoelectron spectroscopy, 141, 142
   of platinum/iridium, 157–159, 168
   of platinum/rhenium, 146–153, 168
   of platinum/tin, 153–156, 167, 168
   by reaction with probe molecules, 40
   with small angle x-ray scattering, 117
   with temperature programmed oxidation, 35
   with temperature programmed reaction, 22, 29
   with transient response method, 22, 38
Chemical equilibrium, 45
   cyclohexane/benzene, 50
   equilibrium constant for, 46
   hydrogenation/dehydrogenation, 56
   paraffin dehydrocyclization, 63, 64
   paraffin isomerization, 57

Index 513

Commercial reforming processes:
  Houdriforming, 428
  Magnaforming, 63, 422–424
  Platforming, 428, 483, 484
  Powerforming, 75, 424
  Octanizing, 421, 422
  Rheniforming, 425, 426
  Ultraforming, 426–428
Crude oil types, 6–8

Electronic effects, 29, 30, 39, 208, 211, 226, 227, 234, 496
EUROPT-1, 25, 30, 33–34, 138

Gasoline:
  desirable properties, 12
  production processes, 181
  vapor pressure reduction, 429, 502
Geometric effects, 29, 30, 71, 208, 227, 234, 302, 327, 496

Hydrocracking, 14, 49, 50, 72–74
  acid function in, 68, 69
  exothermicity, 69, 74
  kinetics of, 69
  products of, 69
Hydrogen:
  effect on aromatization, 36–37
  effect on catalyst stability, 14
  participation in active site on Pt, 38–39
  pressure effect on yields, 39
  types of surface, 36–37
Hydrogenation, 69, 212
Hydrogenolysis, 14, 19, 49, 50, 70, 71, 72–74, 212
  Anderson-Avery mechanism, 22
  with bimetallic catalysts, 71, 227
  effect of hydrogen, 39, 71
  effect of sulfur on, 31, 52, 71

[Hydrogenolysis]
  effect of surface oxygen, 35
  fragmentation factor, 23, 24
  metal catalyzed, 20, 23, 70
  metal dispersion effects on, 71, 227
  suppression by presulfiding, 15, 71, 72, 227
  thermodynamics of, 70
Hydrotreating, 8, 15, 258, 260, 267–272, 410, 411, 418, 501
  catalysts, 263, 266, 268, 272–275
  kinetics, 263–265
  metals removal, 270

Kinetic models:
  of catalytic reformer, 437, 455–468
  for naphtha reforming, 45, 74–77
  reactor profiles, 76, 77
Kinetics of reforming reactions, 51, 52, 56, 57, 72–74

Metals recovery from spent catalyst, 395–406
  chemistry of, 403–405
Metal surface sites:
  catalyzed hydrocarbon reactions, 19, 22
  for chemisorption, 25, 226
  interaction with acidic sites, 27, 237, 238
  interaction with support, 25, 211, 238, 321, 327
  reactive, 24, 25
  single atom, 25, 28
Monofunctional catalyst for aromatization of hexane, 494, 495
  Aromax, 496
  characterization of, 161–165

[Monofunctional catalyst]
  confinement model, 27, 28, 500
  geometric constraint in, 27, 496, 499
  importance of basic sites in, 28, 498
  Pt/alkyl/zeolite, 27, 28, 161, 495
  Pt/K/L zeolite, 28, 34, 79, 495
    characterization of, 161–165
    hindered deactivation in, 28, 499
    sensitivity to sulfur, 79, 164, 496, 498
    stabilization of small crystallites in, 28, 162, 500
  Te/X zeolite, 79
Multimetallic catalysts, 31, 103, 104, 485, 487

Napththenes:
  alkylcyclopentane dehydroisomerization, 53
    kinetics of, 54, 72
  dehydrogenation, 13, 49–51, 72, 212
    kinetics of, 51
    mechanism of, 52, 53
  general properties, 5, 14
  ring opening, 49, 69
Nitrogen compounds:
  basic, 6
  effect of activity and selectivity by, 32, 333
  nonbasic, 6
  poisoning by, 333

Octane boosting processes:
  erionite, 489
  M-forming, 490
  Selectoforming, 489
  ZSM-5, 489–491

Octane number, 9–12
Olefins:
  dehydrogenation to, 55–57
  hydrogenation of, 260, 264, 265, 501
  intermediates, 53, 59, 69
  isomerization, 60
  polymerizing to carbonaceous deposits, 57, 226, 227, 240

Paraffin dehydrocyclization, 13, 19, 49, 50, 61–67
  effects of sulfur on, 67, 68, 301
  ensemble size for, 24, 27–28, 226
  kinetics of, 63, 72–74
  mechanism of, 63–67
  metal catalyzed, 20, 23, 27, 28, 67, 68
  thermodynamics of, 62–64, 72–74
Paraffin dehydrogenation, 49, 50
  effect of sulfur on, 56, 67
  ensemble size for, 24, 226
  kinetics of, 56
  mechanism of, 57
  thermodynamics of, 55, 56
Paraffin isomerization, 13, 16, 19, 49, 50, 57–59
  $C_5$ cyclic mechanism of, 22, 24, 30
  ensemble size, 24
  effect of acid strength on, 59
  effect of sulfur on, 32
  kinetics of, 57, 72–74
  metal catalyzed, 20, 23, 27, 38, 60
  thermodynamics of, 57, 72–74
Paraffins, general properties, 5, 14
Petroleum composition, 2–6
  distillation as measure of, 8
Physical mixture studies:
  of metal and acid, 27
  of Pt/Al$_2$O$_3$ and Re/Al$_2$O$_3$, 31, 228

Index 515

Platinum black catalysts, 29, 30, 33, 35, 36, 39
Platinum/magnesium oxide for aromatization, 28, 29, 80, 500
Process licensors:
  Amoco, 426–428
  Chevron, 425, 426
  Engelhard, 415, 422–424
  Exxon, 415, 424, 425
  Houdry, Division Air Products, 428
  IFP, 417, 421
  UOP, 417, 418, 483, 484
Pt/Al$_2$O$_3$:
  carbon coverage on, 33
  irreversibly held sulfur on, 31
  kinetics of cyclohexane dehydrogenation on, 51, 52
    effect of sulfiding on, 52
  for MCP dehydroisomerization, 55
    effect of sulfur on, 55
  in paraffin dehydrocyclization, 66
  for paraffin isomerization, 58
    effect of fluoride on, 58
Pt/Re/Al$_2$O$_3$:
  irreversibly held sulfur on, 31
  for MCP dehydroisomerization, 55
  in paraffin dehydrocyclization, 66
  selectivity for hydrogenolysis, 71, 96
  selectivity in naphthene dehydrogenation, 52
Pt/Re/S/Al$_2$O$_3$:
  effect of acidity on paraffin isomerization, 59
  impact of coke on isomerization on, 57, 58
  impact of hydrogenolysis, 71
  irreversibly held sulfur on, 31

[Pt/Re/S/Al$_2$O$_3$]
  for MCP dehydroisomerization, 55
  selectivity in naphthene dehydrogenation, 52
  sulfur tolerance of, 31

Reaction mechanisms:
  bond shift, 20, 30, 61
  bond shift isomerization, 22, 23
  C$_5$ cyclic, 20, 30, 61
  1,5 ring closure, 21, 66
  1,6 ring closure, 21, 66, 499
  thermal cyclization, 21
  triene, 21, 29, 67
  two dimensional, 21, 27
Redispersion of metals, 137, 139, 182, 220, 340, 350–356, 389
  model, 357–359
Reformer feeds:
  boiling range, 16, 239, 240, 258–260
  cracked naphthas, 240, 258–260, 267, 410, 500
  impurities in, 208, 240, 257–260, 411
  olefins in, 15, 240, 259, 260, 500
  pyrolysis naphtha, 240, 501
  straight run naphtha, 15, 240, 258, 267, 410
Reforming catalysts:
  chromium oxide, 80, 484
  commercial, 428–434, 485
  molybdenum oxide, 80, 484
Regeneration, 182, 365, 388–391, 414, 415, 417, 418
  adsorbed hydrogen as ignitor, 377, 380
  with hydrogen, 370–375
  model compound, 366
  with oxygen, 376–380

[Regeneration]
  sintering, 237–239, 340–350, 488
    influence on catalytic reactions, 352–353
    migration mechanisms in, 341, 343, 348, 349
    model, 357–359
  of sulfur-contaminated catalyst, 380–388

Single crystal studies, 39
  comparison with Pt/Al$_2$O$_3$, 24, 25, 223–225
  iridium, 24, 218, 219
  nickel, 25
  platinum/rhenium, 33
    effect of sulfur on reactivity of, 33
  reactivities of Pt corners, steps, kinks, ledges, and terraces, 24, 25, 212–218, 220, 221
  reconstruction of platinum, 39, 40, 219, 220
  sulfur-induced reconstruction of platinum, 33, 220, 221
  of surface geometries for platinum, 24, 212–217

[Single crystal studies]
  surface oxygen effects and reconstruction of platinum, 35, 36, 219
Skeletal rearrangement units, 29, 30
Structure-insensitive reactions, 212, 221–224, 326, 334
Structure-sensitive reactions, 212, 221–223, 326, 334
Sulfur:
  activity and selectivity affected by, 234, 325, 327–333
  effect on acidity, 234, 320
  effect on hydrogenolysis activity, 31, 232, 233, 327
  effect on skeletal isomerization, 32, 233
  guardbed, 423
  nature of adsorption, 31, 317–327
  as poisons for metallic catalysts, 6, 15, 31, 232, 260, 314, 316–327, 329–333
Surface unsaturated species, 21, 22, 29

Thermodynamics, 45
  of reforming reactions, 45–50, 54–57, 70, 72–74